R00069 91453

ORGANIC MOLECULAR STRUCTURE

A GATEWAY TO ADVANCED ORGANIC CHEMISTRY

LLOYD N. FERGUSON
California State University, Los Angeles

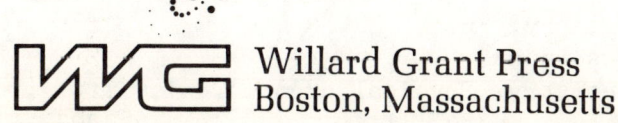

Willard Grant Press
Boston, Massachusetts

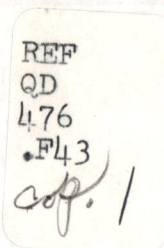

© Copyright 1975 by Willard Grant Press, 20 Newbury Street, Boston, Massachusetts 02116.

All rights reserved. No part of this book may be reproduced or transmitted in any form or by any means, electronic or mechanical, including photocopying, recording, or any information storage and retrieval system, without permission in writing from the publisher.

Printed in the United States of America

Library of Congress Cataloging in Publication Data

Ferguson, Lloyd N
    Organic molecular structure.

    Bibliography: p.
    Includes index.
    1. Chemistry, Physical organic. I. Title.
QD476.F43   547'.1'3   74-32027
ISBN 0-87150-708-0

To Lisa, my daughter

PREFACE

One way of dividing the subjects that fall under the general heading of organic chemistry is in two parts: natural product chemistry and physical organic chemistry (see Figure 1). The former is concerned with structure determination and synthesis of compounds resembling those of natural origin. Physical organic chemistry involves a study of reaction mechanisms and theories of chemical bonding. In recent years, these two aspects of organic chemistry have overlapped more and more. Thus, natural products have provided a fertile field for the testing and application of chemical theories developed by physical organic chemists; in turn, the methods and principles of physical organic chemistry have been invaluable in the practice of natural product chemistry.

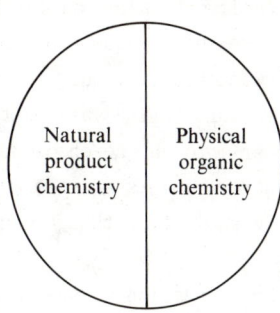

Figure 1. One subdivision of organic chemistry.

An alternate overview of organic chemistry has three divisions: structure, dynamics, and synthesis (see Figure 2). Each has a relationship to the other two. This view of the subject is more in line with the interests and functions of organic chemists. In particular, it defines the scope of this book, namely, organic molecular structure.

This monograph is a synopsis of structural organic chemistry at an intermediate level. As such, it embraces reaction mechanisms only to the extent that it is concerned with the structures of reactive intermediates. Consequently, it cannot be called physical organic chemistry. It has little to do with natural product chemistry except for an occasional illustration of a correlation between structure and physical properties.

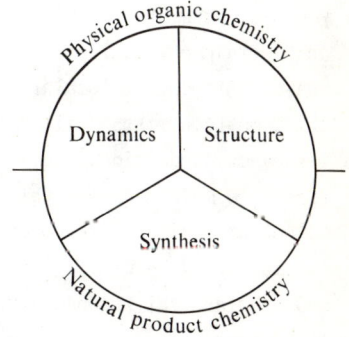

Figure 2. Another subdivision of organic chemistry.

PREFACE

No attempt is made to give a rigorous treatment to any subtopics such as molecular orbital theory, kinetics and reaction mechanisms, linear free-energy relationships, acid-base theories, optical and megnetic spectroscopy, and other. Indeed, several books covering any one of these topics alone are available. An exhaustive treatment of any one topic tends to obscure the over-all hierarchic structure of the subject. However, by drawing together all of these topics briefly, one can get a unified overview of organic molecular structure. Additional educational benefits would accrue by answering the questions posed throughout the text. Proposed solutions follow each set of questions. There is a generous supply of references to the literature on the various topics discussed to enable the reader to delve much deeper into any he chooses. The book is directed toward the student, not the instructor, and is written in a rather informal tone. It offers a unified view of the subject, although many of the explanations could have other interpretations.

Structural theory, of course, provides the foundation or systemization of organic chemistry necessary for industrial and biological applications, the results of which serve as a measure of the success of organic chemists. Just look around you at the ubiquitous synthetic organic kingdom chemists have given us.

During the first year of organic chemistry, you were given a few reaction mechanisms and spectra-structure relationships as explanations for some of the functional group properties and for use in structure determination. In the second year you will examine the bases for these reaction mechanisms and take u ways of getting experimental evidence in support of a proposed mechanism. Also you will attempt to find a rationale for observed relationships between structu and certain physical properties. This monograph is concerned with this latter phase. Although molecular orbital theory offers more powerful means for doing this, most chemists still rely upon contemporary structural theory (i.e., resonance, induction, etc.) for interpreting molecular properties. Often they resort to molecular orbital calculations for support, and occasionally only the molecular orbital model provides a logical explanation. Nevertheless, the initial study is usually in terms of modern structural theory in which classica and nonclassical resonance theory play a central role. It is this general practice which provides justification for the appearance of this monograph at a time when molecular orbital theory is approaching maturity and wide applicability.

It can be envisioned that a good advanced organic chemistry course would cover selected topics of organic molecular structure, dynamics, and synthesis. The course would provide experience in the use of molecular orbital methods for obtaining information about bonding, energies, charge densities, spectra, and other molecular characteristics. Each of these subtopics would be preceded by a cursory view based on modern structural theory and available data. This monograph would be useful at this point in the course. It can serve as a bridge between elementary and advanced organic chemistry. In particular, it gathers many scattered data from the literature pertaining to a given concept or phenomenon, and provides a simple physical model for reasoning by analogy in unstudied cases.

I wish to express my appreciation to the University of Nairobi, Kenya, and the California State University, Los Angeles, for clerical assistance in the preparation of the original and revised drafts of the manuscript for this book. I am grateful for the many helpful criticisms of several reviewers, including Michael Siklosi, Marcel Boegelli, and Professor Robert E. Davis of Purdue University, and Professor Norman L. Allinger of the University of Georgia. Above all, I express my deep appreciation to my wife for her forbearance and patience shown during the many hours of seclusion spent on this project.

Lloyd N. Ferguson

TABLE OF CONTENTS

Part I: MOLECULAR FORCES	1
Chapter 1, CHEMICAL BONDS	1
1. Covalent bonds	2
Questions	18
2. Hydrogen bonds	19
Questions	34
Chapter 2, VAN DER WAALS FORCES	36
1. van der Waals forces	36
2. Inclusion complexes	37
a. Dimensional characteristics, 38; b. General properties, 40; c. Uses of inclusion complexes, 41	
Chapter 3, MOLECULAR POLARIZATION	43
1. Electronegativity	43
a. Atomic electronegativities, 44; b. Relative group electronegativities, 49	
2. Electric dipole moments	55
Questions	61
3. Substituent electrical effects	64
a. Induction, 65; b. Field effects, 72; c. Electrical effects of alkyl groups, 77	
Chapter 4, INTERNAL ENERGIES	91
1. Electromagnetic energy	91
2. Molecular absorption of energy	99
a. Infrared spectra, 99; b. Raman spectra, 113; c. Ultraviolet spectra, 116; d. Nuclear magnetic spectroscopy, 129	
Questions	147
e. Mass spectra, 148	
Questions	165
Part I: REFERENCES	178
Part II: MODERN STRUCTURAL THEORY	179
Chapter 5, RESONANCE THEORY	181
1. The concept of resonance	181
2. Writing resonance structures	183
3. Hyperconjugation	185
Questions	195
4. Resonance energies	196

CONTENTS

Chapter 6, APPLICATIONS OF RESONANCE THEORY	200
1. Structure	
a. Molecular geometry, 200; b. Resonance and polar character, 205	
Questions	214
c. Resonance and aromaticity, 215	
Questions	230
d. Charge-transfer and metal π complexes, 237	
Questions	245
e. Metallocenes, 246	
2. Dynamics -- equilibrium and kinetics	252
a. Resonance and equilibria, 252	
Questions	267
b. Resonance and reaction mechanisms, 273	
Chapter 7, HAMMETT-TYPE TREATMENTS	276
1. Applications of Hammett-type constants	283
Questions	292
2. The Q parameter	293
Chapter 8, NONCLASSICAL RESONANCE	296
Questions	307
Chapter 9, RESONANCE VS. MOLECULAR ORBITAL THEORY	308
Part II: REFERENCES	310

Part III: SPECTRA-STRUCTURE CORRELATIONS   321

Chapter 10, SPECTRA AND ELECTRON DELOCALIZATION	322
Questions 323, 334, 341,	361
Chapter 11, SPECTRA AND STERIC EFFECTS	363
Questions 378,	386
Chapter 12, SPECTRA AND H BONDING	389
Questions	402
Chapter 13, SPECTRA, ELECTRONEGATIVITY, AND POLARIZABILITY	405
Questions	409
Chapter 14, SPECTRA AND CONSTITUENT CONSTANTS	410
Chapter 15, SPECTRA AND INTRAMOLECULAR GEOMETRY	418
1. Dihedral angles	418
2. Carbon bond angles	422
3. H bond distances	424
Chapter 16, SPECTRA AND EXCITED STATES	426
Questions	430
Part III: REFERENCES	433

Part IV: INTRAMOLECULAR FORCES AND MOLECULAR PROPERTIES   441

Chapter 17, STATIC CONFORMATIONAL ANALYSIS   442
    Questions   463

Chapter 18, DYNAMIC CONFORMATIONAL ANALYSIS   469

Chapter 19, RING STRAIN AND MOLECULAR PROPERTIES   484
    1. Ring strain energies   485
       Questions   487
    2. Ring strain and bonding   489
    3. Ring strain and reactivity   497
       Questions   508

Chapter 20, MISCELLANEOUS STERIC EFFECTS   511
    1. Restricted rotations   511
       Questions   518
    2. Steric hindrance to H bonding   520
    3. Base strengths of aliphatic amines   521
    4. Acidity of hindered acids   525
    5. Hindrance to conventional reactions   527
       Questions   533

Part IV: REFERENCES   535

INDEX   545

AUTHOR INDEX   561

Part I: MOLECULAR FORCES

Chapter 1

CHEMICAL BONDS

There are several methods used to describe bonding in organic molecules. There are the electrostatic views, which include Lewis' covalent bond theory, the valence electron-repulsion theories, and the molecular mechanics or force-field methods. The covalent bond theory primarily accounts for *coordination numbers* of atoms and is well known to all chemistry students. The valence electron-repulsion theories [1] are concerned chiefly with molecular *geometry*. Valence shell electrons are treated not as electron pairs but as an array of two spin sets of opposite spins. An arrangement results which achieves minimum intra-electrostatic repulsion, with electrons of like spin exerting more repulsion than those of unlike spin. The molecular mechanics or force-field methods are really ways of calculating geometries and energies of molecules [2]. A molecule is viewed as a system of particles held together by classical forces. The energy differences between different molecular geometries are then calculated by classical mechanical methods and the results yield relative rather than absolute energies. It is particularly suited for computing relative strain energies or relative rates of reactions (from relative activation energies).

There are also theoretical approaches to a description of chemical bonding. The wave mechanical methods are impractical for all but very simple molecules and are too mathematical and complex for most chemists even if complete solutions to the equations could be found. The molecular orbital approximation methods [3] offer much more promise. These methods are variously known as the Hückel MO method, the Mulliken self-consistent field MO method, the Parrison-Pople method, the CNDO methods, and others. Each makes slightly different assumptions and approximations for the calculation of bond characteristics, spectra, reactivities, dipole moments, and thermodynamic stabilities, and one or another will be best suited for a particular molecular system. In many cases, molecular properties can be calculated in agreement with experimental data. Then, once the validity of the method is established it can be used to estimate properties not yet observed or difficult to measure.

Ch. 1   CHEMICAL BONDS

Between the electrostatic and semitheoretical MO approaches for describing molecular structure is the approach of *modern structural theory*. It is an interim, qualitative compromise which describes molecules in terms of structural formulas. It incorporates the symbolism of classical structural theory with the principles of resonance theory and the geometric and energetic concepts of MO theory. This monograph will use this approach.

1.1.   Covalent bonds

1.1a.   Molecular orbital theory

You will recall from physics that a plucked guitar string or a vibrating rope fastened at both ends generates *stationary waves*. As a particle follows in the path of the wave, its *amplitude* outlines opposite *phases* of the wave, arbitrarily designated + and - in Figure 1.1, and

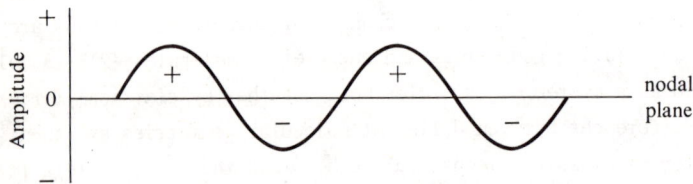

Figure 1.1.

passes through the *nodal plane* -- perpendicular to the plane of the paper -- as it changes from one phase to the next.

The wave can be described by a differential equation, the *wave equation*, whose solution gives the amplitude ϕ as a function $f(x)$ of the distance along the wave. This function, then, is a *wave function*. A wave equation has an infinite *set* of solutions, i.e., wave functions, each corresponding to a different energy level.

Since rapidly moving particles have a wave-like character, it is not surprising that there was an attempt to describe the properties of electrons by a wave equation. This attempt resulted in the Schrödinger equation, written in its simplest form as

$H\psi = E\psi$

where an infinite set of wave functions, called orbitals (ψ), exists. However, only certain orbitals have any physical significance. Each orbital associated with an energy E is a mathematical expression which describes a possible distribution of the electron in three-dimensional space. The left side of the Schrödinger equation H is the Hamiltonian operator, which is a mathematical symbol telling us to perform a manipulation on the quantity which follows it. We must integrate the equation in order to find its solution. As you may know, an exact solution has been found only for such species as the hydrogen atom or the helium molecule ion.

Nevertheless, very good *approximate* solutions to the Schrödinger equation have been developed for many molecules [3]. Thus, we have what are called *ab initio* calculations -- those involving rigorous solutions of approximate formulations of the Schrödinger equation. The rationale is the assumption that the *molecular orbital* of the hydrogen molecule, for example, is a combination of the atomic orbitals of each atom, i.e.,

$$\psi = \phi_A \pm \phi_B.$$

What is important to us are the shapes and relative stabilities of atomic orbitals as calculated by quantum mechanics [4]. These are the so-called *s, p, d*, and *f* atomic orbitals. A *p* orbital, for instance, has two lobes of opposite phase distinguished by + and - signs or by shading. (See Figure 1.2.) The signs do not imply positive and negative charges, merely that the amplitudes in the two lobes are of opposite algebraic sign.

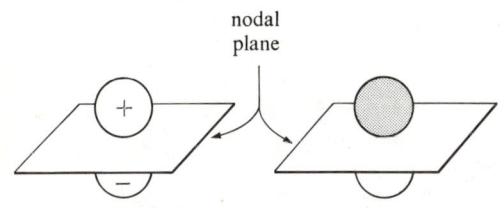

Figure 1.2.

Thus, we visualize a chemical bond as an overlapping of two atomic orbitals such that each electron may then interact with both nuclei. The ensuing increased electrostatic interaction represents the strength of the bond. However, the two orbitals must overlap in phase, i.e., have the same symmetry about the bond axis. For example, if the 1*s* orbital of hydrogen overlaps the $2p_x$ orbital of fluorine (*x*-coordinate taken as the H F axis), the arising molecular orbital leads to bond formation. (See Figure 1.3.) If the $2p_z$ orbital of fluorine were used, however, the

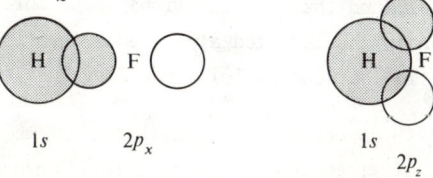

Figure 1.3.   Bond formation   No bond formation

4   Ch. 1 CHEMICAL BONDS

shaded and unshaded lobes would cancel each other, resulting in no net stabilization or bond formation.

The mathematics of the linear combination of two functions works out such that the combination of n atomic orbitals results in the formation of n molecular orbitals. Some will be more stable than the corresponding atomic orbitals (bonding orbitals) and some of the molecular orbitals will be less stable than the component atomic orbitals, called *antibonding* orbitals [5]. The shapes of some bonding and antibonding molecular orbitals are shown in Figure 1.4. (See Figure 6.1 for f orbitals.) Note that an

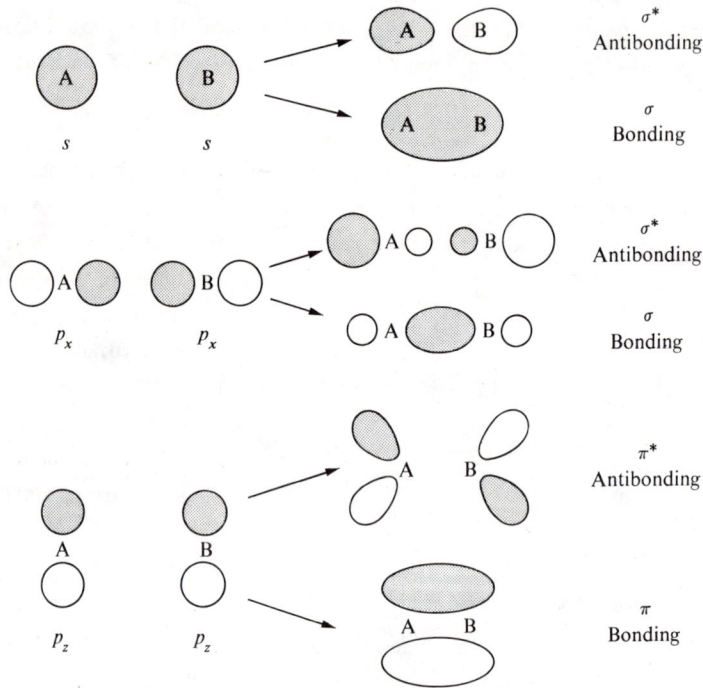

Figure 1.4.

antibonding orbital has a nodal plane perpendicular to the bond axis and between the two nuclei, and that the electrons spend more time farther from the nucleus than in the isolated atom. Consequently, there is less electrostatic attraction between nuclei, resulting in a separation of atoms.

Sec. 1.1 COVALENT BONDS 5

In working out the electronic configurations of molecules, we adopt the rules we used for determining electronic configurations of atoms, e.g., (1) only two electrons, of opposite sign, in each orbital; (2) fill lower energy orbitals first; (3) place an electron into each of two degenerate orbitals before filling either one. For example, the π-electron configurations of three basic systems are shown in Figures 1.5 to 1.7. The

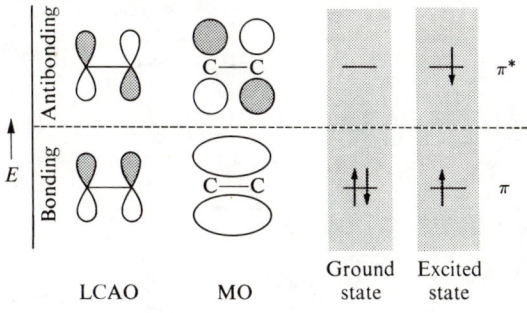

Figure 1.5. Configuration of π electrons in ground and excited states of ethylene.

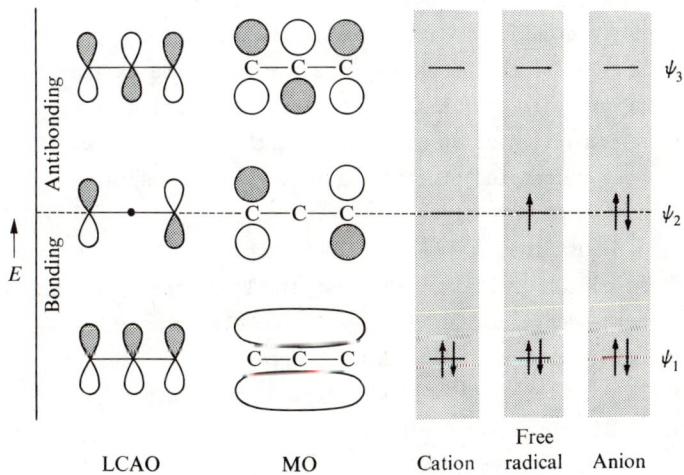

Figure 1.6. Configuration of π electrons in the cation, free radical, and anion of the allyl system.

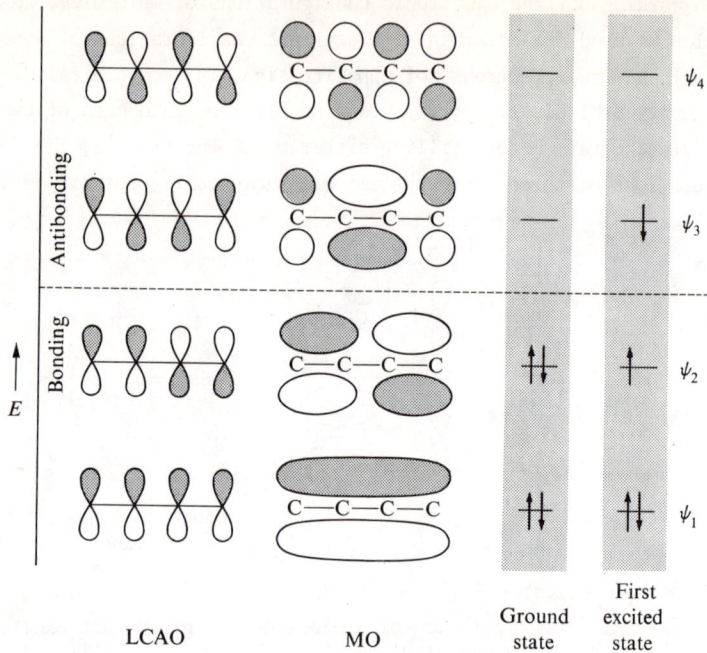

Figure 1.7. Configuration of π electrons in the ground and first excited states of 1,3-butadiene.

broken line in each figure indicates the nonbonding energy level; below it are the energies of the bonding orbitals π and above it lie the antibonding orbitals π*. Excitation (e.g., absorption of light energy) involves a transition of an electron from the highest occupied orbital to the lowest unoccupied orbital, still of the same spin. Light might be emitted in the opposite process.

For 1,3-butadiene, we combine four atomic p orbitals; hence there are four molecular orbitals. The most stable, representing the ground state, has the configuration $\psi_1^2\psi_2^2$, with two electrons in each of the ψ_1 and ψ_2 bonding orbitals. Although ψ_2 represents two isolated π orbitals, it is still more stable than four isolated atomic orbitals. Only ψ_1 encompasses all four carbon atoms.

In the allyl system

$$(CH_2 \cdots CH \cdots CH_2)^+ \qquad (CH_2 \cdots CH \cdots CH_2)^\cdot \qquad (CH_2 \cdots CH \cdots CH_2)^-$$

Cation   Free Radical   Anion

irrespective of the number of π electrons there are three contributing atomic p orbitals giving rise to three molecular orbitals, ψ_1, ψ_2, and ψ_3. As shown in Figure 1.6, ψ_1 is bonding, encompassing all three atoms; ψ_3 is antibonding; and ψ_2 encompasses only the end carbons (with a node at the middle carbon) and has the same energy as an isolated p orbital, i.e., it is nonbonding. Accordingly, the cation has the largest delocalization stabilization, i.e., the largest benefit resulting from atomic orbital overlap.*

In the case of benzene, there are six molecular orbitals, the three most stable of which are shown in Figure 1.8. Orbitals ψ_2 and ψ_3 are of

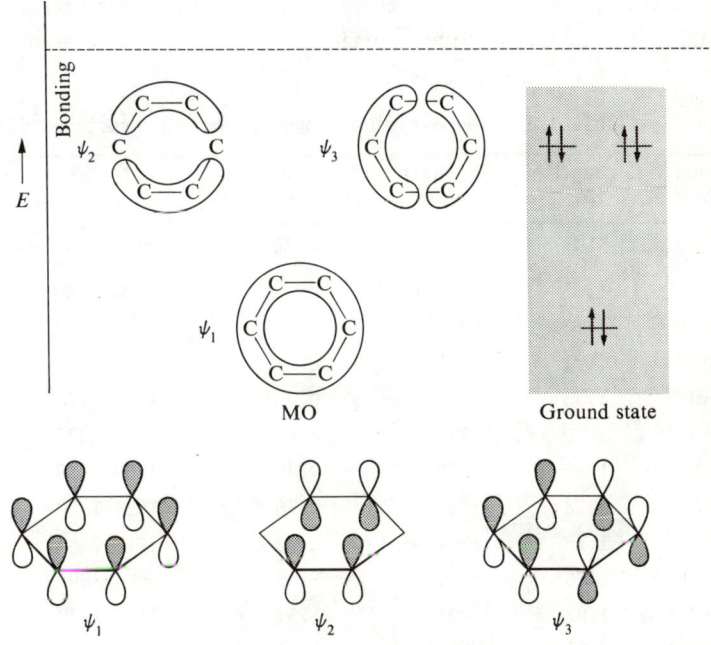

Figure 1.8. Configuration of π electrons in benzene (above) and its three most stable molecular orbitals (below).

*The correct delocalization energy for the allyl radical is still a matter of dispute; see, for example, Doering and Beasley [6].

Ch. 1 CHEMICAL BONDS

different shape but of equal energy (degenerate). More will be said about the aromatic character of benzene in Section 6.1c.

1.1b. Covalent bond correlations

Chemists are interested in molecular species whose structures and compositions are reproducible and are of sufficient stability to permit observations of their properties under chosen conditions; hence we shall use the following as a working definition of a chemical bond:

> A chemical bond exists between two or more atoms when the forces acting between them are of sufficient strength that the group maintains its integrity for a finite time under specified or assumed conditions.

On this basis, the minimum energy of a chemical bond at room temperature is about 2 kcal/mol.

You will recall that a good analogy for a covalent bond is a metal spring. Each bond type has a characteristic range of length, strength, force constant, ir stretching frequency, dipole, and other properties (see Table 1.1). Indeed, many empirical relationships have been drawn

Table 1.1. Some typical covalent bond characteristics

Bond type	Length (Å)	Strength (kcal/mol)	Dipole moment (D)	IR stretching frequency (cm^{-1})	UV λ_m (nm)
$H-C_{sp^3}$	1.12	99	0.30	2850-2960	
C-C	1.53	83	0	1100-1150	
C=C	1.33	145	0	1620-1680	180-190
C≡C	1.21	198	0	2100-2260	190
C-O	1.43	84	0.86	1060-1270	
C=O	1.20	179	2.40	1650-1825	280
C-N	1.47	70	0.40	1030-1230	
C=N	1.28	135	0.90	1615-1700	240
C≡N	1.15	204	2.93	2215-2260	
C-Cl	1.76	79	1.56	600-800	
N-H	1.03	93	1.31	3300-3500	
O-H	0.97	111	1.53	3590-3650	

among these bond parameters [7]. Occasionally, such a correlation is useful for estimating a specific property value for a bond which has not been measured or is difficult to measure.

Bond lengths and bond angles are determined by diffraction (X-rays, electrons, or neutrons) and spectral (infrared, microwave, or Raman) techniques [8]. For a given set of atoms, there are four factors which have a pronounced effect on specific internuclear distances:
1. Hybridization.
2. Electronegativity and polarizability.
3. Electron delocalization (see Section 6.1).
4. Steric conditions (see Section 20.1) [9].

Since electrons can get closer to the nucleus in an s orbital than in a p orbital, the nucleus has a stronger pull on electrons in an s orbital. For a hybrid orbital, then, the larger the s-character, the greater the nuclear attraction for the orbital electrons will be, i.e., the greater the electronegativity of the atom. Hence, the decreasing order for carbon is $C_{sp} > C_{sp^2} > C_{sp^3}$. **As the electronegativity of an atom increases,** whether because of inherent electronegativity or orbital hybridization, its bonding electrons are drawn closer to the nucleus. This decreases its effective atomic radius and shortens its bonds. Examples are given in Table 1.2. However, since the four factors above are sometimes in

Table 1.2. Structural effects on bond lengths (Å)

Effect of carbon hybridization on bond lengths [10]				Effect of inherent electronegativity and polarizability			
	C-H	C-Cl	C-Br		C-H		C-Cl
C_{sp^3}	1.102	1.771	1.927	CH_3CH_2-H	1.102	H_3C-Cl	1.780
C_{sp^2}	1.071	1.711	1.875	H_3C-H	1.094	H_2ClC-Cl	1.772
C_{sp}	1.058	1.631	1.795	Cl_3C-H	1.06	HCl_2C-Cl	1.763
						Cl_3C-Cl	1.755
						F_3C-Cl	1.72
					C-F		C=C [11]
				H_3C-F	1.391	H_2C=CH_2	1.335
				F_3C-F	1.323	F_2C=CF_2	1.313

opposition, the effect of any one is not always clear.

In the absence of delocalization and steric repulsion, the length of a covalent bond A-B is the sum of the normal covalent radii, $r_A + r_B$, of the atoms A and B. For instance, in ethane, chlorine, and carbon tetrachloride, where carbon and chlorine exhibit normal valences, the bond lengths are given by the respective sums

$$r_{C-C} = 0.763 + 0.763 = 1.53 \text{ Å}$$
$$r_{Cl-Cl} = 0.99 + 0.99 = 1.98 \text{ Å}$$

hence,

$$r_{C-Cl} = 0.763 + 0.99 = 1.753 \text{ Å}$$

which agrees with the observed C-Cl bond length in CCl_4. Some useful covalent radii are listed in Table 1.3.

Table 1.3.
Some useful covalent radii (Å)

Atom	Single	Atom	Single	Double	Triple
H	0.37	C	0.763	0.743	0.691
Li	1.34	N	0.74	0.62	0.55
B	0.81	P	1.10	1.00	
Si	1.17	O	0.74	0.62	0.55
F	0.72	S	1.04	0.94	
Cl	0.99				
Br	1.14				

Two common ways of expressing bond strengths are dissociation energies (12) and bond energy values (13). The dissociation energy of a bond R-X is the energy change for the reaction

R-X → R· + X·

carried out in the ideal gas state with the products in their ground states. The data, after correction for readjustment of experimental conditions, include rehybridization energies. In the case of methane for example, $CH_4 \rightarrow \cdot CH_3 + \cdot H$, carbon uses sp^3 hybrid orbitals in the molecule and sp^2 in the radical. Although instantaneous bond dissociation energies, i.e., dissociation energies without orbital rehybridization as dissociation takes place, have been reported (14) most data include the effects of rehybridization (15). Again, the four factors which affect bond length (hybridization, electronegativity and polarizability, delocalization, and steric crowding) have an effect on bond dissociation energies (16) as shown in Table 1.4.

Table 1.4.
Structural effects on dissociation energies

	Effect of hybridization		
	D_{C-C} * (kcal/mol)		D_{C-H} † (kcal/mol)
HC≡C-C≡CH	150	HC≡C-H	125
H_2C=CH-CH=CH$_2$	100	H_2C=CH-H	104
H_5C_6-C_6H_5	100	C_6H_5-H	103
H_5C_2-C_2H_5	87	H_5C_2-H	98
...	...		
HC≡C-CH$_3$	117		
H_5C_2-CH$_3$	85		

	Effect of electronegativity and polarizability		
	D_{C-C} * (kcal/mol)		D_{C-H} † (kcal/mol)
H_3C-CH$_3$	88	Me-H	104
H_3C-C_2H_5	85	Et-H	98
H_3C-CH(CH$_3$)$_2$	84	i-Pr-H	95
H_3C-C(CH$_3$)$_3$	80	t-Bu-H	91
(H$_3$C)$_3$C-C(CH$_3$)$_3$	67.5		...
		F_3C-H	104
		Cl_3C-H	96

* Data apply to the C-C bond shown. † Data apply to the C-H bond shown.

In general, dissociation energies decrease with smaller s-character of the bond orbitals, with decreasing electronegativities, and with increasing degree of alkylation. The analysis is not clear-cut, however, because rehybridization simultaneously involves a change in electronegativity. From relative rates of abstraction of iodine atoms from alkyl iodides by phenyl radicals we infer that the C-I bond dissociation energies decrease in the order Me to *t*-butyl (17). Also, the larger the size of X (X = OH, alkyl, phenyl, etc.), the greater is the difference between Me-X and *t*-Bu-X dissociation energies (18). In addition, the C-O dissociation energy of *t*-Bu$_2$O is only 69 kcal/mol compared to about 78 in most dialkyl ethers (12), and the N-H bond dissociation energy of ammonia and of ammonium ions decreases with increasing degree of alkylation for the same alkyl groups (19). Although it is tempting to attribute these trends to steric repulsion, it is really not clear to what extent σ induction, polarization, steric repulsion, and hybridization changes are responsible for these differences. Delocalization effects on dissociation energies are discussed in Section 6.2a.

The other method of assessing bond energies is by a bond energy term. This is the quantity of energy assigned to each bond type as determined from heats of combustion of simple molecules. For example, the heat of combustion of methane yields a value for the C-H bond. Then values for the C-C, C=C, and C≡C bonds may be obtained from the heats of combustion of ethane, ethylene, and acetylene, respectively, assuming of course that the C-H bond energy remains constant. Extension of this procedure to simple ethers, ketones, and amines would provide values for C-O, C=O, and C-N bonds. The premise is that the sum of the bond energy terms for all bonds in a molecule will give its heat of atomization. However, since two chain isomers may have different potential energies owing to different strain energies (bond angle, torsional, etc.), it is apparent that only an average or best-fit bond energy term can be assigned to each bond type. Although the values do not distinguish heats of atomization for 1,4-dimethylcyclohexane and cyclooctane, for instance, they can be used for estimating relative stabilities of functional isomers. Thus, there is a difference of from 15 to 20 kcal/mol between the bond energy summations for a simple ketone and an enol. This is consistent with the observation that simple ketones are so much more stable than their tautomeric enols that one detects the presence of enols by chemical or physical methods only with difficulty, if at all.

A set of bond energy terms is given in Table 1.5. As might be expected, electronegativity, polarizability, hybridization, electron delocalization, and steric strain can markedly affect bond energy terms.

Table 1.5

Some covalent bond energies

Single bond energies (kcal/mol)							
Elements		Hydrides		Chlorides		Miscellaneous	
H-H	104	H-H	104	H-Cl	103	C-F	105
C-C	83	C-H	99	C-Cl	79	C-Br	66
N-N	38	N-H	93	N-Cl	48	C-I	52
O-O	33	O-H	111	O-Cl	49	C-N	70
F-F	37	F-H	135	F-Cl	61	C-S	66
Si-Si	42	Si-H	76	Si-Cl	86	C-Si	69
P-P	51	P-H	76	P-Cl	79	Si-O	88
S-S	51	S-H	82	S-Cl	60	Si-F	129
Cl-Cl	58	Cl-H	103	Cl-Cl	58	Si-Br	69
Br-Br	46	Br-H	88	Br-Cl	52	N-F	65
I-I	36	I-H	71	I-Cl	50	O-F	44
						I-Br	43

Multiple bond energies (kcal/mol)			
Bond	Single	Double	Triple
C-C	83	147	199
N-N	38	98	225
O-O	33	96	
P-P	51		
S-S	51	101	
C-N	70	140	204
C-O	86	179	250
S-O		120	

Some relevant heats of atomization (kcal/mol)			
Compound	ΔH_a	Compound	ΔH_a
O_2	146	H_2O (liquid)	245
N_2	278	CO_2	411
Br_2 (gas)	46.2	SO_2	297
I_2 (solid)	51.2	HF	148

Bond-type	Effect of hybridization on bond energies (21)
$C_{sp^3}-C_{sp^3}$	85 (kcal/mol)
$C_{sp^3}-C_{sp^2}$	90
$C_{sp^2}-C_{sp^2}$	98
$C_{sp^3}-O$	92
$C_{sp^2}-O$	94
$C_{sp^3}-N$	72
$C_{sp^2}-N$	76

Thus, the higher the *s*-character of the carbon bonding orbital, the stronger is the bond.

So far, we have discussed the length and strength of covalent bonds. Other properties such as electric moments and infrared stretching frequencies will be discussed in later sections. Some comment should be made here, however, regarding bond angles.

You already know that bond angles about a given atom in a covalent molecule depend upon the type of bonding orbitals used by that atom. The configurations for some hybrid bond orbitals are listed in Table 1.6. Accordingly, atoms which use pure *p* orbitals, such as elements in the oxygen and nitrogen families, should form bond angles of 90°. Indeed, the bond angles in H_2S and PH_3 are 92.3° and 93.3°, respectively, not far from the theoretical angle. However, the bond angles in H_2O and NH_3 are 105° and 106.8°, respectively. Several explanations have been offered for these discrepancies. One is in terms of hybridization, another is electron-pair repulsion, and a third is repulsion between atoms attached to the central atom. Actually, each of these effects is a factor but their relative importance varies in different cases. These various explanations are attempts to develop a simple model for assessing these effects in different molecules. In an *ab initio* calculation, all of these factors are taken into account and the results generally agree with experiment.

Table 1.6.

Configurations of some hybrid bond orbitals

Coordination number	Orbital type	Configuration	Illustrative species
2	sp	linear	$Ag(NH_3)_2^{\oplus}$, $HgCl_2$
	sp or dp	linear	$I_3^{\ominus}$
	p^2	angular	H_2O, H_2S
3	sp^2	trigonal	BCl_3, SO_3^{-2}
	p^3	trigonal pyramid	PCl_3, NH_3
4	sp^3	tetrahedral	CH_4, $SiCl_4$, $Ni(CO)_4$
	dsp^2	square planar	$Ni(CN)_4^{-2}$
	d^2p^2	square planar	$ICl_4^{\ominus}$
5	dsp^3	trigonal bipyramid	PCl_5, $Fe(CO)_5$
6	d^2sp^3	octahedral	SF_6, most Werner octahedral complexes
8	d^4sp^3	dodecahedral, with triangular faces	$Mo(CN)_8^{-4}$

In H_2O, for example, the O-H bonds have considerable polar character due to the large difference between the electronegativities of O and H. This produces a repulsion between the partially charged hydrogens to spread the H-O-H angle. The angle in H_2S is smaller because sulfur, which is less electronegative than oxygen, forms less polar bonds and the larger sulfur atom spaces the hydrogens farther apart. Consequently, the nonbonded H,H repulsion is smaller in H_2S than in H_2O because the hydrogens carry less positive charge and are farther apart. This view accounts for the decreasing bond angles, approaching the theoretical $90°$, as one descends the periodic table along these families:

NH_3 $106.8°$ H_2O $105°$
PH_3 $93.3°$ H_2S $92.3°$
SbH_3 $91.5°$ H_2Se $91°$

16   Ch. 1 CHEMICAL BONDS

Interestingly, two factors -- nonbonding interatomic distance and bond dipole charge -- have opposite trends in the compounds NF_3, PF_3, and AsF_3. The net effect is a constant bond angle of 102° about the central atoms.

An alternate view in accounting for bond angles is the electron-pair repulsion theory (1). The theory proposes that the arrangement of bonds around a given atom depends upon the number of *electron pairs* in its valence shell, both shared and lone pair. The bonding electrons diffuse to maximize the distance between bonds, which for coordination numbers of 2 to 6 are

AX_2	linear
AX_3	trigonal
AX_4	tetrahedral
AX_5	trigonal bipyramidal
AX_6	octahedral

Lone-pair electrons (E) hold a position just as a bonding pair; hence, AX_4, AX_3E, and AX_2E_2 are all close to tetrahedral.

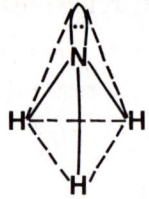

	Bond angle
CH_4	109.5°
NH_3	106.8°
H_2O	105°

NH_3 H_2O

Bond angles can be accounted for fairly well by using three working assumptions:

1. <u>A lone pair takes up more space in the valence shell of an atom than a bonding pair.</u> Hence, the angles in CH_4, NH_3, and H_2O decrease in that order because the lone pairs take more space and "squeeze" the bonds together. As a hypothetical model, BrF_6 would have an octahedral structure with 90° angles. The lone pair in BrF_5 squeezes the bonds together to decrease the F-Br-F angles (22).

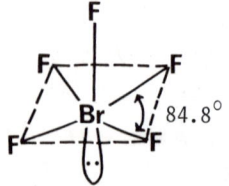

Sec. 1.1 COVALENT BONDS   17

2. <u>The "volume" of bonding electrons decreases with increasing electronegativity of the ligands.</u> For example,

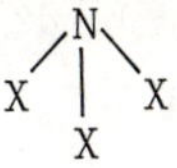

$$\begin{array}{c} N \\ X\ |\ X \\ X \end{array} \quad \begin{array}{l} X = H, 107° \\ F, 102° \end{array} \qquad \begin{array}{c} O \\ X\ \ X \end{array} \quad \begin{array}{l} X = H, 105° \\ F, 100° \end{array}$$

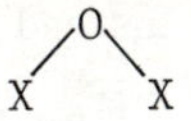

The greater the electronegativity of the ligands, the more the cross section of the bond orbital shrinks at the central atom thereby decreasing the bonding electron-pair repulsion, which results in smaller angles. Additional examples are as follows:

X	∠X-C-X (CH₂=C)	X	∠X-C-X (O=C)	X	∠X-S-X (O=S)	X	∠X-P-X (O=P)
H	116°	H	115.8°	CH₃	100°	Br	106°
Cl	114°	Cl	111.3°	C₆H₅	97.3°	Cl	103.6°
F	110°	F	108°	Br	96°	F	102.5°
				F	92.8°		

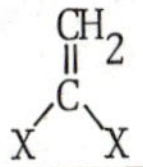

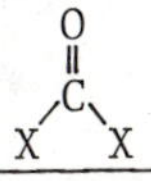

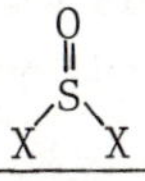

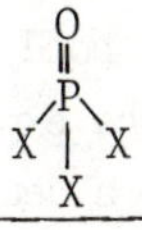

However, inverting cases such as those below are not rare.

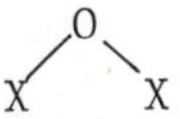

$$\begin{array}{c} O \\ X\ \ X \end{array} \quad \begin{array}{l} X = H, 105° \\ Cl, 110.8° \end{array} \qquad \begin{array}{c} S \\ X\ \ X \end{array} \quad \begin{array}{l} X = H, 92.3° \\ Cl, 102° \end{array}$$

3. <u>Electron pairs of double and triple bonds take up more space than single bonds.</u> In olefins, angle <u>a</u> is always smaller than 120° and angle <u>b</u> is always larger than 120°.

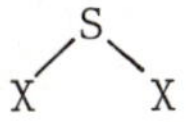

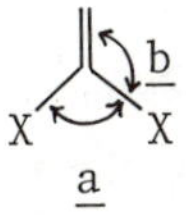

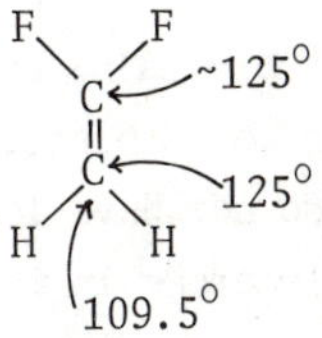

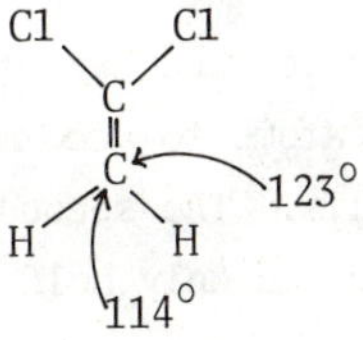

(diagrams: general olefin with angles a and b; $CF_2=CH_2$ with F–C–F ~125°, C=C–H 125°, H–C–H 109.5°; $CCl_2=CH_2$ with Cl–C–Cl 123°, H–C–H 114°)

QUESTIONS

1. What is an argument against the proposal that carbon uses four sp^3 hybrid bonds in ethylene instead of three sp^2 and one π bond?
2. Explain the occurrence of SiF_6^{-2}, PF_6^{-}, and SF_6 but not the corresponding CF_6^{-2}, NF_6^{-}, or OF_6 species.
3. Draw a picture of the orbitals of allene, $CH_2=C=CH_2$.
4. Do you expect the C-Cl bond in CF_3Cl to be shorter than that in CCl_4, and on what basis?
5. What C-C bond length would you expect for $H-C{\equiv}C-CN$?
6. What bond angles would you predict for the hydronium ion, H_3O^+? If the experimental angle were about $110°$, would this help you make a decision?
7. Offer an explanation for the relative C-C and C-H bond lengths (in Å) shown below.

CH_3-CH_3

$\begin{matrix} Cl \\ \diagdown \\ CH-CH_2-H \\ H\nearrow \end{matrix}$

d_{C-H}
d_{C-C} 1.545

1.089 1.092
1.52

Answers

1. Chiefly, the bond angles found in ethylene are not those of four tetrahedral bonds around each carbon.

2. The first three make use of three d orbitals, and these enable the central atoms to accommodate up to 18 electrons in their valence shells. The second three do not have low-energy d orbitals and can only hold eight electrons in their valence shells.

3.

4. On the basis that the carbon of CF_3Cl is more electronegative than that in CCl_4, the C-Cl bond in CF_3Cl should be shorter. Experimentally, it is found to be 0.017 Å shorter (L. S. Bartell and L. O. Brockway, *J. Chem. Phys.*, 23: 1860 (1955)).

5. From Table 1.3, $2 \times 0.691 = 1.38$ Å. This is the experimental value (A. A. Westenberg and E. B. Wilson, Jr., J. Am. Chem. Soc. 72: 199 (1950)).

6. One possibility would be for oxygen to use p orbitals with expanded bond angles near those of H_2O. Another possibility would be for oxygen to use sp^3 bond orbitals. Since sp^3 bonds are stronger than p bonds, the tetrahedral structure would be predicted. An experimental bond angle of about 110° supports this prediction.

7. The chlorine atom increases the electronegativity of the attached carbon over that of the β carbon or a carbon atom of ethane. This shortens the C-C bond of ethyl chloride and shortens the α C-H bond length compared to the β C-H.

1.2 Hydrogen bonds (See also Chapter 12 and Section 20.2.)

You are already quite familiar with hydrogen bonds (H bonds), especially the O-H···O and N-H···N bonds in alcohols, acids, and amines, intramolecular as well as intermolecular. Actually they are fairly ubiquitous and play an important role in nature. For example, H bonds are largely responsible for the helical structure of proteins and nucleic acids and, hence, are deeply involved in controlling many biological processes such as cell development and arrangement of genetic material. H bonds bring about the emulsifying action of nonionic detergents, the adhesion of some glues, the binding of many dyes, and are responsible for the liquid state of water at room temperature. H bonds contribute to all these effects even though the strength of most H bonds is in the range of only 2 to 7 kcal/mol. The energies of some bond types and of some specific H bonds are given in Tables 1.7 and 1.8.

1.2a The nature of H bond forces

The use of the concept of H bonds (23) was slow to gain prominence but with the instrumentation renaissance after World War II a great many studies were made of the nature and scope of H bonding. Although all interatomic forces are electrostatic, it is convenient to partition the H bond energy into four categories (24, 25):

Ch. 1 CHEMICAL BONDS

Table 1.7.
Energies of some types of H bonds

Type of bond	Typical energy (kcal/mole)	Type of bond	Typical energy (kcal/mole)
O-H···N	7	N-H···O	2-3
O-H···O	6	N-H···N	2-4
C-H···O	2-3	N-H···F	5
O-H···π electrons	1-4		

Table 1.8
H bond energies in some specific systems

Bond	Substance and State	Energy (kcal/mole)
F⁻···H⁺···F⁻	Me_4NHF_2, solid	37
F-D···F	$(DF)_6$, vapor	6.85
F-H···F	$(HF)_6$, vapor	6.8
O-D···O	$(CH_3CO_2D)_2$, vapor	7.95
O-H···O	$(CH_3CO_2H)_2$, vapor	7.64
	H_2O, ice	6.4
	$\underline{o}$-HO-$C_6H_4CO_2H$, intramolecular	4.7
	$(CH_3OH)_4$, vapor	6.05
O-H···Cl	$\underline{o}$-Cl-C_6H_4OH, intramolecular CCl_4, solution	1.7
N-H···O	$(C_6H_5)_2CO:H_2NC_6H_5$, CCl_4 solution	2.0
C-H···N	$(HCN)_3$, vapor	4.4
C-H···O	$(CH_3)_2CO:HCCl_3$	2.7

1. Electrostatic attractions, i.e., Coulombic forces arising when the charges on the atoms are regarded as point charges without any delocalization from polarization.
2. Delocalization effects, arising from mutual polarization of valence electrons.

3. Dispersion forces resulting from instantaneous charge fluctuations of the valence shell electrons of the bridged atoms.
4. Repulsive forces between inner shell electrons.

Let us look at some of the data in terms of these four parameters.

1. <u>Electrostatic interactions</u>. The fact that H bonds always involve highly electronegative atoms suggests that dipole-dipole forces are a major component of the strength of H bonds. However, logic dictates that there must be some repulsive forces operating to prevent the H bond from collapsing. Furthermore, if dipole-dipole interactions were the only attractive forces, one would expect a correlation with dipole moments. This is not the case, as shown by the fact that CH_3NO_2 and CH_3CN have large dipole moments but form only weak hydrogen bonds as H-bond acceptors. Nevertheless, H bonds, A-H$\cdots$B, are formed only when atom A is highly electronegative, thereby giving the A-H bond a substantial polarity, $^{\delta-}A$-$H^{\delta+}$. For example, ordinary C-H bonds do not form H bonds with other atoms. However, if the electronegativity of C is increased by <u>sp</u> hybridization or by attaching other highly electronegative atoms, then the arising polar C-H bonds will form H bonds. Some illustrations are:

$$\begin{array}{c} Cl \\ | \\ HC\text{-}Cl\cdots H\text{-}CCl_3 \\ | \\ Cl \end{array} \qquad HC\equiv N\cdots H\text{-}CN\cdots H\text{-}CN$$

$$\begin{array}{c} O \\ \| \\ H_3C\text{-}C\text{-}H\cdots O=CHCH_3 \end{array} \qquad HC\equiv C\text{-}H\cdots OR_2$$

2. <u>Delocalization effects</u>. It is advantageous to use the resonance hybrid <u>1.1</u> as a model for H bonds, where <u>1.1(a)</u> and <u>1.1(b)</u> reflect electrostatic attractions (category 1 above) and <u>1.1(c)</u> the delocalization or charge-transfer effect (category 2).

$$\left[^{\delta-}A\text{-}H^{\delta+}\cdots B^{\delta-}, \quad ^{-}A\cdots \overset{+}{H}\cdots B, \quad ^{-}A\cdots \overset{+}{H\text{-}B} \right]$$

  (a)   (b)   (c)
                   1.1

As Coulson has pointed out, atoms H and B are closer together in some H bonds than the sum of their van der Waals radii. Therefore, there must be some covalent bonding between H and B, otherwise there would be a large repulsion to prohibit formation of the H bond. Indeed, as the overall AB distance decreases, the strength of the bond increases and the A-H and H···B distances approach equality. Thus, the symmetrical bonds are unusually strong, such as the ⁻F···Ḧ···F⁻ bond (37 kcal/mol) in bifluoride salts, (26), whereas the strength of the unsymmetrical F-H···F⁻ bond is in the normal range (6.8 kcal/mol). For O-H···O bonds, the bond becomes symmetrical at an O···O distance of ~2.4 Å. On this basis, the H bond in hydrogen maleate ion is much stronger than that in the diacid.

6-Hydroxy-2-formylfulvene

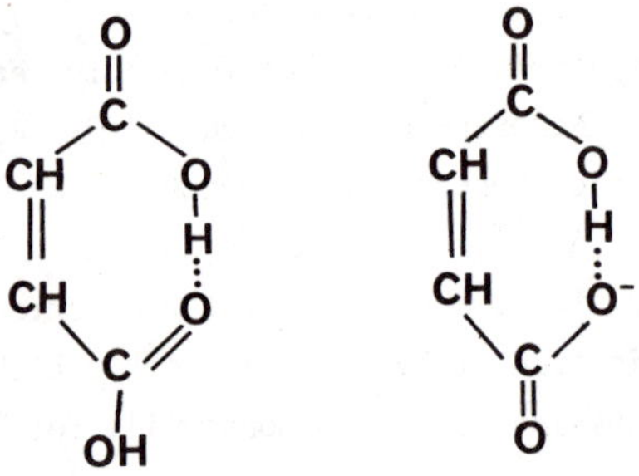

O-H···O distance 2.75 Å   2.44 Å

Symmetrical H bonds are also found by neutron diffraction in calcium bis-dihydrogen arsenate, $Ca(H_2AsO_4)_2$, with an O-H···O distance of 2.436 Å (27), and by microwave spectroscopy in 6-hydroxy-2-formylfulvene (28). Interestingly, associated liquids generally have larger densities and smaller molar volumes than nonassociated liquids. Owing to H bond formation, molecules of the associated liquids are more closely packed.

Additional evidence for the contribution of 1.1(c) to H bonds is the observation that the greater the electron donor character of B the stronger are H bonds. Thus, OH as a proton donor forms stronger bonds with nitrogen or sulfur than with oxygen atoms (see also Chapter 12). This is shown by $\Delta\nu_{OH}$ shifts in ir spectra ($\Delta\nu_{OH} = \nu_{OH} - \nu_{OH\cdots B}$) (29). The stronger the H bond, the larger is $\Delta\nu_{OH}$. Similarly, the C-H···N bond is stronger in $Cl_3CH:N(C_2H_5)_3$ than in $Cl_3CH:NH_3$, since Et_3N is a stronger base than NH_3 in nonaqueous solvents or the gas phase toward small acids (Section 4.2b) (30).

Structure	$\Delta\nu_{OH}$ (cm^{-1})	Structure	$\Delta\nu_{OH}$ (cm^{-1})
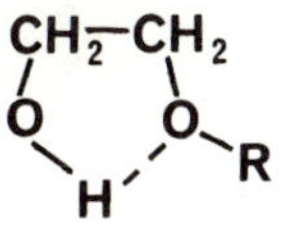 CH₂–CH₂ / O···H / O–R (3-membered)	31	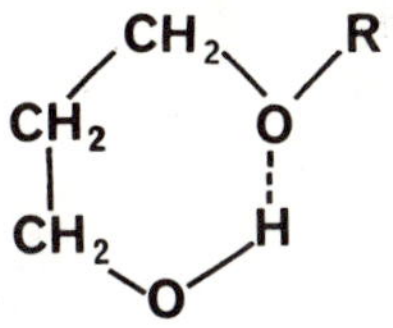 6-membered ring with O···H–O–R	87
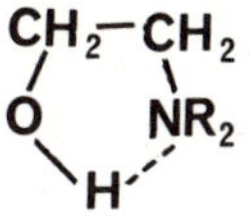 CH₂–CH₂ / O···H / NR₂	138		
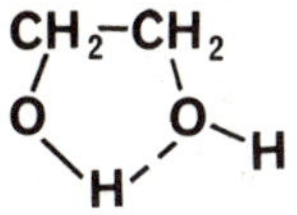 CH₂–CH₂ / O···H / O–H	32	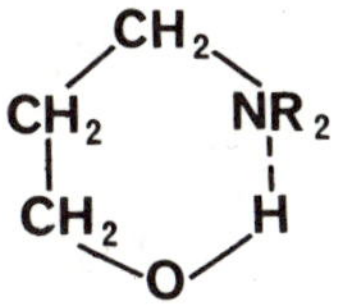 6-membered ring with O···H–NR₂	327
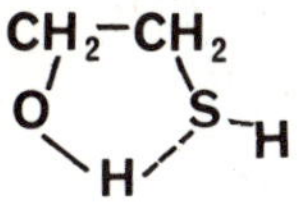 CH₂–CH₂ / O···H / S–H	93		

As might be expected, the relative strengths of the O-H···O and O-H···N bonds are reversed in a β-hydroxyaniline because the lone-pair electrons of nitrogen are now shared with the ring rather than with the oxygen atom.

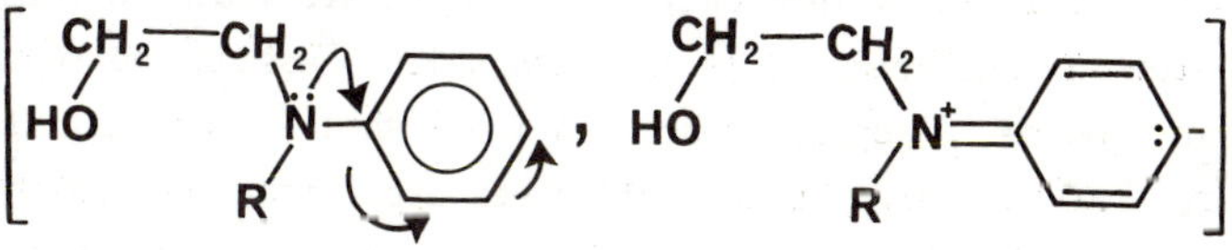

Moreover, as the negative charge on atom A increases, the covalent character of the A-H bond decreases [31].

The electrons involved in structure 1(c) are lone-pair electrons of B. However, they may also be π electrons of a molecular orbital. The most common way of detecting such bonds is by ir spectroscopy, in which $\Delta\nu_{OH}$ ranges from 15 to about 80 cm^{-1} and O-H···π bond strengths range from 1 to 4 kcal/mol [32].

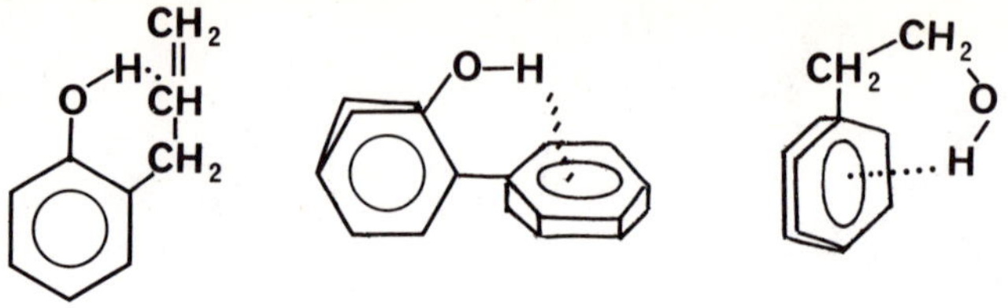

$\Delta\nu_{OH}(cm^{-1})$: 63   45   30

Groups which increase K_a of the phenol, increase $\Delta\nu_{OH}$.

Groups in ring which increase basicity of the π electrons, increase $\Delta\nu_{OH}$.

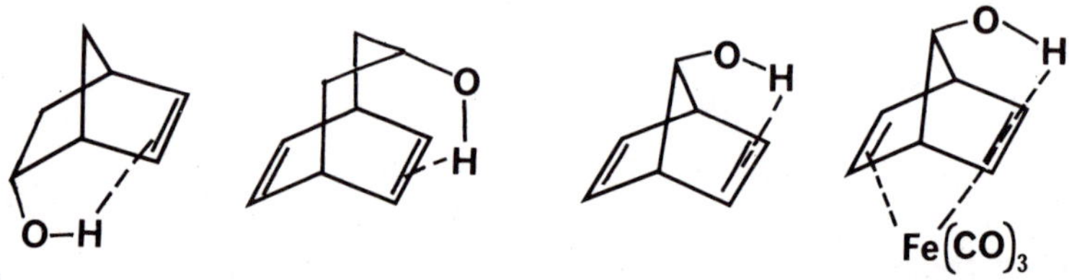

$\Delta\nu_{OH}(cm^{-1})$: 30   32   73   46

3. **Dispersion forces.** Dispersion forces between inner shells of nonbonded atoms are of significant magnitudes and affect dissociation of diatomic molecules [33]. These are dipole-dipole attractions between dissimilar instantaneous charges of oscillating inner shell electrons.

4. **Repulsive forces.** As stated earlier, there must be some force counteracting the three types of attractive forces in an H bond to prevent its collapse. Since only D, H, and Li are found to form the center atom in such bridges, it is reasonable to assign the repulsive forces in H bonds to repulsion between inner shell electrons. That is, such forces are so large between all other atoms that they inhibit the formation of such a bridge. Although sodium, for instance, has a low electronegativity and an O-Na bond is very polar, $\bar{O}$-$\overset{+}{Na}$...$\bar{O}$ bonds would be very weak owing to strong inner shell repulsion. Lithium, however, has a small kernel and has a vacant low-lying 2p orbital for π bonding with a donor atom (i.e., the promotional energy 2s to 2p for hybridization is smaller for Li than for 1s to 2p for hydrogen). As a result, $\bar{O}$-$\overset{+}{Li}$...$\bar{O}$ and $\bar{C}$-$\overset{+}{Li}$...$\bar{C}$ Li bonds are formed [34]. For example, lithium alkoxides are polymeric in solution and in the solid state,

Sec. 1.2 HYDROGEN BONDS 25

$$\overset{+}{Li}-\overset{-}{O}\cdots\overset{+}{Li}-\overset{-}{O}\cdots\overset{+}{Li}-\overset{-}{O}\cdots \quad \text{and} \quad R-O\overset{Li}{\underset{Li}{\cdots}}O-R,$$
  \R \R \R

and lithium alkyls are polymeric in solution and the vapor phase,

$$\overset{+}{Li}-\overset{-}{C}\cdots\overset{+}{Li}-\overset{-}{C}\cdots\overset{+}{Li}-\overset{-}{C}\cdots$$
  \R \R \R

Lithium t-butyl, for instance, is a tetramer [35].

The contribution of the four types of forces just discussed to the over-all H bond strength in O-H···O bonds has been estimated to be as follows [25]:

Electrostatic	6 kcal/mol
Delocalization	8
Dispersion	3
Repulsion	-9
Net	8
Experimental	6

At least the total estimate is not far from experimental values. The individual components, of course, may vary for different H bonds. Experimentally, it is found that the strengths of H bonds A-H···B vary with the electronegativity of A and with the electron-donor character of B. The effect of the electronegativity of A has been identified with correlations between the polarity of A-H, the acidity of A-H, or the electron affinity of A; and the electron-donor character of B has been noted by correlations with the basicity of B and the ionization potential of B. For example, electron-donating groups in the 5 position of structure 1.2 increase the basicity of the OCH_3 and increase the H-bond strength. This lowers the O-H···O ir frequency by 26 cm^{-1}. Such groups in the 4 position, however, decrease the acidity of the COOH and decrease the H-bond strength. This increases the OH frequency by ~20 cm^{-1}. Similarly, electron-withdrawing groups in the 4 position of structure 1.3 decrease the basicity of the carbonyl oxygen, weaken the H bond, and cause the OH nmr chemical shift to move upfield, whereas electron-withdrawing groups in the 5 position strengthen the H bond and cause the chemical shift to move downfield.

Since deuterium is slightly less electronegative than hydrogen, the O-D bond is more polar and the O-D···O bonds are slightly stronger than O-H···O bonds. This order has been observed for the self-association of water, $D_2O > H_2O$; for H bonding of phenols and fluoroform with Lewis bases, $\phi OD > \phi OH$ and $DCF_3 > HCF_3$; and for H bonding of alkylperoxides with aniline, t-BuOOD > t-BuOOH. Consistent with effects caused by electronegativity, the strongly acidic hexafluoroisopropanol forms such a strong H bond with THF that the complex can be distilled (ν_{OH} 3141 cm^{-1}). One of the strongest O-H···O bonds, based on the OH ir frequency, is found in the hydrated hydroxytropylium ion. We also expect the ion to be quite acidic. In contrast to alcohols, phenols [36], and acetylenes [37], intermolecular H bonds of carboxylic acids [38] <u>decrease</u> in strength with increasing acid dissociation constants. The decreasing order of strength of intramolecular H bonds in <u>ortho</u>-substituted phenols, <u>o</u>-X-C_6H_4OH, approximates the decreasing basicity of X: N > S > I > Br > Cl > O > F.

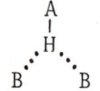

B.p. 100°

ν_{OH} 2390 cm^{-1}

Bifurcate hydrogen bonding, in which an A-H group is coordinated by two proton acceptor atoms B, has been the subject of some dispute for decades. There are some results in support of such bonding but the data are not strongly conclusive [39].

1.2b Effects of H bonds on molecular properties

Much has already been written on the effects of H bonds on molecular properties [40]. For the sake of accessibility, however, a few comparisons will be made here. It is usually well established in the first year of organic chemistry that intermolecular H bonding raises transition temperatures and increases solubility in water. For example, carboxylic acids have higher boiling points and are much more soluble in water than isomeric esters (see Table 1.9). Similarly, as the OH groups of diols and triols are covered by alkyl groups, H bonding is diminished and the boiling points decrease as follows in spite of the larger molecular weights of the ethers.

Table 1.9

Boiling points and water solubilities of some isomeric acids and esters

Molecular Formula	Acid	Ester	Boiling Point Acid	Boiling Point Ester	Water solubility (% near room temperature) Acid	Water solubility Ester
$C_2H_4O_2$	Acetic	Methyl formate	118°	32°	∞	30
$C_3H_6O_2$	Propionic	Methyl acetate	141°	57°	∞	33
$C_3H_6O_2$	Propionic	Ethyl formate	141°	54°	∞	11
$C_4H_8O_2$	n-Butyric	n-Propyl formate	168°	81°	∞	2.2
$C_4H_8O_2$	n-Butyric	Ethyl acetate	168°	77	∞	9
$C_4H_8O_2$	n-Butyric	Methyl propionate	168°	80°	∞	0.5

	$\begin{array}{c}CH_2OH\\|\\CHOH\\|\\CH_2OH\end{array}$	$\begin{array}{c}CH_2OEt\\|\\CHOH\\|\\CH_2OH\end{array}$	$\begin{array}{c}CH_2OEt\\|\\CHOH\\|\\CH_2OEt\end{array}$	$\begin{array}{c}CH_2OEt\\|\\CHOEt\\|\\CH_2OEt\end{array}$
Boiling point	290°	230°	191°	185°

An important biological aspect of intermolecular H bonding is the fact that fibrous proteins are constrained to the alpha-helical configuration because of solvation by water via H bonds. When a polypeptide is placed in D_2O instead of H_2O, more of it assumes a random form [41]. This can be attributed to less H bonding between D_2O and the polypeptide because D_2O is self-associated to a greater extent than is H_2O [42].

Compounds which can form intramolecular H bonds, however, have lower transition temperatures and water solubilities than isomers which form predominantly intermolecular H bonds. Thus, many ortho-substituted phenols and 1,2-glycols have lower melting and boiling points than their isomers.

	Intramolecular	Intermolecular
	$\begin{array}{c}CH_3\text{-}CH\text{-}\text{-}CH\text{-}CH_2CH_3\\ \quad |\quad\quad |\\ \quad O\quad\quad O\\ \quad\quad\diagdown H\diagup H\end{array}$	$\begin{array}{c}CH_2\text{-}CH_2CH_2CH_2\text{-}CH_2\\ |\quad\quad\quad\quad\quad\quad |\\ HO\quad\quad\quad\quad OH\\ O\text{-}\quad\quad\quad\quad\text{-}O\end{array}$
Boiling Point	185°	238°

Ch. 1 CHEMICAL BONDS

	Melting Points	
	Ortho isomer	Para isomer
Nitrophenols	44°	114°
Hydroxybenzaldehydes	-7°	116°
Dihydroxybenzenes	104°	169°
Fluorophenols	16°	48°
Chlorophenols	9°	43°
Bromophenols	5°	64°

Whereas nitration of a substance usually raises its boiling point, nitration of resorcinol in the 2 position and of catechol in the 3 position lowers their boiling points. These decreases can be attributed to intramolecular H bonding in the nitrophenols.

2-Nitroresorcinol   3-Nitrocatechol

Another property strongly affected by H bonding is the acid dissociation constant. If the H bond stabilizes the ion, it promotes dissociation, whereas if the H bond impedes ionization or stabilizes the undissociated form, the dissociation constant is decreased. Thus, on the one hand o-halophenols are weaker acids than the p-isomers owing to H bonding in the acid form of the former. On the other hand, o-hydroxybenzoic acids are stronger acids than the para isomers or the o-methyl ethers owing to stabilization of the ion by an H bond.

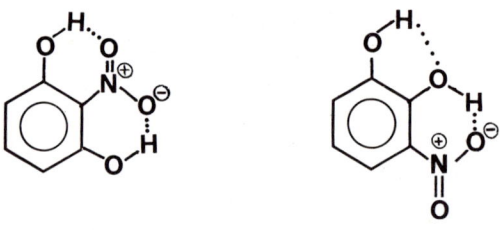

	$10^5 K_a$		
	ortho	meta	para
HOC_6H_4COOH	105	8.3	2.9
$MeOC_6H_4COOH$	8.1	8.2	3.4

The dissociation of o-hydroxybenzoic acid is 17 times that of benzoic acid, and that of 2,6-dihydroxybenzoic acid is 800 times. Similarly, the pK_a of the alcohol in structure 1.4 is the smallest known for an aliphatic alcohol. For comparison, the pK_a of the alcohol $(F_3C)_3COH$ is 9.52. Therefore the unusual acid strength of 1.4 is more than can be attributed to an inductive effect.

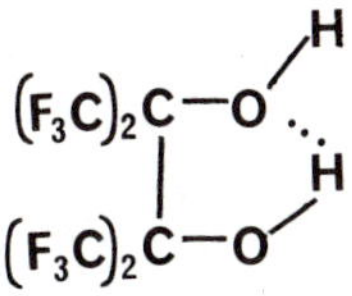

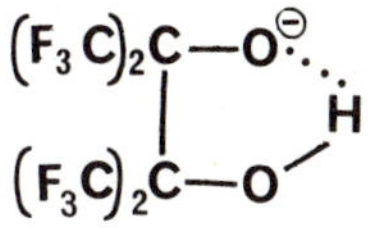

$$(F_3C)_2C-O\cdots H \cdots O-C(CF_3)_2 \qquad (F_3C)_2C-O^- \cdots H-O-C(CF_3)_2$$

1.4

pK_a 5.95

The mono ion is stabilized by a strong intramolecular H bond which is probably symmetrical.

Likewise, structure $1.5^{\oplus}$ is a slightly stronger acid than $1.6^{\oplus}$ because the third ring of $1.6^{\oplus}$ holds the two N atoms in a plane to form a stronger $^{\oplus}$N-H···N bond than in $1.5^{\oplus}$. Also, $1.6^{\oplus\oplus}$ is a stronger

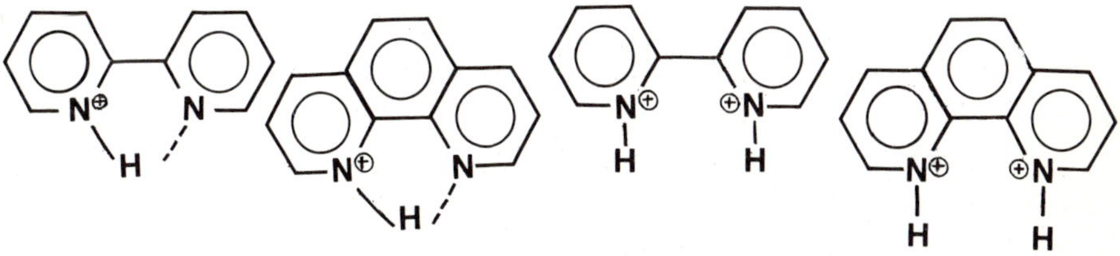

$1.5^{\oplus}$	$1.6^{\oplus}$	$1.5^{\oplus\oplus}$	$1.6^{\oplus\oplus}$
pK_a 4.4	5.0	-0.52	-1.55

acid than $1.5^{\oplus}$ because steric repulsion between $^{\oplus}$N-H groups cannot be relieved in $1.6^{\oplus\oplus}$ as it can in $1.5^{\oplus\oplus}$ by rotation of the pyridine rings. This causes a greater tendency for dissociation than in $1.5^{\oplus\oplus}$, and the resulting ion $1.6^{\oplus}$ is more stable than the resulting ion in the case of dissociation of $1.5^{\oplus\oplus}$ to give $1.5^{\oplus}$ [43]. Similarly, 1.7 is an unusually strong proton acceptor because the ion can form a very strong intramolecular H bond [44].

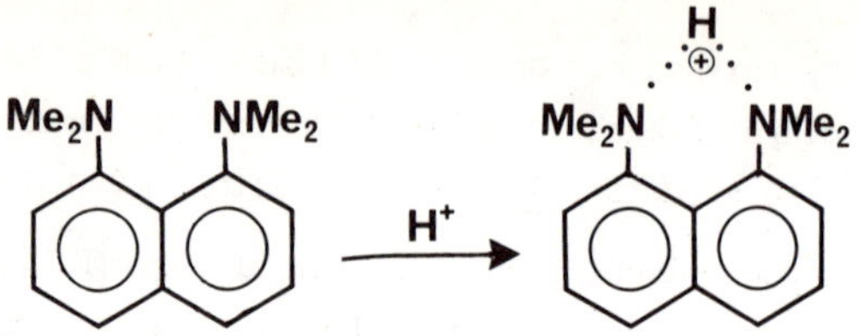

$pK_a = 12.34$

The relative acidity of several pairs of isomeric enols is affected by intramolecular H bonding:

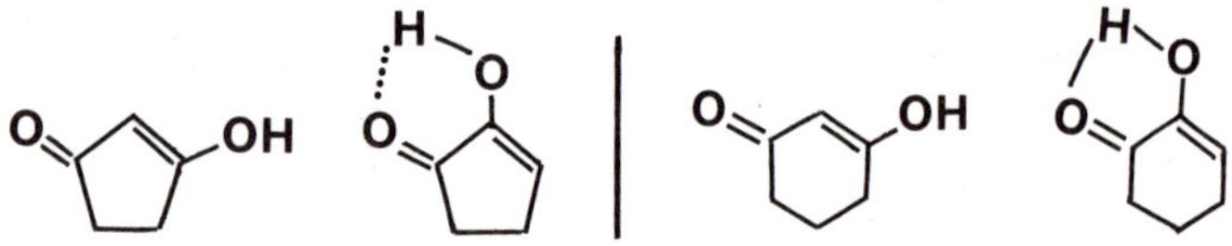

pK_a [45]   4.5   9.1   5.25   10.3

Not only does the H bond lower the dissociation of the 1,2-isomers, but the anions of the 1,3-isomers are resonance stabilized as follows:

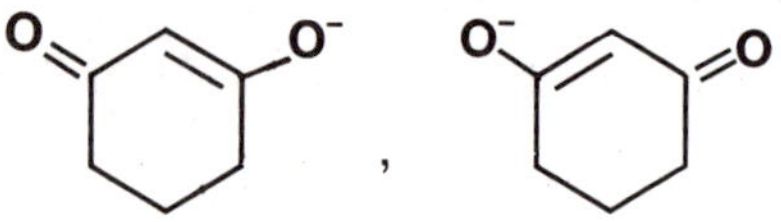

Both effects make the 1,3-isomer the stronger acid.

H bonding is largely responsible for the large K_1/K_2 ratios of some dicarboxylic acids. As the two carboxyl groups get closer to each other in molecules, they exert electrostatic effects on each other. The first $CO_2^\ominus$ group of the fumarate ion, for example, has a mildly retardant electrostatic effect on the dissociation of the second COOH group and decreases K_2 relative to a long-chain dicarboxylic acid. By contrast, the $CO_2^\ominus$ group of the acid maleate ion inhibits dissociation of the second COOH by H bonding to give a very small K_2, and energetically stabilizes the mono ion relative to the undissociated acid, thereby increasing K_1. The K_1/K_2 ratio is therefore very large.

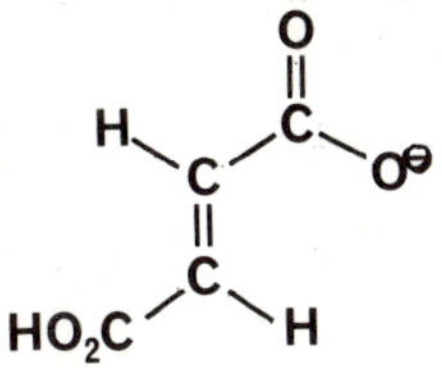

Acid fumarate ion	Acid maleate ion	cis-3,3-Diphenylcyclopropyl-1,2-dicarboxylic acid ion (1.8)
$K_1/K_2 = 32$	26,000	5.5×10^6

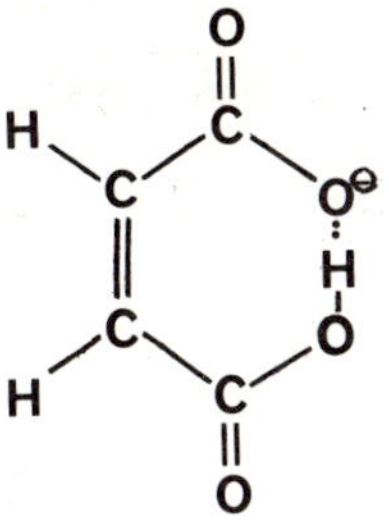

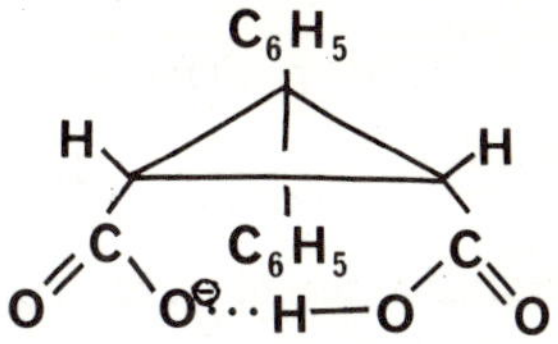

The H bond strength for the ion **1.8** is about 18 kcal/mol, which results in a very large K_1/K_2 ratio [46].

H bonding affects many other physical properties too numerous to discuss separately, such as color [47], wet-melting point depressions [47], heat of mixing, heat of vaporization [48], dipole moments [49], refractive index [50], viscosity [51], polarographic reduction potentials [52], titration curves in nonaqueous solvents [53], conformations, heats of combustion [54], photoreactivity [55], mass spectra [56], and others. H bonding is a major force in the association of molecules, particularly as portrayed in chromatographic processes, and these interactions are frequently studied by spectroscopic techniques [56a]. Presumably, H bonding affects all physical properties, although sometimes it is difficult to determine quantitatively the contribution of H bonding among other effects. General effects of H bonding upon some physical properties are listed in Table 1.10. The influence of H bonding upon spectra, and, in particular, the use of spectra to study H bonding are discussed in Chapter 12.

Table 1.10

Qualitative effects of H bonds on some physical properties

Property	Intramolecular H bonds	Intermolecular H bonds
Transition temperature	Close to normal	Increased
Vapor pressure	Close to normal	Decreased
Mixed melting points	Slightly depressed	Greatly depressed
Molecular weight	Close to normal	Increased
Water solubility	Decreased	Increased
Solvent power	Close to normal	Increased
Ionization	Decreased	Increased
Isomer stability	Increased	Close to normal
Adsorption on polar surfaces (chromatography)	Close to normal	Increased
Infrared absorption		
Stretching frequency	Decreased	Increased
Bending frequency	Increased	Decreased
Effect of environment	Small	Large
Electronic absorption	Large increase	Small increase or decrease
Nmr chemical shift	To lower field	To lower field
Other properties		
Molar volume	Small decrease	Large decrease
Viscosity	Close to normal	Higher
Dielectric constant	Relatively low	Relatively high
Surface tension	Close to normal	Higher
Phototropy	Observed	Not observed

H bonding may alter chemical properties too. They are an important factor in many gas phase reactions, and can be studied by ion cyclotron resonance spectroscopy (57). H-bonded ketones often do not exhibit typical protonic or ketonic properties because the two groups are tied up by H bonding. Thus diaroylmethanes, which are 100 per cent enolic, are not acetylated with acetic anhydride, do not give methane when treated with methyl Grignard, and do not give nitrogen with diazomethane. Neither do they form Schiff bases or add HCN, as expected of a ketone. Likewise, ortho azo- and ketophenols, unlike the para isomers, are insoluble in alkali and inert toward diazomethane, acetic anhydride, and phenylisocyanate. Keto acid amides undergo enzymatic deamination; however, those which can assume a chelated structure are not deaminated.

The fairly strong H bond in tropolone accounts for its different behavior from that of tropone. Tropolone has an O-H···O band at 3100 cm^{-1} and the frequency of its C=O band is 20 cm^{-1} lower than that of its ethers. Tropolone consists of an equilibrium mixture of two indistinguishable tautomers. Whereas tropone is a typical unsaturated ketone -- i.e., it is easily oxidized, undergoes addition reactions, and forms the usual carbonyl derivatives -- tropolone undergoes substitution reactions and the carbonyl group, which is tied up by the H bond, is inert toward semicarbazide, hydroxylamine, and 2,4-dinitrophenylhydrazine.

Resists enzymatic deamination

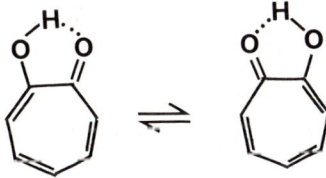

Tropolone

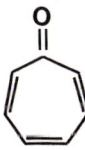

Tropone

An interesting reaction involving H bonds is the degenerate rearrangement $\underline{1.9} \rightleftarrows \underline{1.10}$ (58).

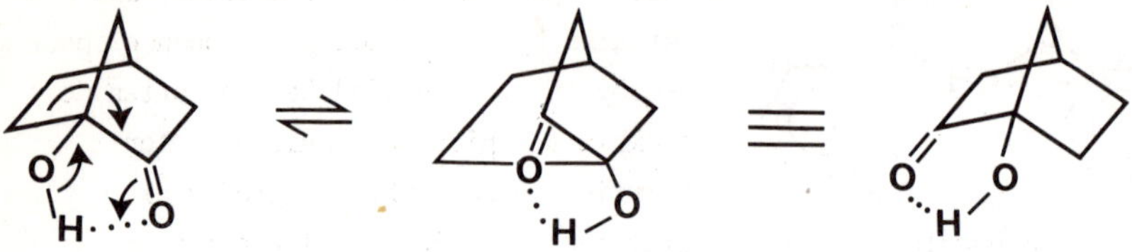

  1.9   1.10   1.11

When the 3 position is tagged with methyl groups, the two isomers $\underline{1.9}$ and $\underline{1.11}$ are no longer equivalent.

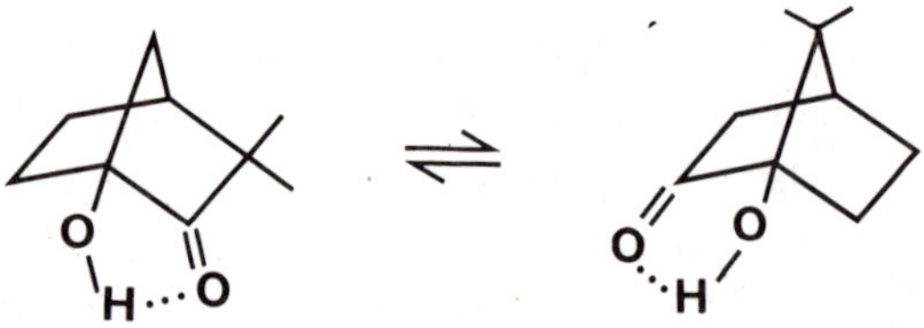

   1.12   1.13

The equilibrium lies about 1:2 in favor of $\underline{1.13}$.

Examples of steric hindrance to H bonding are given in Section 20.2.

QUESTIONS

1. What type of H bonding, i.e., O-H···O or O-H···π, would you expect in the following two compounds?

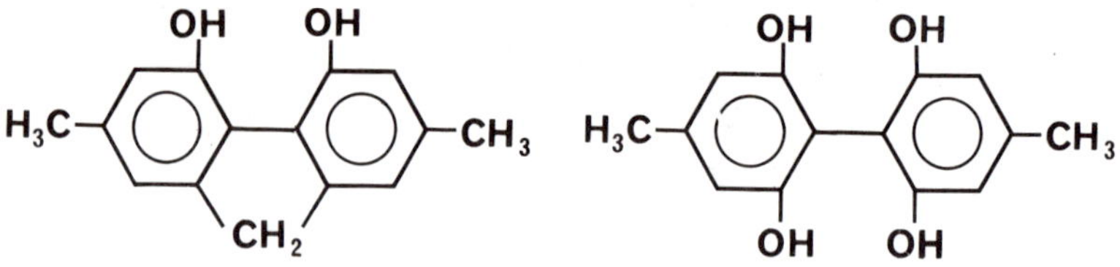

2. Why does the H bond strength in dimeric carboxylic acids <u>decrease</u> with increasing acid strength of the monomeric acids whereas H bond strength <u>increases</u> with greater acidity of alcohols and phenols?

3. In a dilute solution, compound $\underline{A}$ exhibits a free and an H-bonded OH band in its ir spectrum. Which OH do you think is forming the H bond?

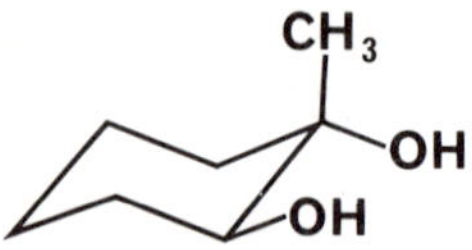

ANSWERS

1. O-H···O in the first and several O-H···π in the second. The CH_2 bridge in the first holds the phenyl rings coplanar whereas steric hindrance prevents them from being coplanar in the second.

2. Lowering the electron density on the OH oxygen atom increases the H bond strength as in alcohols, and an electron decrease on the carbonyl oxygen decreases the H bond strength. With increasing electronegativity of R in RCOOH, there is an increase in acid strength and a decrease in electron densities on the oxygen atoms. Since polarization of π electrons is easier, the change in charge is greatest for the carbonyl oxygen and hence, the observed effect is a decrease in H bond strength with increasing acidity.

3. In solution, the $3°$ alcohol is the weaker acid and the stronger base, which is opposite to the order in the gas phase (Section 3.2). Similar inversion of gas phase and solution properties are observed for H bond strengths. Toward acetone, the relative H bond strengths are MeOH > EtOH > i-PrOH > t-BuOH (determined by ir spectroscopy; see A. Balsubramanian and C. N. R. Rao, Spectrochim. Acta 18: 1337 (1962)), whereas in the gas phase, EtOH forms a stronger H bond to ⁻OEt than does MeOH (measured by ion cyclotron resonance spectroscopy; see R. T. McIver, Jr., J. A. Scott, and J. M. Riveros, J. Am. Chem. Soc. 95: 2706 (1973)).

In the case of A, which is in solution, we will use the solution order and say that the $2°$ OH probably acts as the proton donor. (See A. R. H. Cole and P. R. Jeffries, J. Chem. Soc. 4391 (1956).)

Chapter 2
VAN DER WAALS FORCES

2.1 van der Waals Forces

Although the negative charges of the electrons in a neutral molecule are balanced by the positive charges of the nuclei, the electrons are in constant motion and the center of density of the negative charges (a point which may be regarded as the time-average position for all of the negative charges) does not coincide continuously with the center of density of the positive charges. This situation produces small instantaneous local dipoles; that is, regions with positive and negative ends. When molecules are close enough, these dipoles attract each other like magnetic bars, causing the molecules to cling to one another. Such attraction is called *van der Waals attraction*. The greater the area over which molecules may come in close contact, the greater is the over-all van der Waals attraction. For hydrocarbons, the forces are about 1.0 kcal per CH_2 unit.

Branching of a chain reduces the area over which a molecule may "touch" other molecules. Accordingly, the boiling points of isomers decrease with the degree of branching. For example, the boiling points of normal, iso-, secondary, and tertiary butyl alcohols are 117°, 107°, 100°, and 83°, respectively. van der Waals forces become weaker in proportion to the seventh power of the distance between molecules: $\underline{F} \propto (1/\underline{d}^7)$. A small separation of molecules markedly reduces the van der Waals forces of attraction; hence, they are effective only for short intermolecular distances.

It is fair to ask why only <u>attractive</u> van der Waals forces are considered. The forces are dipolar in character and one can expect that like-charged poles have as much chance of meeting and producing repulsive forces as unlike-charged poles have of meeting to produce attractive forces. However, when the like-charged poles meet and repel each other, they tend to separate and thereby decrease the interatomic repulsion, whereas the unlike-charged poles tend to get closer to each other and thereby increase the attraction. Also, when the dipoles approach each other with similar poles together, they tend to depolarize each other, i.e., decrease the separation of charge in the other dipole. When oriented with dissimilar poles together, however, each dipole tends to increase the charge separation in the other dipole. Consequently, these two effects make the repulsive forces self-diminishing and the attractive forces self-increasing.

Thus, all molecules have a certain attraction for other molecules, which increases as the molecules get closer, up to a certain point. Each atom in a molecule has a definite volume within which it resists penetration by other nonbonded atoms. The radii of these volumes are called <u>van der Waals radii</u>. Large repulsive forces arise when any atom or molecule gets within the van der Waals radius of another atom. The van der Waals radii of atoms or groups of atoms are not known precisely. One approach to measuring them was to determine "distances of closest approach" for a number of elements from diffraction measurements on crystals [59]. Known as "Pauling's van der Waals radii" (Table 2.1), each value is empirically close to the covalent radius plus 0.80 Å. However, the values are usually increased by 10 per cent or 0.2 to 0.3 Å when they are used for molecular considerations.

Table 2.1. van der Waals radii of some atoms (Å)

H	1.2-1.5				
N	1.5	O	1.4	F	1.35
P	1.9	S	1.85	Cl	1.80
As	2.0	Se	2.0	Br	1.95
Sb	2.2	Te	2.2	I	2.15
CH_3	2.0	Half thickness of benzene nucleus 1.85			

The repulsion that results from intramolecular crowding of non-bonded atoms approaching within van der Waals radii has a strong effect on bond angles, conformations, and many chemical and physical properties of molecules. These are discussed in Part IV. van der Waals forces are also primarily responsible for the formation of <u>inclusion complexes</u>, discussed in the next section.

2.2 Inclusion complexes

For decades, the choleic acids, the blue color of starch-iodine mixtures, and the crystalline ice mushes formed in natural gas lines were chemical oddities. Suddenly, following Bengen's accidental discovery of the crystalline urea complex of octyl alcohol in 1940, there was a tremendous burst of activity and interest in these substances by oil companies. It was soon discovered that these compounds, referred to as <u>inclusion complexes</u>, have several valuable commercial and laboratory uses.

Inclusion complexes are formed when certain substances, called <u>host</u> molecules, crystallize and entrap other compounds, the <u>guest</u> molecules, in open crystal lattice cavities. There is no apparent chemical affinity between the host and guest molecules, e.g., even inert gases may be guest molecules. Complex formation depends upon whether the guest molecule has the proper size and shape to match the cavities created by the host molecules. To a certain degree, the couples of many inclusion complexes cannot be predicted but must be determined by trial and error.

There are three general types of inclusion complexes for which the hollow cavities of the crystal lattice differ as follows:

Channel	Clathrates	
	Cage	Layer
Urea and thiourea		
Cyclodextrins	Quinols	Werner type complexes
Choleic acids	Gas hydrates	
Biphenyls, etc. [60]		

In the channel type, host molecules crystallize in helixes or coils and hold the guest molecules in the central channel. The hollow cavities in the cage clathrates may have a variety of polygonal shapes. The host molecules in the layer clathrates are in layers and hold the guest molecules between planes. Usually the imprisoned guest molecules of inclusion complexes can be liberated by any process which destroys the crystal framework such as melting or solvent action.

2.2(a). <u>Dimensional characteristics</u>

<u>Urea complexes</u>. Guests: <u>n</u>-alkanes and their derivatives longer than C_6. In the presence of certain guest molecules, urea molecules link together, head-to-tail by H bonds, in a step-wise manner to form helixes. The diameter of the central channel is about 5 Å which is

just the right size to hold straight chain hydrocarbon derivatives. Branched chains are too large to fit in the channel and do not form complexes, although 2-methylalkanes with chains longer than 10 carbons will do so.

Thiourea complexes. Guests: Cycloalkanes, branched chain aliphatics, $CHCl_3$, durene, 2-bromoethane. Thiourea forms complexes similar in structure to the urea complexes, except that the diameter of the channel is 7 Å. This is too large to hold straight chain compounds but branched chain aliphatics and cycloalkanes do form complexes.

Cyclodextrins [61].

	Channel Diameter	Guests
α	6 Å	Cl_2, Br_2, I_2, fatty acids, substituted benzenes with small groups
β	7-8 Å	Br_2, I_2, miscellaneous organics
γ	10-11 Å	I_2, large organics

Amylose and some of its degradation dextrins form hexagonal helixes or stacked doughnuts and exhibit channel diameters of three sizes. Bromobenzene, for instance, is too large to fit the α-cyclodextrin channel but will form complexes with β- or γ-cyclodextrin. The cyclodextrins will form complexes even in solution, as in the well-known dark blue starch-iodine color reaction used in analytical chemistry.

Choleic acids. Guests: Variety of organics. Certain steroids, of which desoxycholic acid is best known, form inclusion complexes by stacking around a hollow channel in which guest molecules are trapped. Complexes are formed with a wide variety of organic compounds.

Quinols. Guests: SO_2, CH_3OH, CH_3CN, C_2H_4, HCO_2H, CO_2, O_2, HCl, H_2S, the rare gases. Three phenolic molecules trimerize via their OH groups by H bonding to form a cup and two such cups form a hollow cavity in which small guest molecules may be held. Again, complex formation depends upon whether the guest molecule fits the hollow cavity. For example, hydroquinone forms *quinols* with MeOH and argon, but not with EtOH (too large) or He (too small).

Gas hydrates [62]. Guests: CH_4, CO_2, SO_2, Cl_2, Kr, Xe. Water molecules form polyhedra of various sizes through H bonding and trap guest molecules. One prominent type, for instance, is a hexakaidecahedron formed from 28 H-bonded water molecules.

Werner type complexes. Guests: Selected organics. Several complex metal ions plus amines or ammonia form layer clathrates with a variety of organic compounds in which the salts crystallize in planar arrays with the amine molecules bonded to the ions between planes. Complex formation is governed by an unpredictable reciprocal spatial relationship between guest and crystal structure. For example, $Ni(CN)_2$ and ammonia form inclusion complexes [62] as follows.

Complex formed with:	Molecular volume (cc/mole)	No complex formed with:	Molecular volume (cc/mole)
Benzene	88.6	Toluene	105.6
Aniline	90.5	Nitrobenzene	103.6
Phenol	88.8	C_6H_5X (X = F, Cl, Br, I)	94-110
Pyridine	80.0	Toluidine	107
Pyrrole	69.3	Cresol	102
Thiophene	78.5	Naphthalene	111
Furan	72.0		

The complexes above apparently are formed only with guest molecules having molecular volumes smaller than 93 cc/mole [63], although the molecular shape must also be a factor. Somewhat related to this, experiments and molecular models indicate that a methylene chain will pass in close contact through a ring of 22 methylenes and that the limiting ring size for passage of a triphenylmethyl group is about C_{29} [64].

2.2(b). General Properties

Inclusion complexes are usually prepared by recrystallizing the host substance in the presence of guest molecules, either in an inert solvent or with the guest compound as a solvent. For instance, n-butyl 3,5-dinitrobenzoate forms inclusion complexes when recrystallized from long chain hydrocarbon derivatives [65]. Generally inclusion complexes melt near the melting point of the host molecule. Some are soluble in water without decomposition; others may be formed and kept in solution.

Some physical properties of guest molecules are unaltered in the complex whereas other properties are changed. Magnetic character is often unchanged and therefore inclusion complexes provide a convenient means of studying certain compounds, e.g., NO_2, which otherwise would dimerize at low temperatures. Color, redox potential, and other properties frequently undergo substantial changes upon complex formation [66].

2.2(c). Uses of inclusion complexes [67]

The general uses of inclusion complexes fall into three categories: (1) resolution of mixtures, (2) storing substances, and (3) catalysis.

(1) Resolution by clathration has become a useful technique for the resolution of hard-to-separate mixtures such as benzene plus thiophene, isomeric alkylbenzenes [68] and racemic mixtures [69]. For example, $Ni(SCN)_2$ and 4-picoline selectively form complexes with alkylbenzenes by which a mixture of the o-, m-, and p-xylenes may be resolved. The urea complexes are used in the petroleum industry to upgrade fuels by removing straight chain hydrocarbons in the form of their complexes. Urea complexation is also used for the purification of fats, oils, and waxes by complexing and removing fatty acids. Thiourea is commonly used to resolve squalene to an all _trans_ isomer.

(2) Substances may be stored in the form of their inclusion complexes for protection against oxidation or other deterioration. For instance, vitamin A palmitate is stabilized when stored in choleic acid or cyclodextrin complexes. In other cases, guest molecules are complexed so they may be held in a convenient form for study. For example, urea complexes of free radicals have been prepared for X-ray and esr studies [70]. If they were not complexed, the radicals would polymerize.

(3) Some substances react remarkably fast when in an inclusion complex. The oxidation of α-hydroxyketones, the hydrolysis of phenyl pyrophosphates and phenyl acetates [71], and the selective halogenation of aromatics [72] are catalyzed by inclusion complex formation with cyclodextrins.

Inclusion complexes may play important roles in biological systems. For example, one theory on the mechanism of fat assimilation in the intestines is that bile acid salts form soluble inclusion complexes with fats, carry them across the cell membranes, and release the fats after reaching the blood or lymph stream. It has been shown that bile acid salts are good solubilizers of lipophilic substances.

A recently discovered class of complexes is that of the so-called <u>crown</u> <u>ethers</u>. Macrocyclic polyethers have the ability to form complexes with alkali metal cations which are soluble in nonpolar solvents [73]. For instance, potassium permanganate, which has no detectable solubility in benzene, readily dissolves in benzene in the presence of dicyclohexyl-18-crown-6. This reagent is a useful oxidant. Whereas the permanganate oxidation of alpha-pinene proceeds in a 40-60% yield in aqueous medium, the crown ether complex in benzene ("purple benzene") achieves this reaction in a 90% yield. The solubilizing power of crown ethers offers

Dicyclohexyl-18-crown-6

many possibilities in catalysis, analytical and electrochemistry, and in biological transport processes. Crown ethers of the type in structure <u>2.1</u> have been effective agents for the resolution of racemates by differential complexation in liquid-liquid chromatography [74].

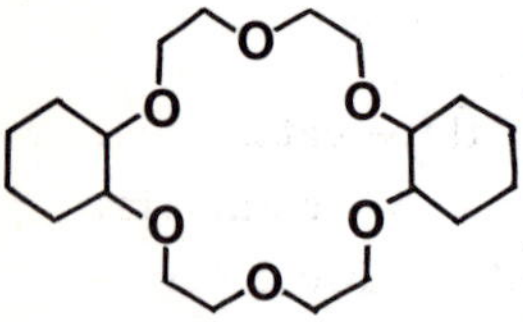

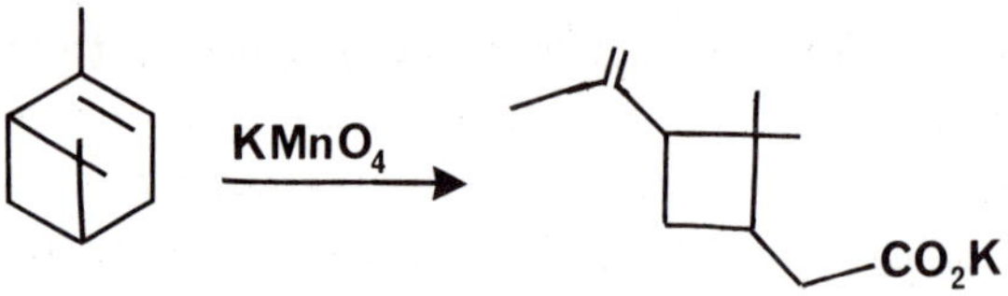

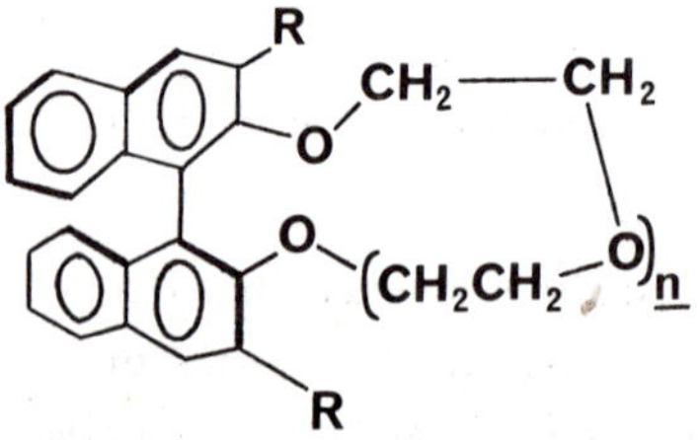

<u>2.1</u>

R = H, CH_2OCH_2COOH
$\underline{n}$ = 1-5

Chapter 3
MOLECULAR POLARIZATION

3.1 Electronegativity

Two aspects of the scientific method are inductive and deductive thought. The former involves reasoning from a particular to the general, marshalling facts to prove a point, i.e., theorizing. Deductive reasoning proceeds from the general to the particular, i.e., by analogy. An organic chemist uses primarily a deductive approach. Thus, synthesis is by analogy and structure determination, based on spectral or chemical functional group analysis, is by analogy. Chemists are always seeking, or are quick to observe, correlations between molecular properties and chemical constitution with the objective of producing a molecule that will have any special desired property. These correlations are usually developed by semitheoretical or empirical methods, again by analogy. Many reliable and useful generalizations have been found which relate properties such as color, acidity or basicity, shock or thermal stability, chemical reactivity, and numerous biological activities (chemotherapy) to structure. One of the most fundamental correlations has been the periodic table. Another widely used relationship is relative electronegativity.

The electronegativity of an atom in a molecule is defined as the intrinsic power of the atom to attract electrons within the molecule. This concept makes it possible to interpret many properties which otherwise would be anomalous. For instance, trichloroacetic acid is as strong as mineral acids whereas most carboxylic acids are weak; or the ether $(CF_3)_2O$, unlike most ethers, is not hydrolyzed in hydriodic acid and is a poor solvent for Grignard reagents. Indeed, we shall see that electronegativity, as applied through the concepts of induction and field effects, is necessary to account for variations in most physical properties such as acid and base strengths, bond characteristics, dipole moments, spectra, H bond strengths, redox potentials, as well as many chemical and biological properties.

3.1a. Atomic electronegativities

Several methods [75] have been offered for establishing a scale of atomic electronegativities. The most widely used is that of Pauling based on thermochemical data [76]. For a pure covalent bond A-B, the energy E_{AB} should be related to the energies of A_2 and B_2 by the equation

$$E_{AB} = \tfrac{1}{2}(E_{A_2} + E_{B_2}). \tag{3.1}$$

If AB has a partial ionic character because of the different electronegativities of A and B, then the bond A-B will be stronger than that computed from equation (3.1) by the quantity Δ_{AB}. Hence,

$$\Delta_{AB} = E_{AB} - \tfrac{1}{2}(E_{A_2} + E_{B_2}). \tag{3.2}$$

Since Δ_{AB} arises from the difference in electronegativities of A and B, Δ_{AB} should be related to this difference, i.e.,

$$\chi_A - \chi_B \propto \Delta_{AB}$$

where χ is the electronegativity of an atom. Pauling used the equation

$$\chi_A - \chi_B = \sqrt{\frac{\Delta_{AB}}{23.06}} = \sqrt{\frac{E_{AB} - \tfrac{1}{2}(E_{A_2} + E_{B_2})}{23.06}} \tag{3.3}$$

where E values are expressed in kcal/mole.

Notice that Δ_{AB} in equation (3.2) is also equal to the heat of the reaction.

$$\tfrac{1}{2}A_2 + \tfrac{1}{2}B_2 \rightarrow A\text{-}B \tag{3.4}$$

when all the substances are in the gaseous state. Thus, Δ values may be obtained experimentally from heats of reactions.

Since equation (3.4) gives only the difference in electronegativities a value has to be assigned arbitrarily to one atom as a reference, in order to get values for individual atoms. Pauling chose a value of 2.1 for hydrogen, and values for all other atoms are relative to this.

By using later data, Pauling's original table was revised and extended (Table 3.1) [77]. The major difference between Pauling's table and the revised table is the alternation in values for certain groups. For example, there is a fluctuation of values for B, Al, Ga, In, Tl, for C, Si, Ge, Sn, Pb, and for N, P, As, Sb, although both tables have a monotonic decrease for the alkali metals, the alkaline earth metals, and the halogens. Rochow [78] and earlier Sanderson [79], showed that a large number of observations which otherwise would be anomalous are consistent with this variation of electronegativity. For example, Ph_3C-H and Ph_3Ge-H react with PhLi to form Ph_3CLi and Ph_3GeLi, respectively, whereas Ph_3Si-H and Ph_3Sn-H yield Ph_4Si and Ph_4Sn, respectively. This implies that the C-H and Ge-H bonds are more polar than the Si-H and Sn-H bonds with the H positive, and this would be expected if the electronegativities of C and Ge are greater than those of Si and Sn. Similarly, SiH_4 is hydrolyzed and oxidized much more readily than are CH_4 and GeH_4. Compounds such as R_3C-Br and R_3Ge-Br are readily reduced by zinc in concentrated hydrochloric acid to the corresponding hydrocarbons whereas the same Si and Sn compounds are inert.

$$\left.\begin{matrix} C-Br \\ Ge-Br \end{matrix}\right\} \xrightarrow{Zn,\ HCl} \left\{\begin{matrix} C-H \\ Ge-H \end{matrix}\right.$$

but

$$\left.\begin{matrix} Si-Br \\ Sn-Br \end{matrix}\right\} \xrightarrow{Zn,\ HCl} \quad \text{No reaction}$$

Although Pauling's table does not show this alternation, its use in organic chemistry has not been appreciably affected because few organic compounds contain elements beyond those in the first two rows and the halogens, for which there is no alternation.

Several other methods have been proposed for assigning atomic electronegativities, only one of which will be presented here. Rochow [78] calculated the force of attraction between the nucleus of an atom A and an electron from a bonded atom at the covalent boundary $\underline{r}$ of A as follows:

$$\text{force} = \frac{e^2 Z_{eff}}{\underline{r}^2} \qquad (3.5)$$

The effective nuclear charge Z_{eff} is smaller than the actual charge Z owing to partial screening by the electrons in the atom. Rochow identified the force in equation (3.5) with the electronegativity of the atom A, which is based on electrostatic data. A plot of the quantity $Z_{eff}/\underline{r}^2$ against Pauling's electronegativity values yields equation (3.6) for electronegativity.

$$\chi = 0.359 \frac{Z_{eff}}{\underline{r}^2} + 0.744 \qquad (3.6)$$

A complete table of electronegativities of the elements based on thermochemical and on electrostatic data [80] has been published. This table is reproduced in Table 3.1

There are some significant differences in the two sets of electronegativity values given in Table 3.1. Neither set can be preferred exclusively over the other. On the one hand, the electrostatic set (upper values) includes more elements and shows a variation for N, P, As, and Sb, for which there is some experimental support [79]. On the other hand, the thermochemical set (lower values) has some values which are more realistic, such as the relative values for carbon and iodine.

In any event, the specific electronegativity of an atom in a molecule may be quite different from that implied by Table 3.1 because several structural conditions affect the electronegativity of an atom: (a) hybridization, (b) other atoms attached, and (c) valence state. We saw in Section 1.1 how hybridization affects bond lengths and strengths (Tables 1.2 and 1.4), a reflection of changes in electronegativity.

	Bond length	Dissociation energy (kcal/mole)	Force constant ($\times 10^5$ dynes/cm)	Dipole moment
$H-C_{sp}$	1.058	125	5.9	1.05 D
$H-C_{sp^2}$	1.071	104	5.1	0.63
$H-C_{sp^3}$	1.102	98	5.0	0.31

We also saw there that attachment of highly electronegative atoms to carbon increases its electronegativity. This was also revealed by changes in bond length and strength, and by an ability to form H bonds (Section 1.2a). Chemical as well as physical properties may be affected. For example, the fluorines in perfluoroamines lower the electronegativity of nitrogen to the extent that it no longer is a base.

Table 3.1. Electronegativity values for the elements [77, 80]*

Element	Upper (electrostatics)	Lower (thermochemistry)
H	2.1	2.20
Li	0.97	0.98
Be	1.47	1.57
B	2.01	2.04
C	2.50	2.55
N	3.07	3.04
O	3.50	3.44
F	4.10	3.98
Na	1.01	0.93
Mg	1.23	1.31
Al	1.47	1.61
Si	1.74	1.90
P	2.06	2.19
S	2.44	2.58
Cl	2.83	3.16
K	0.91	0.82
Ca	1.04	1.00
Sc	1.20	1.36
Ti	1.32	1.54
V	1.45	1.63
Cr	1.56	1.66
Mn	1.60	1.55
Fe	1.64	1.83
Co	1.70	1.88
Ni	1.75	1.91
Cu	1.75	1.90
Zn	1.66	1.65
Ga	1.82	1.81
Ge	2.02	2.01
As	2.20	2.18
Se	2.48	2.55
Br	2.74	2.96
Rb	0.89	0.82
Sr	0.99	0.95
Y	1.11	1.22
Zr	1.22	1.33
Nb	1.23	..
Mo	1.30	2.16
Tc	1.36	..
Ru	1.42	..
Rh	1.45	2.28
Pd	1.35	2.20
Ag	1.42	1.93
Cd	1.46	1.69
In	1.49	1.78
Sn	1.72	1.96
Sb	1.82	2.05
Te	2.01	..
I	2.21	2.66
Cs	0.86	0.79
Ba	0.97	0.89
Hf	1.23	..
Ta	1.33	..
W	1.40	2.36
Re	1.46	..
Os	1.52	..
Ir	1.55	2.20
Pt	1.44	2.28
Au	1.42	2.54
Hg	1.44	2.00
Tl	1.44	2.04
Pb	1.55	2.33
Bi	1.67	2.02
Po	1.76	..
At	1.90	..
Fr	..	..
Ra	..	..
La	1.08	1.10
Ce	1.08	1.12
Pr	1.07	1.13
Nd	1.07	1.14
Pm	1.07	..
Sm	1.07	1.17
Eu	1.01	..
Gd	1.11	1.20
Tb	1.10	..
Dy	1.10	1.22
Ho	1.10	1.23
Er	1.11	1.24
Tm	1.11	1.25
Yb	1.06	..
Lu	1.14	1.27
Ac	1.00	..
Th	1.11	..
Pa	1.14	..
U	1.22	1.38
Np	1.22	1.36
Pu	1.22	1.28

*Upper set based on electrostatics, lower set based on thermochemistry.

Perfluoroamines do not form salts even with strong acids. Similarly, the oxygen atom in perfluoroethers no longer is basic so that perfluoroethers cannot serve as solvents for Grignard reagents. The C-I bond and the C=S bond change polarity by attaching fluorine atoms to the carbon.

$$H_3^{\delta+}C-I^{\delta-} + KOH \rightarrow H_3C-OH + KI$$

$$F_3^{\delta-}C-I^{\delta+} + KOH \rightarrow F_3C-H + KOI$$

Thus, negative OH^- replaces $I^{\delta-}$ in H_3CI but positive H^+ replaces $I^{\delta+}$ in F_3CI. Likewise, hexafluorothioacetone, $(CF_3)_2C=S$, has properties which indicate that the C=S bond is polarized $^{\delta-}C-S^{\delta+}$ rather than the usual $^{\delta+}C-S^{\delta-}$. Based on solvent effects on nmr ^{13}C chemical shifts of ketones and $\Delta\nu_{OH}$ of protic solvents, the carbonyl oxygen of hexachloroacetone has lost almost all of its tendency to be a proton acceptor in an H bond. To take another example, ketone hydrates are normally too unstable with respect to the ketones to enable isolation of the hydrates.

$$R_2C=O + H_2O \rightleftharpoons R_2C\begin{smallmatrix}OH\\OH\end{smallmatrix}$$

However, when the electronegativity of the keto carbon is high and opposes the resonance-stabilizing electron shift toward the oxygen

$$R_2C=O \quad , \quad R_2\overset{\oplus}{C}-\overset{..}{\underset{..}{O}}{}^{\ominus}$$

the equilibrium moves toward the hydrate to permit its isolation. Three such cases where the hydrates are more stable than the ketones are:

$$(CF_3)_2C\begin{smallmatrix}OH\\OH\end{smallmatrix} [81] \qquad Cl_3C-CH\begin{smallmatrix}OH\\OH\end{smallmatrix} \qquad \triangle\begin{smallmatrix}OH\\OH\end{smallmatrix}$$

Finally, the electronegativity of an atom increases with its oxidation state. In this respect, it is observed that sulfur has the +6 oxidation state only when it forms bonds with the highly electronegative oxygen, fluorine, or CF_3 group, e.g., $^{\ominus}SO_4$, SF_6, etc.

3.1b. Relative group electronegativities [82]

In view of the variation of atomic electronegativity with molecular environment, particularly for carbon, it is convenient to use relative group electronegativities for correlations with structure. Several methods have been proposed for establishing a scale of group electronegativities, some of which we can examine because they bring up other points of interest.

(i) *Cleavage of organomercurials.* One of the first extensive series of relative group electronegativities was determined by Kharasch [83] from the reaction

$$R'-Hg-R \rightleftharpoons R'Hg^+ + R^\ominus \xrightarrow{HCl} R'HgCl + RH.$$

He assumed that an organomercurial ionizes to a small degree and that the relative amounts of $^\ominus R'$ and $R^\ominus$ formed are proportional to their relative electronegativities which is reflected by the amounts of R'H and RH liberated. By measuring the quantities of hydrocarbons produced, he set up a scale of relative electronegativities for a large number of groups. A partial sequence from his series, for example, is as follows:

$$\underline{p}\text{-MeOC}_6H_5 > \underline{p}\text{-MeC}_6H_5 > C_6H_5 > Me > Et > \underline{i}\text{-Pr} > \underline{t}\text{-Bu} > (C_6H_5)_3C$$

(ii) *From dipole moments.* Brown [84] showed that a series of organic groups, when listed in order of the C-Cl bond moments in the R-Cl compounds, is essentially the same as Kharasch's series. On this basis, he assigned positions to groups not studied by Kharasch which not only extended the list but gave an over-all series based on a chemical and a physical method.

(iii) *Polar substituent constants.* Taft [85] set up a scale of the inductive effect of groups based on relative rates of hydrolysis of esters. The transition states for hydrolysis of ethyl esters in acidic and basic media are thought to resemble structures 3.1 and 3.2.

$$\begin{bmatrix} & \text{OH} & \\ & | & \text{Et} \\ R - & C - \overset{\oplus}{O} & \\ & | & \diagdown \text{H} \\ & \text{OH} & \end{bmatrix} \qquad \begin{bmatrix} & \overset{\ominus}{O} & \\ & | & \text{Et} \\ R - & C - O & \\ & | & \\ & \text{OH} & \end{bmatrix}$$

3.1 In acid   3.2 In base

Except for two protons, the two species have almost the same steric conditions. In addition, the ester carbon is saturated so there is no resonance involving C and R. Consequently, the major effect of R on the stabilities of 3.1 and 3.2 and hence on the rates of hydrolysis of the esters, is the inductive effect of R. Taft related this inductive effect to the rates of hydrolysis by equation (3.7):

$$\sigma^* = 1/2.5 \; (\log \underline{k}/\underline{k}_0)_b - (\log \underline{k}/\underline{k}_0)_a \qquad (3.7)$$

where $\underline{k}$ and $\underline{k}_0$ are the reaction rate constants for a substituted and a reference ethyl ester, $R\text{-}CO_2Et$ and $R_0\text{-}CO_2Et$, respectively. The subscripts refer to rate data in base and in acid, and σ^* is the polar (inductive) substituent constant for the group R (see Table 3.2). Taft chose R_0 = Me for the reference group, for which $\sigma^* = 1$.

Table 3.2. Taft polar substituent constants for some groups

Group	σ^*	Group	σ^*
$(CH_3)_3\overset{\oplus}{N}CH_2-$	1.90	H-	0.490
O_2NCH_2-	1.40	$C_6H_5CH_2-$	0.260
$NCCH_2-$	1.30	$CH_2=CH-CH_2-$	0.226
FCH_2-	1.10	$c\text{-}C_3H_5CH_2-$	0.011
$ClCH_2-$	1.05	CH_3-	0.00
$BrCH_2-$	1.00	C_2H_5-	-0.10
ICH_2-	0.85	$(CH_3)_2CH-$	-0.190
C_6H_5-	0.600	$(CH_3)_3C-$	-0.300

The order of groups with respect to σ^* is quite similar to that of Kharasch, and indeed, dipole moments or ionization potentials [86] of alkyl halides and other classes of compounds and $\underline{pK_a}$s of alcohols may be computed from σ^* with high accuracy, e.g.,

$$\mu_{RCl} = 1.823 - 1.416 \, \sigma^*$$

$$\underline{pK_a}(RCH_2OH) = 15.74 - 1.316 \, \sigma^*.$$

The latter equation is for alcohols in which C-1 of R is $\underline{sp}^3$ or $\underline{sp}$ hybridized σ^* (87).

σ^* also provides semiquantitative comparisons of group electronegativities. For example, σ^* values show that across each single bond the inductive effect decreases to about a third of the initial value, the same fraction proposed earlier by Branch and Calvin [88] from acid dissociation constants.

	CF_3-	CF_3CH_2-	$CF_3CH_2CH_2-$	$CF_3(CH_2)_3-$
σ^*	2.7	0.92	0.32	0.12

Justification for Taft's assumption that steric effects are equivalent in the transition states for acid and alkaline hydrolysis of esters lies primarily in the excellent correlations of σ^* with various physical and chemical properties. σ^* may also be determined from $\underline{pK_a}$ data for the respective carboxylic acids, RCOOH, or from the carbonyl stretching frequencies of esters ($RCH_2O\text{-}COCH_2CH_2\phi$) of the corresponding alcohols RCH_2OH [87]. This means that the electronegativity of R has a related influence on the dissociation constants of RCOOH and RCH_2OH and on the carbonyl ir frequency of $RCH_2O\text{-}COR'$. How much of this influence of R is due to sigma induction or to field effects is not clear at this time (see Section 3.3).

(iv) From infrared absorption intensities. The intensity of ir absorption is a function of the change in dipole moment of the bond during excitation, which is related to the polar character of the bond. The latter, of course, is largely the result of the electronegativity differences of the bonded atoms. Accordingly, the relative intensities [89] of ir bands for a bond type can be used to set up a scale of relative electronegativities. This has been done for the O-H bond of alcohols [90] and the P=O bond of phosphones $R_3P\text{=}O$ [91], among others. Part of the series obtained from ir intensities is

CCl_3CH_2- > $HC\equiv CCH_2-$ > $C_6H_5CH_2-$ > $H_2C=CH-CH_2-$ >

CH_3- > $CH_3CH_2CH_2-$ > cyclohexyl- > $(CH_3)_3C-$.

(v) <u>From infrared absorption frequencies</u>. The ir frequency of a bond is directly proportional to the bond force constant, and the latter, through its dependence upon the strength of the bond, is related to the bond polarity. Thus, ir frequencies roughly reflect electronegativity differences, and series of relative electronegativities have been set up based on ir frequencies of several bond types:

C=O   in RCO_2Et and $R_2C=O$ [92]

P=O   in R_3PO and $R-P(OEt)_2=O$ [93]

C≡N   in RC≡N

H-X   in HX [94]

One sequence (87) from $\nu_{P=O}$ in $R_3P=O$ compounds, for example, exhibits very good self-consistency:

	$\nu_{P=O}$		$\nu_{P=O}$
F_3PO	1400 (cm^{-1})	Cl_3PO	1293 (cm^{-1})
F_2ClPO	1358	Cl_2BrPO	1285
F_2BrPO	1350	$ClBr_2PO$	1275
FCl_2PO	1331	Br_3PO	1263
$FClBrPO$	1319	$(C_6H_{11})_3PO$	1218
FBr_2PO	1303	$(CH_3)_3PO$	1173

Several other physical properties have been proposed for assessing relative group electronegativities. In some cases the resulting series are too short to mention and in other cases the property measured is influenced by other effects. For instance, nmr chemical shifts are very sensitive to electronegativities and series have been proposed based on such data. However, δ values are frequently affected by anisotropic effects of π electrons and therefore too many exceptions occur. Since the effects of substituents on the dissociation constants of aliphatic acids, RCOOH, was formerly attributed to the inductive effect of R, it is not surprising that the $\underline{K_a}$s of these acids were used as a measure of relative group electronegativities. Nonetheless, we shall see in the next section that several factors affect dissociation constants, most notably solvation, H bonding, and field effects, so that this property is a poor basis on which to determine group electronegativities.

Dissociation constants of deutero acids and phenols [95] and rates of solvolysis of deuterated benzhydryl chlorides [96] are interpreted in terms of the fact that hydrogen is more electronegative than deuterium. This view is supported by the observation that the ^{17}O nmr chemical shift for D_2O is upfield from that of H_2O, indicating that the electron density about the oxygen of H_2O is lower than that in D_2O and hence, that H is more electronegative than D.

From the variety of methods used to assess relative group electronegativities, a "generally expected" order can be established as follows:

$F > CN \stackrel{\sim}{=} F_3C > Cl > Cl_3C > Br > Br_3C > Cl_2HC > Br_2HC >$

$ClH_2C > HC\equiv C > C_6H_5 > H_2C=CH > H > D > Me > Et >$

$\underline{i}\text{-Pr} > \underline{t}\text{-Bu} > (C_6H_5)_3C > (C_6H_5)_3Si$

A set of empirical group electronegativities has been worked out and is given in Table 3.3 [97]. We observe, for instance, that a single directly-attached halogen is more electron withdrawing than are three halogens separated by a carbon (X_3C), and that two of any halogen is more electronwithdrawing than is one halogen of the next higher row, i.e., $Br_2HC > ClH_2C$ and $Cl_2HC > FH_2C$. Experimental attempts to determine the relative acidities of benzene and ethylene have given contradictory results [98].

Although most of the foregoing methods indicate that the relative electronegativities of hydrogen and the alkyl groups is H > Me > Et $\underline{i}$-Pr > $\underline{t}$-Bu, it is shown in Section 3.3c that the order is just the reverse in the gas phase. Obviously, differential solvation is a factor. It is difficult to discern whether the solution phase difference in electron-releasing ability, alkyl > H, is due to relative electronegativities or to relative polarizabilities. That is, Me can be less

Table 3.3 Empirical group electronegativities

Group	Electronegativity	Group	Electronegativity
F	3.95	C_6H_5	3.0
$^+NH_3$	3.8	$CH_2=CH$	3.0
OH	3.7	$N(CH_3)_2$	3.0
OCH_3	3.7	Cl	3.03
NO_2	3.4	COOH	2.85
CF_3	3.35	Br	2.8
NH_2	3.35	$CHCl_2$	2.8
CN	3.3	SH	2.8
$HC\equiv C$	3.3	I	2.47
CCl_3	3.0	PH_2	2.3
		SiH_3	2.2

electron releasing than t-Bu because of a greater electronegativity of Me or a greater polarizability of t-Bu. Consider, for instance, the dipole moments of some simple compounds. Microwave gas phase dipole moments, accurate to ±0.001 D, show that the replacement of CH_3 by CD_3 in methyl fluoride increases the moment whereas the moment is decreased when CH_3 in propane is replaced by CD_3 [99].

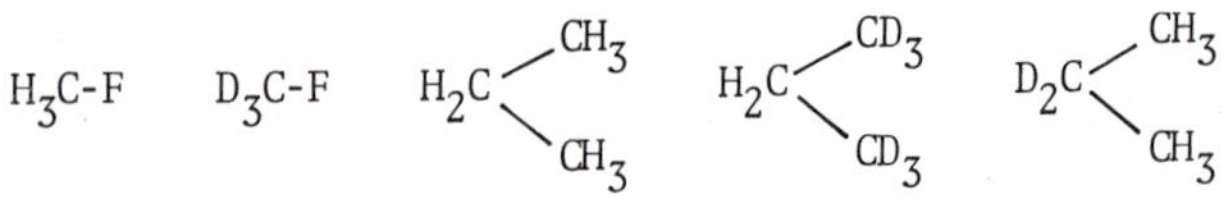

$$H_3C-F \quad D_3C-F \quad H_2C\begin{matrix}CH_3\\CH_3\end{matrix} \quad H_2C\begin{matrix}CD_3\\CD_3\end{matrix} \quad D_2C\begin{matrix}CH_3\\CH_3\end{matrix}$$

$$\underline{DM} = \overrightarrow{1.847} \quad \overrightarrow{1.858} \quad \overrightarrow{0.085} \quad \overrightarrow{0.076} \quad \overrightarrow{0.095} \text{ D}$$

If we make the reasonable assumption that Me is at the positive end of the dipole in methyl fluoride, the changes in moments upon deuteration indicate that Me is at the negative end of the dipole in propane and therefore Me is more electronegative than H.

Assessment of the relative importance of electronegativity and polarizability of alkyl groups is a tenuous matter, but generally, Me is more electronegative than t-Bu in solution and the reverse is true in the gas phase.

3.2 Electric dipole moments [100]

When a body has a negative and a positive end, that is, when the center of positive charge does not coincide with the center of negative charge, it is called a dipole. The product of the charge q at either end by the distance d between centers of charge gives the dipole moment [101].

$$\underline{DM} = \underline{q} \times \underline{d}$$

When a molecule, polar or nonpolar, is placed in an electric field, the electrons are attracted away from their normal positions by the external positive pole, thereby creating a dipole within the molecule, called an induced dipole. Even a charge at one site in a molecule may polarize another portion of the molecule in this fashion. The magnitude of the induced polarization depends upon the external field strength (or local charge in a molecule) and the polarizability α of the molecule or molecular fragment. The polarizabilities of a few bonds and groups are given in Table 3.4.

Table 3.4 Polarizabilities of some groups (10^{-25} cc/mol) [102]

H	0.42	Me	27[103]	C-C	4.74	C-H	6.45
F	0.38	Et	46[103]	C=C	~10	C-F	6.33
Cl	2.28	i-Pr	65[103]	C≡C	18.85	C-Cl	26.0
Br	3.34	t-Bu	84[103]	C-O	5.95	C-Br	37.5
I	5.11	C_6H_5	9.38	C=O	12.71	C-I	57.5

Symmetrical molecules do not have net dipole moments even though they contain highly electronegative atoms, e.g.,

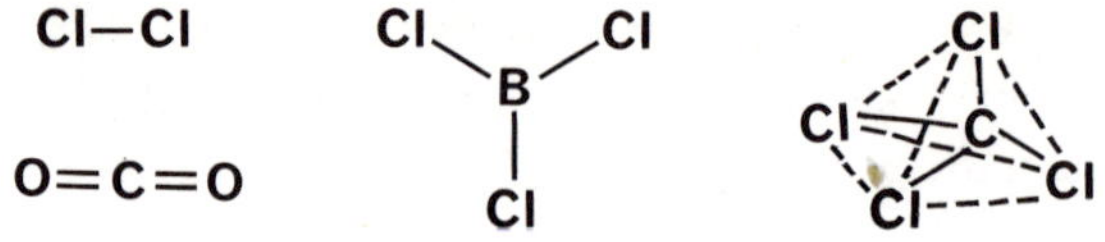

Cl—Cl   Cl\\B/Cl   Cl⋯C⋯Cl
             | (tetrahedral)
O=C=O   Cl
Linear   Trigonal   Tetrahedral

whereas unsymmetrical molecules do have dipole moments, e.g.,

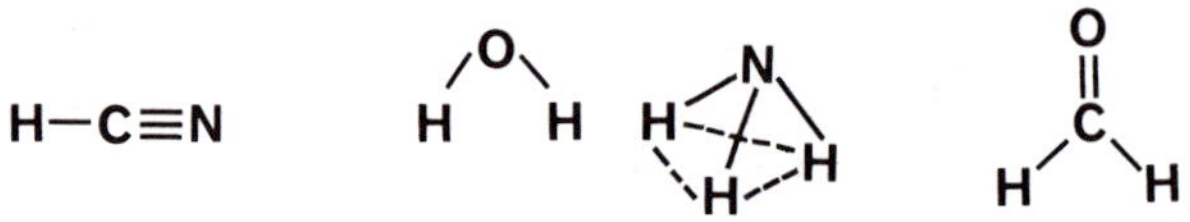

H—C≡N   H–O–H   H–N(H)–H   O=C(H)(H)

DM   2.93   1.84   1.6   2.27 D

     Linear Angular Pyramidal   Trigonal

This observation led to the view that dipole moments can be regarded as vector sums of bond moments. For example, since methane has no dipole moment, the CH_3 group moment must equal the C-H bond moment with a change in sign. If a moment is assigned to the C-H bond, the CH_3 group moment is thereby set.

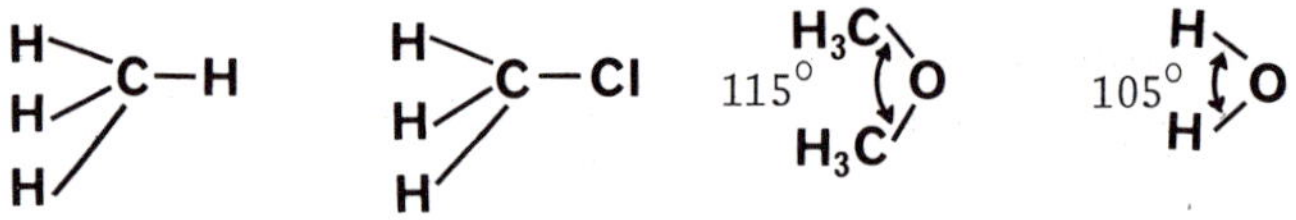

H₃C–H   H₃C–Cl   115° H₃C–O–CH₃   105° H–O–H

Then, from this CH_3 group moment and the measured dipole moments of CH_3Cl and $(CH_3)_2O$, for instance, we can (using the bond angle in $(CH_3)_2O$) calculate the C-Cl and C-O bond moments. The O-H moment can be obtained from the dipole moment of H_2O. In this manner, Smyth set up a table of bond moments, some of which are given in Table 3.5. The contributions of lone pairs to dipole moments are included in the assigned bond moments.

There are divergent views on the direction of the H-C_{sp^3} bond moment. There is no doubt that C_{sp^2} and C_{sp} hybridized carbons are more electronegative than H, but the electronegativity of C_{sp^3} is close to that of H. Because of structural factors such as ring strain and electronegativity of other atoms in the molecule, and because of the polarization of alkyl groups, various experimental attempts to determine whether CH_3 is electron withdrawing or electron donating with respect

Table 3.5. Some bond dipole moments [104]

Bond (+ -)	Bond moment	Bond (+ -)	Bond moment
H−C$_{sp^3}$	0.30	C−I	1.29
H−N	1.31	C−N	0.40
H−O	1.53	C=N	0.90
H−F	1.98	C≡N	2.93
H−Cl	1.03	C−O	0.86
H−Br	0.78	C=O	2.40
H−I	0.38	C−S	2.95
H−S	0.68	C=S	2.80
C−F	1.51	N−O	0.30
C−Cl	1.56	N→O	3.20
C−Cl(2Cl)	1.20	N=O	2.00
C−Cl(3Cl)	0.85	S→O	2.50
C−Br	1.48		

to hydrogen lead to conflicting deductions [105]. It seems reasonable to recognize that the difference is small and that different molecular environments could easily reverse the direction of the H-C$_{sp^3}$ bond moment. In terms of elemental electronegativities, carbon is slightly more electronegative than hydrogen so the H-C$_{sp^3}$ bond dipole was assigned to the negative polo of carbon. Since all other bond moments are based on the original assignment, consistent results are obtained using the dipole direction as given in the table.

In the case of C-C bonds, the moments for differently hybridized carbons are as follows [106].

$C_{sp^3}-C_{sp^2}$ 0.68 D

$C_{sp^2}-C_{sp}$ 1.15

$C_{sp^3}-C_{sp}$ 1.48

For an illustration of the use of bond moments, let us estimate the dipole moment of methyl alcohol using the law of parallelograms.

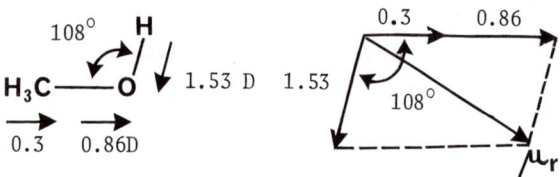

The arrows point to the negative pole by convention.

$$\mu_{resultant} = \sqrt{\mu_1^2 + \mu_2^2 \pm 2\mu_1\mu_2 \cos\theta}$$
$$= \sqrt{(0.3 + 0.86)^2 + (1.53)^2 + 2 \cdot 1.16 \cdot 1.53 \cos 108°}$$
$$= 1.64 \text{ D}$$

$\mu_{observed} = 1.67$ D

Thus, calculated and observed moments are in good agreement.

In the case of aromatic molecules, more use is made of group moments. For example, the moment of a p-disubstituted benzene is close to the vector sum of the moments of the respective monosubstituted benzenes:

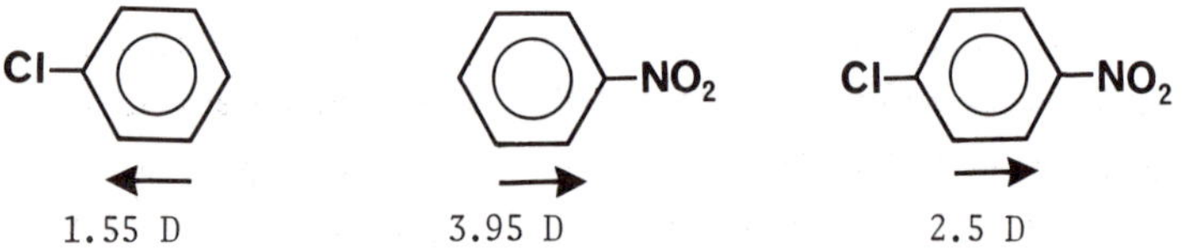

Cl—⟨⟩   ⟨⟩—NO₂   Cl—⟨⟩—NO₂

1.55 D   3.95 D   2.5 D

We see that the moment of p-ClC₆H₄NO₂ (2.5 D) is fairly close to the difference (since the moments are opposed) between the moments of the monosubstituted compounds (3.95 - 1.55 = 2.4 D). Other examples are given in Table 3.6. Equation (3.8) may also be used for ortho and meta isomers; for groups whose moments are not in the plane of the ring, e.g., for the NH₂ group, equation (3.9) is used:

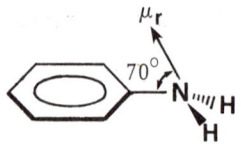

Sec. 3.2 ELECTRIC DIPOLE MOMENTS   59

$$\mu_r = \sqrt{\mu_1^2 + \mu_2^2 + 2\mu_1\mu_2 \cos \phi \cos c_1 \cos c_2} \tag{3.9}$$

where μ_r = resultant moment of the disubstituted benzene;

μ_1, μ_2 = moments of the respective monosubstituted benzenes;

ϕ = 180° for <u>para</u>, 120° for <u>meta</u>, 60° for <u>ortho</u> isomers;

c_1, c_2 = angles which the group moments make with their axes of rotation.

The discrepancies between calculated and observed dipole moments in Table 3.6 are largest for <u>ortho</u> isomers because steric repulsion spreads the groups apart. This change in bond angle is not recognized in the calculations.

The applications of dipole moments are many. They are used to distinguish isomers, confirm conformations, support resonance hybrids, provide information about polar character of molecules, and in a host of other ways. For instance, <u>cis</u> and <u>trans</u> isomers can quickly be distinguished by their measured moments:

Table 3.6. Calculated and observed moments of some disubstituted benzenes [107]

X	Moment of C_6H_5X	Y	Moment of C_6H_5Y	Moment of C_6H_4XY					
				Ortho		Meta		Para	
				Obs.	Calc.	Obs.	Calc.	Obs.	Calc.
CH_3	0.4 D	CH_3	0.4	0.5	0.7	0.4	0.4	0	0
F	1.43	F	1.43	2.4	2.48				
Cl	1.55	Cl	1.55	2.3	2.67	1.48	1.55	0	0
Br	1.52	Br	1.52	2.0	2.63	1.5	1.52	0	0
I	1.3	I	1.3	1.7	2.25	1.3	1.30	0	0
NO_2	3.95	NO_2	3.95	6.0	6.83	3.79	3.95	0	0
CH_3	0.4	Cl	1.55	1.3	1.39	1.78	1.79	1.90	1.95
CH_3	0.4	NO_2	3.95	3.66	3.76	4.17	4.16	4.4	4.35
Cl	1.55	NO_2	3.95	4.3	4.91	3.4	3.44	2.5	2.4
Cl	1.55	Br	1.52	2.2	2.67	1.5	1.54	0	0.03

Ch. 3 MOLECULAR POLARIZATION

	cis	trans
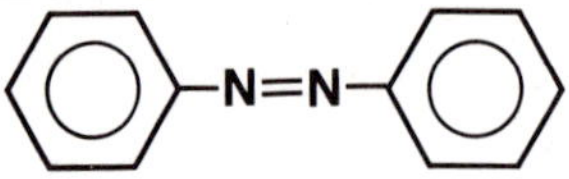 C₆H₅—N=N—C₆H₅	3.0 D	0 D
F—CH=CH—F	2.42	0
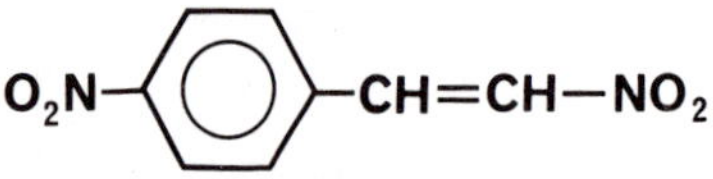 O₂N—C₆H₄—CH=CH—NO₂	7.38	0.50

The direction of the dipole moment of aniline or of anisole might be hard to predict, but from the moments of their para derivatives and the direction of the moment of the para substituent we can quickly deduce that the moments are toward the ring:

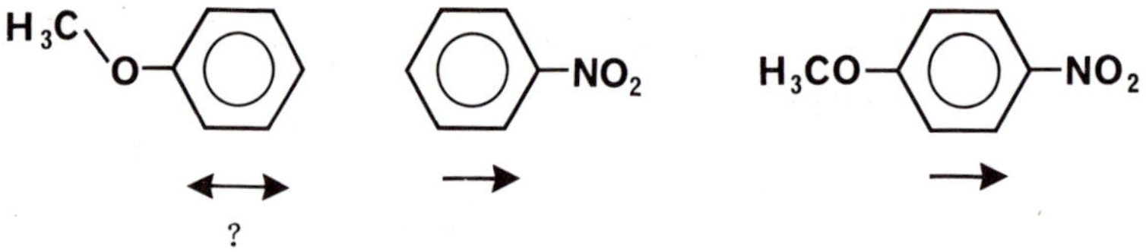

H₃C—O—C₆H₅   C₆H₅—NO₂   H₃CO—C₆H₄—NO₂
  ←→   →   →
  ?
 1.38 D   4.27 D   5.26 D

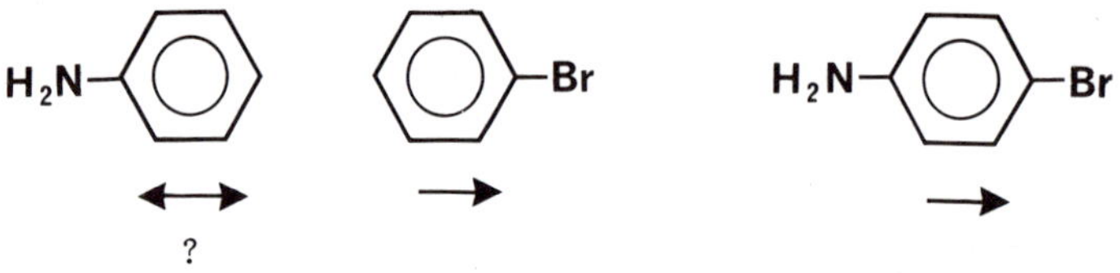

H₂N—C₆H₅   C₆H₅—Br   H₂N—C₆H₄—Br
  ←→   →   →
  ?
 1.53 D   1.52 D   2.91 D

The decrease in moment of alkylamines as the number of alkyl groups increases indicates that the nitrogen pyramid is flattening out or that the alkyl groups are more electron withdrawing than hydrogen, or both.

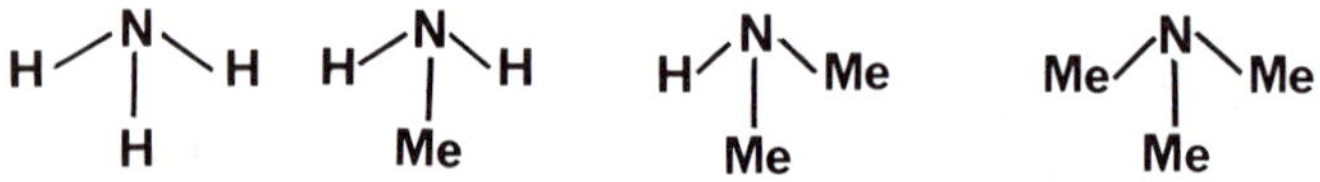

	H–N(H)(H)–H	H–N(H)(Me)–H	H–N(Me)(Me)–H	Me–N(Me)(Me)–Me
DM	1.47	1.33	1.03	0.61 D

Surprisingly, many molecules which could assume a symmetrical planar configuration do not, e.g.,

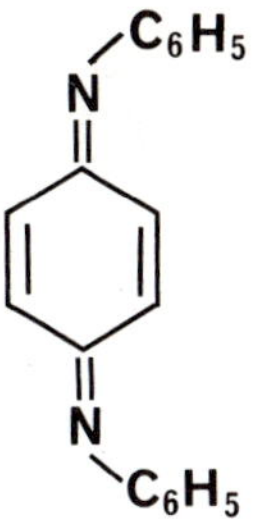

Sec. 3.2 ELECTRIC DIPOLE MOMENTS   61

DM   1.67 D   1.73 D

The fact that these molecules have a moment indicates that they do not
have a symmetrical configuration. The use of dipole moments to
estimate the fraction of conformations present is discussed in
Chapter 18.

Questions

1. Match the respective isomers with the measured dipole moments
   for the following isomeric pairs:

   (a)   and
                        cis   trans

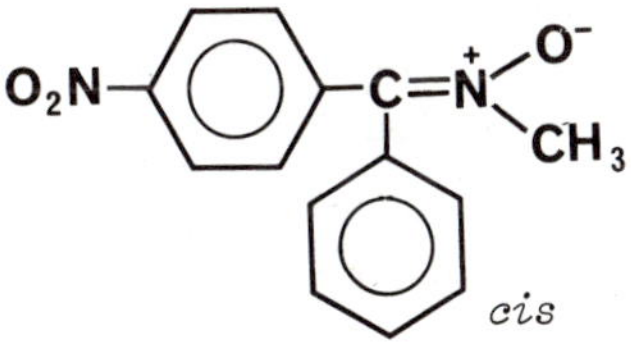

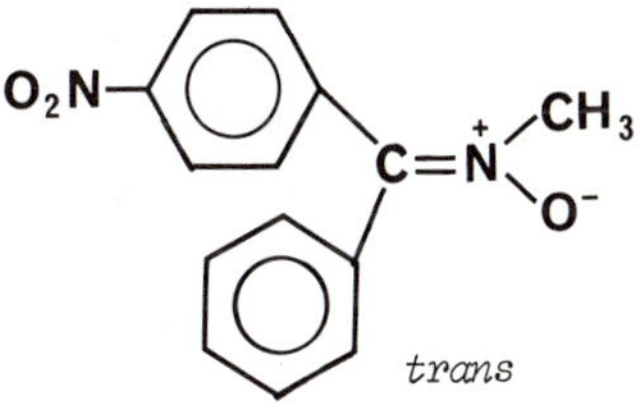

   μ_{obs} = 1.09 and 6.20 D
   (L. E. Sutton and T. W. J. Taylor, J. Chem. Soc. 2190 (1933))

   (b)   and
              trans   cis

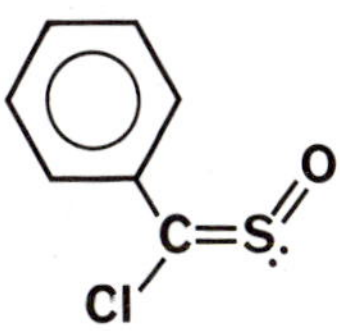

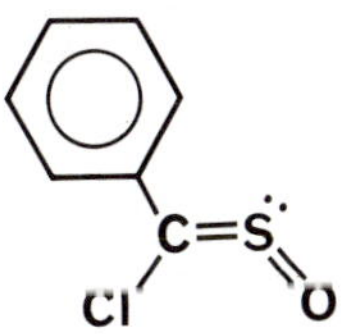

   μ_{obs} = 2.63 and 3.97 D

2. The dipole moments of some mono- and disubstituted benzenes in
   CCl_4 solution are given below. From the approximate additivity
   of moments observed, predict the dipole moment of $C_6H_5-CCl_3$ in
   CCl_4 (R. J. W. LeFevre, et. al., J. Chem. Soc. (B): 1120
   (1970); 159 (1969)).

Ch. 3 MOLECULAR POLARIZATION

| Substituent | | Moment of | | Moment of |
X	Y	C_6H_5X	C_6H_5Y	p-X-C_6H_4-Y
CH_3	F	0.34	1.38	1.76
"	Cl	"	1.59	1.88
"	Br	"	1.51	1.87
"	I	"	1.39	1.72
"	NO_2	"	3.95	4.35
"	CN	"	4.02	4.33
$C(CH_3)_3$	Br	0.3	1.51	1.91
"	CH_3	"	0.34	0
CCl_3	Cl	??	1.59	0.50

3. The dipole moment of phenylacetylene is 0.78 D. In which direction is the dipole?

4. The dipole moment of pentafluorobenzene is smaller than that of fluorobenzene, and the moment of 2,3,5,6-tetrafluorotoluene is larger than that of toluene (H.-H. Huang, J.C.S. Chem. Comm. 723 (1973). From the discussion in this chapter, these differences are unexpected. Propose an explanation for the differences.

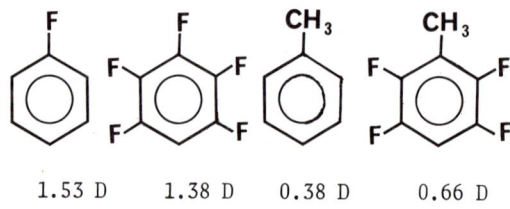

μ_{obs} 1.53 D   1.38 D   0.38 D   0.66 D

Answers

1. (a) <u>trans</u>, μ= 1.09 D.

   (b) <u>trans</u>, μ= 2.63 D.

2. Since the moments of the disubstituted compounds are within 0.04 D of the sums of the moments of the respective monosubstituted compounds, the moment of $C_6H_5CCl_3$ can be expected to be 1.59 - 0.50 = 1.09 ± 0.04 D.

Sec. 3.2 ELECTRIC DIPOLE MOMENTS   63

2. (Cont'd)

$$\xleftarrow{\dfrac{CCl_3C_6H_5}{??}} \qquad \xrightarrow{\dfrac{C_6H_5Cl}{1.59}}$$

$$\xrightarrow{\dfrac{Cl_3C-C_6H_4-Cl}{0.50}}$$

3. From the relative group electronegativities, the dipole is directed away from the ring toward the C≡C bond. This is expected on a basis of the hybridization of the phenyl and acetylenic carbons and is supported by most data. For example, in benzene solution the dipole of p-nitrophenylacetylene (3.45 D) is less than that of nitrobenzene (3.96 D) (A. J. Boulton, et. al., J. Chem. Soc. (B) 822 (1966)).

4. Neighboring dipoles mutually polarize each other so that the observed dipole is less than that calculated when the dipoles are parallel and the observed dipole is greater than that calculated when the dipoles are antiparallel. For example,

μ_{obs}	4.29 D	3.17 D	2.38 D
μ_{calc}	4.53 D	2.78 D	2.53 D

Thus, in pentafluorobenzene, the dipoles associated with the three center fluorines polarize each other to reduce the observed moment to less than that of fluorobenzene, whereas the opposite effect is observed in the case of toluene and tetrafluorotoluene. Another example is as follows:

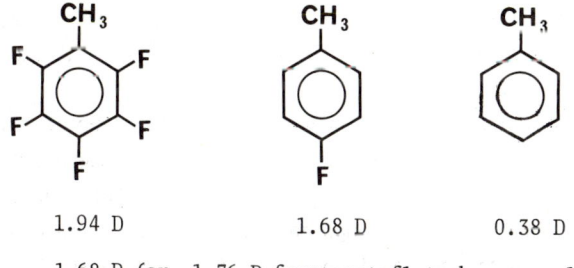

μ_{obs}	1.94 D	1.68 D	0.38 D
μ_{calc}		1.68 D (or, 1.76 D from pentafluorobenzene plus toluene)	

3.3   Substituent electrical effects

Chemists have found it useful to partition the electrical effects of substituents on the physical and chemical properties of a molecule into at least five modes of action:

(1)   σ induction

(2)   π induction

(3)   Field effects

(4)   Delocalization

(5)   Orbital rehybridization

Modes (1) and (2) refer to the successive polarization of bonding electrons along a chain of atoms, (1) for σ electrons and (2) for π electrons. The field effect is the electrical effect of a substituent on the reaction site transmitted through space (including the solvent) in accord with classical electrostatics. Mode (4) operates only when there is a conjugated π system (including homo- and spiroconjugation) and is discussed in Chapters 6, 7, and 8. Steric conditions which alter bond angles may produce changes in orbital hybridization of the reacting atom and therefore affect its electronegativity. Substituents may thereby exert an indirect electrical effect on the reactivity or physical characteristics of a molecule.

All these modes of action of electrical effects of substituents often operate collectively. Not only is there no satisfactory quantitative theory to assess the magnitude of each but it is even difficult to discuss each one separately. Consequently, we will merely try to pick out structural situations which justify the application of these five modes or concepts of electrical effects. A discussion of the electrical effects of alkyl groups is included because much work has been done in this area and it offers a chance to discuss the composite effects of the five modes of action.

Several modifications of former views will emerge in this section. Thus, the inductive electron-withdrawing ability of the alkyl groups in the gas phase will be found to be $\underline{t}$-Bu > Me > H rather than the reverse order; the effects of alkyl groups on acid strengths of alcohols and base strengths of aliphatic amines is best rationalized on the basis of polarization of the alkyl groups rather than inductive effects; and most electrical effects of substituents formerly attributed to σ-induction can be assigned to field effects.

3.3a. Induction

G. N. Lewis first used the notion of chain induction to explain the relatively high acid strengths of the chloroacetic acids:

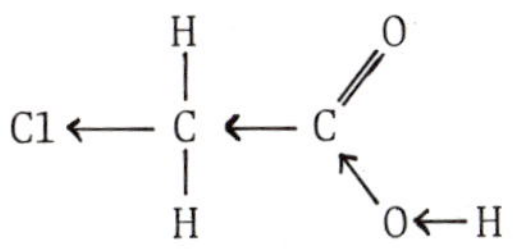

$$Cl \leftarrow \underset{H}{\overset{H}{\underset{|}{\overset{|}{C}}}} \leftarrow C \overset{\nwarrow}{\underset{O \leftarrow H}{\overset{O}{\diagup}}} \quad \text{relative to} \quad H - \underset{H}{\overset{H}{\underset{|}{\overset{|}{C}}}} - C \overset{O}{\underset{O - H}{\diagup}}$$

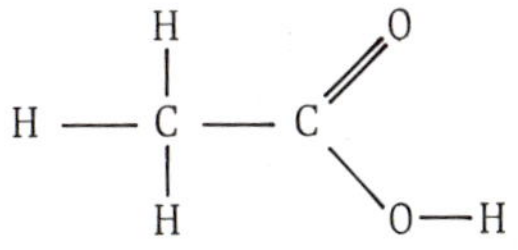

It was argued that the chlorine atom of chloroacetic acid induces a positive charge on the adjacent carbon, owing to the greater electronegativity of chlorine, which then polarizes the C-C bond, and in turn the hydroxyl oxygen becomes slightly more positive than the corresponding oxygen of acetic acid. It therefore pulls in the electrons from the hydrogen, to facilitate dissociation of the proton as H^+. The more chlorine atoms there are on the methyl carbon, the greater is this acid strengthening inductive effect. As a result, trichloroacetic acid is a strong acid like mineral acids ($10^3 K_a$ for CH_3COOH = 0.0175).

$$10^3 K_a \ (25°C)$$

X	XCH_2COOH	$X_2CHCOOH$	X_3CCOOH
Br	1.3	33	85
Cl	1.4	50	220
F	2.6	56	590

To make other comparisons:

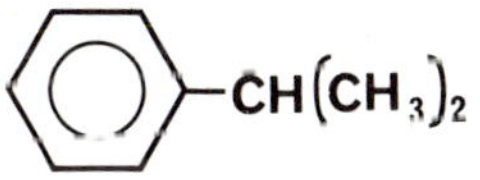

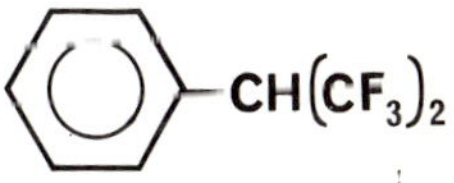

$\text{C}_6\text{H}_5-CH(CH_3)_2$ $\text{C}_6\text{H}_5-CH(CF_3)_2$

pK_a (benzylic proton)   37   17.9

Relative acid strength

CH_3CO_2H	10^{-5}	CH_3SO_3H	17
CF_3CO_2H	1	CF_3SO_3H	427

Actually, CF_3SO_3H is the second strongest monobasic acid known (FSO_3H is the strongest [108]), even stronger than H_2SO_4 and $HClO_4$ which have relative acid strengths of 30 and 397, respectively.

The σ-inductive effect is negligible beyond two single bonds from the reaction site. For example, there is a sharp fall in acid strengths of aliphatic acids when an electronegative substituent is moved to the β carbon atom (Table 3.7). The substituent effect then becomes a field effect.

Table 3.7. $10^5 \underline{K_a}$ of some aliphatic acids, $25^\circ C$

X	XCH_2COOH	$X(CH_2)_2COOH$	$X(CH_2)_3COOH$
H	1.75	1.34	1.50
I	71	9.0	2.3
CF_3	100	7	
Br	125	9.8	2.6
Cl	136	10.1	3.0
COOH	149	6.4	4.5
CN	342	10.2	3.7

NMR chemical shifts are quite sensitive to chain inductive effects, however, which is often reflected beyond two or three C-C bonds. For example, the ^{13}C chemical shifts (in ppm relative to external benzene) in some aliphatic acids were assigned as follows [109]:

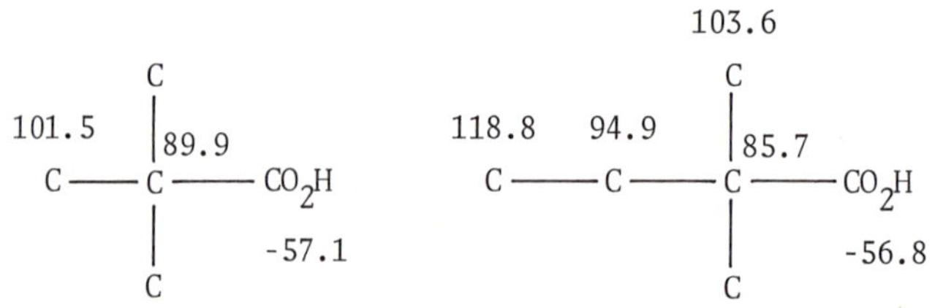

```
                C                         103.6
                                           C
   101.5   |89.9              118.8  94.9  |85.7
    C —— C —— CO₂H              C —— C —— C —— CO₂H
         |         -57.1              |         -56.8
         C                            C
```

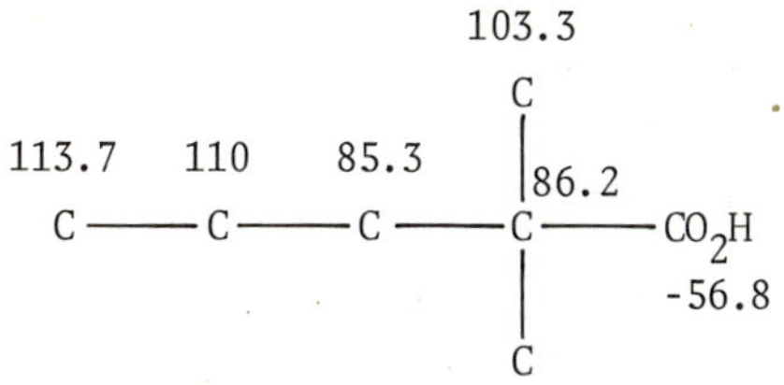

```
                       103.3
                         C
 113.7    110    85.3    |  86.2
  C ——— C ——— C ——— C ——— CO₂H
                         |  -56.8
                         C
                                               107.8
                                                 C
                   113.6  110.3  87.1      82.2  | 96.5  119.4
                    C ——— C ——— C ——— C ——— C ——— C
                                             |
                                            CO₂H
                                            -56.2
```

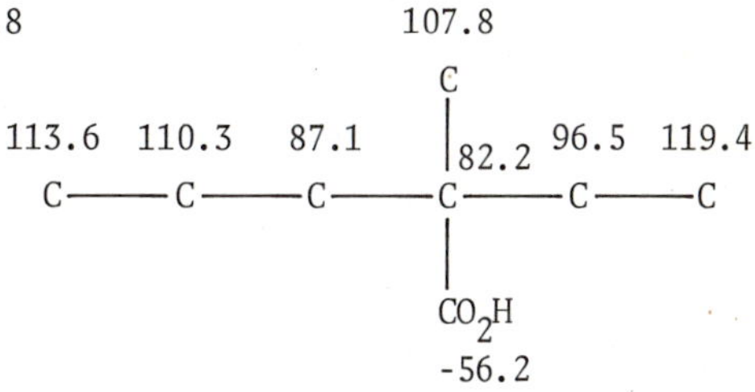

Thus, the ^{13}C signals move steadily upfield, inferring increased electron density, as the carbons are farther removed from the oxygen atoms.

Photoelectron spectral data showing inductive effects on orbital ionization potentials are presented in Chapter 10.

It must be pointed out here that it is only by chance that the foregoing explanation of the dissociation constants of substituted acetic acids can be made. There would be numerous exceptions if data obtained at temperatures other than 25°C had been used because relative acid strengths change with temperature (Figure 3.1).

It is common to identify the strength of an acid with the equilibrium constant $\underline{K_a}$ for the reaction

$$HA_{aq} = H^{\oplus}_{aq} + A^{\ominus}_{aq}$$

where the usual thermodynamic constants apply.

$$\Delta G° = -RT \ln \underline{K_a} = \Delta H° - T\Delta S°$$

Surprisingly, for this simple reaction, the $T\Delta S°$ term varies considerably for different substituted acetic acids and actually has a greater

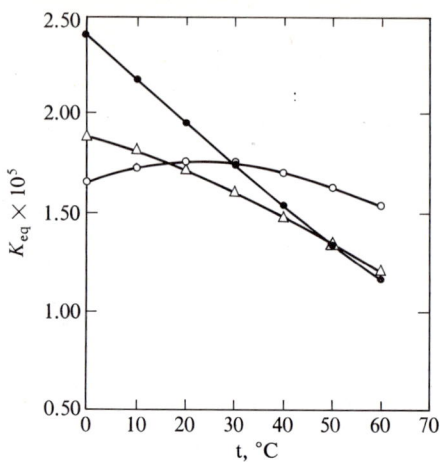

Figure 3.1. The ionization constants of some alkyl acetic acids as a function of temperature. O, acetic acid; Δ, isopropylacetic acid; and ■, diethylacetic acid.

influence on $\underline{K}_a$ than the enthalpy, ΔH^o (Table 3.8) [110]. Negative ΔS^o values are indicative of electrostriction of the solvent by the ions, an effect which usually exceeds all others in magnitude. Apparently there is extensive positioning of solvent molecules about the ionic products of dissociation which lowers the entropy of the system and retards ionization. Not only does this entropy effect outweigh enthalpy changes with structural modifications, but it varies with temperature [111]. As a result, relative acid strengths of acids fluctuate with temperature.

Moreover, the enthalpy of ionization usually changes with temperature, and even changes *sign* for acetic and benzoic acids near room temperature [112]:

ΔH^o (cal/mol)

t°C	Acetic	Benzoic
5	657	736
15	275	460
25	-137	118
35	-430	-272
45	-671	-657

Table 3.8. Enthalpy and entropy terms for the ionization of some acetic acids at infinite dilution in water at 25°C [110]

Acid	ΔH^o (cal/mol)	$T\Delta S^o$ (cal/mol)	$10^5 \underline{K}_a$
HCOOH	-41	-3440	17.6
CH_3COOH	-137	-6570	1.76
Me_3CCOOH	-690	-7540	0.93
ICH_2COOH	-1420	-5750	66.8
$BrCH_2COOH$	-1240	-5180	125.3
$ClCH_2COOH$	-1120	-5040	135.6
FCH_2COOH	-1390	-4920	259.2
$Br_2CHCOOH$	-500	-2380	3300
$Cl_2CHCOOH$	-100	-1790	5000
$F_2CHCOOH$	0	-1790	5600
Br_3CCOOH	-800	-595	8500
Cl_3CCOOH	+1000	+595	22000
F_3CCOOH	0	-298	59000

Accordingly, in aqueous solution (1) σ **induction** does not adequately account for the relative enthalpy of ionization of weak acids and (2) relative dissociation constants may vary with temperature. The fact that at 25°C relative electronegativities of substituents correlate aqueous ionization constants of acetic acids quite well suggests that the sequence of relative $\underline{K}_a$s is the result of a field effect (solvation), not of σ **induction**. The X-C bond dipole in the XCH_2COO^- anion can exert a field effect upon the ordering of solvent molecules about the ionic species. This view is confirmed by the observation that there is a reversed sequence in the gas phase, i.e., acidities decrease in the order $FCH_2COOH < ClCH_2COOH < BrCH_2COOH$ [109a].

One view that has appeared from time to time is that the charges along a chain of saturated carbon atoms alternate rather than decrease from a point of high charge. This notion has been revived by some recent

CNDO calculations [113] and finds experimental support in the observation that there is an alternation in substituent effects on nmr J_{HH} and J_{HF} coupling constants [114]. For example, J_{HF} for the system X-C-C-C-F *increases* with the electronegativity of X but *decreases* with the electronegativity of X for the fragment X-C-C-F.

Since π electrons are more polarizable than σ electrons, we expect and observe that π induction is larger than σ induction. This explains, for example, the low reactivity of pyridine toward electrophilic substitution and relative acid strengths when resonance delocalization effects are minor [115].

	pK_a		pK_a
C_6H_5OH	9.9	C_6H_5COOH	4.21
C_6F_5OH	5.5	C_6F_5COOH	3.38

We saw earlier that relative group electronegativities of substituents R can be determined from ir C=O frequencies in $\overset{R}{\underset{X}{>}}$C=O systems, where X = Cl, OEt, or OH. This is because π induction is greater than σ induction and the C=O bond character is more sensitive to the inductive effects of R. Polarization of a C=O bond affects the basicity of the oxygen, which is revealed by various properties such as pK_a, polarographic reduction potential, $\Delta\nu_{OH}$ of proton donors, C=$\overset{\oplus}{O}$H pmr chemical shift of the protonated ketones, C=O ir frequencies, etc. (see Section 7.1). For one example, the basicities of α-haloketones decrease with halogenation [116].

	pK_a
$CH_3-CO-CH_3$	-7.5
$BrCH_2-CO-CH_3$	-10.7
$ClCH_2-CO-CH_3$	-10.7
$FCH_2-CO-CH_3$	-10.8
$FCH_2-CO-CH_2F$	-12.9
$Cl_3C-CO-CH_3$	-14.8
$F_3C-CO-CH_3$	-14.9
$F_2CH-CO-CHF_2$	~ -17

Sec. 3.3 SUBSTITUENT ELECTRICAL EFFECTS

We saw in Section 3.1a that hydrates and hemiacetals of trihalomethyl ketones, $(CX_3)_2C(OH)_2$, are readily isolable because the halogens inhibit the $[R_2C=O, R_2\overset{\oplus}{C}-O^{\ominus}]$ carbonyl resonance and destabilize the ketones. Also, many properties of hexafluorothioacetone indicate that the polarity of the $\overset{\sigma+}{C}=\overset{\sigma-}{S}$ bond has been reversed owing to π induction.

A comparison of substituent effects in aryl and bicyclo-[2.2.2]-octyl systems has shown that π-**inductive** effects are distinctly larger than σ-**inductive** effects. For example, the ^{19}F SCS (substituent-induced chemical shifts) in DMF for NO_2 and CO_2Me in compounds **3.3** - **3.6** are as follows (117).

R = NO_2	^{19}F SCS	R = CO_2Me	^{19}F SCS
3.3	-10.72	3.3	-5.69
3.4	-5.51	3.4	-2.80
3.5	-1.83	3.6	-1.01

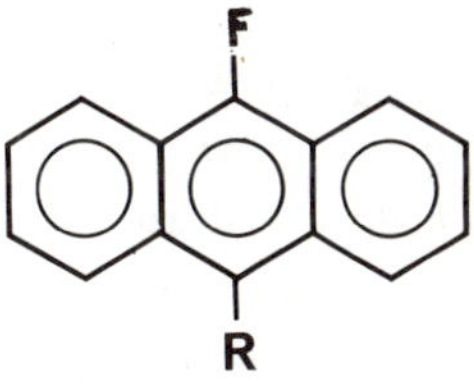

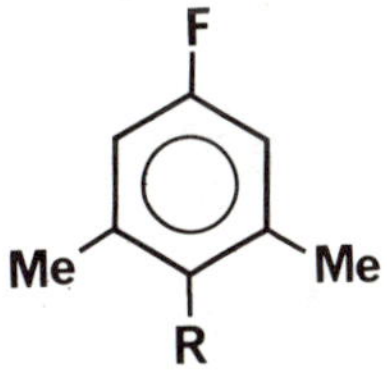

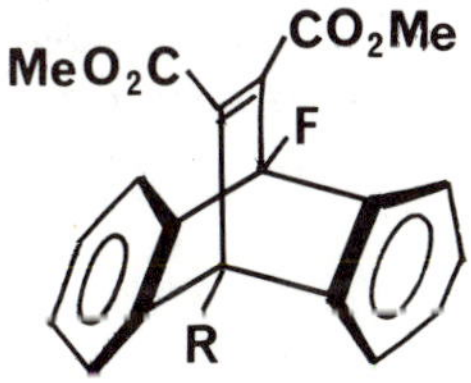

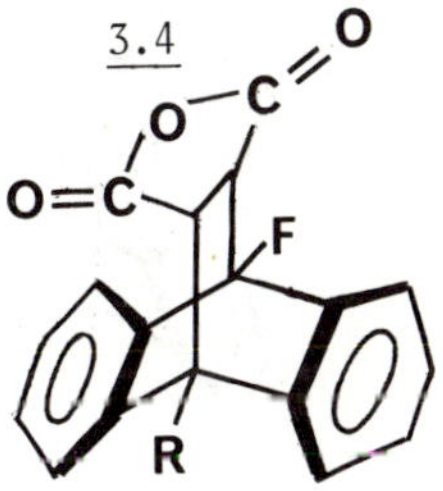

In **3.5** or **3.6**, the ^{19}F SCS is due to a σ-inductive effect (plus a possible field contribution) and is rather small. In **3.3** and **3.4** where resonance is sterically inhibited (by the peri H's in **3.3** (see Section 20.1) and the 3,5-dimethyl groups in **3.4**) the ^{19}F SCS is due primarily to π induction (plus the same possible field contribution). It can be seen that π induction is the dominant effect here and is larger in the anthracene ring than in the benzene ring.

3.3b. Field effects

The field effect has been the elusive factor in the transmission of electrical effects of substituents. It certainly is necessary to explain the difference between the first and second dissociation constants of saturated dicarboxylic acids [118]. In the absence of any good information on the relative magnitudes of the σ-inductive, the π-inductive, and the field effect, most chemists have lumped them together as an inductive effect. This has not been a serious problem in correlating inductive effects because the manner in which the parameter constants have been determined has adopted this same assumption. Thus, excellent quantitative correlations have been obtained in terms of σ_I and σ^* constants.

There have been numerous attempts to separate inductive from field effects but there is no general agreement on their relative importance [119]. Convincing arguments have recently been offered to show that σ induction is insignificant beyond two bonds, and that in the absence of π induction, field effects rather than σ induction are responsible for most observed inductive effects [120,121]. A few supporting observations from various sources can be presented here.

Some of the most persuasive evidence for the field effect is the angular dependence of substituent effects. For example, the inductive model would predict that compounds **3.7a** and **3.7b** would have the same $\underline{pK_a}$, and the fact that they do not supports the notion of a field effect rather than σ induction. Nevertheless, since **3.7a** is a stronger acid than **3.7c**, the substituents must exert more than a simple field effect.

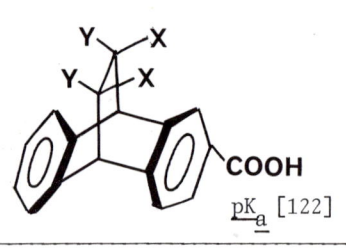

$\underline{pK_a}$ [122]

3.7a, X = Cl, Y = H	6.07
3.7b, X = H, Y = Cl	5.68
3.7c, X = Y = H	6.27

Sec. 3.3 SUBSTITUENT ELECTRICAL EFFECTS   73

Substituents separated from a halogen atom by three or more intervening carbon atoms have a negligible effect on the nuclear quadrupole resonance frequency of the halogen, indicating that inductive effects fall off rapidly along a carbon chain [121]. The nmr substituent-induced chemical shifts across four bonds are very small (Tables 3.9 and 3.10) [119a]. Values for the transmission coefficient ε of inductive effects across single bonds have ranged from 0.16 to 0.6 [119] but these data and work of others [121,123] indicate that a value near the lower end of this range is most appropriate.

Table 3.9. ^{19}F SCS values* for compounds 3.8 and 3.9 in dioxane [119a]

R	3.8	3.9
H	0.00 ppm	0.00 ppm
OCH_3	-0.14	0.26
$OCOCH_3$	-0.36	0.03
CN	-0.49	-0.41
Br	-0.71	-0.87
NO_2	-1.62	-0.65

*SCS = δ^F_H - δ^F_R . A positive value is a shift upfield

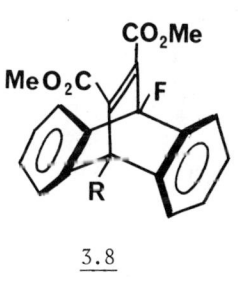

    3.8   3.9   3.10

Table 3.10. Proton SCS values* for 3.10 in CCl_4 [119a]

R	SCS
H	0.00 ppm
$COCH_3$	-.022
Br	-.037
NO_2	-.063
CN	-.072

*$SCS = \delta_H^H - \delta_R^H$

Inductive effects should increase with the number of transmission paths. Thus, the effect of $CO_2^\ominus$ on $\underline{K}_1$ for the acids 3.11-3.13 (R = $CO_2^\ominus$) should increase in the order 3.11 < 3.12 < 3.13 in a ratio of 1:3:6 (120).* The fact that the statistically corrected ratios of the first and second dissociation constants, log ($\underline{K}_1/4\underline{K}_2$), are virtually the same argues against an inductive effect. Since the C_1 to C_4 distances in 3.12 (2.59 Å) and 3.11 (2.69 Å) are not greatly different, these results support the notion of a field effect [120].

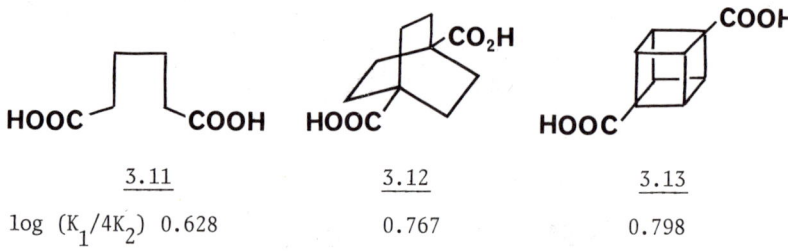

3.11	3.12	3.13
log ($K_1/4K_2$) 0.628	0.767	0.798

---

*Even if one argues that the effect of R should *diminish* rather than *increase* because its effect is more diffused with an increasing number of transmission paths, at least there should be a difference for compounds 3.11-3.13.

Sec. 3.3 SUBSTITUENT ELECTRICAL EFFECTS   75

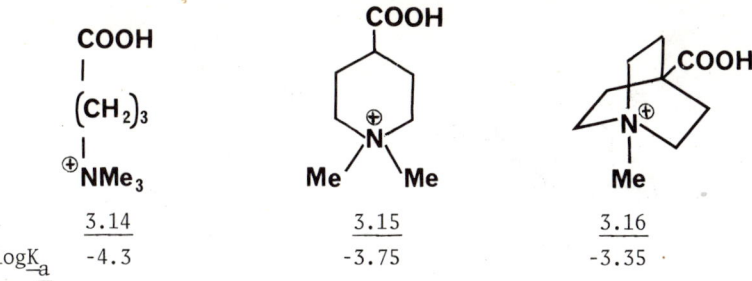

	3.14	3.15	3.16
logK_a	-4.3	-3.75	-3.35

In the cations 3.14-3.16 the number of bonds between $\overset{\oplus}{N}$ and the COOH remains constant whereas the distances are certainly different. Grob argued that since the distance between $\overset{\oplus}{N}$ and the COOH has a greater influence on the dissociation than the number of bonds, this is evidence for a field effect rather than induction [124].

The dissociation constants for the acids 3.17-3.19 are correlated well by the equations

$$\log K \ (3.17) = 1.63 \ \sigma_I - 6.88$$
$$\log K \ (3.18) = 1.75 \ \sigma_I - 6.49$$
$$\log K \ (3.19) = 1.15 \ \sigma_I - 5.75$$

where σ_I is the Taft induction substituent constant (see Chapter 7).

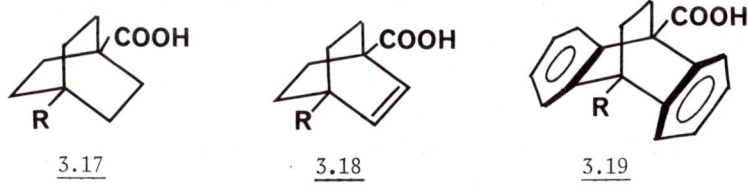

| 3.17 | 3.18 | 3.19 |

If the effect of R on log K were operating via σ induction, one would expect the influence of R (the coefficient for σ_I) to decrease in the order 3.19 > 3.18 > 3.17 [125], which is not observed. Again, a field mode of transmission of the effects of R best accounts for the data [120]. The acidities of another series of acids, 3.20, are shown below [126].

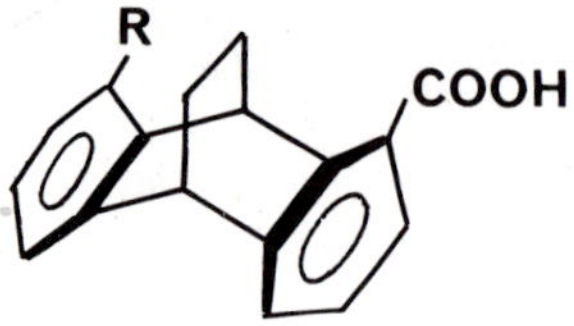

Ch. 3 MOLECULAR POLARIZATION

R	pK_a
H	5.99
F	6.01
Cl	6.24
CN	6.05
CO_2Me	6.16
OCH_3	6.15
CH_3	6.07
OH	5.98
COOH	5.97
CO_2^-	6.89

3.20

The acids with an electronegative group, e.g., R = F, Cl, etc., are weaker rather than stronger than the parent acid (R = H), which is evidence that the electrical effect of the substituent is transmitted by a field effect. The log ($K_1/4K_2$) ratio is normal, 0.92, (log ($K_1/4K_2$) is 0.73 for glutaric acid and 1.30 for succinic acid) and indicates that the acid-weakening effect of an electronegative R is not due to intramolecular H bonding [127].

In conclusion, there are several strong arguments to show that many of the inductive effects of substituents are transmitted mostly by a field mode of action rather than by chain induction. Thus, in several systems, such as spirane carboxylic acids 3.21 (128), calculations of polar effects

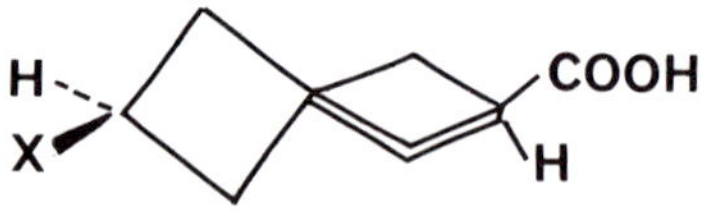

3.21

based on a field model are closer to experimental values than those based on chain induction (129). This applies to chemical reactivities too. For example, axial esters solvolyze more slowly than the corresponding equatorial isomers in certain systems. The results can be explained in terms of the transition state dipole-dipole interaction (field effect) between the substituent and the leaving group, which is more unfavorable for the axial isomers.

	3.22	3.23
$k_{ax/eq}$ (acetolysis)	0.539 (130)	0.333 (131)

Several pairs have been studied. Not only does the field effect provide a logical explanation, since an inductive effect would predict identical rates, but a Kirkwood-Westheimer treatment based on a field model accounts very well for the observed rates of acetolysis of 3.22 (131). When there is also a π system involved, π induction may be as large as the field effect.

3.3c. Electrical effects of alkyl groups

Interest in the electronic properties of alkyl groups remains high and a steady flow of related publications continues unabated. Electronic theory of organic chemistry has considered the three major modes of relay of polar effects of alkyl groups to be induction, polarization, and hyperconjugation, and until recent years, to exert a +I inductive effect. This results in the following order of electron release by the "big four" [132], called the inductive order:

t-Bu > i-Pr > Et > Me > H.

Relative polarizabilities are also in this order. When attached to a π system, particularly an sp^2-hybridized carbon, alkyl groups usually exhibit the Baker-Nathan order of electron release (see Section 5.3):

Me > Et > i-Pr > t-Bu.

The situation is quite complex because there are many reactions of alkylbenzenes for which the inductive order is observed and an equal number which follow the Baker-Nathan order [132,133]. Several hypotheses have been proposed to explain the Baker-Nathan order, including hyperconjugation, steric inhibition of solvation [134], steric hindrance to bond contraction [135], and H bonding involving the alkyl group protons [136].

These two orders are based on much data observed for systems in which it is difficult to separate inductive (or field) and hyperconjugative effects. Recently, a wide variety of properties has been studied and two modifications of past views have evolved [137]. One is that polarizability rather than induction often provides an explanation for the observed properties, and the other is that frequently alkyl groups exert a $-I$ instead of a $+I$ effect.

The observation which shows that the importance of polarization is greater than that of induction is that alkyl groups stabilize positive as well as negatively charged ions [123,138]. It is well established experimentally and theoretically that tertiary carbenium ions are more stable than primary carbenium ions, although it is not universally attributed to hyperconjugation [139,140].

Heats of formation (kcal/mol) for some alkyl carbenium ions [123]

Primary		*Secondary*		*Tertiary*	
Ethyl	219	2-Propyl	192	$\underline{t}$-Butyl	167
1-Propyl	208	2-Butyl	183		
1-Butyl	201				
Isobutyl	199				

Likewise, in the gas phase, the stabilities of acylium ions parallel the order of polarizabilities [141].

Heats of formation in the gas phase (kcal/mol)

R in R-$\overset{+}{C}$=O	Experimental	Calculated*
Me	153	157
Et	143	145
$\underline{i}$-Pr	133	133
$\underline{t}$-Bu		121

*By the MINDO approximation method

Recent experimental [142] and theoretical [143,144] studies of the acidities of alcohols and alkanes have shown that alkyl groups also stabilize negatively charged ions. Calculations indicate that the C-H bonds in the alkyl groups of RO$^-$, for example, are strongly polarized $^{\delta+}$C-H$^{\delta-}$. This view explains the fact that the gas phase basicities of amines [145], alcohols [146], and ethers, and the gas phase acidities of alcohols [147], increase with greater polarizabilities of the respective alkyl groups. For example, in the gas phase the relative basicities of

(1) primary amines:

$$\underline{t}\text{-BuNH}_2 > \underline{i}\text{-PrNH}_2 > \text{EtNH}_2 > \text{MeNH}_2 > \text{NH}_3;$$

(2) secondary and tertiary amines:

$$\text{Et}_2\text{NH} > \text{Me}_2\text{NH}$$

$$\text{Et}_3\text{N} > \text{Me}_3\text{N};$$

(3) multiple alkylated amines:

$$\text{Me}_3\text{N} > \text{Me}_2\text{NH} > \text{MeNH}_2 > \text{NH}_3$$

$$\text{Et}_3\text{N} > \text{Et}_2\text{NH} > \text{EtNH}_2$$

$$\text{Me}_3\text{N} > \underline{t}\text{-BuNH}_2$$

all indicate that the inherent basicity of the nitrogen of ammonia increases with degree of alkylation and with the polarizability of the alkyl groups [145]. The results are remarkably self-consistent, and indicate that the net polarizability of three methyls is about the same as two ethyls and the effect of two methyls is equivalent to one isopropyl. Of course, large steric effects such as F strain or solvation energies [148] can outweigh the small effects of polarization and completely different sequences may be observed (Section 20.3). For instance, the polarizability order is observed in the gas phase for the acidity [142, 147] as well as basicity [146] of alcohols and the stability of acylium ions [141], whereas the Baker-Nathan order is found for these same species in aqueous solution. Also, the relative acidity of acetylenes and water are reversed for the gas phase and in aqueous solution [149]. The electron-withdrawing or electron-donating effect of the alkyl groups relative to hydrogen also depends upon whether the attached system is electron rich or electron poor. Thus, the energy of the principal electronic transition in the uv of p-substituted alkylbenzenes decreases with increasing polarizability of the alkyl group for electron-donating and electron-accepting substituents (see Chapter 13).

There are many measurements--dipole moments, ir spectra, ionization potentials, and pmr chemical shifts--which indicate that the order of electron release in p-substituted alkylbenzenes coincides with the polarizabilities of the alkyl groups:

$\underline{t}$-Bu > $\underline{i}$-Pr > Et > Me > H

(Table 3.11). The same relative order for $\underline{t}$-Bu and Me is observed for the integrated intensity (A) of the ir band in the 1600 cm^{-1} region for $\underline{p}$-alkylbenzaldehydes, pyridines, pyridinium iodides, and triphenylmethyl carbocations [150].

R	R-C$_6$H$_4$CHO	R-C$_5$H$_4$N	R-C$_5$H$_4$NI$^{\oplus}$$^{\ominus}$	(R-C$_6$H$_4$)$_3$CBF$_4$$^{\oplus}$$^{\ominus}$
H	1308	1360	1821	6336
Me	3549	2555	4150	14280
$\underline{t}$-Bu	3722	2900	4723	17315

Integrated ir intensities are proportional to the square of the change in dipole moments during excitation, which in this case is related to the polar character of the alkylbenzene ring. The greater the electron displacement, the larger the value of A.

The order of ionization potentials of the alkylbenzenes and *resistance* to one-electron reduction ($\underline{K}$ in Table 3.11) indicates that the electron densities of the rings follow the polarizabilities of the

Table 3.11.

Electronic effects of alkyl groups in $\underline{p}$-substituted alkylbenzenes

Property		R			
	H	Me	Et	i-Pr	t-Bu
μ_{gas}, C$_6$H$_5$R (D) (151)	0	0.37	0.58	0.65	0.70
μ, $\underline{p}$-NC-C$_6$H$_4$R (152)	4.08	4.42	4.53	4.60	4.64
IP (ev) (153)	9.24	8.92	8.75	8.6	8.5
$\delta(^1H)_{cyclohexane}$ (para) (154)	~7.8	5.9	4.4	3.8	~1.1
A$_{CCl_4}$,* C$_6$H$_5$R (155)		275	289	332	677
$\underline{K}^{\dagger}$ (C$_6$H$_5$R$^-$ → C$_6$H$_5$R) (156)		4.4	22	40	110

* Integrated intensity of the 1600, 1585 cm^{-1} bands.
† Equilibrium constant for the reaction
$$C_6H_6 + RC_6H_5^{\cdot -} \rightleftharpoons C_6H_6^{\cdot -} + RC_6H_5.$$

alkyl groups. Thus, it is easiest to strip an electron from the ring of t-butylbenzene and hardest to force an electron into its ring.

Hence, the electronic effects of alkyl groups on a wide variety of aliphatic and aromatic systems can be attributed to the relative polarizabilities of the alkyl groups. In many cases, the trend could be the result of induction but this view would not be valid where alkyl groups stabilize electron-demanding as well as electron-repelling centers.

Numerous isolated observations are consistent with the polarizability (or inductive) effect of the alkyl groups. One, for instance, is the relative ionization potentials of the "big four" alcohols, determined from λ_m of the iodine charge-transfer complexes [157]:

	IP
MeOH	10.85 eV
EtOH	10.48
i-PrOH	10.16
t-BuOH	9.93

Another simple example is the dipole moments of the corresponding alkyl chlorides:

	MeCl	EtCl	i-PrCl	t-BuCl
DM =	1.86	2.01	2.1	2.15 D

The trend in each case is the result of increased electron release by t-Bu over Me. Note also that the carbonyl ir frequency of saturated aliphatic ketones follows the polarizability rather than the hyperconjugative order,

	$\nu_{C=O}$ (cm^{-1})
Me-CO-Me	1719
Me-CO-Pr-i	1718
n-Pr-CO-Pr-n	1716
i-Pr-CO-Pr-i	1713
Me-CO-Bu-t	1710

Ch. 3 MOLECULAR POLARIZATION

although the changes in mass and C-CO-C bond angle (due to repulsion between groups) are also in the direction to reduce $\nu_{C=O}$. Similarly, the C=O ir frequencies of substituted acetic acids correlate with the electronegativities as well as the polarizabilities of the substituents:

Acid	$\Delta\nu$ (cm^{-1}) (for the 1721-cm^{-1} band)
Trichloroacetic	43
Dichloroacetic	30
Chloroacetic	15
Iodoacetic	0
Acetic	0
Dimethylacetic	-6
Diethylacetic	-15
Trimethylacetic	-17

Since Taft polar substituent constants σ^* (Section 3.1b) and inductive constants σ_I (Chapter 7) reflect electron-releasing ability, irrespective of mode, i.e., induction or polarization, many properties of alkyl derivatives have been correlated with one or the other of these substituent constants (Section 7.1). Taking basicity constants as one example, $\underline{pK^+_{BH}}$ for alcohols, defined by the reaction

$$ROH + H_3O^+ \rightleftarrows ROH_2^+ + H_2O,$$

can be related to σ^* by the equation [158]

$$\underline{pK^+_{BH}} = 2.35\ \sigma^* - 2.18.$$

The second striking development to emerge from recent studies interpreting the electrical effects of alkyl groups is the indication that alkyl groups exert -$\underline{I}$ inductive effects (electron-withdrawing relative to hydrogen) in many cases [159-161]. The most convincing evidence comes from nmr data. Nmr chemical shifts for an atom are related to the electron density at that atom such that reduced electron density produces a downfield shift in the resonance signal. Although there may be unrecognized factors affecting chemical shifts, the downfield shift of the ^{14}N signal of alkylamines and ammonium ions with

Sec. 3.3 SUBSTITUENT ELECTRICAL EFFECTS

increasing alkylation on nitrogen as well as at the α carbon, can apparently be attributed to a -$\underline{I}$ effect of alkyl groups [162].

	R	δ_N (ppm from CH_3NO_2)		δ_N (ppm from CH_3NO_2)
RNH_2	H	+383	$MeNH_2$	+378
	Me	+378	Me_2NH	+371
	Et	+355	Me_3N	+365
	$\underline{i}$-Pr	+334		
	$\underline{t}$-Bu	+319	$EtNH_2$	+355
			Et_2NH	+332
RNH_3^+	Me	+351	Et_3N	+327
	Et	+336		
	$\underline{t}$-Bu	+314		

A simple interpretation of substituent effects can be given to ^{13}C and ^{17}O shifts observed for the "big four" alcohols listed in Table 3.12. Replacement of H by Me results in a decrease in electron density, not only at the α and β carbons, but also at the oxygen atom. Additional

Table 3.12. ^{13}C and ^{17}O chemical shifts of some aliphatic alcohols [161]

	$\delta(^{13}C)$*		$\delta(^{17}O)$†
	α	β	
MeOH	143.7		37
EtOH	135.2	174.5	-6
$\underline{i}$-PrOH	129.4	167.8	-38
$\underline{t}$-BuOH	124.5	161.7	-70

*ppm upfield from CS_2. †ppm from H_2O.

84   Ch. 3 MOLECULAR POLARIZATION

examples are as follows:

$$\begin{array}{cccc} & H_2O & Et_2O & i\text{-}Pr_2O \\ \delta(^{17}O)\ [163] & 0 & -15 & -62\ \text{ppm} \end{array}$$

$$\triangleright\!-\!\overset{\oplus}{C}HMe \qquad \triangleright\!-\!\overset{\oplus}{C}Me_2$$

$$\delta(^{13}C)\ [164] \quad\quad -59.1 \quad\quad\quad -86.8\ \text{ppm relative to } CS_2$$

CMR shifts, ppm relative to CS_2

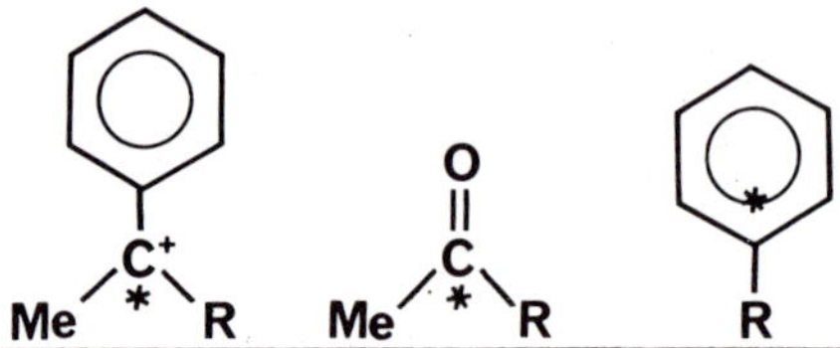

	Ph-C⁺(Me)(R)*	Ph-C(=O)(R)* with Me	Ph(R) ring
Me	-25	-12	56
Et	-30	-14	50
i-Pr	-33	-17	46
t-Bu	-44	-18	42

* Denotes carbon tabulated.

In all cases here, replacement of H by alkyl groups has a deshielding effect on the α or β positions. The reduced electron densities (ρ) of the charged carbon in the alkyl carbenium ions [165, 166]

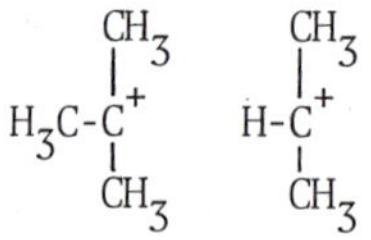

$$H_3C\!-\!\overset{CH_3}{\underset{CH_3}{\overset{|}{\underset{|}{C^+}}}} \qquad H\!-\!\overset{CH_3}{\underset{CH_3}{\overset{|}{\underset{|}{C^+}}}}$$

δC^+	-135.4	-125 (ppm from CS_2, in $SO_2ClF\text{-}SbF_5$)
$\delta C+$	0.692	0.611

and of the carbon and oxygen atoms in the alcohols and ethers with increasing alkylation are hard to explain unless one assigns a -I effect to the alkyl groups. MO calculations have shown that the successive α-methyl substitution in the methyl cation increases the positive charge on the central carbon, the stability of the cation, and total

$\pi_{C_\alpha - C^+}$ overlap. The third effect reflects increased hyperconjugation, although the double-bond character of $C_\alpha - C^+$ bonds diminishes in going from ethyl to isopropyl to t-butyl cations. Thus, hyperconjugation increases π−electron density while induction decreases σ-electron density at the central carbon, with the net effect that the positive charge increases with successive alkylation of the methyl cation.

Similarly, the chemical shift of the protons attached to α carbon atoms in alcohols move downfield with increased alkylation [161].

$CH_3-\underline{CH}-CH_3$ with HO	1.145*	$Et-\underline{CH}-CH_3$ with OH	1.140
$CH_3-\underline{CH}-CH_2-CH_3$ with OH	1.481	$Et-\underline{CH}-CH_2-CH_3$ with OH	1.397
$CH_3-\underline{CH}-\underline{CH}(CH_3)_2$ with OH	1.580	$Et-\underline{CH}-CH(CH_3)_2$ with OH	1.49

*ppm relative to TMS

This is the opposite trend that would be observed from anisotropic shielding by the methyl groups. Similar -I effects are observed for ^{13}C chemical shifts in hydrocarbons [167].

	$\delta^{13}C^*$ (relative to CS_2)
C^*Me_4	165.9 ppm
C^*Et_4	159.9
$MeCH=C^*HEt$	60.4
$MeCH=C^*HPr-\underline{i}$	54.9
$MeCH=C^*HBu-\underline{t}$	50.8

In the case of ketones [168] and similarly constituted alkenes [167], the chemical shifts for the carbonyl and α carbons [161] move downfield with α alkylation but that of the oxygen [163] moves upfield.

Ch. 3 MOLECULAR POLARIZATION

	$\delta(^1H)$*	$\delta(^{13}C)$†		$\delta(^{17}O)$**
	(αH)	(C=O)	(αC)	
Me-CO-CH$_3$	2.09	-11.31	164.7	-572
Me-CO-CH$_2$CH$_3$	2.44	-13.75	157.3	-563
Me-CO-CH(CH$_3$)$_2$	2.59	-16.80	157.3	
Me-CO-C(CH$_3$)$_3$		-17.9	149.1	
Et-CO-CH$_2$CH$_3$	2.43	-16.2	157.5	-548
Et-CO-CH(CH$_3$)$_2$	2.61	-19.3		
Et-CO-C(CH$_3$)$_3$		-20.58	150.1	

* ppm, relative to TMS. † ppm, from CS$_2$. ** ppm from H$_2$O.

The opposite trend in chemical shift values for the carbonyl carbon and oxygen atoms is understandable. The inductive effect alone would cause both resonance signals to move downfield, or at least in the same direction. However, a combined inductive and hyperconjugative effect can account for the observed changes. A -I effect produces a decrease of electron density at the carbonyl carbon and the oxygen atom but a much larger hyperconjugative effect <u>increases</u> the electron density at oxygen without any appreciable effect on the carbonyl carbon. Similar changes occur upon alkylation of α, β-unsaturated esters or acids. Upon replacement of H by Me, $\delta^{13}C$ for the attached carbon moves downfield, reflecting a -I effect, and that for the vicinal carbon moves upfield, the result of hyperconjugation:

$\delta^{13}C$ (ppm from TMS) [169] $\delta^{13}C$ (ppm from TMS) [169]

```
H₂C    129.9    125.0    CH₂          H₂C    130.8    126.2    CH₂
  ‖                         ‖            ‖                         ‖
 HC     128.7    136.2    C-Me         HC     129.2    136.3    C-Me
  |                         |            |                         |
 CO₂Me                    CO₂Me        COOH                     COOH

H₂C    129.9    144.5    HC-Me        H₂C    130.8    146.2    HC-Me
  ‖                         ‖            ‖                         ‖
 HC     128.7    120.4    HC           HC     129.2    121.0    HC
  |                         |            |                         |
 CO₂Me                    CO₂Me        COOH                     COOH
```

Sec. 3.3   SUBSTITUENT ELECTRICAL EFFECTS   87

In support of the -I effect at the attached carbon and the delocalization +R effect at the vicinal carbon, changes in the same direction are observed upon the introduction of a chlorine atom:

$\delta^{13}C$ (ppm from TMS) [169] $\delta^{13}C$ (ppm from TMS) [169]

H$_2$C 129.9   133.3   HC-Cl   MeCH 144.5   145.7   MeC-Cl
‖   ‖   ‖   ‖
HC 128.7   121.5   HC   HC   120.4   117.2   CH
| | | |
CO$_2$Me   CO$_2$Me   CO$_2$Me   CO$_2$Me

However, the picture is clouded by the fact that substitution of bromine or iodine produces shifts reversed from those of methyl and chlorine.

Other studies also infer -I effects of alkyl groups. For one example, the t-butyl cation is believed to be a harder carbocation than methyl cation, implying that Me groups are more electronegative than hydrogen [170].

      3.24   3.25

Alkyl groups exert a rate-retarding -I effect on the solvolysis of 3-substituted-1-bromo-adamantanes (3.24, X = Br) and 4-substituted-bicyclo-[2.2.2]-octyl tosylates (3.25, X = OTs) [159]. There is, however, a +I effect of methyl on the radical abstraction of hydrogen atoms from the adamantyl system [171]. Apparently, solvation is again inflicting a reversal of the electrical effects of alkyl groups. To add to the confusion, the Baker-Nathan order is found for the effect of alkyl groups on the ^{19}F nmr chemical shifts in series 3.24 (X = F) [172] and 3.25 (X = F) [173]. Since the -I effect of Me is almost as large as the electron-withdrawing effect of CO$_2$R, the effect of alkyl groups may be more than inductive. One possibility [173] is that substituents modify the C-C bond orbital hybridization as a result of alterations in bond angles. Increasing bond angle strain, for instance, increases the p character of the C-C bonds and allows for greater electron delocalization by the F atom. This would change the shielding environment of the F nucleus and thereby affect the nmr chemical shift.

Even in simple systems, methyl appears to exert a -I effect. For instance, we saw in Section 3.3b that Me exerts a -I effect in propane. When Me is attached to a π system however, as in methylacetylene, it can exert an electron-releasing hyperconjugative effect.

	$CH_3C\equiv CH$	$CD_3C\equiv CH$	$CH_3C\equiv CD$	$CD_3C\equiv CD$
DM =	0.780	0.784	0.767	0.772 D

Thus, deuteration (CD_3 for CH_3) increases the dipole moment when methyl is at the positive end of the dipole. Since the H-C-H angle does not equal the D-C-D angle, this could be due to a change in hybridization of carbon. MO calculations on methylacetylene indicate that the π system is polarized $H\bar{C}-\overset{+}{C}-\bar{C}-\overset{+}{H}_3$, attributable to hyperconjugation, and the acetylenic σ system is polarized in an opposite direction, ascribed to an -I effect of methyl [174].

^{13}C Nmr chemical shifts of alkyl cations indicate that the charges on the central carbons decrease (are deshielded and $\delta(^{13}\overset{\oplus}{C})$ moves downfield) as H is replaced by Me, and decrease further when substituted by Et [166]. Simultaneously, the chemical shift for an attached methyl group moves upfield:

	Et-$\overset{\underset{\mid}{Me}}{\overset{\oplus}{C}}$-Me	Me-$\overset{\underset{\mid}{Me}}{\overset{\oplus}{C}}$-Me	H-$\overset{\underset{\mid}{Me}}{\overset{\oplus}{C}}$-Me	
$\delta(^{13}\overset{\oplus}{C})$	-139.4	-135.4	-125.0	(ppm from CS_2)
$\delta(^{13}CH_3)$	150.1	146.3	132.8	

Hence, the alkyl groups are exerting a -I effect on $\overset{\oplus}{C}$, with Et > Me. But, with increasing charge on $\overset{\oplus}{C}$, the polarization of Me and $^{\delta-}C-H^{\delta+}$ bonds becomes greater, to move the Me carbon chemical shifts upfield. A similar pattern is observed for the pmr spectra of protonated ethers and alcohols [175].

δC_α-H		$\delta \overset{\oplus}{O}H$	δC_α-H		$\delta \overset{\oplus}{O}H$
4.35	$CH_3-\overset{\oplus}{O}\diagdown^H_H$	9.2	4.49	$CH_3-\overset{\oplus}{O}\diagdown^{Me}_H$	9.03
4.86	$^{Me}\diagdown CH_2-\overset{\oplus}{O}\diagdown^H_H$	9.2	4.73	$^{Me}\diagdown CH_2-\overset{\oplus}{O}\diagdown^{Et}_H$	8.61
5.2	$^{Me}_{Me}\diagdown CH-\overset{\oplus}{O}\diagdown^H_H$	8.92	5.18	$^{Me}_{Me}\diagdown CH-\overset{\oplus}{O}\diagdown^{Pr-i}_H$	7.88

(δ in ppm relative to TMS, all in $FSO_3H-SbF_5-SO_2$ at -60°)

As α-hydrogens are replaced by methyls, which exert a -$\underline{I}$ effect, the charge on C is decreased and the chemical shift for an α-hydrogen moves downfield. At the same time, the chemical shift for the atom attached to the charged atom moves upfield.

Dipole moments of methyl, phenyl, and mesityl ketones decrease generally as the other alkyl group is changed from methyl to $\underline{t}$-butyl [176].

Ketone	Dipole moment
CH_3CO-Me	2.84 D
CH_3CO-Et	2.86
CH_3CO-Pr-$\underline{i}$	2.83
CH_3CO-Bu-$\underline{t}$	2.70
C_6H_5CO-Me	3.05
C_6H_5CO-Et	2.90
C_6H_5CO-Pr-$\underline{i}$	2.93
C_6H_5CO-Bu-$\underline{t}$	2.58
2,4,6-$Me_3C_6H_2CO$-Me	2.81
2,4,6-$Me_3C_6H_2CO$-Et	2.76
2,4,6-$Me_3C_6H_2CO$-Pr-$\underline{i}$	2.64
2,4,6-$Me_3C_6H_2CO$-Bu-$\underline{t}$	2.50

Although the trend might imply greater electronegativity going from Me to $\underline{t}$-Bu, the changes in dipole moments can also be due partly to changes in C-CO-C bond angles and partly to slight differences in orbital hybridization of the carbonyl carbon owing to strain.

In summary, it is reasonable to say:

(1) There are three properties of alkyl groups commonly considered to explain their electrical effects in molecular systems, namely, induction plus field effect, hyperconjugation, and polarizability.

(2) When attached to a saturated atom with low polar character, alkyl groups will usually exert a -$\underline{I}$ effect, $\underline{t}$-Bu > Me.

(3) When attached to a polar atom, the polarizabilities of alkyl groups generally govern their electron-donating or electron-withdrawing effects, $\underline{t}$-Bu > Me.

(4) When attached to a π system, most often the electron-donating ability of alkyl groups parallels their C-H hyperconjugative capacity, Me > $\underline{t}$-Bu.

(5) The variation in electrical effects of alkyl groups is so small that it can easily be inverted or canceled by other effects such as hydration, steric crowding, or orbital hybridization changes. For instance, thermodynamic parameters for the ionization of 2,3-dimethylsuccinic acid and tartaric acid not only exhibit different entropy effects, indicative of differential solvations, but also differ for dl and meso isomers of the same acid [177].

Chapter 4
INTERNAL ENERGIES

You are probably familiar with the fundamentals of electromagnetic radiation and the general techniques for measuring absorption spectra. Hence, in this chapter we will give a cursory treatment of the theory of absorption spectroscopy to refresh your memory, but we will not take up the interpretation of spectra (178, 179). In Part III we will discuss correlations between spectra and molecular structural parameters, i.e., how spectra are affected by resonance, electronegativity, reversible rearrangements, H bonding, etc., and how these generalizations may be used to infer related properties and structures of molecules.

4.1 Electromagnetic Energy

We know from experience that X-rays penetrate flesh but not bones, and darken a photographic plate to reveal bone structure; that long exposure to sunlight produces painful skin burns; that heat lamps (infrared radiation) can warm us inside without heating up the air around us, and can bring relief from certain body aches and pains. These various forms of radiation -- heat, visible light, X-rays -- are all parts of the total electromagnetic spectrum and differ merely in wavelengths. The electromagnetic spectrum is an ordered arrangement of radiation according to wavelength. Our present-day knowledge extends from cosmic rays (short waves) to radio waves (long waves; see Table 4.1). The range primarily used by chemists is from the ultraviolet up to the microwave region.

The energy of a molecule in the gaseous state may be expressed as the sum of its translational and its internal energy. The translational energy is not quantized and therefore a molecule may gain or lose any finite quantity of this form of energy. The internal energy electronic, vibrational, rotational -- is quantized (180), however. Consequently, molecules may gain or lose only integral amounts of these three forms of energy, for instance through collisions or the absorption or emission of light energy. The internal energy of a molecule is conveniently represented by diagrams such as Figure 4.1. The position of a molecule on such a scale is called its *potential energy* or *energy state*.

Table 4.1. Regions of the Electromagnetic Spectrum

Approximate wavelength	Region	Some special uses
10^{-4} Å	Cosmic rays	
0.1 Å	Gamma rays	Nuclear chemistry
100 Å	X-rays	Medicine
4000 Å	Ultraviolet*	Luminescent lights
8000 Å	Visible light	Colorimetry, human sight
1 µ	Near infrared*	Physical therapy
100 µ	Far infrared	Jet engine thermometry, missiles, serial reconnaisance
3 cm	Microwave	Airborne radar, microwave ovens
30 m	Radar	FM radio, ground radar, Tacan, television, military communications
300 m	Radio waves	AM radio, long-range communications

*The term *ultraviolet* once meant "extreme *violet* end" of the visible range and the term *infrared* meant "beyond the *red* end" of the visible range.

The potential energy of a substance is defined as the negative value of the energy that would be required to dissociate it into its atoms to infinite distances. Therefore, a compound is considered more stable as its potential energy moves lower on the negative scale. This is best explained by the diagram in Figure 4.2. The isomers A and B of potential energy X and Y, respectively, on an arbitrary scale would require amounts of energy represented by the distances XG and YG for dissociation into atoms. If they were burned they would liberate amounts of energy indicated by the distances XZ and YZ. Since A and B are isomers, the combustion products are identical. Thus, B is more stable than A, has the greater energy of formation from its atoms, and has the smaller heat of combustion. The lower line at Z could represent the products of a different exothermic reaction, such as hydrogenation. Hence of two isomers, the one with the smaller heat of hydrogenation (liberated) would be the more stable. Strictly, this comparison applies only to isomers, but in general it holds in a semiquantitative way for closely similar compound

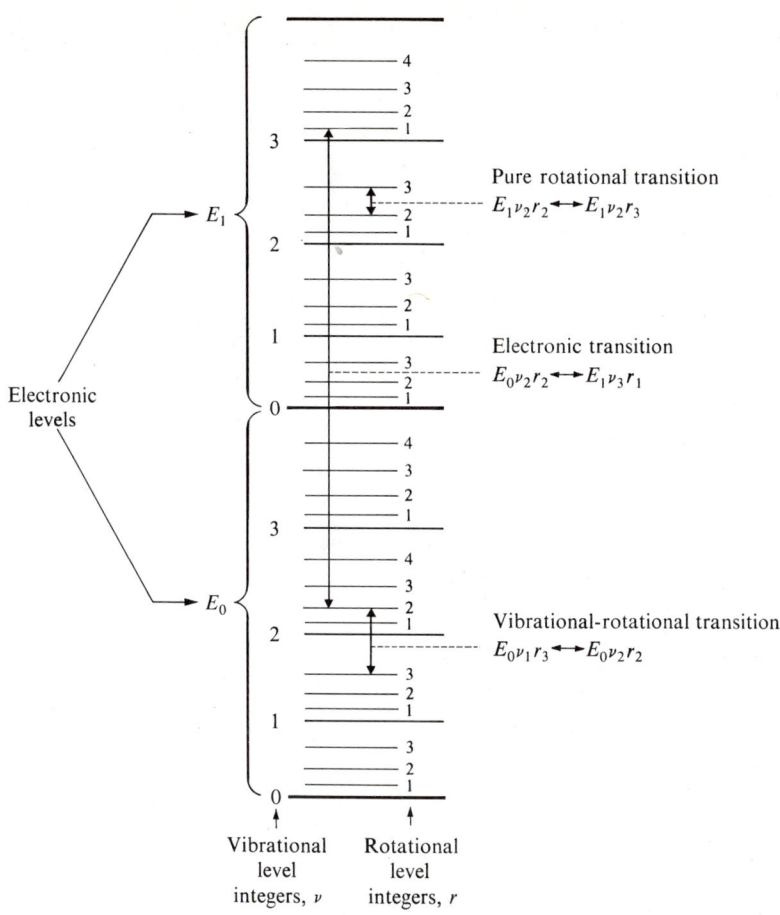

Figure 4.1. Schematic energy level diagram for a simple molecule.

If a molecule gains a small amount of energy, say on the order of magnitude of energy from a light source in the far infrared region of the spectrum (about 1 kcal/mol), this will be only enough energy to bring about a transition between rotational levels, no matter in which vibrational state or electronic state the molecule may be. The result will be a change in the modes of rotation of groups about a given axis in the molecule. If the light source is of greater energy, such as that of the near infrared (see Table 4.2), the quanta of energy are sufficiently large (about 10 kcal/mol)

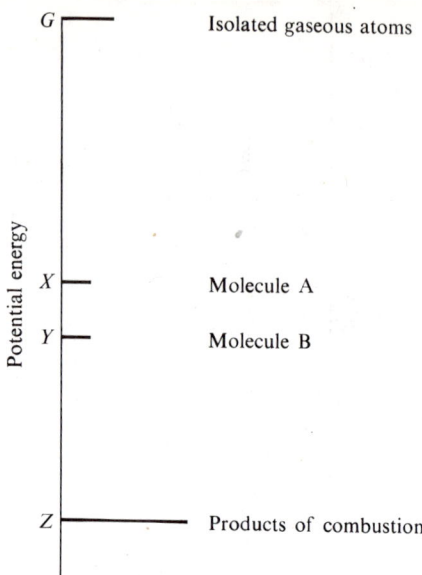

Figure 4.2. Schematic diagram of the relative stabilities of two isomeric molecules, their products of combustion, and their constituent atoms.

to produce vibrational and rotational transitions. The molecule will then undergo changes in mode of atomic vibration and rotation. Finally, if the energy gain is greater (10 to 250 kcal/mol), such as absorption of ultraviolet or visible light, changes in all three forms of internal energy may occur. Changes in electronic energy correspond to alterations in electron configurations of molecules which usually differ by one or more valence bonds. The electronic energy level of a molecule under normal conditions is called its ground state, and the succeeding higher electronic levels represent the first, second, ..., excited states, with energies E_0, E_1, E_2, ..., respectively. Similarly, at each electronic plateau, there are the ground and the first, second, ..., excited vibrational states, and for each vibrational level there are the ground and excited rotational states (see Figure 4.1).

Spectroscopy is one of the best methods for studying internal energy changes in molecules. The use of spectroscopy mushroomed after World War II when commercial instruments became available, and now it is the most widely used tool in chemistry. Almost every theoretical or practical problem

Table 4.2. Energy conversion chart for spectroscopy*

Electron volts	Ergs per molecule	Calories per mole	Wave numbers (cm^{-1})
1	1.602×10^{-12}	23,053	8067
6.242×10^{11}	1	1.439×10^{16}	5.036×10^{-5}
4.34×10^{-5}	6.95×10^{-17}	1	0.35
1.2395×10^{-4}	1.986×10^{-16}	2.86	1

Spectral region	Approximate wavelength (Å)	Approximate energy (cal/mol)
Microwave	10^7-10^8	10
Far infrared	10^5-10^7	10^3
Near infrared	10^4-10^5	10^4
Visible	4000-8000	5×10^4
Near ultraviolet	2000-4000	10^5
Far ultraviolet	1-2000	10^6

* The wavelength of light corresponding to the peak of an absorption band is designated λ_{max} or λ_m, and the intensity is expressed as molar extinction coefficient, ε,

$$\varepsilon = \frac{\log(I_0/I)}{c \times l}$$

where I_0 and I are the intensities of incident light and transmitted light, c is the molar concentration of solute, and l the path length of the solution in centimeters. Log $(I_0/I) = A$, the absorbance (formerly, optical density), is normally recorded directly by the spectrophotometer.

There is no uniform system of recording data. A common practice is to report the intensity of absorption as ε (or log ε), to express the region of the spectrum in the uv and visible region in nm (formerly in mμ or Å), to use wavelength (in μ) for the near infrared (0.5 to 2.0 μ); and to use wave numbers (in cm^{-1}) for the regular infrared (200 to 4000 cm^{-1}).

Wavelength (λ) 1 nm = 1 mμ = 10 Å

Wave number, ν, = $1/\lambda$

Energy (kcal/mol) = $\frac{1239.5}{\lambda \text{(nm)}} \times 23.06$

That is, $\underline{E}$ for λ_{400}nm = $\frac{1239.5}{400} \times 23.06 = 71.5$ kcal/mol.

in chemistry can use one or more forms of spectroscopy to advantage. The British physicist and mathematician Lord William Kelvin (1824-1907) said, "If it cannot be expressed with numbers, we do not know much about it." In the same spirit chemists say, "If it has not been studied with spectroscopy, there is much more to be learned about it."

Spectroscopy has provided quick, accurate analyses for guidance in research and process control; it has contributed to fundamental knowledge about chemical bonding; and it has facilitated the unraveling of complex molecular structures so that chemists could synthesize useful substances such as the ubiquitous plastics, synthetic textiles, rubbers, drugs, detergents, dyes, lubricants, cosmetics, perfumes, and a host of other commodities. It is standard procedure now, when reporting the preparation of a new compound not only to give its melting or boiling point, but also to give some of its spectral characteristics as a means for future identification. Two different compounds may have the same melting point or two others may have the same boiling point, but to the author's knowledge, no two compounds have the same melting and boiling points, except optical isomers. We can see that two compounds would not have the same set of ε values at the hundreds of wavelengths at which measurements are made. Consequently, the set of ε values (the absorption spectrum) for a given compound is unique for that compound and serves as a much stronger characterizing property than any single-valued physical constant such as boiling or melting point, index of refraction, viscosity, etc. Thousands of reference spectra are on file for making comparisons in the identification of compounds or for chemical analysis. The literature abounds with material on spectroscopy and there are entire books devoted to each type of spectroscopy (181).

From classical physics we have learned that a negative charge moving in a closed orbit is equivalent to a current in a coil of wire and it will produce a magnetic field. When an atom, with its attendant electrons moving in closed orbits, is placed in a magnetic field, the external field superposes an angular velocity of precession on the electronic motions. This induces in the system of randomly oriented atoms a small magnetic moment in opposition to the external field. This magnetism, called *diamagnetism*, will be found in all atoms whether or not they also possess a permanent magnetic moment (182). Diamagnetism increases with the effective radius of the electronic orbits and is essentially temperature independent. For example F^-, Ne, Na^+, and Mg^{+2} are isoelectronic, but their diamagnetism decreases in this order because the increasing nuclear charge decreases the orbital radii.

Sec. 4.1 ELECTROMAGNETIC ENERGY

	F^-	Ne	Na^+	Mg^{+2}	
Mag. susc./g-ion	-11	-7.6	-5	-3	$\times 10^{-6}$ esu

On this basis, we can expect the magnetic susceptibility* of a covalently bonded atom to increase as the negative formal charge on the atom increases. Similarly, the diamagnetic susceptibility will be smaller as the oxidation state of an atom increases:

Cr^{+2} -15 Mn^{+2} -14 Pb^{+2} -28

Cr^{+3} -11 Mn^{+3} -10 Pb^{+4} -26

Cr^{+4} -8

P. Pascal measured the diamagnetic susceptibilities of many compounds and showed that their susceptibilities χ_m can be closely approximated by the expression $\chi_m = \Sigma n_i x_i + \lambda$ where n_i is the number of atoms of susceptibility x_i in the molecule and λ is a structural parameter (sometimes called a constitutive constant or __exaltation__) which depends upon the nature of the bonding between the atoms. Pascal empirically established a set of atomic susceptibilities x_i and structural parameter constants λ, some of which are given in Tables 4.2 and 4.4. C_Δ, etc., in Table 4.4 refer to alicyclic rings. C_3^α, etc., refer to carbon atoms in α, β, γ, δ, or ε positions with respect to oxygen atoms attached to three, etc., other atoms, excepting hydrogen. Pascal's constants may be used to provide supporting evidence for a proposed structure for a compound. For example, if we assume that cyclooctatetraene has an aromatic ring, we would calculate its magnetic susceptibility to be

$$10^6 \chi_m = 8x_C + 8x_H + 8\lambda_{arom}$$
$$= 8(-6.00) + 8(-2.93) + 8(-0.24)$$
$$= -73.36 \text{ cm}^3/\text{mol};$$

whereas, if we assume a nonaromatic structure, we have

$$10^6 \chi_m = 8x_C + 8x_H + 2\lambda_{C=C-C=C}$$
$$= 8(-6.00) + 8(-2.93) + 2(10.6)$$
$$= -50.24 \text{ cm}^3/\text{mol}.$$

This latter value is in good agreement with the observed value of

---

* A measure of the magnetization of a substance when it is placed in a magnetic field.

Table 4.3. Pascal's constants (x_i) for some elements (10^{-6} cgs units)

H	-2.93	F	-11.5	Li	-4.2
C	-6.00	Cl	-20.1	Na	-9.2
N (open chain)	-5.57	Br	-30.6	K	-18.5
N (ring)	-4.61	I	-44.6	Si	-13.0
N (monoamide)	-1.54	S	-15.0	B	-7.0
N (diamide, imide)	-2.11	Se	-23.0		
O (alcohol, ether)	-4.61	Te	-37.3		
O (aldehyde, ketone)	1.72	P	-10.0		
O (carboxyl, ester, etc.)	-3.36	As	-21.0		

Table 4.4 Pascal's structural parameter constants, λ (10^{-6} cgs units)

C=C	+5.5	C-Cl	+3.1
C≡C	+0.8	C-Br	+4.1
C=C-C=C	+10.6	C-I	+4.1
$H_2C=CH-CH_2-$	+4.5	$C_3^\alpha, C_3^\gamma, C_3^\delta, C_3^\varepsilon$	-1.3
N=N	+1.85	$C_4^\alpha, C_4^\gamma, C_4^\delta, C_4^\varepsilon$	-1.55
C=N	+8.15	C_3^β, C_4^β	-0.5
C≡N	+0.8	$C_\triangle$	+4.1
C_{arom}(in one ring)	-0.24	$C_\square$	+3.05
C_{arom}(in two rings)	-3.1	$C_{\pentagon}$	-0.98
C_{arom}(in three rings)	-4.0	$C_{\hexagon}$	+0.86

-51.9 × 10^{-6} cm^3/mol, which is consistent with its nonaromatic physical and chemical properties.

The constants in Tables 4.3 and 4.4 should not be used to distinguish structures with calculated magnetic susceptibilities closer than several units because they are not that accurate. There are several structural factors for which correction constants are not given, such as ionic character and ring strain. Note that aromatic carbon atoms and, therefore, aromatic rings have large negative structural constants. This provides evidence that the π electrons circulate around the ring and thereby generate diamagnetic moments.

In addition to diamagnetism, some substances exhibit a magnetic character of much larger magnitude than diamagnetism and of opposite sign. Since a rotating charge creates a magnetic field, a spinning electron will produce a magnetic dipole and cause the atom to have a permanent magnetic moment called paramagnetism. The spinning electrons of an atom may or may not pair up with opposed dipoles. Obviously any atom with an odd number of electrons will be paramagnetic. A very large paramagnetism is called ferromagnetism, and is confined to the solid state. Neither diamagnetism nor paramagnetism is observed in the absence of an external magnetic field because diamagnetism is an induced magnetism and, owing to random orientation of molecules, paramagnetic substances do not have an excess magnetism in any one direction.

4.2 Molecular Absorption of Energy

As mentioned in the previous section, absorption of energy from the far infrared region of the spectrum produces molecular rotations, absorption of near infrared radiation produces rotations plus atomic vibrations, and absorption of ultraviolet and visible light produces not only molecular rotations and atomic vibrations but also electron oscillations, i.e., electronic excitations. Although useful, subtle, or sometimes unique information can be obtained from electron spin resonance, photoelectron, ion cyclotron resonance (183), and other forms of spectroscopy, this discussion will be limited to the types most commonly used to determine molecular structure, namely infrared (ir) and Raman, ultraviolet-visible (uv), nuclear magnetic resonance (nmr), and mass spectrometry. This discussion is not intended to teach the student how to interpret spectra. It merely draws from treatments given in spectroscopy books those features pertinent to the general topic of this book.

4.2a Infrared spectra

The Danish physical chemist Niels Bjerrum first showed in 1915 that the ir spectra of molecules can best be understood if we think of atoms as small charged balls and of covalent bonds as metal springs holding the balls together. Each spring-like bond has its own natural frequency of vibration and when a ray of electromagnetic energy with this same frequency passes the molecule, the molecule absorbs some of the incident light energy. This is recorded by the spectrometer. The frequency ranges associated with each bond type were identified over

Table 4.5. Some typical ir frequency ranges

Group	ir frequency cm^{-1}	Group	ir frequency cm^{-1}
H-C$_{sp^3}$	2850-2960	O-H	3610-3650
H-C$_{sp^2}$	3010-3040	S-H	2500-2600
H-C$_{sp}$	3250-3300	N-H	3300-3500
C-C	1100-1150	π-C=C	1600-1630
C=C	1620-1680	π-C=O	1680-1700
C≡C	2100-2260	C=O···H-O	1605-1640 and 2500-3590
C-O	1060-1270	Phenyl ring	~1450 and ~1580
C=O	1700-1825		675-730
C-N	1030-1230	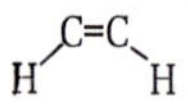 H₂C=CH₂ (cis)	
C=N	1640-1700	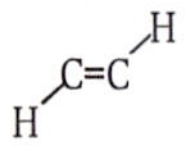 HC=CH (trans)	965-975
C≡N	2215-2260		
C-F	1000-1400	C=CH$_2$	890-910
C-Cl	600-800		
C-Br	500-600		

several years, following the pioneering work of W. W. Coblentz and J. Lecomte. Some of these data are listed in Table 4.5.

In a similar fashion, a molecule may be regarded as a mechanical system of atoms all interconnected by springs of different tensile strengths. When one bond is set into vibration, the entire system is affected and the specific resonant frequency of a given bond type will depend upon the entire molecule (184). Thus, the C-H bond does not absorb at an exactly identical frequency in all molecules. Fortunately the range of frequencies for each bond type is sufficiently narrow that usually an observed absorption band can be attributed to the presence of a certain bond type.

Furthermore, a given bond may undergo several different types of vibratory motions. For example, some modes for the C-O-H group are shown in Figure 4.3. As another example, the C-H bonds of a CH$_2$ group may stretch or bend, sometimes in unison and sometimes out of phase to

Sec. 4.2 MOLECULAR ABSORPTION OF ENERGY

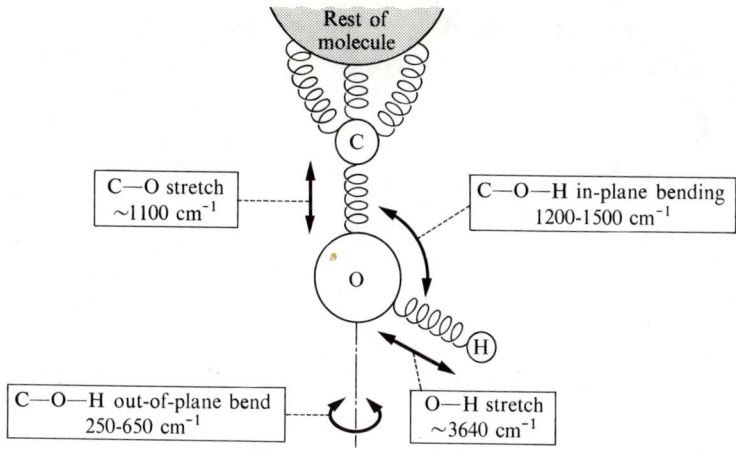

Figure 4.3

give rise to several different modes of molecular distortions. Some of these are given in Figure 4.4.

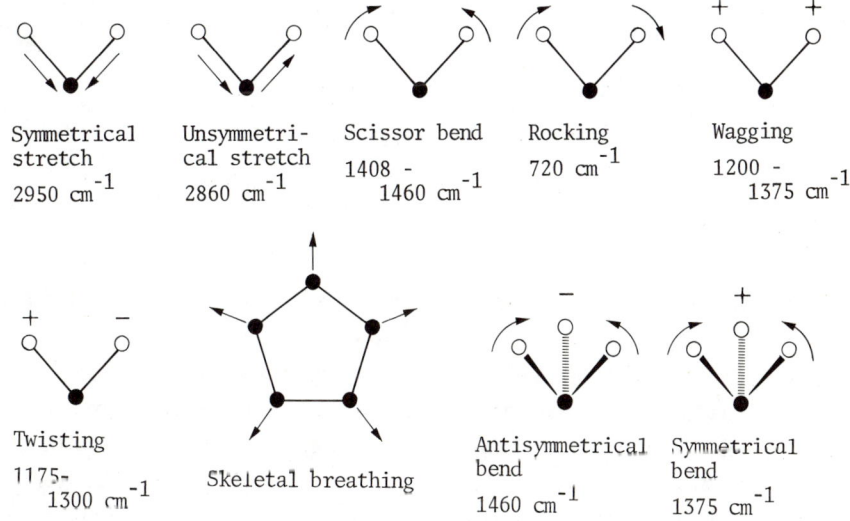

● = C atom; o = H atom; + and - refer to vibrations perpendicular to page.

Figure 4.4. Several modes of CH_2 group vibrations.

Each mode of vibration requires a different amount of energy and absorbs radiation of different frequencies accordingly. Hence, each bond type gives rise to several absorption bands and the absorption spectrum of a molecule is quite complex. Nevertheless, most of the bands in a spectrum can be assigned to certain chemical bonds. An example is shown in Figure 4.5 for the spectrum of 1,3,5-heptatriene.

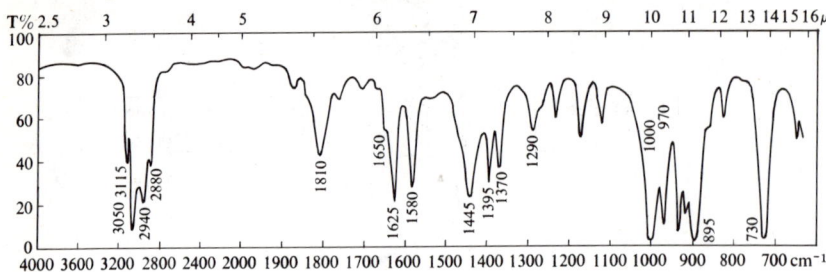

Figure 4.5. Frequency assignments in the ir spectrum of trans,cis,1,3,5-heptatriene ⌇⌇⌇ (185)

3115 and 3050 cm^{-1}, ν_{C-H} (=CH)

2940 and 2880 cm^{-1}, ν_{C-H} (CH$_3$)

1810 cm^{-1}, overtone of =CH$_2$ (895 cm^{-1})

1650, 1625, 1580 cm^{-1}, $\nu_{C=C}$

1445, 1370 cm^{-1}, γ_{CH_3} bend

1395, 1290 cm^{-1}, γ_{C-H} in-plane bend (=CH$_2$)

1002, 895 cm^{-1}, γ_{C-H} out-of-plane bend (=CH$_2$)

971 cm^{-1}, γ_{C-H} out-of-plane bend, trans $\overset{H}{}$C=C$\overset{}{H}$

732 cm^{-1}, γ_{C-H} in-plane bend, cis $_H$⟋C=C⟍$_H$

In quantum mechanical concepts, the infrared spectra of substances arise from transitions to higher vibrational energy levels of the molecules involved. For a molecule consisting of n atoms, there will be 3n - 6 fundamental or normal modes of vibration, although they may not all be different. The expression 3n - 6 is based on the number of coordinates required to establish the position of n atoms relative to some fixed origin. For n atoms, 3n Cartesian coordinates are needed for each atom. Since each of the 3n coordinates may change with time, the molecule has 3n degrees of freedom. However, some of the 3n degrees of freedom correspond to certain motions of the molecule as a whole, which are the three coordinates which fix the center of mass of the molecule and the three which specify its orientation with respect to the origin. Hence, the number of possible modes of vibration for an n-atom molecule will be 3n - 6. A linear molecule needs only two coordinates to establish its orientation and it will therefore have an additional mode of vibration. For example, $\underline{n} = 3$ for the linear CO_2 molecule; hence there are 9 - 5 = 4 modes of vibration.

$$\begin{array}{ccc} \leftarrow \quad \rightarrow & \rightarrow \leftarrow \rightarrow & \downarrow \uparrow \downarrow \quad\quad + \; - \; + \\ O-C-O & O-C-O & O-C-O \;\; \text{or} \;\; O-C-O \\ \text{Symmetrical} & \text{Unsymmetrical} & \text{Bending, } \nu_2 \\ \text{stretch, } \nu_1 & \text{stretch, } \nu_3 & \end{array}$$

Because of the symmetry of the molecule, the two bending modes are equivalent, i.e., doubly degenerate. As a second example, the molecule CH_3Cl should have 3 × 5 - 6 = 9 vibrational frequencies. If we regard the molecule as a set of diatomic vibrators, then we expect one C-Cl stretching vibration and three C-H stretching vibrations of which two are degenerate. This leaves three discrete frequencies which must be of the bending type. The observed frequencies are given in Table 4.6.

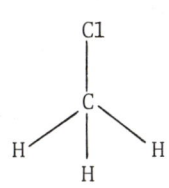

Table 4.6. Infrared frequencies for CH_3Cl (186)

Non-degenerate	C-H	stretch	2966 cm^{-1}
	C-H	bend	1355
	C-Cl	stretch	732
Doubly degenerate	C-H	stretch	3042
	CH_3	bend	1455
	CH_3	rock	1015

The energy of absorbed ir light, $\underline{\Delta E}_{ir}$, is related to the frequency of the absorption band by equation (4.1):

$$\underline{\Delta E}_{ir} = \underline{h}\nu_0(\ell + 1/2) \quad (\ell = 0,1,2,\ldots) \quad (4.1)$$

in which ν_0 is the fundamental frequency and ℓ is the vibrational quantum number. For a transition of $\Delta\ell = 1$,

$$\Delta E_{ir} = h\nu_0((\ell + 1) + 1/2) - h\nu_0(\ell + 1/2)$$
$$= h\nu_0.$$

However, the vibrations of chemical bonds are not pure harmonic vibrations, and transitions take place in which $\Delta\ell$ is more than ± 1. The corresponding absorption bands are called <u>harmonic overtones</u>, or frequently, just overtones. The frequencies of the first and higher overtones are

$$\nu_1 = (1 - 2\underline{x})\nu_0$$
$$\nu_2 = (1 - 3\underline{x})2\nu_0$$
$$\nu_3 = (1 - 4\underline{x})3\nu_0$$

where $\underline{x}$ is the anharmonicity constant. It is usually very small, e.g., $\underline{x}$ for the C-H bond is 0.01 to 0.05; hence, the first and second overtones occur at approximately twice and three times the frequency of the fundamental band. For instance, ν_0 and ν_1 for the C=O bond of cyclohexanone are 1720 and 3420 cm^{-1}. The fact that the C=O first overtone falls in the region of the H-bonded hydroxyl group, O-H···O, has sometimes led to a misinterpretation of spectra. For example, 4-hydroxycyclohexanone exhibits bands at 3626 (I) and 3427 cm^{-1} (II). Band II was simply attributed to an H bond. However, when the OH was replaced by OD, the deuterio compound gave bands at 2678 (I') and 3427 cm^{-1} (II') (187). Obviously, the band at 3427 cm^{-1}, which appears in both spectra, is not that for an H bond, and can be assigned to the C=O first overtone. This assignment is confirmed by the spectrum of cyclohexanone which has a first overtone at 3420 cm^{-1}. Compounds of type <u>4.1</u> however have two bands near 3600 and 3400 cm^{-1}. In one instance, the 3400-cm^{-1} band was simply assigned to a C=O overtone, although it is fairly strong. Later, it was found that if OH is replaced by OD, the two bands move to 2662 and 2514 cm^{-1} (188). Therefore, the 3400-cm^{-1} band in this case can be attributed to an H bond. Since its molar extinction coefficient is concentration dependent, it is

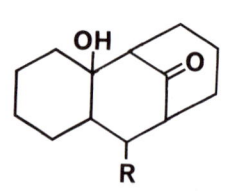

<u>4.1</u>   R = H, Me, Et, <u>i</u>-Pr

due to intermolecular rather than intramolecular <u>H</u> bonding. Often a check is made by examining the ratio ν_{OH}/ν_{OD}, which is normally about 1.35 (i.e., close to the classical value of $\sqrt{2}$ for the H-D shift) (189).

The intensities of overtone bands are only 1/100 to 1/10 that of the fundamental band, so overtones are often hidden by much stronger fundamental bands. Since most fundamentals fall below 4000 cm^{-1}, it is convenient to work with overtones above this frequency, called the near infrared or overtone region. The approximate frequencies in this region for some bonds are given in Table 4.7.

Table 4.7. Approximate frequencies of overtones of some bond types

Bond	Overtone*	Frequency (cm^{-1})
C-H	First	5900
	Second	9000
O-H	First	7100
	Second	10,500
N-H	First	7100
S-H	First	5100
Si-H	Second	6700
C-C	Third	5700
C-O	Third	5400
C-Cl	Sixth	5300

* Fundamentals are given in Table 4.5.

Hydrogen bonds, for example, have been studied extensively in this region (190). The C-O bands for alcohols are in the 1050 to 1150 cm^{-1} region and they are often difficult to distinguish from fundamental C-C bands. If one works in the OH first overtone region, primary, secondary, and tertiary alcohols give bands at

$1°$ 7090-7115 cm^{-1}

$2°$ 7067-7078 cm^{-1}

$3°$ 7042-7053 cm^{-1}

from which one may characterize the class of the alcohol (191).
Cyclopropyl groups have three prominent bands:

(1)   3090 cm^{-1} C-H stretch

(2)   6250 cm^{-1} First overtone

(3)   1020 cm^{-1} Ring deformation

Although band (3) is often used for the detection of a cyclopropyl group, many studies have been made in the overtone region (192). For instance, the effects of substituents on this band are found to be well correlated with the polar substituent constant, σ^*.

In addition to overtones and different modes of vibration, two other effects which complicate ir spectra are <u>combination tones</u> and <u>Fermi resonance</u>. The former are simply cases where two bands, ν_x and ν_y, combine to give new bands with frequencies $\nu_x + \nu_y$ or $\nu_x - \nu_y$. Fermi resonance is a splitting of a fundamental band or overtone by coupling with another (193). It is known from classical physics that if two balls are hanging by strings from a single bar which itself is free to swing, and if one ball is set to swinging, the swinging ball transmits some of its motion to the second ball and the amplitude of oscillation of the first ball gradually decreases. Similarly, when two vibrating bonds in a molecule are directly connected or attached to the same third atom, there is some mechanical interaction or coupling between them. If they have vibrational frequencies fairly close together, an overtone, normally forbidden and very weak, becomes more intense at the expense of the fundamental and two bands are observed in the same region.

For instance, the frequencies for the first overtone of the H-C_{sp^2} bend ($\nu \approx 860$ cm^{-1}, hence first overtone at ~1720 cm^{-1}) and C=O stretch ($\nu = 1720$ cm^{-1}) are close together and often exhibit Fermi resonance (194). Thus, 4-pyrone has a doublet C=O band as a result of coupling with the adjacent C-H bonds, but if the H's are replaced by D, the C=O band is a singlet (1648 cm^{-1}). When two coupled bonds are equivalent and therefore have the same frequency, the ir band is a doublet, often observed for

Doublet $\nu_{C=O}$ Singlet $\nu_{C=O}$

1872, 1796 cm^{-1} 1770, 1789 cm^{-1} 1729, 1686 cm^{-1}

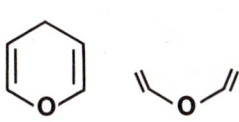

Doublet C=C bands

anhydrides and imides. Similarly, the C=C band of vinyl ethers often appears as a doublet (195). For weak coupling, the bands merely separate further. A good example is found in cyclopropene derivatives (196). The C=C stretch is coupled with the =C-X bending vibration. If the C-X bend is lower than the C=C stretch, the C=C frequency is increased, and if higher, the C=C frequency is lowered. Thus, $\nu_{C=C}$ is raised for X = CH_3 and lowered for X = D.

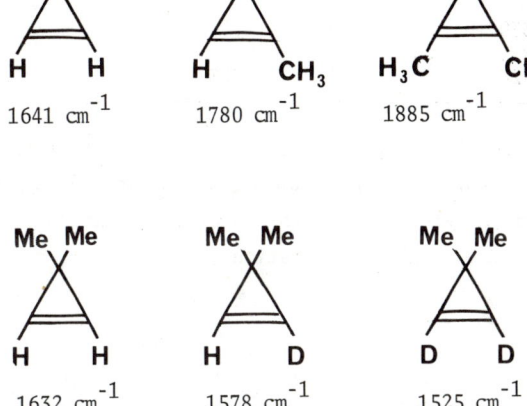

Because of coupled oscillations, the notion of group frequencies is not always valid in certain molecular systems. This appears to be the case in strong chelates (197), many amides, and some cyclanones (198). Whether or not it accounts for the alternation in certain bond frequencies with position in an alkane chain, it is fascinating to see the alternation shown by some groups on the octadecane chain (Table 4.8.)

The motions of a simple pendulum, of a mass suspended by a spring, and of a diatomic molecule are similar in that each involves a mass whose motion is resisted by a force proportional to the distance from an equilibrium position. That is, each approximates a harmonic oscillator for which Hooke's law applies, $F = -kd$, where $\underline{F}$ is the force, $\underline{d}$ the displace-

Table 4.8. Ir frequencies for some groups on the octadecane chain (199)

Position of OH	$\nu_{OH}(cm^{-1})$	Position of C=O	$\nu_{C=O}(cm^{-1})$	Position of C=NOH	$\nu_{C=N}(cm^{-1})$
1	3290				
2	3360	2	1712	2	1655
3	3325	3	1710	3	1656
4	3400	4	1717	4	1658
5	3320	5	1702	5	1655
6	3330	6	1718	6	1657
7	3320	7	1705	7	1654
8	3330	8	1725	8	1657
9	3325	9	1723	9	1656

ment from the equilibrium position, and $\underline{k}$ is a constant called the <u>restoring force constant</u>. For a system of two masses $\underline{m_A}$ and $\underline{m_B}$ moving in a potential field $\underline{V} = 1/2 \underline{k} r_{AB}^2$, where $\underline{r_{AB}}$ is the interatomic distance and $\underline{k}$ is the bond force constant, the resulting vibration will have a frequency

$$\nu_0 = \frac{1}{2\pi \underline{c}} \sqrt{\underline{k/M}}$$

where $\underline{c}$ is the velocity of light and $\underline{M}$ is the effective mass for the combined vibration. Usually $\underline{M}$ can be replaced by the reduced mass μ

$$\mu = \frac{\underline{m_A} \underline{m_B}}{\underline{m_A} + \underline{m_B}}$$

which leads to equation (4.2):

$$\nu = 1307\sqrt{\underline{k/\mu}} \; cm^{-1}. \tag{4.2}$$

One can estimate the absorption frequency of a given bond with equation (4.2). For example, for the C-H bond in methane, $\underline{k}$ and μ have values close to 5×10^5 dynes/cm and 1, respectively. Hence, $\nu_{C-H} = 1307\sqrt{5/1} = 2920$ cm^{-1}. For the C=O bond in acetone, $k = 12 \times 10^5$ dynes/cm and $\mu = 6.8$, so that $\nu_{C=O} = 1307\sqrt{12/6.8} = 1730$ cm^{-1}. In the ir spectra of these two compounds there are strong bands at 2915 and 1720 cm^{-1}, respectively.

Inasmuch as the force constant $\underline{k}$ is related to the strength of a bond, which in turn is affected by the electronegativities of the bonded atoms and the interatomic distance, the bond ir frequency is a function of all these bond parameters,

$$\nu = \underline{f}(\frac{k\chi D}{r\mu}) \tag{4.3}$$

where $\underline{k}$ is the force constant, $\underline{D}$ the dissociation energy, χ is related to the electronegativities of the bonded atoms, $\underline{r}$ is the bond length, and μ the reduced mass. Equation (4.3) indicates how ν is quantitatively related to $\underline{k}$ and μ. Thus, any effect which reduces $\underline{k}$ lowers ν. We saw in Section 1.2b that H bonding of O-H to C=O, O-H···O=C, helps each bond to stretch, i.e., lowers $\underline{k}$, and is accompanied by lower ν_{OH} and $\nu_{C=O}$ frequencies. Similarly, acid-base or charge-transfer complexation with a C=O group, $R_2C=O$, lowers the stretching frequencies and raises bending vibrations. Some data for I_2 complexes with cyclohexanone and acetone are given in Table 4.9. In another study, it was found that the enthalpy of complex formation, $-\Delta\underline{H}$, is linearly related to the change of the C=O frequency of ethyl acetate (200):

Complexing agent	$-\Delta\underline{H}$ (kcal/mol)	$\Delta\nu_{C=O}^{CCl_4}$ (cm^{-1})
I_2	2.5	26
C_6H_5OH	4.8	36
$\underline{p}$-Cl-C_6H_4OH	5.0	33
ICl	6.1	55
BF_3	13.0	107
$SbCl_5$	17.1	138

Table 4.9. IR frequencies for two ketones and I_2-ketone complexes (201)

Deformation	Cyclohexanone (cm^{-1})	Cyclohexanone-I_2 complex (cm^{-1})
C=O stretch	1715	1698
C=O stretch overtone	3412	3389
C-C=O bend	487	493
	Acetone	Acetone-I_2 complex
C=O stretch	1716	1700
C-C=O bend	529	534

Equations (4.2) and (4.3) indicate that ν varies inversely with the reduced mass and this is commonly observed. For example, the ir frequencies for some bonds involving isotopes are as follows (202):

Compound	$\nu_{C=O}$	$\nu_{C=O18}$ (cm^{-1})
Benzophenone	1664	1635
Benzamide	1690	1666
Methyl benzoate	1727	1696
	ν_{C-H}	ν_{C-D}
Methane	3020	2258
	ν_{O-H}	ν_{O-D}
Methanol	3620	2680

	$\underline{i}\text{-Pr}_2\text{C=O}$	$\underline{i}\text{-Pr}_2\text{C}^{13}\text{=O}$	$\underline{i}\text{-Pr}_2\text{C=O}^{18}$
$\nu_{C=O}$	1712	1675	1681
$\nu_{C-C-C \atop \| \atop O}$	1024	1004	1024

The larger the reduced mass, the lower is the frequency. In isopropyl ketone, $\nu_{C=O}$ is lowered by increasing the mass of carbon or of oxygen but the C-C vibration is unaffected by the change in oxygen (203). However, there can be indirect mass effects too. For instance, the $\nu_{C=O}$ bands in $H_2C=O$ and $D_2C=O$ are at 1743 and 1700 cm^{-1}, respectively.

We will see how other structural parameters such as resonance, electronegativity, steric hindrance, and ring strain affect $\underline{k}$, and therefore ν, in Part III.

A few comments should be made here about the effects of solvents. Many quantitative studies have been made, usually in terms of the dielectric constant of the solvent. These have been more successful for nonpolar solvents (204). Qualitatively, the stretching frequency of a **polar bond** decreases in a polar solvent. For example, $\nu_{C=O}$ is distinctly lower in DMSO than in CCl_4:

	DMSO $\nu_{C=O}$	CCl_4 $\nu_{C=O}$
$C_6H_5COCH_3$	1657 cm^{-1}	1664 cm^{-1}
$\underline{p}$-$HOC_6H_4COCH_3$	1645	1661

Given a bond A-B, in which the solvent is sterically able to associate with atom B, if the bond is

polarized $^{\delta+}A$-$B^{\delta-}$ ν_{A-B} will decrease in acidic solvents and undergo little change in basic or neutral solvents.

polarized $^{\delta-}A$-$B^{\delta+}$ ν_{A-B} will decrease in basic solvents and change little in acidic or neutral solvents.

nonpolar ν_{A-B} will suffer little change with changes in solvents.

Some data are listed in Table 4.10 for illustration. Thus, $\nu_{C=O}$ is not markedly different in the neutral and basic solvents but is **distinctly** lower in the acidic solvent. ν_{X-H} is very sensitive to the electron-donor character of the solvent for H bonding but only in the basic solvent is there a large decrease.

The solvent effect is sometimes used as a **method** of identifying the carbonyl band (205). One plots $\Delta\nu/\nu$ for the sample $\underline{vs}$. $\Delta\nu/\nu$ for acetone in a series of solvents, where $\Delta\nu$ is frequency change for each solvent compared to a reference solvent, say hexane. If the band is that of a

Table 4.10. Ir group frequencies (cm^{-1}) in different solvents

Compound	Bond	Vapor	Hexane (nonpolar)	CCl$_4$	CHCl$_3$ (acidic)	Et$_2$O (basic)
C$_6$H$_5$CO$_2$Me	C=O		1735	1730	1720	1730
C$_6$H$_5$COCH$_3$	C=O	1709	1697	1692	1683	1694
C$_6$H$_5$OH	O-H	3654	3622	3611	3595	3344
Pyrrole	N-H	3530	3506	3500	3486	3352
C$_6$H$_5$C≡CH	≡C-H	3337	3323	3313	3309	3250

carbonyl group, there results a linear correlation with a slope of one. For example, 2,6-diaryl-4-pyrones (4.2) exhibit a doublet in the 1600 to 1700 cm^{-1} range and as a means of assigning these to the carbonyl group, Δν/ν values for 4.2 were plotted against the corresponding values for the 1715-cm^{-1} bond of acetone in cyclohexane, isooctane, carbon tetrachloride, tetrachloroethylene, acetonitrile, methylene chloride, and chloroform. Both bands gave a straight line with slope close to unity, indicating that the two bands are for the carbonyl group, i.e., a doublet (206). Tropolones also give two prominent bands in the 1570 to 1650-cm^{-1} range and most authors have assigned the higher band to the carbonyl group. However, if the frequency shifts produced by a series of solvents on the tropolone and acetone bands are compared, only the lower-frequency band is apparantly the carbonyl band (207). Only the 1590-cm^{-1} band exhibits the steady shift shown by acetone:

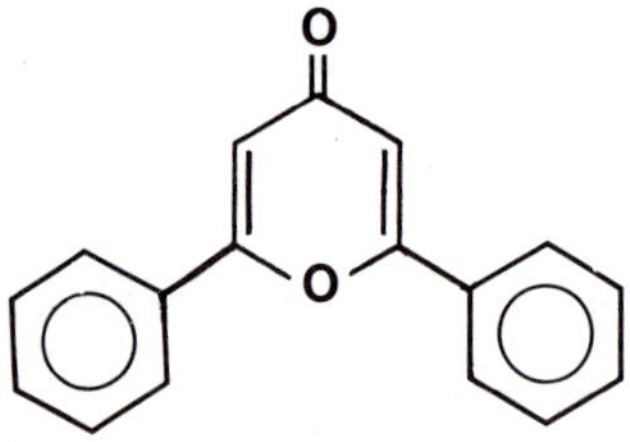

4.2

Solvent	$\nu_{C=O}$ Acetone			$\nu_{C=O}$ Tropolone
Hexane	1726 cm^{-1}	1647	1637	1597 cm^{-1}
Tetrachloroethylene	1722	1646	1633	1597
Benzene	1718	1645	1635	1591
Acetonitrile	1716	1650	1634	1583
Chloroform	1713	1650	1635	1577
1,1,2,2-Tetrabromoethane	1709		1633	1572

The intensity of an ir absorption band is approximately proportional to the square of the change in dipole moment per unit distance at the instant of absorption; accordingly, ir bands are normally observed only for unsymmetrical vibrations. Fortunately, the symmetrical vibrations produce strong Raman lines and their frequencies may be determined in this way.

4.2b. Raman spectra (206a)

Now that instruments are available commercially, particularly laser Raman spectrometers, Raman spectroscopy is taking its place among the corps of spectral techniques for routine analysis and research. The technique, developed by Sir C. V. Raman in 1928, is closely related to infrared absorption spectroscopy. When a beam of light passes through a transparent medium and the scattered light is viewed at right angles to the incident beam, in the spectrum of the scattered light additional lines (which are displaced at regular intervals from the incident lines) will be superimposed upon the original spectrum. Raman found that a similar pattern of lines is always found for a given substance irrespective of the wavelength of the incident beam. He also showed that the frequency shifts are of the same magnitude as the energy differences between vibrational states corresponding to bond vibrations within the molecules. The frequency shifts are called Raman frequencies and the set of Raman frequencies comprises the Raman spectrum. Raman excitation is active for a change in polarizability whereas ir is active for a change in dipole moment. Since polarization is a "volume" property, Raman frequencies are most active for vibrations which bring about a change in volume, usually symmetrical vibrations. Thus, Raman and ir spectroscopy complement each other by producing the strongest bands in the ir spectrum for asymmetrical vibrations and the strongest Raman frequencies for symmetrical vibrations. For example, compound 4.3 has a center of symmetry whereas 4.4 does not. By comparing the ir and Raman spectra of the two samples, the compounds can be identified because 4.3 will have bands in its ir spectrum not found in its Raman spectrum and 4.4 will have bands in its Raman spectrum not found in its ir spectrum (208). This method has also been used to assign conformations to 1,4-dimethylenecyclohexane, dispiro(2.2.2.2)decane, and to cyclohexane-1, 4-dione (209). The first two compounds have a centrosymmetric chair structure and the third has a twist-boat conformation.

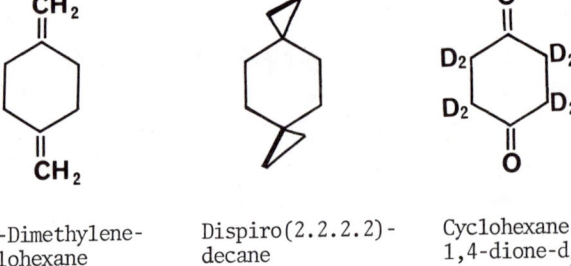

1,4-Dimethylene-cyclohexane   Dispiro(2.2.2.2)-decane   Cyclohexane-1,4-dione-d_8

Since Raman frequencies, $\Delta\nu$, can be related to bond vibrations, we have

$$\Delta\nu = \frac{1}{2\pi c}\sqrt{k/\mu}$$

or $k = 0.05877\ \Delta\nu^2 \mu$ (10^5 dynes/cm), with $\Delta\nu$ in cm^{-1}. Hence, Raman spectra provide a good method of determining bond force constants. Some values are listed in Table 4.11. Note that the values for single, double, and triple bonds are in different ranges:

Single bonds	$<7 \times 10^5$ dynes/cm
Double bonds	7 - 15
Triple bonds	>15

Table 4.11. Some bond force constants (210)

Bond	Substance	$10^{-5}\ \underline{k}$ (dynes/cm)	Bond	Substance	$10^{-5}\ \underline{k}$ (dynes/cm)
H-H	H_2	5.76	C-C	C_2H_6	4.5
F-H	HF	9.7	C-N	CH_3NH_2	4.9
Cl-H	HCl	5.2	C-O	$(CH_3)_2O$	4.5
Br-H	HBr	4.1	C-F	CH_3F	5.6
I-H	HI	3.1	C-Cl	CH_3Cl	3.4
B-H	B_2H_6	3.6	C-Br	CH_3Br	2.9
C-H	CH_4	5.0	C-I	CH_3I	2.3
N-H	NH_3	6.5	$C_{ar}-C_{ar}$	C_6H_6	7.6
O-H	H_2O	7.6	C=C	C_2H_4	9.8
S-H	H_2S	4.0	C=O	H_2CO	12.3
P-H	PH_3	3.1	C≡C	C_2H_2	15.6
Si-H	SiH_4	2.7	C≡N	CH_3CN	17.5
N-N	N_2H_4	3.6	C≡O	CO	18.9
Cl-Cl	Cl_2	3.3	N≡N	N_2	22.8
Br-Br	Br_2	2.5			
I-I	I_2	1.7			

This generalization often provides support for the bond character of a given bond. For example, the bond force constants of the carbon-carbon bonds in benzene are all identical, 7.6×10^5 dynes/cm, which is midway between the values for the C-C of ethane, 5×10^5, and the C=C bond of ethylene, 9×10^5 dynes/cm.

The force constant itself is used to calculate the force necessary to distort a given bond. For small displacements where Hooke's law would apply, we have $F = -\underline{kd}$ and $\underline{E} = 1/2\underline{kd}^2$, where $\underline{E}$ is the energy required and $\underline{d}$ is the displacement. For example, to stretch the C-H bond of methane by 0.1 Å would take a force of

$$0.1(\times\ 10^{-8}) \times 5(\times\ 10^5) = 5 \times 10^{-4}\ \text{dynes}.$$

The energy required to do this would be

$$0.5 k d^2 = 0.5(5 \times 10^5) \times (0.1 \times 10^{-8})^2 = 3.62 \text{ kcal/mol}.$$

In a similar fashion, the force required to bend an H-C-H bond one degree (1/57.3 radians) from its normal bond angle can be estimated with use of the bending force constant (0.5×10^{-11} dynes-cm/radian).

$$\begin{aligned} F &= k\omega = 0.5 \times 10^{-11} \times 1/57.3 \\ &= 8.33 \times 10^{-14} \text{ dyne-cm} \\ E &= 0.5 \; k d^2 = 0.5 \times 0.5 \times 10^{-11} (\tfrac{1}{57.3})^2 \\ &= 7.6 \times 10^{-15} \text{ erg/molecule} \\ &= 0.11 \text{ kcal/mol} \end{aligned}$$

We can see that it is much easier to bend a bond through a small angle than it is to stretch it a short distance. The torsional force constant for ethylene about the C=C bond is 3.6×10^{-12} erg/radian^2. Therefore, to twist the bond 5°, 10°, and 15° requires 0.2, 0.8, and 1.8 kcal/mol, respectively.

One useful relationship involving force constants and bond distances is Badger's rule (211):

$$k = 1.86(r_e - d_{ij})^{-3}$$

where r_e is the equilibrium bond length and d_{ij} is a constant characteristic of a diatomic molecule made up of one element in the ith row and one in the jth row of the periodic table. Some values of the constant d_{ij} were provided by Badger, and later data have extended the applicability of the relationship (212). This rule and equation (4.2) involve three physical properties for a given bond, ν, k, and r_e. If any two are measured, the third may be estimated. Sometimes, for example, certain vibrational frequencies are needed for semitheoretical calculations and the frequencies have not been or cannot be measured. If the bond distance is known and the d_{ij} constant may be obtained, then the force constant may be calculated and the vibrational frequency estimated therefrom.

4.2c. Ultraviolet spectra (213)

Absorption of energy in the ultraviolet-visible region of the spectrum produces an excitation of valence electrons--bonding and lone pair--and a transition of the molecule from the electronic ground state to an electronically excited state. The change in energy $E_1 - E_0$ is related to the wavelength of the absorbed light by equation (4.4):

$$E_1 - E_0 = hc\nu = \frac{hc}{\lambda}. \qquad (4.4)$$

The molecular orbital energy diagrams and electronic transitions for some common bond types are as follows.

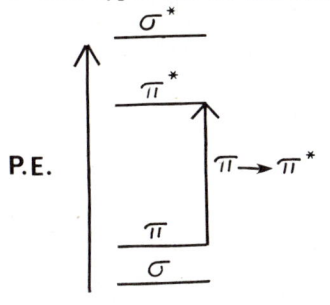

$$C=C \xrightarrow{h\nu} \overset{\mp}{C}-\overset{\pm}{C}$$

$\lambda_m \approx 180$ nm

The excited state resembles the polar structure more than the ground state does, i.e., the former is the more polar. Hence, λ_m shifts to longer wavelengths in more polar solvents, with attached groups which delocalize charge, or with easily polarizable attached groups.

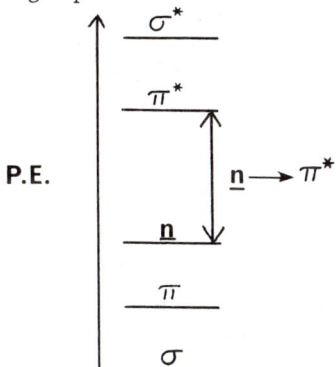

$$C=\ddot{X} \xrightarrow{h\nu} C\dot{=}\dot{X}$$

This is a transition of a lone-pair electron of the hetero atom to an empty antibonding π^* molecular orbital. The excited state is less polar than the ground state (214). Such bands are (1) of low intensity, usually $\varepsilon \leq 1500$, (2) shifted to shorter wavelengths by more polar solvents, and (3) shifted to shorter wavelengths by attached electron-donating groups. Of course, these groups may give rise to $\pi \to \pi^*$ and $n \to \sigma^*$ bands too (215). The positions of the $n \to \pi^*$ and $\pi \to \pi^*$ bands for several groups are as follows:

	$\pi \to \pi^*$		$n \to \pi^*$	
Group	λ_m	ε	λ_m	ε
C=C	185 nm	8000		
C=N	180	8000	245 nm	200
C=O	188	900	279	15
NO$_2$	201	5000	274	17

118   Ch. 4 INTERNAL ENERGIES

The two bands may be distinguished by their extinction coefficients, or by noting the effects of substituents or solvents. For example when groups which can provide electrons (σ, π, $\underline{n}$) are attached to a C=O group there is a stabilization of the lone-pair electrons on the oxygen in the ground state which increases the transition energy (216).

λ_m of the $\underline{n} \rightarrow \pi^*$ band of $CH_3 - C \overset{\displaystyle O}{\underset{\displaystyle Z}{\diagdown}}$

Z	Solvent	λ_m (nm)	ε
H	Hexane	290	17
CH_3	Hexane	279	15
Cl	Hexane	220	100
NH_2	H_2O	220	63
OEt	EtOH	211	57
OH	EtOH	208	32

Electron-withdrawing groups, which are not π- or $\underline{n}$-electron donors, increase λ_m (217):

	CH_3-CO-CH_3	CF_3-CO-CF_3
λ_m	280 nm	300 nm
ε	14 nm	8 nm

It is apparent from these low intensities that these are still $\underline{n} \rightarrow \pi^*$ bands. The effects of such groups on the $\pi \rightarrow \pi^*$ band is just the opposite: λ_m is increased. The use of solvent shifts to distinguish $\pi \rightarrow \pi^*$ and $\underline{n} \rightarrow \pi^*$ bands is discussed later in this section.

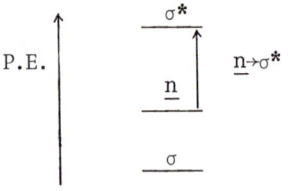

$C-\ddot{\underset{..}{X}} \xrightarrow{h\nu} C-\dot{\underset{..}{X}}\cdot$

X = O, N, S, halogen

This is an excitation of a lone-pair electron from the hetero atom to an antibonding orbital. The positions of this band for several compounds are:

Sec. 4.2 MOLECULAR ABSORPTION OF ENERGY

	λ_m
MeCl	173 nm
Me_2O	185
MeBr	204
Me_3N	227
MeI	258

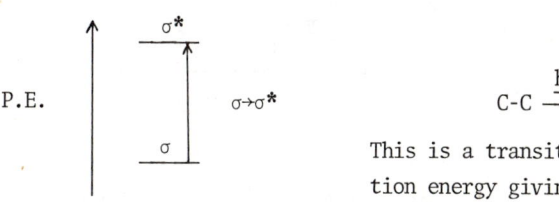

P.E. $\sigma \to \sigma^*$ $C-C \xrightarrow{h\nu} C \cdot C \cdot$

This is a transition of large excitation energy giving bands of $\lambda_m < 150$ nm, e.g., λ_m for $CH_3CH_2CH_3 \approx 135$ nm.

It was early recognized that colored molecules, i.e., molecules which absorb light in the visible spectrum, contain multiple bonds in their structures. The unsaturated groups were given the name <u>chromophore</u> (Greek <u>chroma</u>, color, + <u>phoros</u>, bearer). These groups have different chromophoric powers, in the following order:

$$C=C < C=N < C=O < N=N < C=S < N=O$$

λ_m 185   240   280   347   400   665 nm

Groups which do not confer color to an otherwise colorless substance but do increase the color-producing power of a chromophore are called <u>auxochromes</u> (Greek <u>auxo</u>, increase). Examples are NH_2, NR_2, OH, OR, CH_3. For example, polyaminobenzenes are colorless but one amino group in the nitrobenzene ring changes the nitrobenzene from colorless to yellow.

Benzene derivatives exhibit three prominent bands, all of the $\pi \to \pi^*$ type:

First primary: R—⬡ $\xrightarrow{h\nu}$ $\overset{\pm}{R}$=⬡*

λ_m increases with increasing polarity of solvent and with greater electron interaction between R and the ring (either through polarization or π-electron delocalization).

---

* The ± symbol is used to denote that either sign may be used at a given atom and the opposite sign will occur at the other charged atom. This practice saves space in writing structures.

Secondary: R—⟨○⟩ —hν→ R—⟨⟩

Transitions due to local excitations

The first primary and secondary bands of benzene are both forbidden and the intensity of each would be zero if the molecule did not undergo deformations to break the molecular symmetry. There is also a high intensity second primary band of shorter wavelength than the first primary.* The positions of these bands for some disubstituted benzenes are as follows:

Compound	Second primary λ_m (nm)	ε	First primary λ_m (nm)	ε	Secondary λ_m (nm)	ε
Benzene	180		203.5	7400	254	204
p-H_2N-C_6H_4COOH	216.5	18500	248	3900	327	1940
p-HO-$C_6H_4NO_2$	230	3900	278.5	6600	351	3200
p-H_2N-$C_6H_4NO_2$	245	7000	282.5	5400	412	4500

Criteria which help to identify these bands are (i) the ratio of secondary to primary λ_m's is close to 1.22 (218), and (ii) the ratio of λ_m's of first primary (λ) to second primary (λ') is usually less than 2 and increases as λ increases, and the ratio of the corresponding extinction coefficients, ε/ε', decreases as λ increases (219).

An extensive study of the spectra of mono- and disubstituted benzenes has shown that when the substituents are divivded into ortho-para directing and meta-orienting groups, and arranged in order of increasing displacement ($\Delta\lambda$) of the first primary band of benzene (203 nm), the following sequences

---

*Terminology for uv bands is sometimes confusing:
$\pi\to\pi^*$ has been called charge-transfer (CT) or electron transfer (ET) when a heteroatom is involved, i.e., C=O, C_6H_5-C=N-.
$n\to\pi^*$ is called R, LE (local excitation).
First Primary: K, B, C, or ^1L_a, $1B_{1u}$
Second Primary: ^1B, LE, or $1E_{1u}$
Secondary: ^1L_b, K, ET, B, $1B_{2u}$

($\Delta\lambda = \lambda_m - 203$) occur (218):

o,p-directing	$\Delta\lambda$	m-orienting	$\Delta\lambda$
CH_3	3 nm	$\overset{+}{N}H_3$	-0.5 nm
Cl	6	SO_2NH_2	14
Br	6.5	CO_2^-	20.5
OH	7	CN	20.5
OCH_3	13.5	CO_2H	26.5
NH_2	26.5	$COCH_3$	42.0
O^-	31.5	CHO	46.0
		NO_2	65.0

Each series parallels the amount of electron delocalization occurring in the excited state, irrespective of the direction of electron shift. That is, λ for the primary absorption band, corresponding to the transition

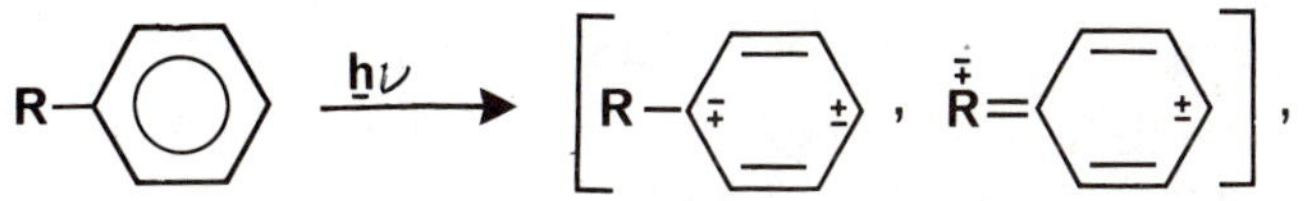

$$R-\bigcirc \xrightarrow{h\nu} \left[R-\overset{\mp}{\bigcirc}{\pm} , \overset{\mp}{R}=\bigcirc{\pm} \right],$$

is at longer wavelengths as the system is more polarizable. For example, the more readily Z donates electrons to the ring in m-substituted nitro-benzenes

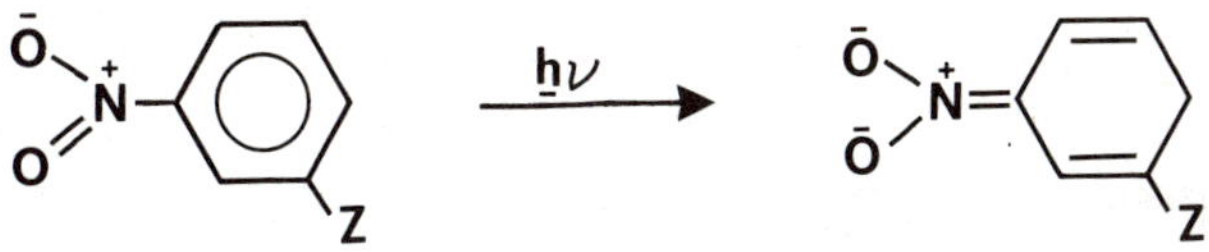

$$\overset{\bar{O}}{\underset{O}{\searrow}}\overset{+}{N}-\bigcirc_Z \xrightarrow{h\nu} \overset{\bar{O}}{\underset{\bar{O}}{\searrow}}\overset{+}{N}=\bigcirc_Z$$

the smaller is the transition energy and the larger the λ_m.

Z	λ_m (nm)
NO_2	234
$^+NH_3$	253
COOH	257
Cl	257
Br	259
H	260
CH_3	265
OH	269
O^-	288

The same trends are observed in the spectra of m-substitutued benzaldehydes, acetophenones, etc.

Whereas the wavelength of absorbed uv-visible light is determined by the energy difference between ground and excited states, the intensity of the absorption band is a function of the probability of the transition and of the length of the polarized chromophore. This is sometimes expressed in terms of oscillator strength (f) given by equation (4.5)

$$\underline{f} = 4.32 \times 10^{-9} \int \varepsilon d\nu \simeq 0.5 \Delta \nu \varepsilon \qquad (4.5)$$

where ε is the molar extinction coefficient and $\Delta \nu$ is the range of wave numbers over which the absorption band extends. The spectra of mono- and dinitrobenzenes illustrate these generalities concerning wavelength and intensity (220).

	λ_m	$10^{-3} \varepsilon$
$C_6H_5NO_2$	260 nm	8.1
$\underline{m}-O_2N-C_6H_4NO_2$	234	17.0
$1,3,5-(O_2N)_3-C_6H_3$	224	26.9

As more nitro groups are introduced into the ring to lower its polarizability, there is an increase in the transition energy of the phenyl-NO_2 chromophore, hence λ_m is smaller. However, each NO_2 attached to the ring serves as a phenyl-NO_2 chromophore. As the number increases, so does the probability of excitation and therefore the intensity (ε). Likewise, λ_m of $\underline{p}$-O_2N-$C_6H_4NO_2$ is a little smaller than that of $C_6H_5NO_2$ but has an ε twice that of $C_6H_5NO_2$ (221).

Solvent effects on uv spectra can be very dramatic. For instance, pyridinium N-phenolbetaine (4.6) is red in methanol and green in acetone. In general, increasing the polarity of the solvent shifts λ_m of $\pi \to \pi^*$ bands to longer wavelengths and $\underline{n} \to \pi^*$ bands to shorter wavelengths. Many kinds of parameters have been used as a measure of the polarity of a solvent (222). For example, there are those based on:

1. An intrinsic property of the solvent:
   dielectric constant, ε; refractive index, $\underline{n}$; solubilizing ability, δ (223a).
2. Solvent effects on rates of reactions:
   $\underline{Y}$ values (223b): relative rates of solvolysis of $\underline{t}$-butyl chloride in different solvents.

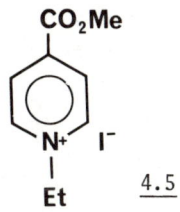

4.5

log $\underline{k}_{ion}$ (223c): relative solvolysis rates of $\underline{p}$-CH_3O-C_6H_4-$C(CH_3)_2CH_2OTs$.
$\underline{X}$ values (223d): relative rate of bromination of tetramethyltin.
Ω values (233e): ratio of endo to exo product in the Diels-Alder addition of cyclopentadiene to methyl acrylate.

3. Solvent effects of spectra:
   $\underline{Z}$ values (233f): based on λ_m of 4.5.

$$\underline{Z} = 2.859 \times 10^{-3} \nu_m \ (\nu_m \text{ in cm}^{-1})$$

$\underline{Z}$ values are known for about 30 solvents.
$\underline{E}_T$ values (233g): based on λ_m of 4.6. λ_m is 810 nm in $(C_6H_5)_2O$ and 453 nm in H_2O, the largest solvent shift known for any compound. $\underline{E}_T$ values are known for 67 solvents.

Pyridinium-N-phenolbetaine
4.6

$\Delta\nu_D$ and $\Delta\nu_A$ values (233h): Based on changes of the O-D band of deuteriomethanol and the C=O band of acetophenone in various solvents compared to benzene.

4. Multiple correlation indexes (based on two or more parameters):
S values (233i): Based on rate coefficients and uv spectral shifts. S values are known for 158 solvents.

Some correlations are better than others but no single scale accurately measures all solvent-solute interactions. The spectral indexes, of course, are best for correlating solvent effects on spectra and the solvoletic indexes correlate solvolyses best. For instance, ^{19}F nmr shifts of fluoropyridines in different solvents do not correlate well with solvent functions which contain the dielectric constant but do correlate with the E_T or Z values (224).

Since the excited states of C=O and C=N bonds for n→π* transitions are less polar than the ground states, e.g., the dipole moment of $H_2C=O$ in the ground and excited states is ~2.5 and ~1.5 D, respectively (225), λ_m undergoes a blue-shift with increasing polarity of solvent; and inasmuch as the excited state of π→π* transitions are more polar than ground states, the corresponding λ_m undergoes a red-shift with increasing solvent polarity. Several examples can be given (see also Table 4.12).

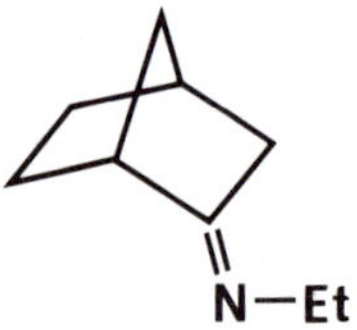

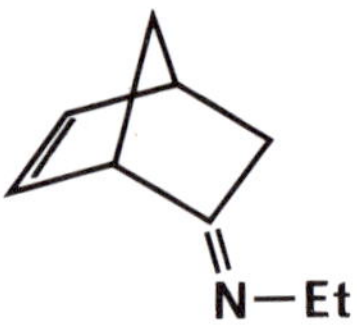

	Isooctane (226) λ_m (nm)	MeOH (226) λ_m (nm)
π→π*	209	210
n→π*	249	236
π→π*	214	215
n→π*	250	242

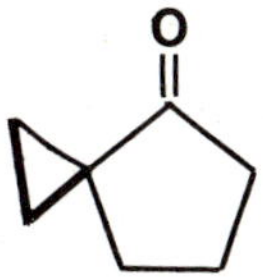

Sec. 4.2 MOLECULAR ABSORPTION OF ENERGY

		λ_m (227)	
Solvent	Z	$\pi \to \pi^*$ $\varepsilon \simeq 3000$	$n \to \pi^*$ $\varepsilon \simeq 20\text{-}40$
Isooctane	60	195 (nm)	293 (nm)
Acetonitrile	71	198	287
Methanol	84	201	281
$F_2CH\text{-}CF_2CH_2OH$	96	205	275

On the basis of the low extinction coefficient (~60) and also from the solvent effect, the 210-nm band of esters can be labeled an $n \to \pi^*$ band (228):

	Isooctane λ_m (nm)	EtOH λ_m (nm)	H_2O λ_m (nm)
CH_3CO_2Me	210	207	203
$\underline{i}\text{-Pr-}CO_2Me$	213	211	206

Adsorption of ketones on silica gel provides a strongly polar medium and there are observed blue-shifts for $n \to \pi^*$ bands and red-shifts for $\pi \to \pi^*$ bands:

		Cyclohexane λ_m (nm)	EtOH λ_m (nm)	Silica gel λ_m (nm)
$(C_6H_5)_2C=O$	$\pi \to \pi^*$	248	251	262
$(CH_3)_2C=O$	$\underline{n \to \pi}^*$	278		266
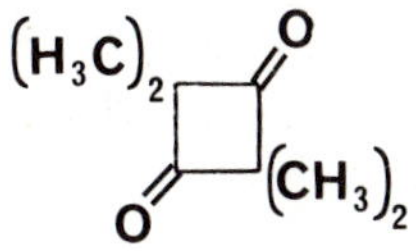 (H₃C)₂ cyclobutanedione (CH₃)₂	$\underline{n \to \pi}^*$	351	344	330
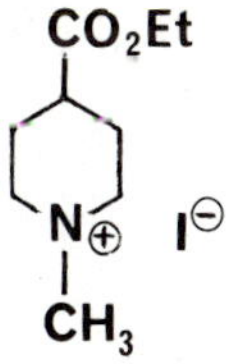 piperidine-CO₂Et N-CH₃ I⁻	$\underline{n \to \pi}^*$ (229)	420		320

	Cyclohexane λ_m (nm)	CF_3CH_2OH λ_m (nm)	Cyclohexane silica gel λ_m (nm)
$\pi \to \pi^*$ (230)	302	318	330

Table 4.12. λ_m of mesityl oxide in various solvents

Solvent	Z value	E_T value (kcal/mol)	ε	λ_m $\pi \to \pi^*$ (nm)	λ_m $n \to \pi^*$ (nm)
Isooctane	60		1.94	231	321
Acetonitrile	71	46	37.5	234	314
Ethanol	80	52	24.3	236	311
Methanol	84	55.5	32.6	237	309
Ethylene glycol	85	56.3	37.7	240	307
Water	95	63	78.5	243	302

Temperature has an effect on uv spectra. A uv absorption band is much broader than an ir band. An electronic excitation produces vibrational and rotational changes in addition to electronic transitions. Not all molecules are in the same vibrational level so there are groups of different vibrational transitions, each group giving a narrow peak of slightly different wavelength. The uv band, then, consists of an envelope of these peaks, which are too close for resolution. However, as the temperature is lowered, the molecules fall more and more into lower and fewer vibrational states to give a smaller distribution of transitions upon excitation. This produces fine structure in the uv bands. An example is given in Figure 4.6.

Photoelectron (pe) spectroscopy provides a relatively new and extremely versatile technique for measuring first and higher valence electron ionization potentials (233). A beam of photons is used to ionize a molecule. Since the process occurs much faster (10^{-15} sec) than a molecular vibration (~10^{-12} sec), the nuclei of the ionized molecule do not have time to reorient themselves into the new equilibrium position. This is called a

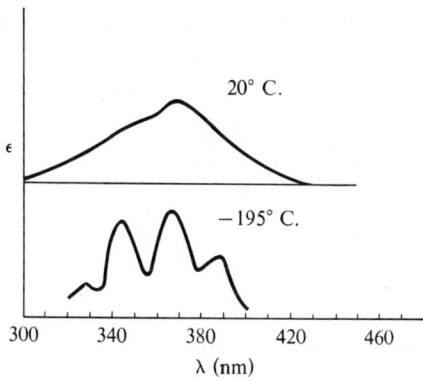

Figure 4.6. Spectra of dodecapentaenic acid at two different temperatures. (231)

vertical ionization. Whereas uv spectroscopy reflects ground and excited state interactions, pe spectroscopy provides a probe into ground electronic states alone. Some progress has been made toward correlating pe and uv spectral data (230). PE spectroscopy is also useful for determining energies of atomic and molecular orbitals. For example, pe spectroscopy shows the following facts. (1) The two lone-pair orbitals of water are not of the same energy. (2) The lone-pair electrons of nitrogen, phosphorus, arsenic, and antimony are in orbitals of virtually s character, all of very nearly the same orbital energies: NH_3, 10.9 eV; PH_3, 10.6 eV; AsH_3, 10.5 eV; and SbH_3, 10.0 eV (234). (3) The relative ionization potentials of vinyl chloride increase from $\pi(C=C)$ (lowest), through non-delocalized $\underline{n}$ and de-localized $\pi(\underline{n})$, to $\sigma(C=C)$ electrons (highest). The $\pi(\underline{n})$ orbital is orthogonal to the molecular plane and therefore is able to overlap in π fashion with the C=C $\underline{p}$ orbitals. (4) With increasing unsaturation, the σ-bonding framework is increasingly responsible for the integrity of a molecule and therefore σ-ionization potentials rise (235). Of the three filled π orbitals of benzene, two ($\pi_{\underline{a}}$, $\pi_{\underline{b}}$) are degenerate (Figure 5.1) and are responsible for the first IP of the compound. In a monosubstituted benzene, π_a has no density at the point of substitutuion whereas π_b has a maximum. The p or π orbitals on groups which are electron donating by resonance (+R groups) mix with π_b and lower the IP, whereas electronegative and -R groups raise the IP but produce only a small splitting of the π_a and π_b orbitals.

PE spectra are especially sensitive to conjugation, homoconjugation, and spiroconjugation (Chapter 10). The orbital overlap between two π orbitals splits their energies into two levels, π_1 and π_2 (see Figure 10.1), and there is a substantial difference (ΔIP) between the first and second ionization potentials. In contrast, ΔIP is distinctly smaller for isomers with isolated π orbitals. For example, ΔIP values for some dienes are as follows:

	ΔIP (eV)		ΔIP (eV)
Butadiene	3.17	1,3-Cyclohexadiene	2.5
1,1,4,4-Tetra-fluorobutadiene	2.66	1,4-Cyclohexadiene	1.0
		Fulvene	1.0
Hexafluorobuta-diene	1.0	<u>4.7</u>	0.5
		<u>4.8</u>	1.7

<u>4.7</u>   <u>4.8</u>

On this basis, we conclude that there is orbital overlap in <u>4.8</u> (<u>trans</u>, <u>trans</u>-1,6-cyclodecadiene) but little if any in <u>4.7</u> (the <u>cis</u>, <u>cis</u> isomer) (236) and that in contrast to butadiene and tetrafluorobutadiene there is little conjugation in hexafluorobutadiene and hence the diene is nonplanar (237). This latter deduction is supported by uv spectral data. Likewise, ΔIP for the conjugated enones below (left column) are larger than for nonconjugated isomers (238). UV spectra confirm the absence of homoconjugation in the two β, γ-enones on the right (239). Similarly, pes has been used to study non-bonding interactions between $\pi_{C=C}$ and $\underline{n}_0$ orbitals in unsaturated ethers (240).

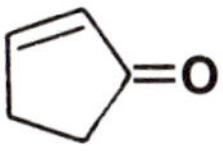

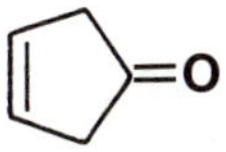

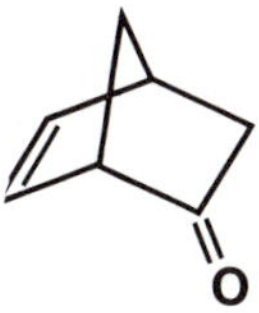

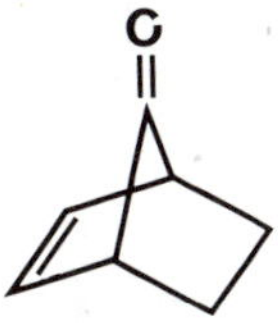

The ionization potentials of nitro- and N-methylnitro-anilines, compared to nitrobenzene and aniline, respectively, are all in the expected directions, although assignments to the correct molecular orbitals are not as readily made (241). For instance, the first IP of p-nitroaniline (8.60 eV) could be that of aniline (8.02 eV) shifted to higher energy by the electron-withdrawing NO_2 group, or that of nitrobenzene (9.99 eV) shifted to lower energy by the electron-donating NH_2 group.

4.2d. Nuclear magnetic resonance spectroscopy

Nmr spectroscopy is presently the organic chemist's most powerful tool, considering the variety and subtlety of structural information it can provide. The technique is based on the behavior of certain nuclei in a magnetic field. Nuclei spin and, like spinning tops in the gravitational field, they precess when placed in a magnetic field (see Figure 4.7). The spinning nucleus will possess an angular momentum based on its spin number, $\underline{I}$. The numerical value

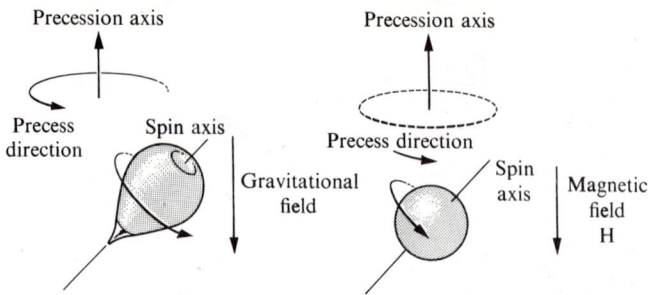

Figure 4.7. Gyroscopic actions of a spinning top in a gravitational field and of a nucleus in a magnetic field tend to tip over the precessing body.

of the spin number is related to the mass and atomic numbers of an atom as follows:

Mass number	Atomic number	Spin number	Examples
Even	Even	0	$^{12}_{6}C$, $^{16}_{8}O$, $^{18}_{8}O$, $^{32}_{6}S$
Even	Odd	1, 2, 3, ...	$^{2}_{1}H$ (D) 1, $^{14}_{7}N$ 1
Odd	Even or odd	1/2, 3/2, 5/2, ...	$^{1}_{1}H$ 1/2, $^{11}_{3}B$ 3/2, $^{13}_{6}C$ 1/2, $^{15}_{7}N$ 1/2, $^{17}_{8}O$ 5/2, $^{19}_{9}F$ 1/2, $^{29}_{14}Si$ 1/2, $^{31}_{15}P$ 1/2

Only nuclei with $\underline{I} \neq 0$ will have an angular momentum, which is quantized. That is, the magnetic dipole will assume only a discrete set of orientations with respect to the applied magnetic field. The number of orientations is $2\underline{I} + 1$, with each orientation corresponding to a given energy state of the nucleus. Thus, for ^{1}H, ^{11}B, ^{13}C, ^{15}N, ^{19}F and ^{31}P, with $\underline{I} = 1/2$, there are $2(1/2) + 1 = 2$ orientations, and for ^{2}H (D) and ^{14}N, there are $2(1) + 1 = 3$ orientations, whereas $\underline{I} = 0$ for ^{12}C, ^{16}O, and ^{18}O, which are nonmagnetic. The energy of each orientation is $-\mu_N \underline{H}_0$, where μ_N is the magnetic moment of the atom and $\underline{H}_0$ is the applied field strength.* For a proton, the energy of the two orientations is $+\mu_H \underline{H}_0$ and $-\mu_H \underline{H}_0$, where the orientation against the field is the higher energy state. If $\underline{H}_0$ is increased

---

* $\mu_N = \dfrac{\gamma_N h}{2\pi} \underline{I}$ where γ_N is a characteristic of the nucleus called the magnetogyric ratio, and $\underline{h}$ is Planck's constant.

Sec. 4.2 MOLECULAR ABSORPTION OF ENERGY

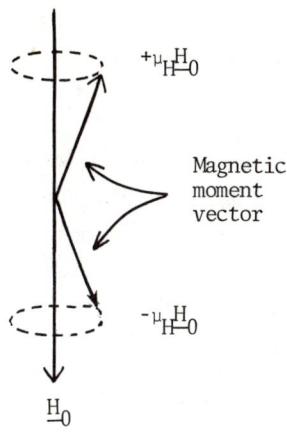

Magnetic moment vector

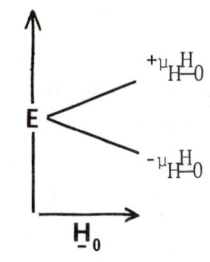

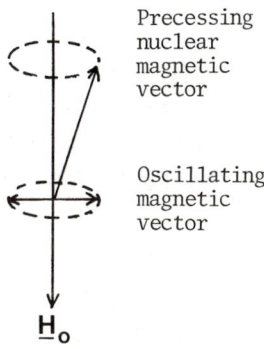

Precessing nuclear magnetic vector

Oscillating magnetic vector

the nucleus is not forced to align with the field; the nucleus merely precesses faster and the difference between the two energy levels increases.

If an oscillating magnetic field is superimposed on the nucleus, perpendicular to the stationary field $\underline{H}_0$, then when the frequency of the oscillating field equals the rate of precession of the magnetic moment of the nucleus, the two are said to be in <u>resonance</u>. Energy may then be transferred to the nucleus and it will undergo a transition from the lower energy state to the upper, i.e., it will flip over. The energy difference between the two states is

$$\mu_H \underline{H}_0 - (-\mu_H \underline{H}_0) = 2\mu_H \underline{H}_0.$$

Hence, when resonance occurs,

$$\Delta \underline{E} = 2\mu \underline{H}_0 = \underline{h}\nu$$

where ν is the frequency of the oscillating field and $\underline{h}$ is Planck's constant. Since μ and $\underline{h}$ are constants, one may alter $\Delta \underline{E}$ by holding $\underline{H}_0$ constant and changing ν until resonance occurs, or by holding ν constant and varying $\underline{H}_0$. Mechanically it is easier to do the latter, so most instruments are built with a fixed ν and the operator slowly changes $\underline{H}_0$ until resonance takes place. Since μ_F for fluorine is smaller than μ_H, for example, $\underline{H}_0$ must be larger to reach resonance for fluorine in molecules at a given value of ν. With fixed oscillatory frequencies of 60, 100, 220 MHz (megacycles per second), increasingly higher $\underline{H}_0$ fields must be used.

132   Ch. 4 INTERNAL ENERGIES

The chemical shift. Electrons rotating about a nucleus produce a magnetic field which shields the nucleus from the full magnetic force of the external field. Hence, H_0 must be increased to get resonance, depending upon the magnitude of the shielding effect. The shielding is directly proportional to the field strength, $= \sigma H_0$, where σ is a shielding constant. It would take an accuracy of one part in 10^8 to measure the exact value of H_0 where resonance occurs. However, one can distinguish two peaks which are separated by 60 ppm. Therefore, in order to raise the accuracy of readings and to have a value independent of field strength, the positions of resonance for a given atom are expressed relative to a reference atom:

$$\delta = \frac{H_r - H_s}{H_r} \text{ ppm} = \frac{\Delta\nu \times 10^6}{\text{oscillator frequency (cps)}}$$

where δ is called the chemical shift (242) for a given atom, H_r and H_s are the field strengths for resonance of particular nuclei in the sample and the reference compound, and $\Delta\nu$ is the difference in absorption frequencies of the sample and the reference in cps. The usual reference compound for proton measurements is tetramethylsilane, which gives a signal at higher field than most protons. As a result, δ decreases with increasing field strength.

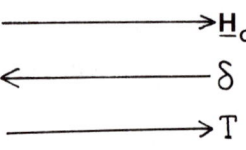

Since it is conventional to express parameters as increasing toward the right on graph paper, the chemical shift is sometimes expressed as τ, where $\tau = 10 - \delta$.*

The shielding effect of valence electrons depends on how close they are to the nucleus. Thus, of two protons attached to different atoms, the one attached to the more electronegative atom will experience less shielding, i.e. will be deshielded with respect to the other. For example, δ decreases (moves upfield) in the order $\underline{a} > \underline{b} > \underline{c} \ldots$.

δ (ppm) = 3.77   3.57   3.1 2.0 1.2   3.56 1.05
               $\underline{a}$ $\underline{b}$ $\underline{a}$ $\underline{b}$ $\underline{c}$ $\underline{a}$ $\underline{b}$
            $Cl-CH_2-CH_2-Br$ $I-CH_2-CH_2-CH_3$ $(CH_3)_2\underset{H}{C}-O-\underset{H}{C}(CH_3)_2$

---

* Not to be surpassed by ir spectroscopy, where two inverse units ν and λ are used, nmr has the two scales which vary in opposite directions.

It is interesting that the CH_2 of the ethyl group in $HC(OCH_2CH_3)_3$ resonates <u>downfield</u> of the CH_3 whereas the CH_2 gives a signal <u>upfield</u> of the CH_3 in $Si(CH_2CH_3)_4$ because of the relative electronegativities of $O > C > Si$.

Equivalent protons resonate at the same field strength and merely increase the height of the signal or what is called the <u>intensity</u> of the absorption peak. A compound such as $\overset{a}{C}H_3\overset{b}{C}H_2Cl$ then, would have two peaks with relative heights 3:2 for <u>a</u>:<u>b</u>. Consequently, so far there are two characteristics we would look for in an nmr spectrum, the chemical shift δ and the relative peak height.*

<u>Anisotropic effects.</u> The chemical shift for a particular atom is a function of the electron density around that atom. This may be altered by the inductive effects of neighboring atoms transmitted through chemical bonds. The magnetic field about an atom will usually vary with direction owing to local magnetic fields from neighboring electron clouds. Such uneven magnetic fluxes are called <u>anisotropic</u> effects. For instance, the revolving π electrons of an aromatic ring produce a diamagnetic field with a flux as pictured. Hence, atoms, above or below the plane are shielded whereas those attached to the ring are deshielded; that is, the magnetic field created by the π-electron ring current augments the field at the periphery of the ring so that $\underline{H}_0$ need not be as large to reach resonance conditions. Such protons, then, give a signal at lower field. There are many excellent examples of this anisotropic effect.

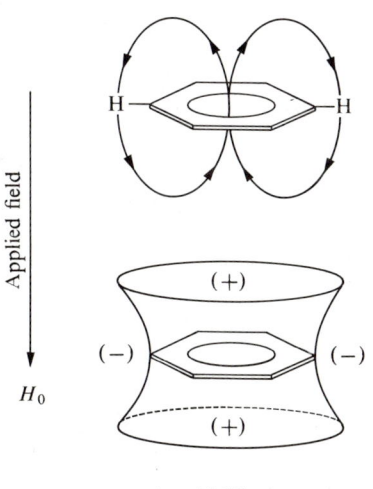

+ = shielding zone

---

* You probably know that modern instruments measure the area of nmr peaks (integrated intensity) which can be expressed in terms of number of protons giving that peak.

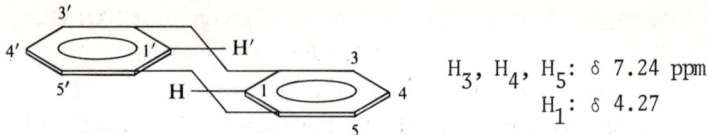

H_3, H_4, H_5: δ 7.24 ppm
H_1: δ 4.27

Thus, H_1 (and H_1') sits over a ring and is partially shielded whereas H_3, H_4, and H_5 in each ring are in the "aromatic range," i.e., the value observed for protons attached to an aromatic ring carbon. The _a_ protons in _4.9_ are deshielded by two benzene rings and are at lower field than the _b_ protons.

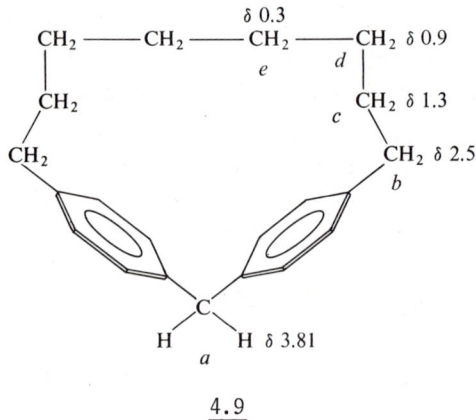

4.9

As the _c_, _d_, and _e_ protons lie more and more over a benzene ring, δ is increasingly shifted upfield.

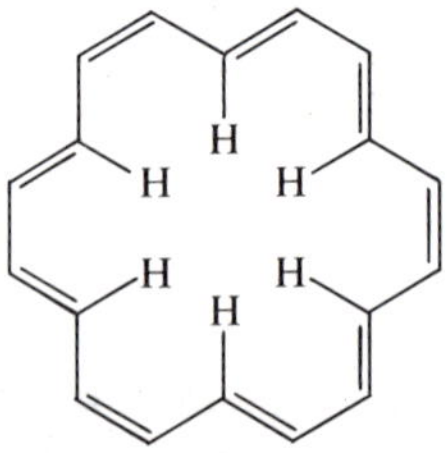

4.10
(18)Annulene
Inner Hs, δ -1.8 ppm
Outer Hs, δ 8.9

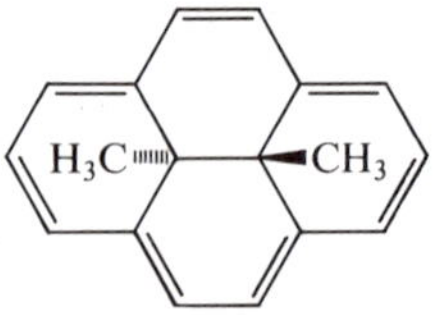

4.11
Outer Hs δ 7.98-8.67 ppm
$\underline{CH_3}$, δ -4.23 (Ar$\underline{CH_3}$ δ 2.3-2.5)

Note that the inner protons of 4.10 and the methyl protons of 4.11, which lie within the shielding zone of the respective aromatic rings, are shifted even beyond the TMS signal. Although carbon atoms are shielded by these aromatic ring currents, the changes in ^{13}C chemical shifts are rather small, e.g., (243)

3.48 ppm   36.9 ppm

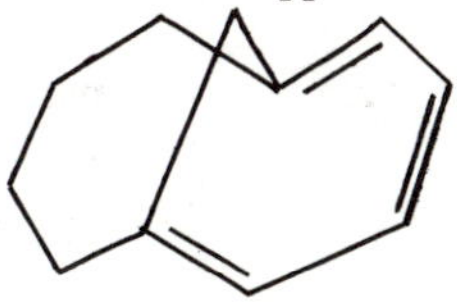

Shielding and deshielding zones of some other unsaturated groups are as follows (244):

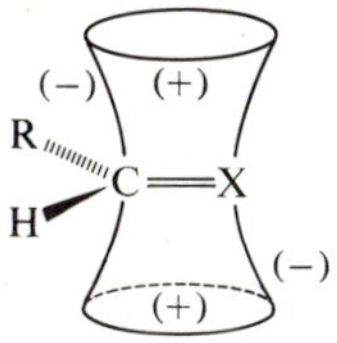

X = O, N, = CR_2

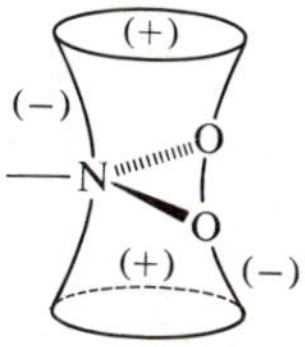

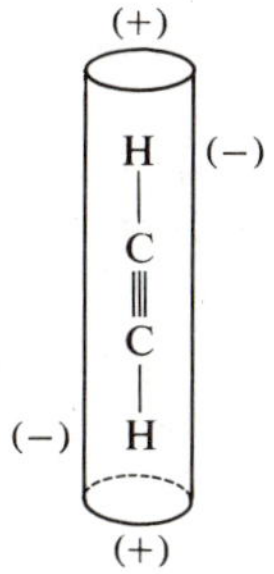

Thus aldehydic and ethylenic protons are **deshielded**.

Thus acetylenic protons are **shielded**.

For example, the CH_3 groups and the protons of 4.12 and 4.13 have different chemical shifts; those cis to the C=O group are in its deshielding zone. When NO_2 groups on the benzene ring are forced to rotate out of the plane of the ring, the adjacent protons fall into the nitro shielding zone and the chemical shift moves upfield (245).

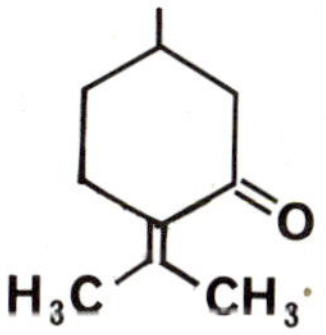

δ = 1.77   1.95   5.98   6.86 ppm
     4.12   4.13

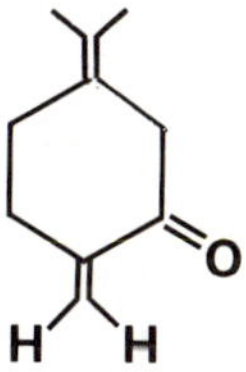

136   Ch. 4   INTERNAL ENERGIES

Spin-spin interactions (246). The magnetic field surrounding a given atom is affected not only by nearby π electron clouds but also by the nuclear magnetic moments of neighboring atoms. Consider the molecular fragment 4.14 in which H_a and H_b are not equivalent protons. The nuclear moment of H_a can be oriented with or against the external field, H_0. Atom H_b, then, can experience two slightly different magnetic fields and will resonate at two correspondingly different values of H_0. Likewise, H_a will exhibit two peaks, corresponding to the two orientations of the nuclear moment of H_b. The sum of the areas under either doublet is equal to the area that there would be under the single peak if there were no splitting, since only one atom is responsible for each doublet. The separation between tips of the doublet is called the spin-spin coupling constant, J, and is expressed in cycles/sec or Hertz. J is independent of the magnitude of H_0.

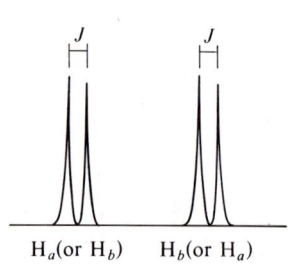

Magnetically equivalent atoms do not split each other. Accordingly, CH_4, C_6H_6, and H_2 have only one line in their spectra. Similarly, H_a and $H_{a'}$ in 4.15 give rise to a single two-proton peak which is split into a doublet by H_b. H_b may however, feel three different fields from H_a and $H_{a'}$, which may be diagrammed as follows:

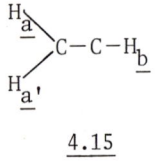

4.15

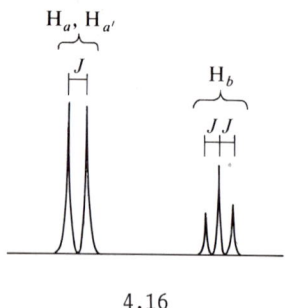

4.16

Spin orientation H_a $H_{a'}$	Total effect
↑ ↑	1/2 + 1/2 = 1
↑ ↓	1/2 − 1/2 = 0
↓ ↑	−1/2 + 1/2 = 0
↓ ↓	−1/2 − 1/2 = −1

Sec. 4.2 MOLECULAR ABSORPTION OF ENERGY   137

Thus, H_b may suffer three different field effects from H_a and $H_{a'}$. Since there is a two-fold chance of the zero effect, the relative intensities of the three peaks are 1:2:1 as shown in 4.16. Likewise, each H in CH_2F_2 is split by each F but is unaffected by the other H. The spectrum, then, has two triplets, one in the proton range and one in the fluorine range (not seen in the proton spectrum). If there were three neighboring equivalent protons H_a, they would have the following effect on H_b:

H_a	H_a	H_a	Total effect
↑	↑	↑	1/2 + 1/2 + 1/2 = 3/2
↑	↑	↓	1/2 + 1/2 - 1/2 = 1/2
↑	↓	↑	1/2 - 1/2 + 1/2 = 1/2
↓	↑	↑	-1/2 + 1/2 + 1/2 = 1/2
↓	↓	↑	-1/2 - 1/2 + 1/2 = -1/2
↓	↑	↓	-1/2 + 1/2 - 1/2 = -1/2
↑	↓	↓	1/2 - 1/2 - 1/2 = -1/2
↓	↓	↓	-1/2 - 1/2 - 1/2 = -3/2

H_b therefore feels four different fields from H_a nuclei with a probability (i.e., relative intensity) of 1:3:3:1, shown in 4.17.

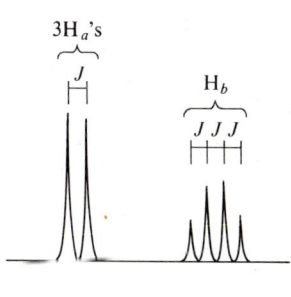

4.17

Based on the number of protons giving the peaks, the relative intensity or integrated ratios of the doublet and the quartet are 3:1.

As a rule of thumb, the number of peaks in the fine structure or multiplicity m of an nmr band is m = $2nI$ + 1, where n is the number of nuclei causing the splitting and I is their spin quantum number. The relative areas under the peaks are given by the coefficients for each term in the expansion formula of $(x + 1)^y$ in which y = m - 1. For example, for n = 2, 3, and 6 coupled protons:

$$m = \frac{2 \times 2}{2} + 1 = 3 \qquad (x + 1)^{3-1} = x^2 + 2x + 1$$

Peak area ratio is 1:2:1.

$$m = \frac{2 \times 3}{2} + 1 = 4 \qquad (\underline{x}+1)^{4-1} = \underline{x}^3 + 3\underline{x}^2 + 3\underline{x} + 1$$
Peak area ratio is 1:3:3:1.

$$m = \frac{2 \times 6}{2} + 1 = 7 \qquad (\underline{x}+1)^{7-1} = \underline{x}^6 + 6\underline{x}^5 + 15\underline{x}^4 + 20\underline{x}^3 + 15\underline{x}^2 + 6\underline{x} + 1$$
Peak area ratio is 1:6:15:20:15:6:1.

We see that the electronegative effect on the chemical shift and spin-spin coupling are each transmitted through bonds. Each, then, is more effective across a double bond and each decreases markedly with distance. For example, $^3J_{\underline{H},\underline{H}}$ (coupling between protons across three bonds) is 2 to 9 Hz across a C-C bond and 13 to 18 Hz across a C=C bond (see Table 4.13). Often $J_{\underline{cis}} \simeq 2/3 J_{\underline{trans}}$. Coupling across more than three single bonds can usually be ignored (see Section 15.1 for exceptions).

Table 4.13. Spin-spin coupling constants

Fragment	J (Hz)	Fragment	J (Hz)
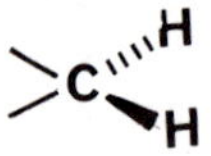 >C(H)(H)	12-15	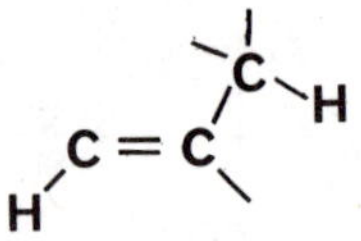 (C)C=C(H)/H (geminal vinyl-allyl)	0.5-2.5
>CH—CH<	2-9	H\C=C/(C)H	0
>C=C<H,H	0.5-3	benzene ortho	6-9
H\C=C/H (cis)	13-18	benzene meta	1-3
H\C=C/H (trans)	7-12	benzene para	0-1

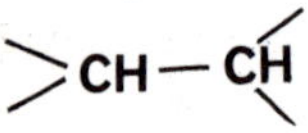

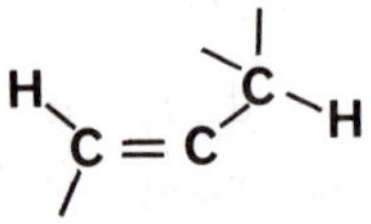

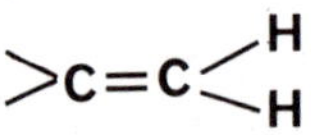

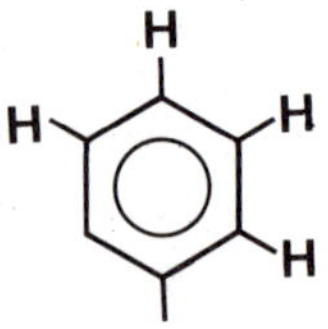

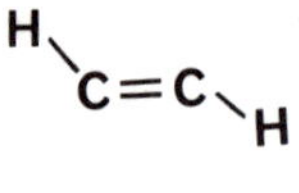

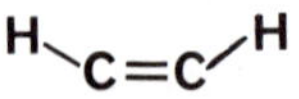

Although it is not necessary to memorize the positions of all chemical shifts or ir absorption frequencies for all bond types, it is valuable to know a few approximately, just to do any modern organic chemistry with dispatch. It is convenient, for instance, to know the range for aromatic, saturated aliphatic, or vinylic protons in the nmr. For this reason, the chemical shifts for protons attached to few common groups are listed in Table 4.14.

Spin-coupled nuclei are designated A, B, ..., X, in order of increasing τ. A,B refers to protons with close chemical shifts, A,M to medium close, and A,X refers to protons with widely separated chemical shifts. Some examples are:

CH_3CH_2Br $HO-CH_2-CH_3$ CH_2F_2 HCl_2C-CH_2Cl $\underset{H_A}{Cl}\diagdown C=C \diagup \underset{H_C}{H_B}$
 B   A   A   B   C A_2X_2 AX_2 ABC

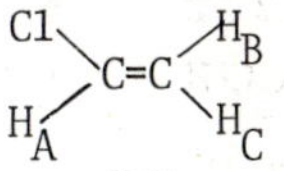

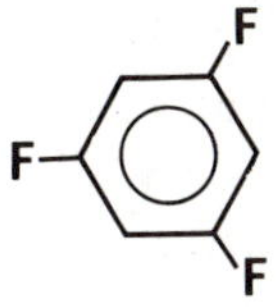

A_3X_3 (1,3,5-trifluorobenzene with F labels)

AA'BB' (O_2N-aryl-OR with $H_{A'}, H_{B'}, H_A, H_B$)

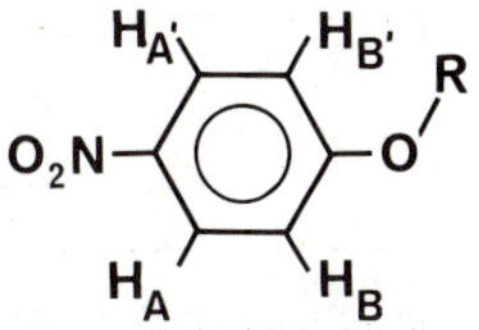

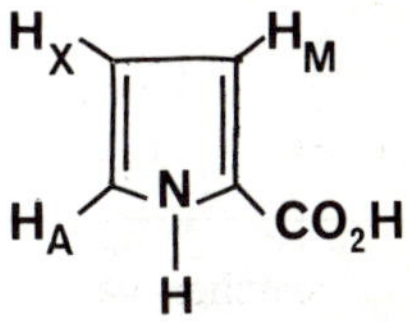

AMX (pyrrole with H_X, H_M, H_A, CO_2H)

Table 4.14. Nmr chemical shifts for protons attached to selected groups

A. BROAD RANGES

Fragment	Chemical shift (ppm)	Fragment	Chemical shift (ppm)
$H-C_{sp^3}$	0-1.8	O-H alcohol	0.5-5.5
$H-C_{sp^3}-X$	1.5-5.0	phenols	4.0-7.7
X = Cl, O, N,		enols	15-17
C_{sp^2}, C_{sp}		acids	10.5-12
≡C-H	2.0-3.0	N-H aliphatic	0.3-2.2
=C-H	4.5-6.0	aromatic	2.6-5.0
Ar-H	6.5-9.5	amide	5.0-8.5
O=C-H	9.0-10	S-H aliphatic	1.2-1.6
		aromatic	3.6

(Table 4.14 continued)

B. SOME HANDY VALUES (δ ppm)

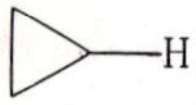 ▷—H	0.5-0.2	Ar-H	6.5-9.5
R-CH$_3$	0.9	Ar-C-H	2.2-3.0
R$_2$CH$_2$	1.3	Ar-C-C-H	1.0-1.3
R$_3$C-H	1.5	O=C-C-H	1.7-2.2
≡C-H	2-3	RO-C-H	3.3-4.0
C=C(H) 4.5-6	C=C-CH$_3$ ~2.0	C=C-CH$_2$- ~2.5-3.0	C=C-CH 3.0-4.0

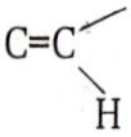

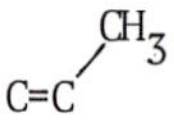

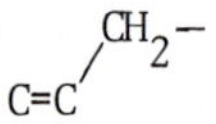

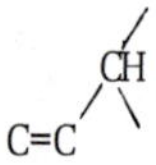

Another way of analyzing a spectrum is by use of spin-spin diagrams. For example, for the system **4.18**:

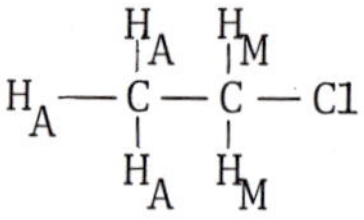

$$H_A-\underset{H_A}{\overset{H_A}{C}}-\underset{H_M}{\overset{H_M}{C}}-Cl$$

4.18

δ_A = 1.48 ppm
δ_M = 3.57
$\underline{J}_{AM}$ = 0.15

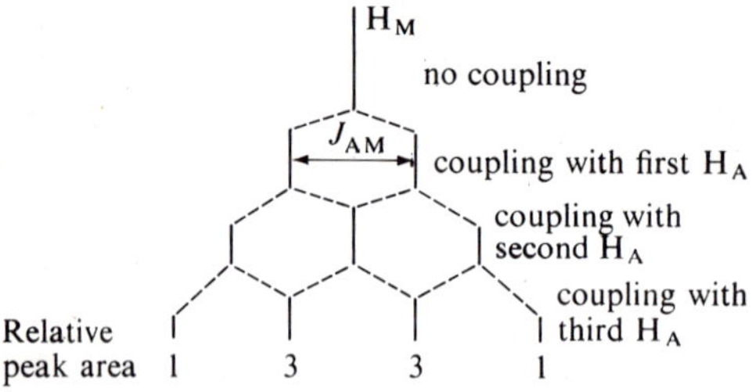

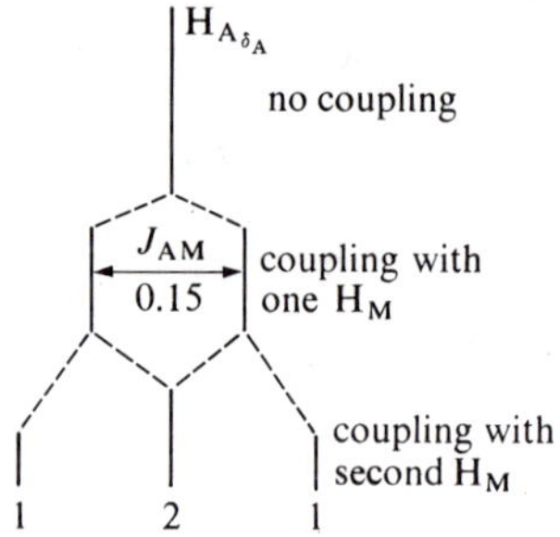

H$_M$	no coupling		H$_{A\delta_A}$	no coupling
J_{AM}	coupling with first H$_A$		J_{AM} 0.15	coupling with one H$_M$
	coupling with second H$_A$			
	coupling with third H$_A$			coupling with second H$_M$

Relative peak area 1   3   3   1   1   2   1

The splitting pattern for H$_M$ in the system $\underset{H_A\ H_M\ H_X}{C-C-C}$ where $\underline{J}_{AM}$ = 12 Hz and J_{MX} = 4 could be:

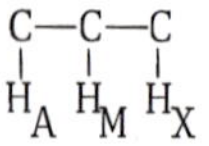

Sec. 4.2 MOLECULAR ABSORPTION OF ENERGY 141

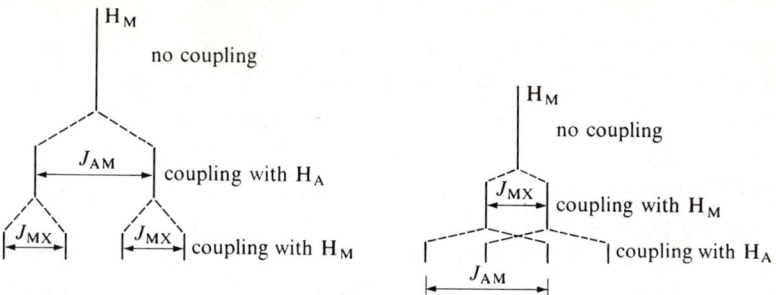

Compound <u>4.19</u> has a complex structure but a relatively simple spectrum which illustrates the foregoing comments (247).

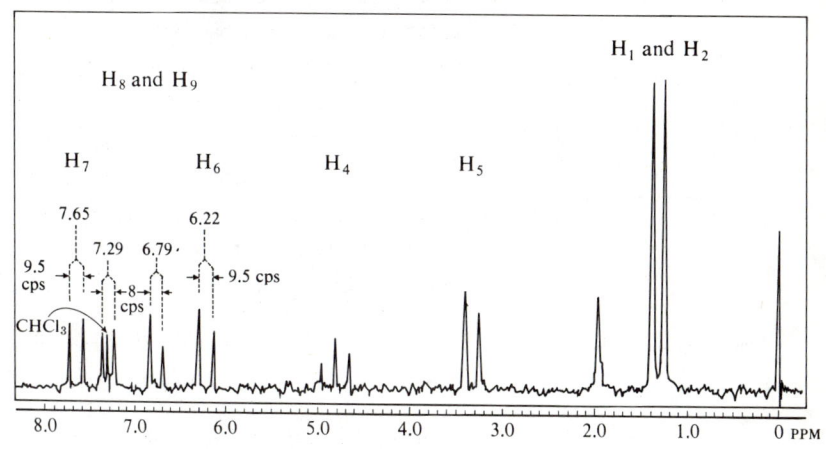

Figure 4.8

Notice that H_6, H_7 and H_8, H_9 form A,B systems. The identical coupling constants help to identify each pair. The methyl groups, in different molecular environments, exhibit different chemical shifts.

Chemically equivalent protons are not necessarily magnetically equivalent, in which case they can couple.

Magnetically and chemically equivalent

Chemically but not magnetically equivalent

Nuclei are magnetically equivalent if they couple to all other nuclei in the system in exactly the same way (248). Since trans and cis coupling are not equal, H and H' in 4.20 do not always experience identical fields from F and F' so H and H' are not magnetically equivalent. As a result, H and H' couple with each other and the proton spectrum of 4.20 has more than eight peaks in it.

We mentioned earlier that the number of peaks to expect in a band is based on the quantity $m = 2I + 1$. The expression applies only when $\Delta\nu/J$ is greater than about 20. This is called first-order coupling, in which case J can be read directly from the spectrum. Such is the case for an AX or AX_2 spectrum. J is independent of H_0, so that as H_0 increases, the ratio $\Delta\nu/J$ increases. When $\Delta\nu/J$ falls below about 20, the nuclei perturb each other and the number of lines observed depends on the specific ratio. This is well illustrated by comparing the bands for the B protons of $HO\text{-}CH_2\text{-}CH_3$ and of $I\text{-}CH_2CH_2CH_3$ (249).

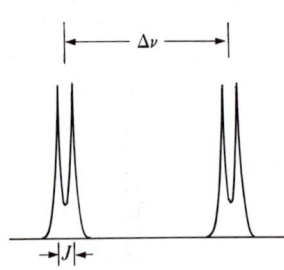

$HO\text{-}CH_2\text{-}CH_3$

  A   B   C

δ 5.28 3.62 1.17 ppm

$J_{AB} = 5$; $J_{BC} = 7.2$ Hz

$\Delta\nu_{BC} = (8.83 - 6.38) \times 60$
$= 147$ c/s

therefore $\Delta\nu_{BC}/J = \dfrac{147}{7.2} \approx 20$

$\Delta\nu_{AB}/J = \dfrac{99.6}{5} \approx 20$

As expected, the A and C protons each give a triplet and the B protons give eight lines:

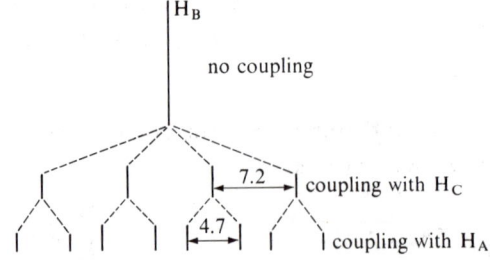

I-CH$_2$-CH$_2$-CH$_3$
A B C
δ 3.17 1.86 1.02 ppm

$\Delta\nu_{AB}/\underline{J} = \frac{78.6}{6.8} \simeq 11$; $\Delta\nu_{BC}/\underline{J} = \frac{50.4}{7.3} \simeq 7$

$\underline{J}_{AB} = 6.8$; $\underline{J}_{BC} = 7.3$ Hz

Again, the A and C protons each give a triplet but the B protons give a complex multiplet.

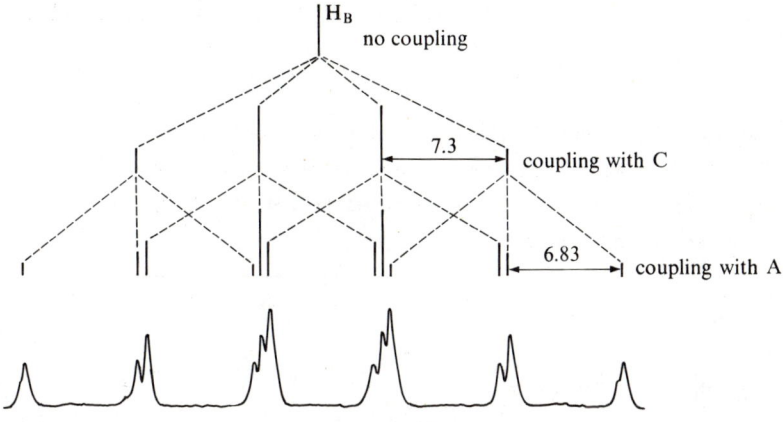

Among the combinations of three interacting nuclei, there are AB$_2$, ABC, AX$_2$, ABX, AMX, and ABM. Some examples are:

AMX

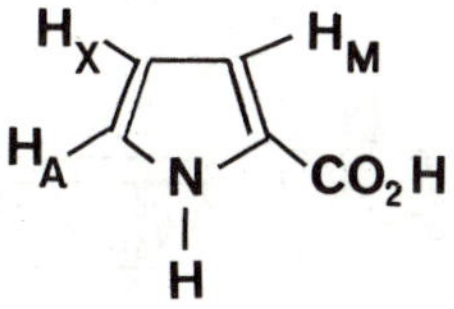

δ_A = 6.99 ppm, $\underline{J}_{AM}$ = 1.6 Hz

δ_M = 6.79 $\underline{J}_{AX}$ = 2.6

δ_X = 6.26 $\underline{J}_{MX}$ = 3.7

AX$_2$

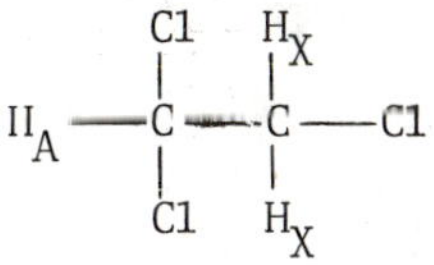

δ_X = 3.95 ppm

δ_A = 5.75

ABX ≡ AXY

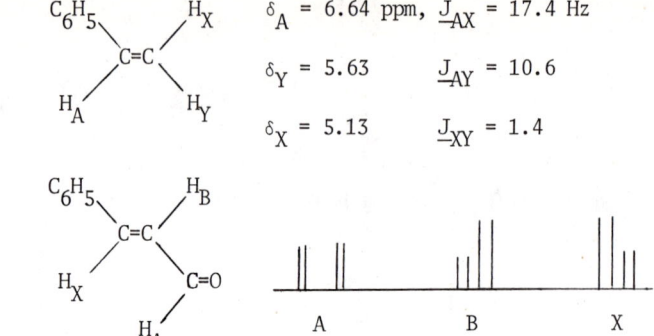

δ_A = 6.64 ppm, J_{AX} = 17.4 Hz
δ_Y = 5.63, J_{AY} = 10.6
δ_X = 5.13, J_{XY} = 1.4

These compounds have roughly the expected first-order spectra.

When $\Delta\nu/J$ is less than about 20 and there results a perturbed splitting pattern, one may use the equations (4.6) to (4.10). I_n is the relative intensity of line n.

$$\Delta\nu_{AB} = \sqrt{(\delta_1 - \delta_4)(\delta_2 - \delta_3)} \quad (4.6)$$

$$J_{AB} = \delta_1 - \delta_2 = \delta_3 - \delta_4 \quad (4.7)$$

$$\frac{I_2}{I_1} = \frac{I_3}{I_4} = \frac{\delta_1 - \delta_4}{\delta_2 - \delta_3} \quad (4.8)$$

$$I_1 = I_4 = 1 - \frac{J_{AB}}{\sqrt{\Delta\nu_{AB}^2 + J_{AB}^2}} \quad (4.9)$$

$$I_2 = I_3 = 1 + \frac{J_{AB}}{\sqrt{\Delta\nu_{AB}^2 + J_{AB}^2}} \quad (4.10)$$

Usually δ_1, δ_2, δ_3, and δ_4 can be read from the spectrum. From them $\Delta\nu_{AB}$ may be calculated and the chemical shifts for the unsplit protons determined.

The number of theoretical peaks from coupled protons can be ascertained by another diagrammatic technique. Consider vinyl chloride:

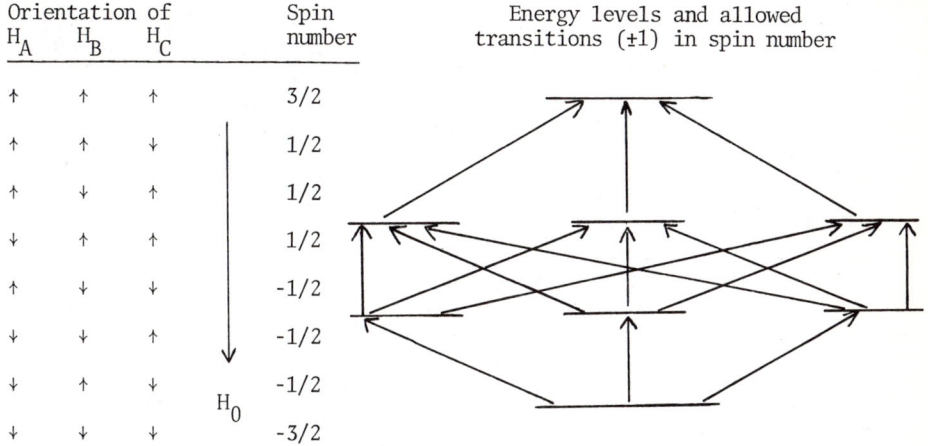

| Orientation of | | | Spin |
H_A	H_B	H_C	number
↑	↑	↑	3/2
↑	↑	↓	1/2
↑	↓	↑	1/2
↓	↑	↑	1/2
↑	↓	↓	-1/2
↓	↓	↑	-1/2
↓	↑	↓	-1/2
↓	↓	↓	-3/2

Hence, 15 lines are predicted, and at least 13 peaks are readily picked out in the spectra of vinyl chloride and acrylonitrile. ABC, ABM, and ABX systems are usually complex.

To illustrate how spectra are dependent upon $\Delta\nu/J$ ratios, consider the hypothetical case in which $\Delta\nu_{AB}$ changes. (Refer to the diagram at the top of the next page.) Accordingly, spectra can be quite complex with only three interacting nuclei and they can be even more so with four or more coupled nuclei (250). One approach is to use computer programs (251) for calculating band positions which can then be matched with the observed lines.

146   Ch. 4 INTERNAL ENERGIES

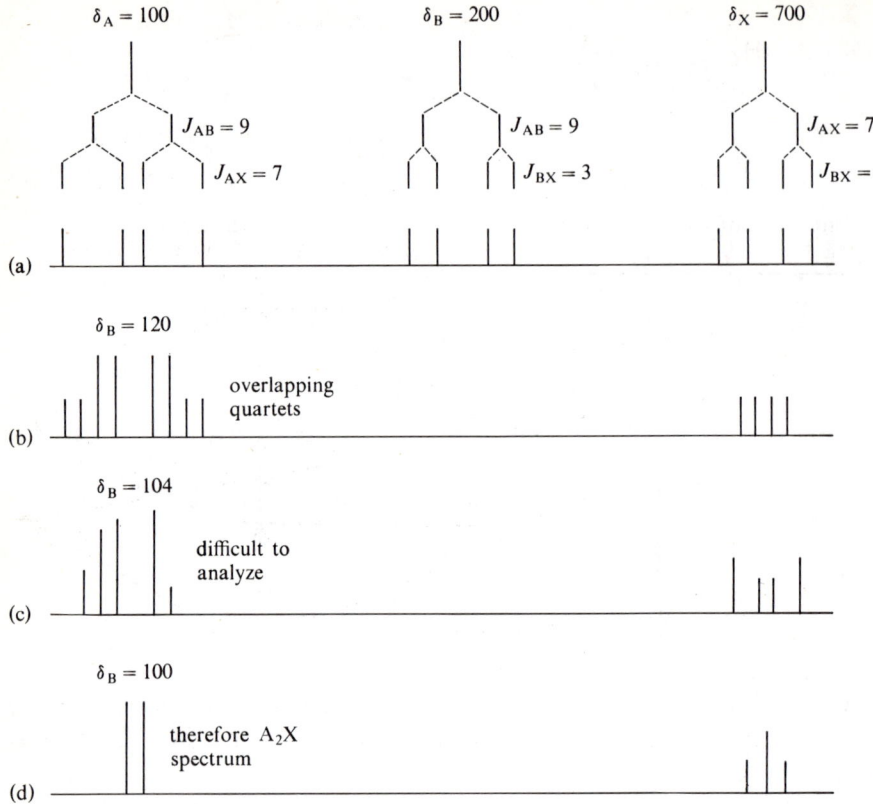

In this section we have considered three parameters concerning nmr spectra: chemical shift, spin-spin coupling constants, and integrated intensities of peaks. Several additional important features could be discussed, such as coupling with other nuclei, exchanging nuclei, and multiple resonance. Effects of geometry on the coupling constant and the alteration of chemical shifts by temperature changes wil be discussed in Chapters 15 and 18, respectively.

We have discussed only pmr spectra. Spectra of other nuclei, such as ^{31}P, ^{11}B, ^{15}N, ^{13}C (252), and others, are equally useful for certain classes of compounds. Much of what we have said about ^{1}H could apply to ^{13}C (252). In this connection, although substituent constants are available for estimating olefinic ^{13}C chemical shifts of substituted ethylenes (252a), the effects of substituents are very complex and remain unclear (252b). Since ^{13}C is only about one per cent of the carbon in organic compounds, and ^{12}C is nonmagnetic, ^{13}C peaks are very small. If one were to take many, many scans, the intensity of ^{13}C peaks could be built up but this would take months. Fortunately, recent instrumental developments enable one to take tens of thousands of scans in a few hours and now routine cmr spectral measurements at natural ^{13}C abundance can be made (253). This has enhanced the use of cmr spectroscopy for structure determinations during the past two years (254).

QUESTIONS

1. (2.2)Metaparacyclophane has proton chemical shifts as follows: 5.24(1H); 5.70(2H); 6.97(2H); 6.63(3H) ppm. Assign these to the various ring protons in the compound.
2. Would dimethylacetylene be a good reference compound to use to ascertain the C≡C stretching frequency in the ir spectrum?
3.  The *endo* proton on C6 of *endo*-5-methylnorbornene resonates in the nmr at abnormally high field. Suggest a reason for this (R. G. Foster and M. C. McIver, Chem. Comm. 380 (1967)).
4. Which halobenzene should exhibit the least fine structure in its uv spectrum?

Answers

1. 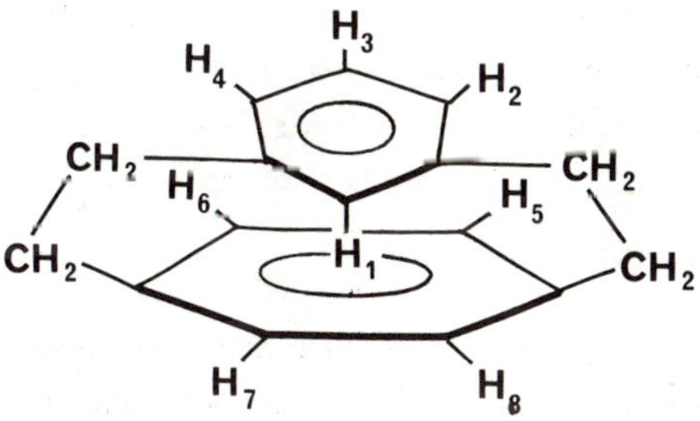 Thus, H_1 sits over the bottom phenyl ring and H_6, H_5 sit almost below the top benzene ring. H_1 is shielded the most, then H_5 and H_6. H_7 and H_8 are equivalent, and H_2, H_3, and H_4 form an approximately equivalent group. The integrated intensities help in making the assignments. The nmr signals, then, are: H_1, δ 5.24; H_5,H_6, δ 5.70; H_2,H_3,H_4, δ 6.63; H_7,H_8, δ 6.97.

2. No. Since the molecule is symmetrical, its C≡C band cannot be seen in the ir spectrum.
3. It is shielded by the coparallel endo C-CH_3 bond. It is also in the shielding zone of the double bond, although this effect is small.
4. Iodobenzene. If some electronic energy is changed into low energy vibrational or rotational energy immediately after electronic excitation occurs, the observed distribution of excited molecules is enlarged and the fine structure diminishes. This energy transformation takes place when groups which can easily vibrate or rotate are attached to the chromophoric system. Normally, rigid molecules have more fine structure in their spectra than conformationally loose molecules. In this case, iodine has the smallest energy of rotation or oscillation and exhibits the broadest absorption bands among the halobenzenes.

4.2e. Mass spectra (255)

The function of a mass spectrometer is to produce ions from a sample, usually by bombardment with a high energy electron beam, to separate the ions according to their mass-to-charge ratios (m/e), and to measure the relative abundance of each ion. The heaviest ion in a spectrum is normally that formed by the loss of an electron from a molecule of the compound being studied and is called the molecular ion, symbolized M^+. Since most ions generated in this way are singly charged, i.e., $e = +1$, mass spectrometry provides one of the most exact methods for measuring the molecular weight of a molecule. The molecular ion may undergo a variety of fragmentations, and a record of the m/e values and abundances for the numerous ions produced constitutes the mass spectrum. The peak representing the molecular ion is called the parent or molecular ion peak, P. Occasionally the molecular ion is too unstable to reach the detector and no parent peak is observed. The most intense peak is called the base peak and is arbitrarily assigned a relative peak height of 100. If the rate of decomposition of an ion m_1^+

$$m_1^+ \rightarrow m_2^+ + \text{neutral fragment}$$

formed in the ionization chamber is very fast, most such ions will decompose prior to reaching the acceleration region and only the m_2^+ ion will reach the detector. Some ions of intermediate stability may decompose into fragments while passing through the accelerator region. Such ions will first

Sec. 4.2 MOLECULAR ABSORPTION OF ENERGY   149

be accelerated with mass $\underline{m}_1$ and after fragmentation with mass $\underline{m}_2$. It will be recorded as a broad peak, called a metastable peak, generally with low intensity. The position of a metastable peak can be calculated from the formula

$$\underline{m}^* = \frac{\underline{m}_2^2}{\underline{m}_1}.$$

Metastable peaks can be helpful in unraveling fragmentation patterns, and their use is demonstrated in example 3 at the end of this section.

Since molecules contain the elements in accordance with the natural proportions of the respective isotopes, the parent peak will be accompanied by small peaks $\underline{P} + 1$, $\underline{P} + 2$, etc., representing ions containing one or more heavy isotopes. Each isotopic species of a particular molecule will have a different $\underline{m/e}$ ratio and produce a separate peak. For instance, the spectrum of methylene chloride has three prominent peaks at $\underline{m/e}$ = 84, 86, and 88 for $CH_2{}^{35}Cl_2$, $CH_2{}^{35}Cl^{37}Cl$, and $CH_2{}^{37}Cl_2$. There are also three smaller peaks at $\underline{m/e}$ = 85, 87, and 89 for the ^{13}C analogs of the above ions.

The natural isotopic percentages for some elements are given in Table 4.15. It is noteworthy that fluorine, phosphorus, and iodine are monoisotopic. The intensities of the $\underline{P} + 1$ and $\underline{P} + 2$ peaks relative to that of $\underline{P}$ depend upon the ratios of natural abundances of heavy isotopes in the molecule. Thus, sulfur-, chlorine-, and bromine-containing compounds have relatively high $\underline{P} + 2$ peaks because of their large abundances of heavy isotopes. To take a simple example, CO and N_2 each have a nominal molecular weight of 28. In the spectrum of CO there is a $\underline{P} + 1$ peak with about 1.12% of the intensity of $\underline{P}$, representing the $^{13}C^{16}O$ and $^{12}C^{17}O$ species.

$$^{13}C = \frac{1.107}{98.89} = 1.08\%\ ^{12}C \text{ and } ^{17}O = \frac{0.037}{99.759} = 0.04\%\ ^{16}O$$

i.e., $^{13}C^{16}O + ^{12}C^{17}O = 1.08 + 0.04 = 1.12\%$ as abundant as $^{12}C^{16}O$. The $\underline{P} + 2$ peak at $\underline{m/e}$ = 30 would have an intensity about 0.2% $\underline{P}$:

$$^{12}C^{18}O = \frac{0.204}{99.759} = 0.2\%\ ^{12}C^{16}O; \quad ^{13}C^{17}O = (0.00108)(0.0004)\ ^{12}C^{16}O$$
$$= 0.000432\%\ ^{12}C^{16}O.$$

The relative intensities for P, $\underline{P} + 1$, and $\underline{P} + 2$ for $\underline{N}_2$, however, would be:

$\underline{P}$ + 1: $\dfrac{0.36}{99.64}$ = 2 × 0.38 $^{14}N_2$ = 0.76 $\underline{P}$

$\underline{P}$ + 2: $(0.0038)^2$ = 0.001296% $\underline{P}$.

Table 4.15. Natural isotopic distributions for some elements

Light isotopes	Natural abundances	Heavy isotopes			
			Natural abundances		Natural abundances
^1H	99.985%	^2H (D)	0.015%		
^{12}C	98.893	^{13}C	1.107		
^{14}N	99.64	^{15}N	0.36		
^{16}O	99.759	^{17}O	0.037	^{18}O	0.204%
^{19}F	100				
^{28}Si	92.2	^{29}Si	4.7	^{30}Si	3.1
^{31}P	100				
^{32}S	95.02	^{33}S	0.76	^{34}S	4.22
^{35}Cl	75.77	^{37}Cl	24.23		
^{79}Br	50.537	^{81}Br	49.463		
^{127}I	100				

Consequently, CO and N_2 could be distinguished from the profile of the $\underline{P}$, $\underline{P}$ + 1, and $\underline{P}$ + 2 peaks.

	$\underline{P}$	$\underline{P}$ + 1	$\underline{P}$ + 2
CO	100	1.12	0.2
N_2	100	0.76	~0

We see, then, that if the molecular formula is known, the intensities of $\underline{P}$ + 1 and $\underline{P}$ + 2 peaks relative to $\underline{P}$ can be calculated. When the molecular formula is not known, an arithmetic trial and error approach or other calculation method (256) is used to find the best fit with the observed ratios. Fortunately, Beynon (257) has constructed a table of masses and isotopic abundance ratios for various combinations of carbon, hydrogen, nitrogen, and oxygen. For example the molecular ion and isotope peaks of

Sec. 4.2 MOLECULAR ABSORPTION OF ENERGY

a substance have the following m/e and relative abundance values:

m/e	relative abundance
178	100
179	15.4
180	1.1

There is a large number of compounds with a nominal molecular weight of 178, a few of which are listed here with their $\underline{P}$ + 1 and $\underline{P}$ + 2 intensity ratios, taken from Beynon's table.

Molecular formula (178)	Relative intensities	
	$\underline{P}$ + 1	$\underline{P}$ + 2
$C_{14}H_{10}$	15.29	1.09
$C_{13}H_{22}$	14.40	0.96
$C_{12}H_{20}N$	13.67	0.86
$C_{12}H_{18}O$	13.29	1.01
$C_{12}H_4NO$	13.45	1.03
$C_{11}H_{14}O_2$	12.19	1.08
$C_{10}H_{14}N_2O$	11.83	0.84
$C_8H_{18}O_4$	9.09	1.17

Clearly, the observed relative peak intensities agree best with those for the molecular formula $C_{14}H_{10}$. In this type of analysis, extreme care must be taken to assure that the experimental relative abundance measurements are recorded accurately. This is usually done by taking the average of several measurements.

Alternately, high-resolution spectrometers can routinely measure mass weights to six and seven significant figures. For such work, McLafferty (258) has composed a table of elemental compositions for various masses. A small segment at mass 58 is given here for inspection.

Elemental composition	Mass
$CHNO_2$	57.993
CH_2N_2O	58.017
CH_4N_3	58.041
$C_2H_2O_2$	58.005
C_2H_4NO	58.029
$C_2H_6N_2$	58.053
C_3H_6O	58.042
C_3H_8N	58.066
C_4H_{10}	58.078

Actually, one obtains elemental compositions not only for the molecular ion, but also for each fragment observed. By noting the manner in which known structures tend to break up, several empirical rules have been developed for guidance in reconstructing the molecular structure of an unknown from its spectrum. Some of these generalizations and fragmentation patterns are summarized here (259).

Some generalizations

1. Factors governing the fragmentation of an ion are:
   (a) The relative strengths of the various bonds in the decomposing ion; e.g., C-C in preference to C-H.
   (b) The relative stabilities of the potential ionic, neutral, or radical fragments which could be formed by the competing degradation pathways.
   (c) Steric conditions, e.g., concerted cleavage of bonds via cyclic transition states.
2. The nitrogen rule: The mass of a molecular ion must be even unless the molecule contains an odd number of nitrogen atoms. Nitrogen (and deuterium) is the only element commonly found in organic compounds with an even atomic mass but an odd-numbered covalency.

    For instance, the P, P + 1 and P + 2 peaks in the spectrum of N-ethylaniline have relative intensities of 100 (m/e 121), 9.2 (m/e 122), and 0.4 (m/e 123). A section of Beynon's table for nominal molecular weight of 121 is shown below.

	P + 1	P + 2
$C_{10}H$	10.82	0.53
C_9H_{13}	9.93	0.44
$C_8H_{11}N$	9.20	0.38
C_8H_9O	8.83	0.54
$C_7H_9N_2$	8.47	0.32
C_7H_7NO	8.10	0.49
$C_7H_5O_2$	7.72	0.66

The formulas with an even number of nitrogen atoms (or no nitrogen) do not conform to ordinary valence rules, and the best fit of $P + 1$ and $P + 2$ intensity ratios are for $C_8H_{11}N$.

Some fragmentation modes

1. **Simple C-C bond cleavages.** Since only ions* are detected, attention is focused on structural features which stabilize the ions, although of course, this is not the only factor. Thus, if other factors are equal, formation of the most stable carbocation directs the favored cleavage of a C-C bond.

$$C_6H_5\overset{+}{C}H_2 > CH_2=CH-\overset{+}{C}H_2 > R_3\overset{+}{C} > R_2\overset{+}{C}H > R\overset{+}{C}H_2 > \overset{+}{C}H_3$$

In a saturated chain, rupture occurs preferably at branched carbons with the positive charge remaining with the branch:

$$(C-C-\underset{C}{C}-C-C)^{+\cdot} \rightarrow C-C-\underset{C}{\overset{+}{C}} + \cdot C-C$$

Generally, the parent peak of branched chain hydrocarbons is absent or of very low intensity.

In an alkylbenzene, cleavage is most probable at the <u>beta</u> bond.

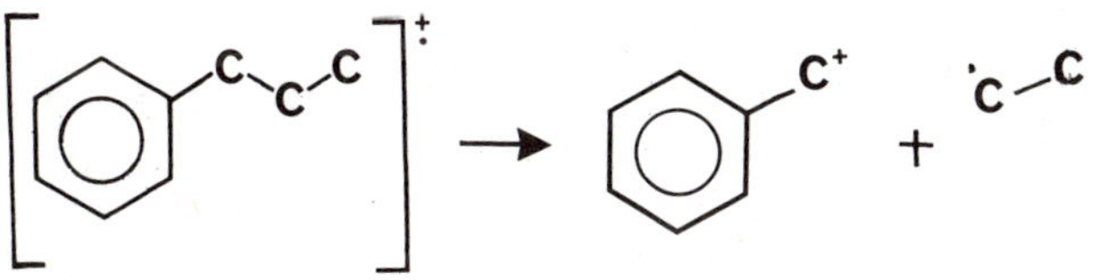

---

* Usually cations, but anions can also be detected and are just beginning to be studied in detail.

Saturated rings tend to lose side chains at the alpha bond.

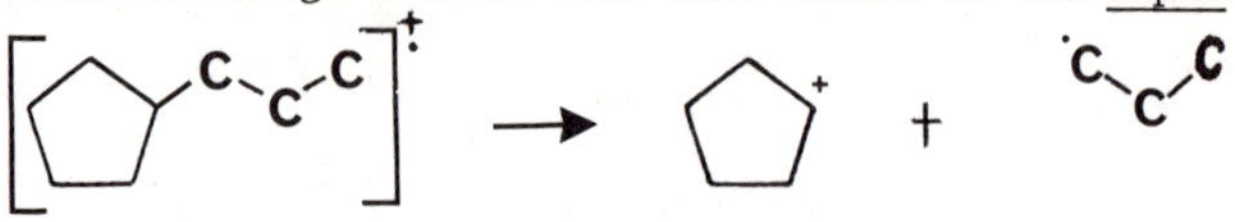

$$\left[\bigcirc\!\!-\!\!C\!-\!C\!-\!C\right]^{+\cdot} \longrightarrow \bigcirc^{+} + C\!-\!C\!-\!C$$

Fission in alkynes occurs mostly at alpha bonds to give an abundance of $(C_nH_{2n-3})^+$ ions.

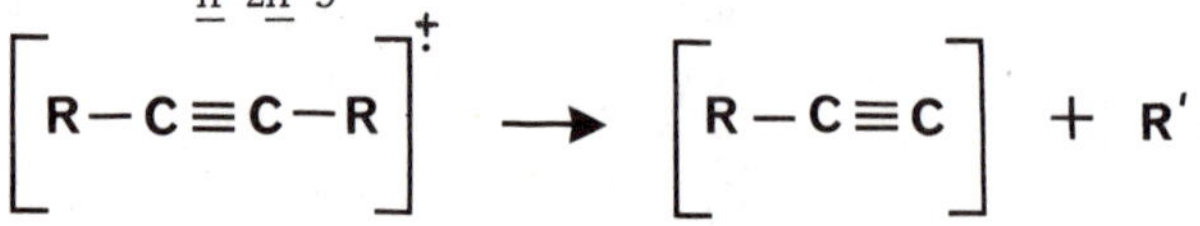

$$[R-C\equiv C-R]^{+\cdot} \longrightarrow [R-C\equiv C]^+ + R'$$

2. Fragmentations involving heteroatoms favor fission of the beta bond.

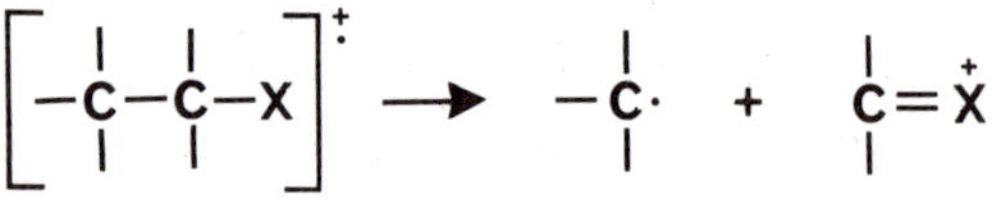

$$\left[-\!\!\underset{|}{\overset{|}{C}}\!-\!\underset{|}{\overset{|}{C}}\!-\!X\right]^{+\cdot} \longrightarrow -\!\!\underset{|}{\overset{|}{C}}\cdot + \underset{|}{\overset{|}{C}}\!=\!\overset{+}{X}$$

X = halogen, OH, OR, NR_2, SR

$$\left[CH_3-CH_2-\underset{OH}{\overset{|}{CH}}-(CH_2)_4-CH_3\right]^{+\cdot} \longrightarrow (C_5H_{11}\overset{+}{C}HOH) +$$
$$\underline{m/e}\ 101$$

$$CH_3CH_2^{+} + (C_2H_5\overset{+}{C}HOH)^+ + C_5H_{11}^{+} + \cdots$$
$$\underline{m/e}\ 59$$

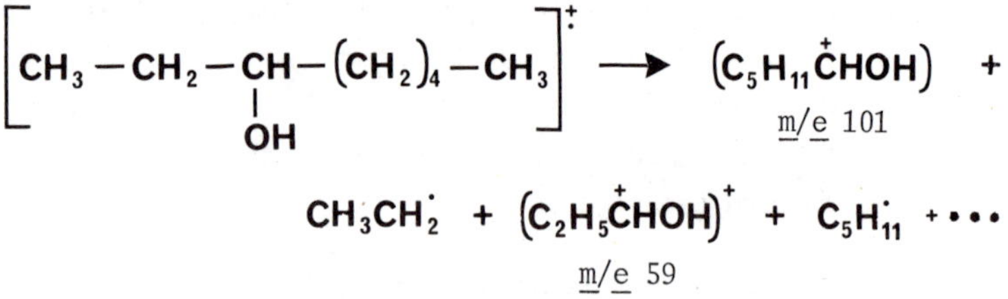

Thus, in the spectrum of 3-octanol there are peaks at $\underline{m/e}$ 101 and 59 corresponding to the loss of C_2H_5 and C_5H_{11} radicals from the molecular ion. The base peak at $\underline{m/e}$ 59 illustrates the preferential cleavage into (and loss of) larger alkyl fragments.

3. There is a tendency to eliminate stable small molecules. For example, the molecular ion of phthalic anhydride expels CO_2, followed by CO, then by C_2H_2.

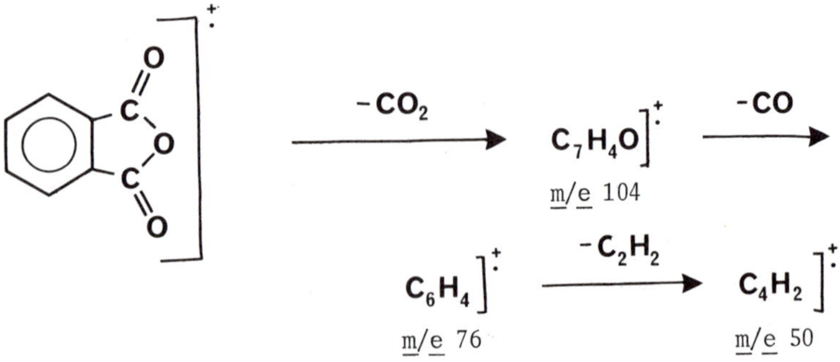

$$[\text{phthalic anhydride}]^{+\cdot} \xrightarrow{-CO_2} [C_7H_4O]^{+\cdot} \xrightarrow{-CO}$$
$$\underline{m/e}\ 104$$

$$[C_6H_4]^{+\cdot} \xrightarrow{-C_2H_2} [C_4H_2]^{+\cdot}$$
$$\underline{m/e}\ 76 \qquad\qquad\qquad \underline{m/e}\ 50$$

The elimination of water, ammonia, or hydrogen halides is prevalent.

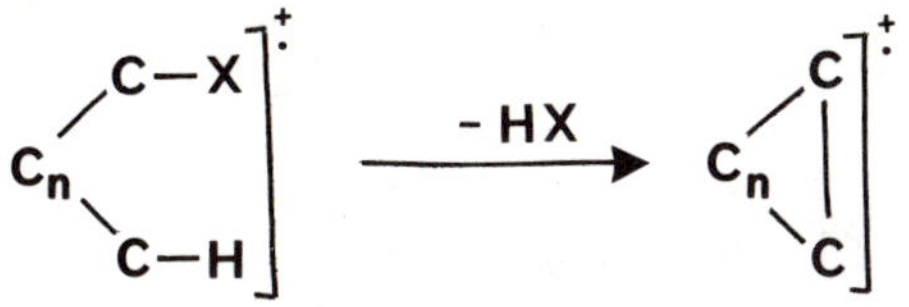

$$\left[C_n \begin{matrix} C-X \\ C-H \end{matrix} \right]^{\ddagger} \xrightarrow{-HX} \left[C_n \begin{matrix} C \\ \| \\ C \end{matrix} \right]^{\ddagger}$$

Owing to relative bond strengths, iodides and bromides tend to yield halide halogen atoms,

$$[R-C-C-X]^{\ddagger} \longrightarrow [R-C-C]^{\ddagger} + \cdot X$$

whereas fluorides and chlorides lose HX.

$$(R-\underset{H}{\overset{|}{C}}-C-X)^{\ddagger} \longrightarrow (R-C=C)^{\ddagger} + HX$$

Further fragmentation then resembles that of an alkane or alkene.

Alcohols and ethers alternately fractionate by modes 2 or 3:

$$(R-CH_2-CH_2-OR')^{\ddagger} \quad \text{or} \quad \begin{matrix} \nearrow (R-CH=CH_2)^{\ddagger} + R'OH \\ \searrow (RCH_2CH_2O)^{\ddagger} + \cdot R' \end{matrix}$$

Primary and secondary alcohols generally lose water, and the spectra resemble those from the corresponding alkenes. Tertiary alcohols usually eliminate the larger alkyl group:

$$R_3COH)^{\ddagger} \longrightarrow R_2COH)^{+} + \cdot R$$

Cyclic alcohols often undergo a series of rearrangements leading to a characteristic peak at m/e 57:

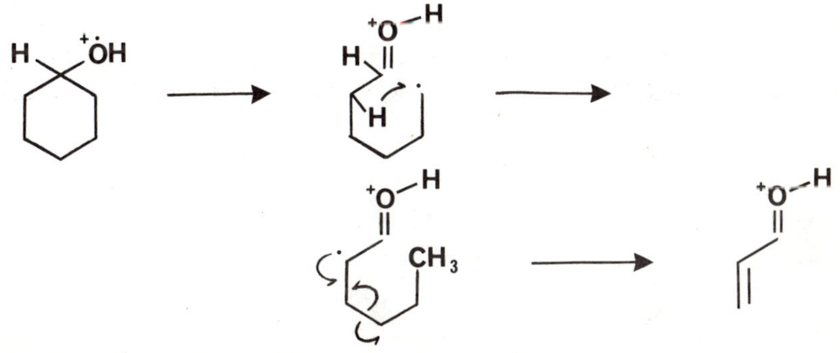

4. Rearrangement via six-membered cyclic transition states is prevalent:

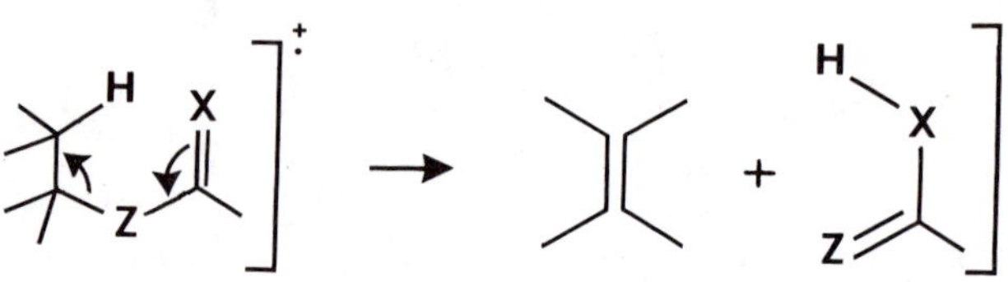

Known generally as the McLafferty rearrangement, it accounts for major peaks in the spectra of many classes of compounds such as olefins, carbonyl and acyl compounds, etc. Thus, cracking of olefins at the site <u>beta</u> to the double bond, i.e., <u>allylic fission</u>, is most likely, which produces ions of general formula $(C_n H_{2n-1})^{+\cdot}$.

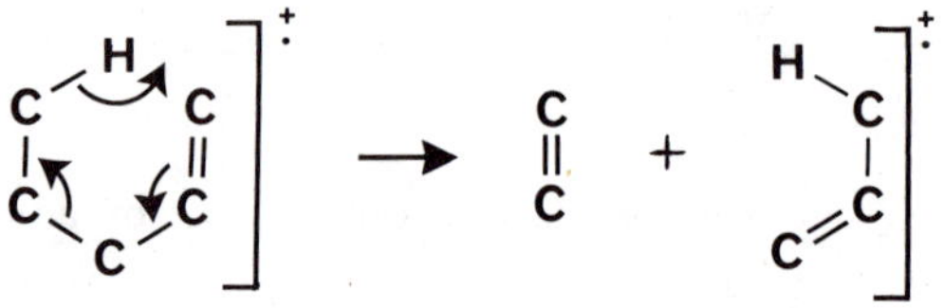

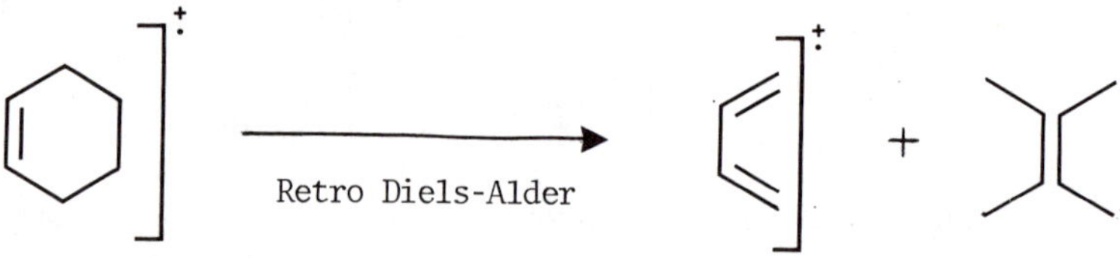

Ethyl esters give large peaks at m/e 88,

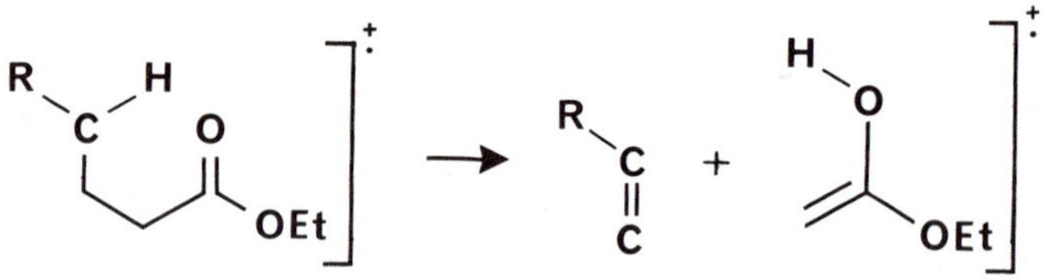

and methyl ketones give peaks at m/e 58. The mechanism of this fragmentation has been supported by using deuterated analogs.

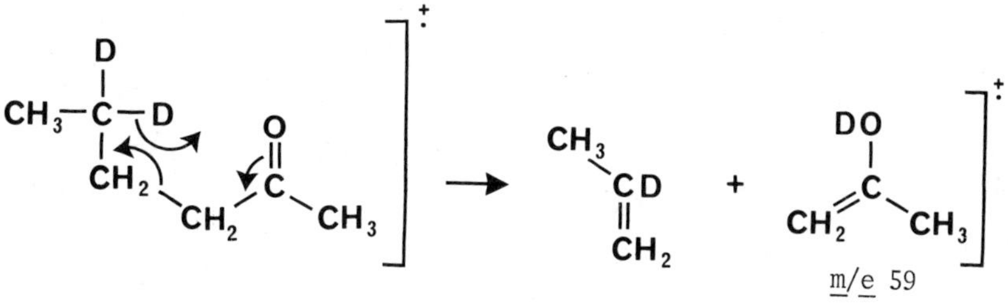

Sec. 4.2 MOLECULAR ABSORPTION OF ENERGY

[Scheme showing α-cleavage mechanism of a deuterated ketone giving CH$_3$-CD=CH$_2$ and the enol cation at m/e 59]

Thus, aliphatic ketones usually undergo α cleavage to give the two sets of alkyl and acyl ions:

$$R\text{-}CO\text{-}R' \rightarrow \underline{RCO^+ + R'} + \underline{R + {}^+COR'}$$

In general, fragments resulting from a rearrangement can be recognized by the simple observation that fragments formed by the cleavage of only one bond from a molecule of <u>even</u> mass will always be of <u>odd</u> mass. When the mass of such a fragment is <u>even</u>, it is the result of a rearrangement. This characteristic cleavage of ketones has facilitated the identification of surface lipids from plant and insect sources (260). Another example is as follows.

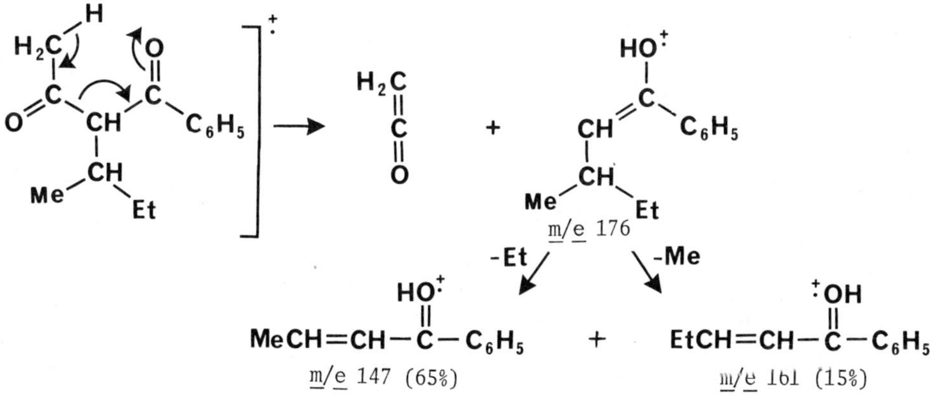

[Scheme showing McLafferty rearrangement of a ketone at m/e 176, losing Et to give MeCH=CH-C(OH)=C$_6$H$_5$ at m/e 147 (65%), and losing Me to give EtCH=CH-C(OH)=C$_6$H$_5$ at m/e 161 (15%)]

As might be anticipated, the m/e 147 peak is of greater intensity than that at m/e 161, since the former represents the loss of a larger fragment.

<u>o</u>-Alkylbenzoic acids readily lose water, unlike their <u>m</u>- or <u>p</u>-isomers (261).

158   Ch. 4 INTERNAL ENERGIES

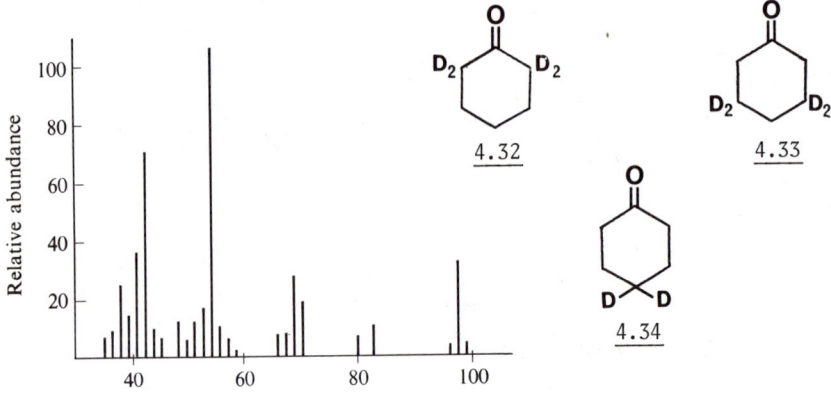

Compound	Relative intensity (base = 100%)	
	Ion: M^+	$(M - H_2O)^+$
Benzoic acid	86	1
o-Methylbenzoic acid	84	100
m-Methylbenzoic acid	87	4
p-Methylbenzoic acid	97	2
2,4,6-Trimethylbenzoic acid	87	100

The mass spectrometric fragmentation of cyclohexanone has been studied extensively. Its spectrum is abbreviated in Figure 4.9.

Figure 4.9. Bar mass spectrum of cyclohexanone.

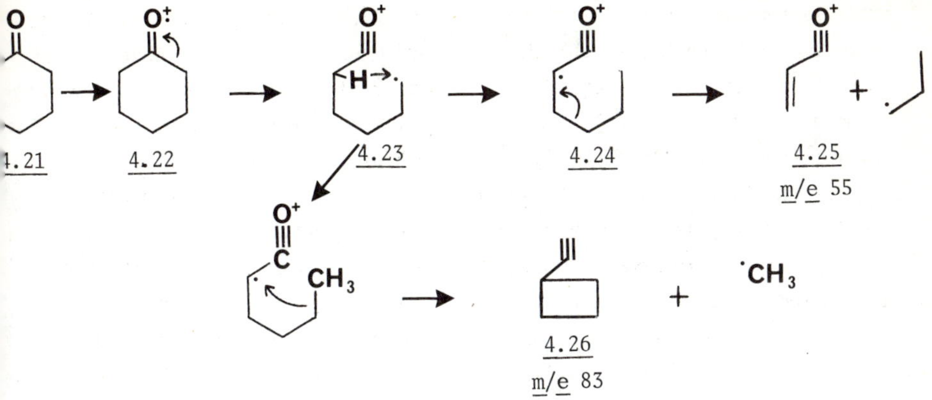

Sec. 4.2 MOLECULAR ABSORPTION OF ENERGY 159

4.21 4.22 4.23 4.24 4.25 m/e 55

4.26 m/e 83

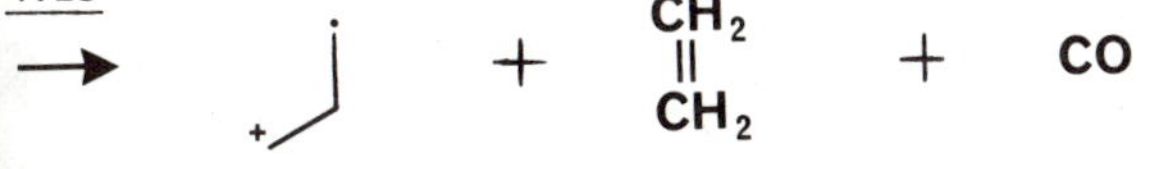

4.23 →

4.27 m/e 42

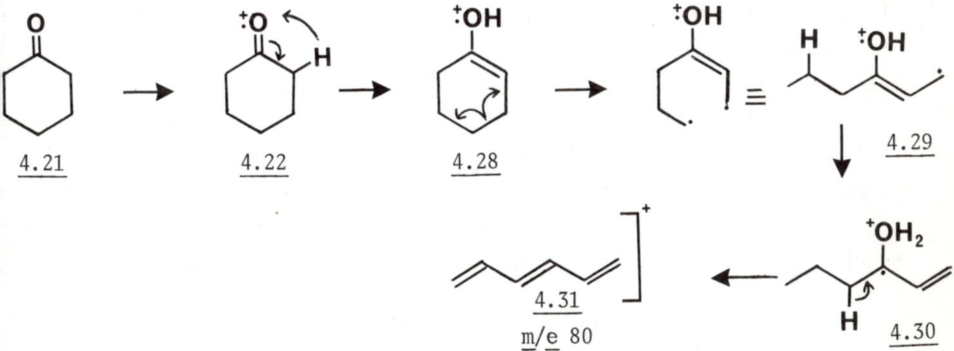

4.21 4.22 4.28 4.29

4.31 m/e 80 4.30

The base peak in the spectrum of the ketone 4.21 (262) is at m/e 55 for ion 4.25. Deuterated analogs 4.32-4.34 show shifts of this peak to m/e 56 and m/e 57, while 4.34 does not shift. The loss of a methyl group (M-15, ion 4.26) is due to C_1-C_2 bond cleavage, transfer of a hydrogen from C_6 to C_1, and expulsion of the methyl radical. A shift of three mass units in the spectrum of the 2,2,6,6-d_4 analog 4.32 is in agreement with this process. A hydrogen fragment containing C_2, C_3, and C_4 which is the result of expulsion of CO and CH_2=CH_2 is the major contributor to 4.27, m/e 42. A minimum of four

bonds is ruptured in the elimination of water (M-18, 4.31) from cyclohexanone
Deuterium labeling has established that hydrogen expelled in a neutral fragment comes from (C_6) and C_3 (C_5) in almost equal amounts. The mechanism proposed by Djerassi (263) predicts that allylic homolytic cleavage (4.28 to 4.29) is energetically favored, as is the hydrogen rearrangement (4.29 to 4.30). The last step (4.30 to 4.31) involves a hydride rearrangement with simultaneous expulsion of water to form the conjugated triene (4.31, m/e 80).

Likely rearrangements must always be kept in mind when interpreting a spectrum. Thus toluene, alkylbenzenes, and even t-butylbenzenes, commonly have a base peak at m/e 91 which is thought to be for the tropylium ion by rearrangement of the benzyl ion.

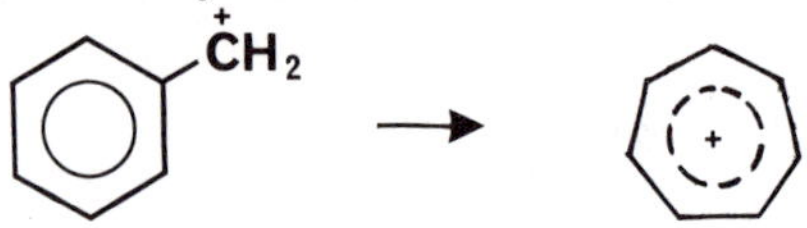

This shows the importance of fragment-ion stability to the ion intensity pattern. Indeed, the mass spectra of toluene and cycloheptatriene are identical.

Three examples of structure determination from mass spectra follow.*
Elegant solutions to a number of such problems are given by Beynon, Saunders, and Williams (264).

Example 1. The mass spectrum of an unknown ketone shows relatively abundant peaks at m/e 128 (M$^+$), 86, 85, 71, 57 (base peak), 43, 29, and 28 (265). The molecular ion at m/e 128 corresponds to a molecular formula of $C_8H_{16}O$. The peak at m/e 86 is 42 mass units less than that of the molecular ion, and since it is even, suggests the possibility that the ketone is one which underwent a McLafferty rearrangement.

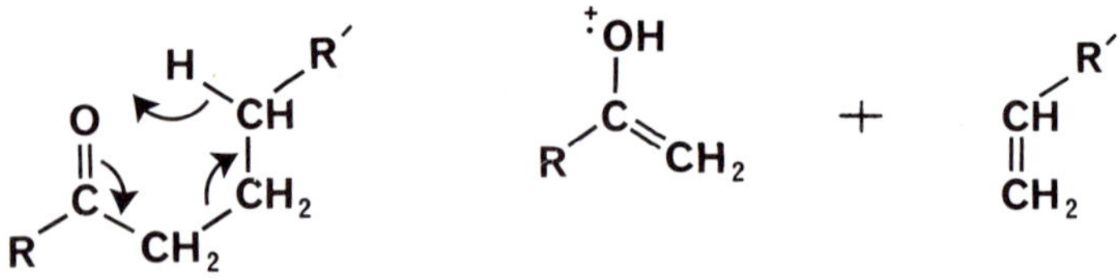

$$M^+ - 42$$

---

* The author is indebted to Professor Costello L. Brown of the California State University, Los Angeles, for pointing out these examples.

Since the eliminated neutral molecule corresponds to 42 mass units, R' must be CH_3. The peak at m/e 85 is 43 mass units less than that of the molecular ion, suggesting the loss of fragments C_3H_7 or C_2H_3O. Ketones usually exhibit α cleavage to give the respective alkyl and acyl ions, and the neutral fragment by a McLafferty rearrangement is $-CH_2CH_2CH_3$, hence speculation on the structure at this time would be

$$C_3H_7-CO-CH_2CH_2CH_2CH_3.$$

4.35

For the McLafferty rearrangement to occur, there must be a three-carbon chain attached to C=O, and since there is no M-29 peak (m/e 99), which would result from such a rearrangement of the C_3H_7 group if it were n-propyl, then C_3H_7 in this case must be isopropyl. The compound is thus thought to be

$$(CH_3)_2CH-\overset{\overset{O}{\|}}{C}-CH_2CH_2CH_2CH_3.$$

Checking the other peaks for confirmation, the peak at m/e 71 (M-57) and the base peak m/e 57 (M-71) represent the expected α-cleavage products from 4.35.

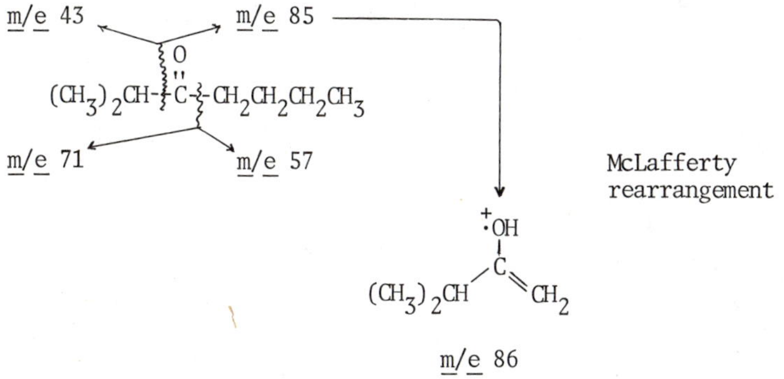

$$(CH_3)_2CH-\overset{\overset{O}{\|}}{C}-CH_2CH_2CH_2CH_3$$

m/e 43 ← → m/e 85
m/e 71 ← → m/e 57

McLafferty rearrangement

$$(CH_3)_2CH\diagdown\overset{\overset{+\cdot OH}{|}}{C}\diagup CH_2$$

m/e 86

Note also:

$$[CH_3-CH-CH_3]^{+\cdot} \longrightarrow [CH_3-CH]^{+\cdot} + \cdot CH_3$$
m/e 43 m/e 28

$$[CH_3CH_2CH_2CH_2]^{+\cdot} \longrightarrow [CH_3CH_2]^{+\cdot} + CH_2=CH_2$$
m/e 57 m/c 29

<u>Example 2</u>. The mass spectrum of an unknown ketone contains a molecular ion

at m/e 162 and major peaks at m/e 106, 105, 91, 77, 71, and 43. The m/e 162 molecular ion corresponds to a molecular formula of $C_{10}H_{12}O$. A large peak at m/e 77 usually indicates the presence of an aromatic ring. Unsaturation is also indicated by the C:H ratio of the molecular formula. The m/e 91 ion could represent a benzyl group because this ion is always present in monosubstituted alkylbenzene derivatives. Thus, we have these two fragments:

m/e 77 $[C_6H_6]^{+\cdot}$ m/e 91 $[C_6H_5-CH_2]^{+\cdot}$

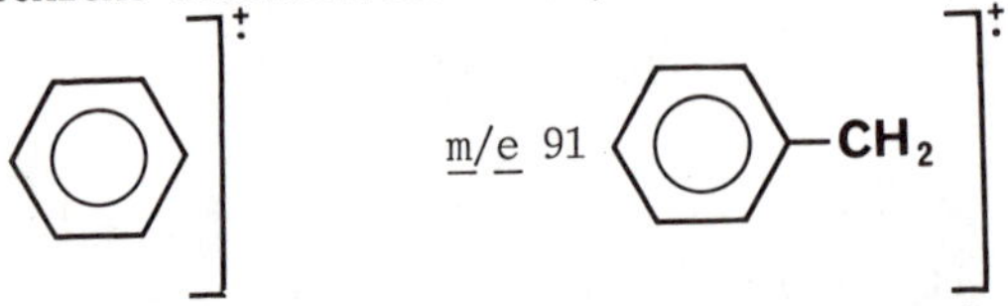

We saw earlier that ketones usually undergo α cleavage to give alkyl ions and we can surmise here that the m/e 43 ion is a propyl group from a ketone. If we add 43 to 91, to give 134, and subtract this mass from that of the molecular ion, we are left with only 28 mass units, supposedly CO. Accordingly, a possible structure for the ketone is 4.36.

$$4.36 \quad C_3H_7-\overset{O}{\underset{\parallel}{C}}-CH_2-C_6H_5$$

$$C_3H_7^+ \qquad C_3H_7C\equiv O^+ \qquad CH_2-C_6H_5$$
$$\text{m/e 43} \qquad \text{m/e 71} \qquad \text{m/e 91}$$

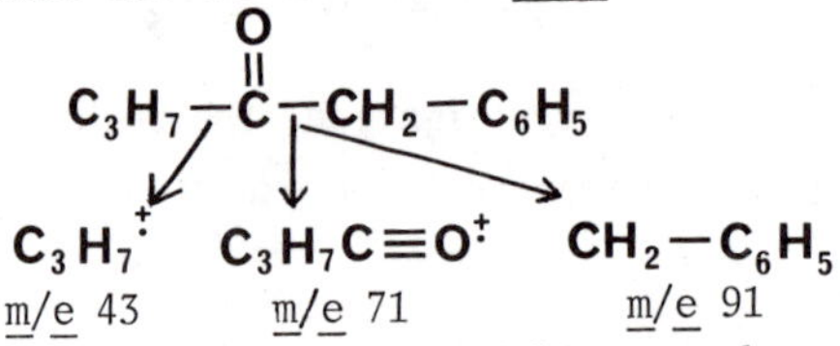

Again, the C_3H_7 group is probably isopropyl because the n-propyl group would give McLafferty rearrangement fragments not present in this spectrum.

It is not clear how the m/e 105 and 106 ions were formed. For example, the m/e 105 ion is probably $C_6H_5C\equiv O^+$, but its origin is not well understood.

Example 3. An example involving several metastable transitions is as follows (266). The mass spectrum of a pyrrolizidine alkaloid, 7-angeloylheliotridene, has peaks at m/e 237 (M^+), 210, 154, 137, 124, 111, 106, 94, and 80. Its chemical and other spectral properties suggest structure 4.37.

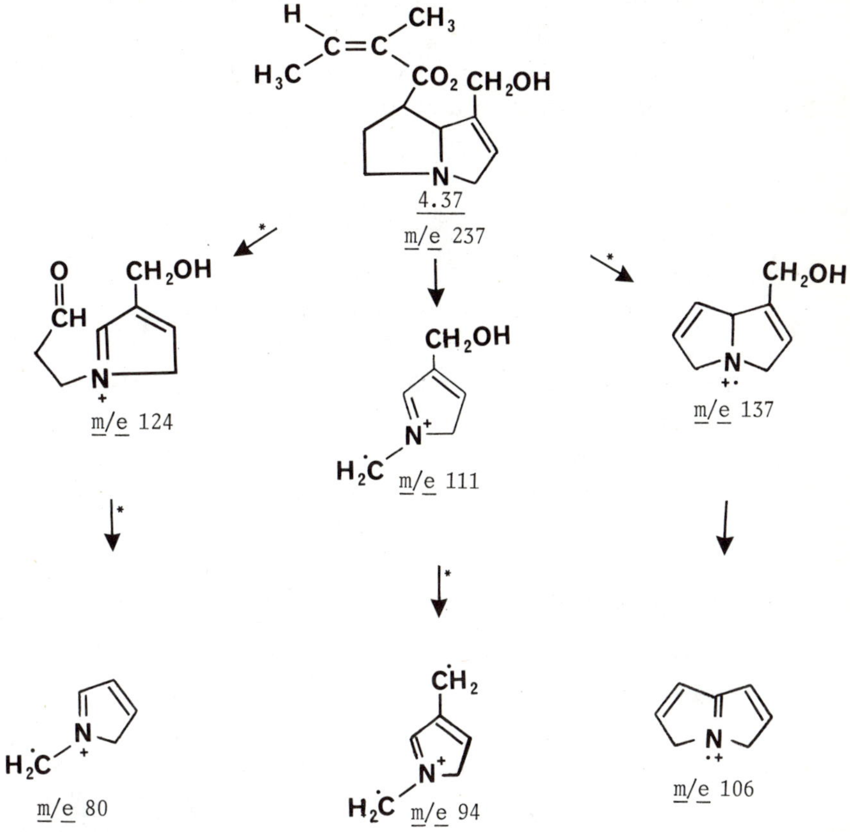

Sec. 4.2 MOLECULAR ABSORPTION OF ENERGY   163

[Structure 4.37, m/e 237: pyrrolizine with substituents including C=C(H)(CH$_3$)/C(CH$_3$), CO$_2$CH$_2$OH, and CH$_2$OH]

Fragmentation pathways lead to:

- m/e 124: pyrrole-N-CH$_2$CH$_2$-CHO with CH$_2$OH substituent
- m/e 111: pyrrole with CH$_2$OH substituent, N-CH$_2$·
- m/e 137: pyrrolizine with CH$_2$OH substituent
- m/e 80: N-methyl pyrrole cation (H$_2$C-N)
- m/e 94: pyrrole with ĊH$_2$ substituent, N-CH$_2$·
- m/e 106: pyrrolizine

The empirical formulas of all of the ions shown were determined by high resolution mass spectrometry. The fragmentation pathways giving each of the ions above can be ascertained by the identification of metastable peaks. Those marked with an asterisk indicate that a metastable peak was found in the spectrum for that process. For example, the metastable ion for m/e 137 from the molecular ion m/e 237 would have the mass

$$m^* = \frac{(137)^2}{(237)} = 79.19.$$

The fact that a metastable peak for this process was found in the spectrum indicates in general that the m/e 137 ion is derived directly from the molecular ion in a one-step process. This means that this loss of 100 mass units from the parent ion cannot be accomplished by the elimination of sustituents at several different positions in the molecule because the possibility of several cleavages occurring in a one-step process is extremely small. A 1,2-elimination of the anhydride group fits the data and could easily occur by a one-step concerted process. The other fragments can be accounted for in a similar manner with the aid of metastable ions.

QUESTIONS

1. Substituted butylbenzenes display two major peaks in their mass spectra at M-42 and M-43 mass units. Propose a likely mechanism for these two peaks.

2. Predict the relative peak heights of the acetyl cation CH_3CO^+ in the mass spectra of p-X-C_6H_4-$COCH_3$ for X = H, NH_2, and NO_2, and give the basis of your prediction.

3. α-Carotene, but not β-carotene, has a prominent peak in its mass spectrum at M-56. How would you account for this peak?

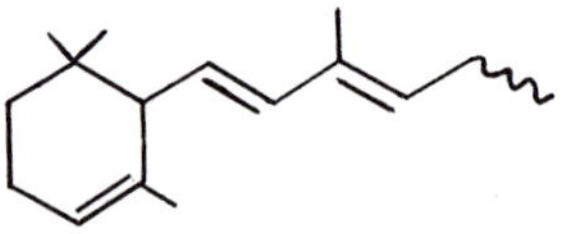

Fragment of α-carotene molecule

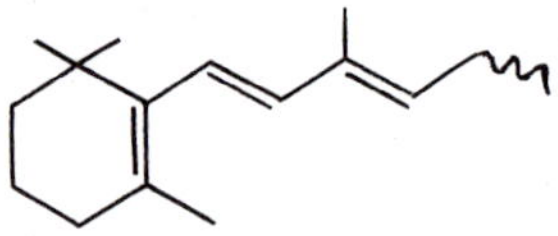

Fragment of β-carotene molecule

4. In the mass spectrum of flavone there are prominent peaks at m/e = 222, 194, 120, 105, 102, 97, and 92. Give reasonable mechanisms for the formation of fragments corresponding to the peaks

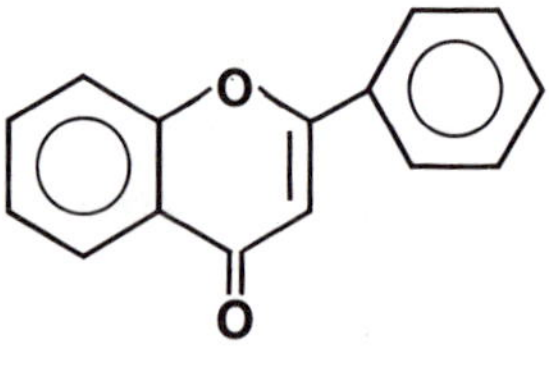

Answers

1. 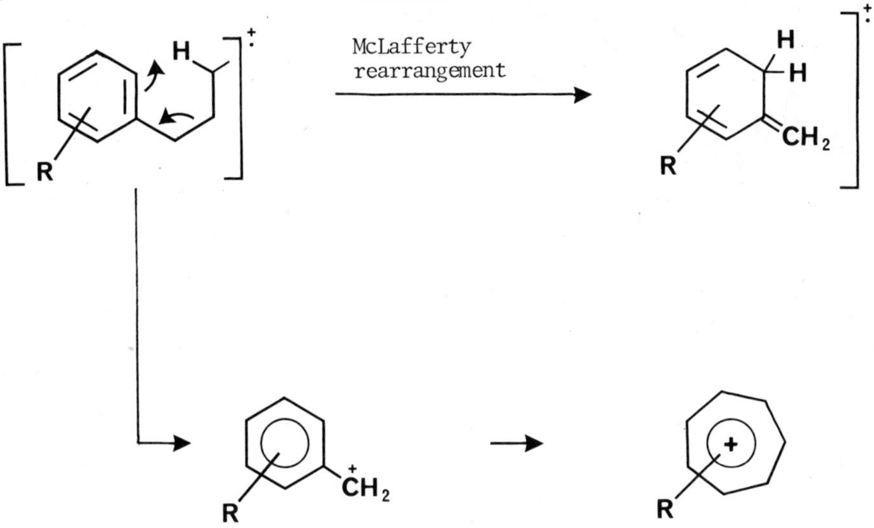 [structural diagram: McLafferty rearrangement of alkylbenzene radical cation to form methylenecyclohexadiene radical cation and CH₂ benzyl cation rearranging to tropylium cation]

(R. Nicoletti and D. A. Lightner, *Tetra. Letters* 4553 (1968))

2. Relative $(CH_3CO)^+$ abundances should be in the order $X = NO_2 > H > NH_2$ because this order relfects the descending order of electron-withdrawing ability of the substituted phenyl group.

$$[X\text{-}C_6H_4\text{-}COCH_3]^{+\cdot} \rightarrow [X\text{-}C_6H_4]^{\cdot} + CH_3CO^+$$

The experimental values are 4.5, 1.00, 0.26 (M. M. Bursey and F. W. McLafferty, *J. Am. Chem. Soc.* **89**:1 (1967)).

3. A retro-Diels-Alder reaction would account for the M-56 peak.

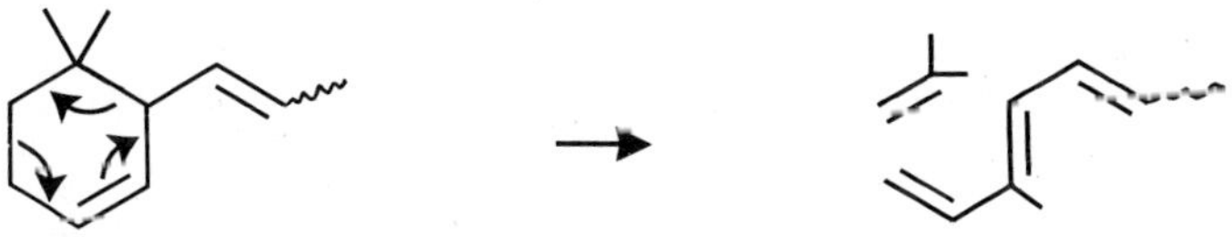

[structural diagram showing retro-Diels-Alder fragmentation yielding M-56]

M-56

4.

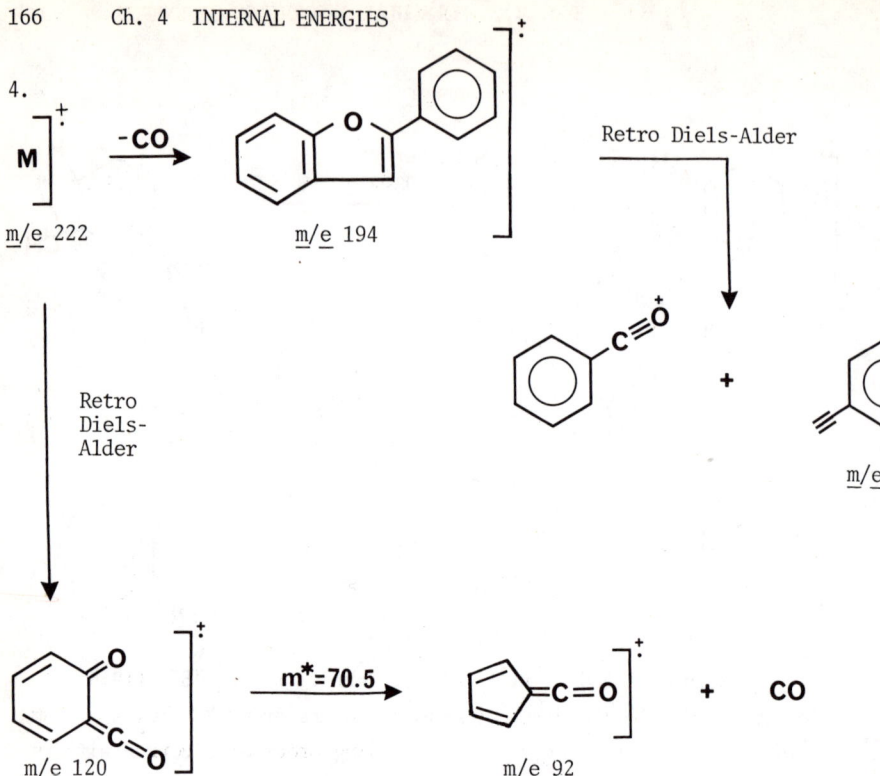

The peak at m/e 97 could be a doubly-charged fragment from the initial extrusion of CO.

REFERENCES

Part I

1. R. J. Gillespie, *J. Chem. Educ.*, 51: 367 (1974); R. A. Firestone, *J. Org. Chem.*, 34: 2621 (1969); W. F. Luder, *J. Chem. Educ.*, 44: 206, 269 (1967); J. W. Linnett, *The Electronic Structure of Matter, A New Approach*, New York: Wiley, 1964; H. A. Bent, *J. Chem. Educ.*, 40: 446 (1963).

2. N. L. Allinger, M. T. Tribble, M. A. Miller, and D. H. Wertz, *J. Am. Chem. Soc.*, 93: 1637 (1971); R. C. Bingham and P. v. R. Schleyer, *J. Am. Chem. Soc.*, 93: 3189 (1971). See also the "consistent force field" method, S. Lifson, *et al.*, *J. Am. Chem. Soc.*, 95: 4121 (1973).

3. O. Sinanoglu and K. B. Wiberg, eds., *Sigma Molecular Orbital Theory*, New Haven, Conn.: Yale Press, (1973); W. C. Herndon, *Progr. Phys. Org. Chem.*, 9: 99 (1972); J. A. Pople and D. L. Beveridge, *Approximate Molecular Orbital Theory*, New York: McGraw-Hill, (1970); M. J. S. Dewar, *The Molecular Orbital Theory of Organic Chemistry*, New York: McGraw-Hill, 1969; L. Salem, *The Molecular Orbital Theory of Conjugated Systems*, New York: Benjamin, 1966; K. Higasi, H. Baba, and A. Rembraum, *Quantum Organic Chemistry*, New York: Interscience, (1965); J. D. Roberts, *Notes on Molecular Orbital Calculations*, New York: Benjamin, (1961); A. Streitwieser, Jr., *Molecular Orbital Theory for Organic Chemists*, New York: Wiley, 1961; R. Daudel, R. Lefebvre, and C. Moser, *Quantum Chemistry*, New York: Interscience, 1959.

4. W. L. Jorgensen and L. Salem, *The Organic Chemist's Book of Orbitals*, New York: Academic, 1973.

5. M. Orchin and H. H. Jaffe, *The Importance of Antibonding Orbitals*, Boston: Houghton Mifflin, 1967.

6. W. v. E. Doering and G. H. Beasley, *Tetrahedron*, 29: 2231 (1973).

7. J. Stals, *Rev. Pure and Appl. Chem.*, 20: 1 (1970).

8. P. J. Wheatley, *The Determination of Molecular Structure*, New York: Oxford Press, 1959. For a table of bond lengths and angles, see "Tables of Interatomic Distances and Configurations in Molecules and Ions," *Chem. Soc.* (London), *Spec. Publ.*, 11 (1958); "Interatomic Distances Supplement," *Chem. Soc.* (London) *Spec. Publ.* 18 (1965); *Molecular Structures and Dimensions*, Vol. A1, Pittsburgh, Pa.: Polycrystal Book Service, 1973.

9. T. G. Traylor, *J. Am. Chem. Soc.*, 90: 74 (1968); L. S. Bartell *Tetrahedron*, 17: 177 (1962).

10. M. G. Brown, *Trans. Farad. Soc.*, 55: 694 (1959).

11. I. Karle and J. Karle, *J. Chem. Phys.*, 18: 963 (1950).

12. S. W. Benson, *J. Chem. Educ.*, 42: 502 (1965).

13. J. D. Cox and G. Pilcher, *Thermochemistry of Organic and Organometallic Compounds*, New York: Academic, (1970).

14. J. M. Hay, *J. Chem. Soc.*, (B) 45 (1970).

15. J.A. Kerr, *Chem. Revs.* 66: 465 (1966).
16. For a recent discussion, see R.T. Sanderson, *Chemical Bonds and Bond Energy*, New York: Academic, 1971.
17. W.C. Danen, T.J. Tipson, and D.G. Saunders, *J. Am. Chem. Soc.* 93: 5186 (1971).
18. C. Rüchardt, *Angew. Chem. Int. Ed.* 9: 830 (1970).
19. J.L. Brauman, J.M. Rivers, and L.K. Blair, *J. Am. Chem. Soc.* 93: 3914 (1971).
20. L.N. Ferguson, *The Modern Structural Theory of Organic Chemistry*, Englewood Cliffs, N.J.: Prentice-Hall, 1963, p. 48; E.G. Lovering and K.J. Laidler, *Canad. J. Chem.* 38: 2367 (1960).
21. J.D. Cox, *Tetrahedron* 19: 1179 (1963).
22. A.G. Robiette, R.H. Bradley, and P.N. Brier, *Chem. Commun.* 1567 (1971).
23. M.L. Huggins introduced the concept of an H bond. He has given an historical development of the nature of H bond forces in *Angew. Chem. Int. Ed.* 10: 147 (1971).
24. For a recent review on the theory of H bonds, see P.A. Kollman and L.C. Allen, *Chem. Revs.* 73: 283 (1973).
25. C.A. Coulson in *Hydrogen Bonding*, eds. D. Hadzi and H.W. Thompson, New York: Pergamon, 1959, p. 339.
26. S.A. Harrell and D.H. McDaniel, *J. Am. Chem. Soc.* 86: 4497 (1964).
27. G. Ferraris, D.W. Jones, and J. Yerkess, *Chem. Commun.* 1566 (1971).
28. H.M. Pickett, *J. Am. Chem. Soc.* 95: 1770 (1973).
29. L. Joris and P.v.R. Schleyer, *J. Am. Chem. Soc.* 90: 4599 (1968); C.H. Robinson and L. Milewich, *J. Org. Chem.* 36: 1814 (1971).
30. M.E. Emerson, E. Grunwald, M.L. Kaplan, and R.A. Kromhout, *J. Am. Chem. Soc.* 82: 6312 (1960).
31. W.B. Pearson, R.E. Erickson, and R.E. Buckles, *J. Am. Chem. Soc.* 82: 29 (1960).
32. D. Bloser, J. Murphy, and J.N. Spencer, *Canad. J. Chem.* 49: 3913 (1971).
33. K.S. Pitzer and E. Catalano, *J. Am. Chem. Soc.* 78: 4844 (1956).
34. P.D. Bartlett, C.V. Goebel, and W.P. Weber, *J. Am. Chem. Soc.* 91: 7425 (1969); P.A. Kollman, J.F. Liebman, and L.C. Allen, *J. Am. Chem. Soc.* 92: 1142 (1970).
35. T.L. Brown, *Adv. in Organometallic Chem.* 3: 365 (1965).
36. R.A. Nyquist, *Spectrochim. Acta* 19: 1655 (1963); M.M. Maguire and R. West *Spectrochim. Acta* 17: 369 (1961).
37. R. West and C.S. Kraihanzel, *J. Am. Chem. Soc.* 83: 765 (1961).
38. H. Fales, *J. Am. Chem. Soc.* 88: 5544 (1966); M.D. Taylor, *et al.*, *J. Am. Chem. Soc.* 78: 2950 (1956); P.M.G. Bavin and W.J. Canady, *Canad. J. Chem.* 35: 1555 (1957).
39. J. Dabrowski, Z. Swistun, and U. Dabrowska, *Tetrahedron* 29: 2257, 2261 (1973).

40. L.N. Ferguson, *J. Chem. Educ.* 33: 267 (1956); I.D. Sadekov, V.I. Menkin, and A.E. Lutskii, *Russ. Chem. Revs.* 39: 179 (1970); G.C. Pimentel and A.L. McClellan, *The Hydrogen Bond*, San Francisco: Freeman, 1960.
41. M. Calvin, J. Hermans, Jr., and H.A. Scheraga, *J. Am. Chem. Soc.* 81: 5048 (1959).
42. G. Dahlgren, Jr., and F.A. Long, *J. Am. Chem. Soc.* 82: 1303 (1960).
43. O.T. Benfey and J.W. Mills, *J. Am. Chem. Soc.* 93: 922 (1971).
44. R.W. Alder, P.S. Bowman, W.R.S. Steele, and D.R. Winterman, *Chem. Commun.* 723 (1968).
45. D.G. Farnum, J. Chickos, and P.E. Thurston, *J. Am. Chem. Soc.* 88: 3075 (1966).
46. J.L. Haslam, E.M. Eyring, W.W. Epstein, G.A. Christiansen, and M.H. Miles, *J. Am. Chem. Soc.* 87: 1 (1965).
47. L.N. Ferguson and I. Kelly, *J. Am. Chem. Soc.* 73: 3707 (1951).
48. A. Bondi and D.J. Simkin, *A. I. ch. E. Journal* 3: 473 (1957); R.W. Taft, Jr., and H.H. Sisler, *J. Chem. Educ.* 24: 175 (1947).
49. W. Hückel, *Theoretical Principles of Organic Chemistry*, Vol. II, Princeton, N.J.: Van Nostrand-Elsevier, 1958, p. 318.
50. C.H. Giles, *et al.*, *J. Chem. Soc.* 72 (1956) and earlier papers.
51. L.A.K. Stavely and P.F. Taylor, *J. Chem. Soc.* 200 (1956).
52. M.J. Astle and W.V. McConnell, *J. Am. Chem. Soc.* 65: 35 (1943).
53. A.T. Shulgin and H.O. Kerlinger, *J. Org. Chem.* 25: 2037 (1960).
54. O. Bastiansen, *Acta Chem. Scand.* 3: 415 (1949).
55. E.J. O'Connell, Jr., *J. Am. Chem. Soc.* 90: 6550 (1968).
56. J.T. Bursey, M.M. Bursey, and D.G.I. Kingston, *Chem. Revs.* 73: 181 (1973).
56a. C.H. Lochmüller and R.W. Souter, *American Laboratory* Nov. 1973, p. 25.
57. D.P. Ridge and J.L. Beauchamp, *J. Am. Chem. Soc.* 93: 5925 (1971).
58. A. Nickon, T. Nishida, and J. Frank, *J. Org. Chem.* 36: 1075 (1971).
59. L. Pauling, *Nature of the Chemical Bond*, 3rd ed., Ithaca, N.Y.: Cornell U. Press, 1960.
60. J. Booth, E. Boyland, and S.F.D. Orr, *J. Chem. Soc.* 598 (1954).
61. For a source of cyclodextrins and a bibliography of uses of their inclusion complexes, write: Corn Products Development, International Plaza, Englewood Cliffs, N.J., 07632.
62. G.A. Jeffrey, *Accts. Chem. Res.* 2: 344 (1969).
63. V.M. Bhatnager, *J. Chem. Educ.* 40: 646 (1963).
64. I.T. Harrison, *Chem. Commun.* 231 (1972).
65. E.H. White, *J. Am. Chem. Soc.* 77: 6081 (1955).
66. F.D. Cramer, *Rev. Pure and Appl. Chem.* 5: 143 (1955).
67. M.M. Hagan, *J. Chem. Educ.* 40: 643 (1963).
68. W.D. Schaeffer, W.S. Dorsey, D.A. Skinner, and C.G. Christian, *J. Am. Chem. Soc.* 79: 5870 (1957); F.V. Williams, *J. Am. Chem. Soc.* 79: 5876 (1957).

69. A.C.D. Newman and H.M. Powell, *J. Chem. Soc.* 3747 (1952).
70. O.H. Griffith and A.L. Kwiram, *J. Am. Chem. Soc.* 86: 3937 (1964).
71. D.L. Vander Jagt, F.L. Killian, and M.L. Bender, *J. Am. Chem. Soc.* 92: 1017 (1970); R. Breslow and L.E. Overman, *J. Am. Chem. Soc.* 92: 1075 (1970); F. Cramer and G. Mackensen, *Angew. Chem. Int. Ed.* 5: 601 (1966).
72. R. Breslow and P. Campbell, *J. Am. Chem. Soc.* 91: 3085 (1969).
73. C.J. Pedersen, *J. Am. Chem. Soc.* 89: 7017 (1967); D.J. Sam and H.E. Simmons, *J. Am. Chem. Soc.* 94: 4024 (1972).
74. R.C. Helgeson, K. Koga, J.M. Timka, and D.J. Cram, *J. Am. Chem. Soc.* 95: 3021 (1973).
75. H.O. Pritchard and H.A. Skinner, *Chem. Revs.* 55: 745 (1955); J. Hinze, M.A. Whitehead, and H.H. Jaffe, *J. Am. Chem. Soc.* 85: 148 (1963).
76. L. Pauling, *Nature of the Chemical Bond*, 3rd ed., Ithaca, N.Y.: Cornell U. Press, 1960, p. 88.
77. A.L. Allred, *J. Inorg. Nucl. Chem.* 17: 215 (1961).
78. E.G. Rochow and A.L. Allred, *J. Inorg. Nucl. Chem.* 5: 264, 269 (1958).
79. R.T. Sanderson, *J. Chem. Educ.* 29: 539 (1952); R.T. Sanderson, *J. Am. Chem. Soc.* 74: 4792 (1952).
80. E.J. Little, Jr., and M.M. Jones, *J. Chem. Educ.* 37: 231 (1960).
81. F.E. Rogers and R.J. Rapiejko, *J. Am. Chem. Soc.* 93: 4596 (1971).
82. P.R. Wells, *Progr. Phys. Org. Chem.* 6: 111 (1968).
83. M.S. Kharasch, *et al.*, *J. Am. Chem. Soc.* 48: 3130 (1926); M.S. Kharasch, *et al.*, *J.Org. Chem.* 3: 405 (1938).
84. H.C. Brown, *J. Am. Chem. Soc.* 61: 1483 (1939).
84a. Y.E. Rhodes and L. Vargas, *J. Org. Chem.* 38: 4077 (1973).
85. R.W. Taft, Jr., *Steric Effects in Organic Chemistry*, ed. M.S. Newman, New York: Wiley, 1956, Chap. 13.
86. L.S. Levitt and B.W. Levitt, *Tetrahedron* 29: 941 (1973).
87. S. Takahashi, L.A. Cohen, H.K. Miller, and E.G. Peake, *J. Org. Chem.* 36: 1205 (1971).
88. G.E.K. Branch and M. Calvin, *The Theory of Organic Chemistry*, Englewood Cliffs, N.J.: Prentice-Hall, 1941, p. 218.
89. For comments on methods of determining intensities, see A. Cabana and C. Sandorfy, *Spectrochim. Acta* 16: 335 (1960).
90. T.L. Brown and M.T. Rogers, *J. Am. Chem. Soc.* 79: 577 (1957); T.L. Brown and M.T. Rogers, *J. Phys. Chem.* 61: 820 (1957).
91. J.V. Bell, J. Heisler, H. Tannenbaum, and J. Goldenson, *J. Am. Chem. Soc.* 76: 5185 (1954).
92. R.E. Kagarise, *J. Am. Chem. Soc.* 77: 1377 (1955); E.T. McBee and D.L. Christman, *J. Am. Chem. Soc.* 77: 755 (1955); W.D. Closson, *J. Am. Chem. Soc.* 86: 2386 (1964).
93. J.K. Wilmshurst, *J. Chem. Phys.* 28: 733 (1958); R.D. Kross and V.A. Fassel, *J. Am. Chem. Soc.* 77: 5858 (1955).

94. E.L. Wagner, *J. Am. Chem. Soc.* 85: 161 (1963).
95. E.V. Halevi, *Tetrahedron* 1: 174 (1957); R.P. Bell and M.B. Jensen, *Proc. Chem. Soc.* (London) 307 (1960).
96. H.S. Klein and A. Streitwieser, Jr., *Chem. and Ind.* (London) 180 (1961).
97. P.R. Wells, *Progr. Phys. Org. Chem.* 6: 127 (1968).
98. See M.J. Maskornick and A. Streitwieser, Jr., *Tetra. Letters* (17) 1625 (1972) and references cited there.
99. V.W. Laurie and J.S. Muenter, *J. Am. Chem. Soc.* 88: 2883 (1966).
100. V.I. Minkin, O.A. Osipov, and Y.A. Zhdanov, *Dipole Moments in Organic Chemistry*, New York: Plenum, 1970.
101. For the measurement of dipole moments, see H.B. Thompson, *J. Chem. Educ.* 43: 66 (1966). For a table of dipole moments, see A.L. McClellan, *Tables of Experimental Dipole Moments*, San Francisco: Freeman, 1963.
102. A. Burawoy and A.R. Thompson, *J. Chem. Soc.* 4313 (1956); J.A.A. Ketelaar, *Chemical Constitution*, Princeton, N.J.: Van Nostrand-Elsevier, 1953, p. 90.
103. Calculated: T.L. Brown, *J. Am. Chem. Soc.* 81: 3229 (1959).
104. C.P. Smyth, *J. Chem. Phys.* 4: 209 (1937).
105. D. Steele, *Quart. Revs.* (London) 18: 21 (1964); F.W. Baker, R.C. Parish, and L.M. Stock, *J. Am. Chem. Soc.* 89: 5677 (1967); S.H. Pine and D.R. Steele, *Spectrochim. Acta* 23A: 1509 (1967); T.L. Brown, *Chem. Revs.* 58: 581 (1958).
106. A.J. Petro, *J. Am. Chem. Soc.* 80: 4230 (1958).
107. G.E.K. Branch and M. Calvin, *The Theory of Organic Chemistry*, Englewood Cliffs, N.J.: Prentice-Hall, 1941, p. 135.
108. E.M. Arnett, T.C. Moriarty, L.E. Small, J.P. Rudolph, and R.P. Quirk, *J. Am. Chem. Soc.* 95: 1492 (1973).
*109. Y. Souma, H. Sano, and J. Iyoda, *J. Org. Chem.* 38: 2016 (1973).
110. G.V. Calder and T.J. Barton, *J. Chem. Educ.* 48: 338 (1971).
111. I.N. Maksimova, *Russ. J. Phys. Chem.* 41: 27 (1967).
112. C.S. Leung and E. Grunwald, *J. Phys. Chem.* 74: 687 (1970); see also P.D. Bolton, K.A. Fleming, and F.M. Hall, *J. Am. Chem. Soc.* 94: 1033 (1972).
*109a. K. Hiraoka, R. Yamdagni, and P. Kebarle, *J. Am. Chem. Soc.* 95: 6833 (1973).
113. G.E. Maciel, J.W. McIver, Jr., N.S. Ostlund, and J.A. Pople, *J. Am. Chem. Soc.* 92: 1197 (1970).
114. K.L. Williamson, S. Mosser, and D.E. Stedman, *J. Am. Chem. Soc.* 93: 7208 (1971).
115. W.A. Sheppard, *J. Am. Chem. Soc.* 92: 5419 (1970).
116. G.C. Leey, J.D. Cargioli, and W. Racela, *J. Am. Chem. Soc.* 92: 6238 (1970); E.M. Arnett, R.P. Quirk, and J.W. Larsen, *J. Am. Chem. Soc.* 92: 3977 (1970).
117. G.L. Anderson, R.C. Parish, and L.M. Stock, *J. Am. Chem. Soc.* 93: 6984 (1971).
118. J.G. Kirkwood and F.H. Westheimer, *J. Chem. Phys.* 6: 506, 513 (1938).

119. (a) G.L. Anderson and L.M. Stock, *J. Am. Chem. Soc.* 91: 6904 (1969) and references cited therein. (b) C.G. Swain and E.C. Lupton, *J. Am. Chem. Soc.* 90: 3328 (1968); P.R. Wells and W. Adcock, *Australian J. Chem.* 18: 1365 (1965); E.J. Grubbs, D.J. Lee, and A.G. Bellettini, *J. Org. Chem.* 31: 4069 (1966) and references cited therein; W.T. Simpson, *J. Am. Chem. Soc.* 73: 5363 (1951); C.F. Wilcox and C. Leung, *J. Am. Chem. Soc.* 90: 336 (1938). (c) C.L. Liotta, W.F. Fisher, E.L. Slightom, and C.L. Harris, *J. Am. Chem. Soc.* 94: 2129 (1972) and references cited therein.

120. F.W. Baker, R.C. Parish, and L.M. Stock, *J. Am. Chem. Soc.* 89: 5677 (1967)

121. P.E. Peterson, *et al.*, *J. Am. Chem. Soc.* 89: 5902 (1967); M.J.S. Dewar and P.J. Grisdale, *J. Am. Chem. Soc.* 84: 3548 (1962); M.J.S. Dewar and A.P. Marchand, *J. Am. Chem. Soc.* 88: 354 (1966).

122. E.J. Grubbs and R. Fitzgerald, *Tetra. Letters* (47) 4901 (1968).

123. L. Radom, J.A. Pople, and P.v.R. Schleyer, *J. Am. Chem. Soc.* 94: 5935 (1972); S.-L. Chong and J.L. Franklin, *J. Am. Chem. Soc.* 94: 6347 (1972).

124. C.A. Grob, E. Renk, and A. Kaiser, *Chem. and Ind.* (London) 1222 (1955).

125. K. Bowden, *Canad. J. Chem.* 41: 2781 (1963).

126. R. Golden and L.M. Stock, *J. Am. Chem. Soc.* 94: 3080 (1972).

127. K. Bowden, M.J. Price, and G.R. Taylor, *J. Chem. Soc.* (B) 1022 (1970).

128. C.L. Liotta, W.F. Fisher, G.H. Greene, Jr., and B.L. Joyner, *J. Am. Chem. Soc.* 94: 4891 (1972).

129. L.M. Stock, *J. Chem. Educ.* 49: 400 (1972).

130. H. Tanida, S. Yamamoto, and K. Takeda, *J. Org. Chem.* 38: 2077 (1973).

131. D.S. Noyce and B.E. Johnston, *J. Org. Chem.* 34: 1252 (1969).

132. E.M. Arnett and J.W. Larsen, *J. Am. Chem. Soc.* 91: 1438 (1969).

133. E. Berliner, *Tetrahedron* 5: 202 (1959); J.W. Larsen, *Tetrahedron* 92: 5136 (1970); T. Sorenson, *Tetrahedron* 89: 3794 (1967).

134. W.M. Schubert, R.B. Murphy, and J. Robins, *J. Org. Chem.* 35: 951 (1970) and references therein.

135. A. Burawoy and E. Spinna, *J. Chem. Soc.* 3752 (1956).

136. W. Hanstein, H.J. Berwin, and T.G. Taylor, *J. Am. Chem. Soc.* 92: 829 (1970); V.J. Shiner, Jr., and C.J. Vervanic, *J. Am. Chem. Soc.* 79: 373 (1957).

137. J.F. Sebastian, *J. Chem. Educ.* 48: 97 (1971).

138. M. Davies, *Some Electrical and Optical Aspects of Molecular Behavior*, New York: Pergamon, 1965, p. 160.

139. J.E. Williams, V. Buss, and L.C. Allen, *J. Am. Chem. Soc.* 93: 6867 (1971)

140. A. Streitwieser, P.H. Owens, and R. Wolf, *Tetra. Letters* (38) 3385 (1970); W.J. Hehre, R.F. Stewart, and J.A. Pople, *J. Chem. Phys.* 51: 2657 (1969); R.B. Hermann, *J. Am. Chem. Soc.* 92: 5298 (1970); J.E. Huheey, *J. Org. Chem.* 36: 204 (1971).

141. J.W. Larsen, P.A. Bouis, M.W. Grant, and C.A. Lane, *J. Am. Chem. Soc.* 93: 2067 (1971).

142. J.I. Brauman and L.K. Blair, *J. Am. Chem. Soc.* 90: 6561 (1968); R.T. McIver, Jr., J.A. Scott, and J.M. Riveros, *J. Am. Chem. Soc.* 95: 2706 (1973).

143. N.C. Baird, *Canad. J. Chem.* 47: 2306 (1969).
144. T.P. Lewis, *Tetrahedron* 25: 4117 (1969); S. Fliszar, *J. Am. Chem. Soc.* 94: 1068 (1972).
145. J.I. Brauman, J.M. Riveros, and L.K. Blair, *J. Am. Chem. Soc.* 93: 3914 (1971).
146. W.B. Nixon and M.M. Bursey, *Tetra. Letters* 4389 (1970).
147. M.S.B. Munson, *J. Am. Chem. Soc.* 87: 2332 (1965); D.K. Bohme, E. Lee-Ruff, and L.B. Young, *J. Am. Chem. Soc.* 94: 5153 (1972).
148. M.S. Nozari and R.S. Drago, *J. Am. Chem. Soc.* 94: 6877 (1972).
149. J.I. Brauman and L.K. Blair, *J. Am. Chem. Soc.* 93: 4315 (1971).
150. T.J. Broxton, L.W. Deady, A.R. Katritzky, A. Liu, and R.D. Topson, *J. Am. Chem. Soc.* 92: 6845 (1970).
151. J.W. Baker and L.G. Groves, *J. Chem. Soc.* 1144 (1939).
152. T.L. Brown, *J. Am. Chem. Soc.* 81: 3232 (1959).
153. W.C. Price, *Chem. Revs.* 41: 257 (1947).
154. R.B. Moodie, T.M. Coonor, and R. Stewart, *Canad. J. Chem.* 38: 626 (1960); F.A. Bovey, F.P. Hood, E. Pier, and H.E. Weaver, *J. Am. Chem. Soc.* 87: 2060 (1965).
155. R.T.C. Brownlee, R.E.J. Hutchinson, A.R. Katritzky, and R.D. Topson, *J. Am. Chem. Soc.* 90: 1757 (1968).
156. R.G. Lawlah and C.T. Tabit, *J. Am. Chem. Soc.* 91: 5671 (1969).
157. M.J. Kurylo and N.B. Jurinski, *Tetra. Letters* (12) 1083 (1967).
158. L.S. Levitt and B.W. Levitt, *Tetrahedron* 27: 3777 (1971).
159. P.v.R. Schleyer and C.W. Woodworth, *J. Am. Chem. Soc.* 90: 6528 (1968).
160. C.G. Swain and W.P. Langsdorf, *J. Am. Chem. Soc.* 93: 2813 (1971). As Schleyer (see ref. 159 above) has said, all possibilities are observed: $-\underline{I}$, ~ 0, and $+\underline{I}$.
161. L.M. Jackman and D.P. Kelby, *J. Chem. Soc.* (B): 102 (1970).
162. M. Witanowski and J. Januszewski, *Canad. J. Chem.* 47: 1321 (1969).
163. H.A. Christ, P. Diehl, H.R. Schneider, and H. Dahn, *Helv. Chim. Acta* 44: 865 (1971).
164. G.A. Olah, C.L. Jeuell, D.P. Kelly, and R.D. Porter, *J. Am. Chem. Soc.* 94: 146 (1972).
165. Compiled by G.A. Olah, R.D. Porter, C.L. Jeuell, and A.M. White, *J. Am. Chem. Soc.* 94: 2044 (1972).
166. G.A. Olah and A.M. White, *J. Am. Chem. Soc.* 91: 5801 (1969).
167. R.A. Friedel and H.L. Retcofsky, *J. Am. Chem. Soc.* 85: 1300 (1963).
168. G.E. Maciel, *J. Chem. Phys.* 42: 2746 (1965).
169. H. Brouwer and J.B. Stothers, *Canad. J. Chem.* 50: 601 (1972).
170. R.G. Pearson and J. Songstad, *J. Am. Chem. Soc.* 89: 1827 (1967).
171. P.H. Owens, G.J. Gleicher, and L.M. Smith, Jr., *J. Am. Chem. Soc.* 90: 4122 (1968).

172. G.H. Wahl, Jr., and M.R. Peterson, Jr., *J. Am. Chem. Soc.* 92: 7238 (1970).
173. G.L. Anderson and L.M. Stock, *J. Am. Chem. Soc.* 91: 6804 (1969).
174. M.D. Newton and W.N. Lipscomb, *J. Am. Chem. Soc.* 89: 4261 (1967).
175. G.A. Olah and D.H. O'Brien, *J. Am. Chem. Soc.* 89: 1725 (1967); G.A. Olah and E. Namanworth, *J. Am. Chem. Soc.* 88: 5327 (1966).
176. A.G. Pinkus and H.C. Custard, Jr., *J. Phys. Chem.* 74: 1042 (1970).
177. N. Purdue and M.B. Tomson, *J. Am. Chem. Soc.* 95: 48 (1973).
178. C.J. Creswell and O. Runquist, *Spectral Analysis of Organic Compounds*, Minneapolis, Minn.: Burgess, 1970, is excellent in this respect for uv, ir, and nmr.
179. R.M. Silverstein and G.C. Bassler, *Spectrophotometric Indentification of Organic Compounds*, 2nd ed., New York: Wiley, 1967, is very good for mass spectroscopy.
180. See J.L. Hollenberg, *J. Chem. Educ.* 47: 2 (1970).
181. For a good, recent bibliography, see P. Laszlo and P.J. Stang, *Organic Spectroscopy*, New York: Harper and Row, 1971. For a useful method of determining molecular weights by nmr, see S. Barcza, *J. Org. Chem.* 28: 1914 (1963).
182. P.J. Wheatley, *Molecular Structure*, 2nd ed., New York: Oxford U. Press, 1968.
183. M.L. Gross and C.L. Wilkins, *Anal. Chem.* 43, No. 14: 65A (1971).
184. See N.B. Colthup, "Vibrating Molecular Models," *J. Chem. Educ.* 38: 394 (1961).
185. K. Alder and H.v. Brachel, *Ann.* 608: 195 (1957).
186. W.T. King, J.M. Mills, and B.L. Crawford, *J. Chem. Phys.* 27: 455 (1957).
187. R.D. Stolow, *J. Am. Chem. Soc.* 84: 686 (1962).
188. J. Pitha, *J. Org. Chem.* 35: 2411 (1970).
189. R.A. Nyquist, *Spectrochim. Acta* 19: 1655 (1963).
190. I.C. Kogon, *J. Am. Chem. Soc.* 79: 2253 (1957); O.R. Wulf and V. Liddell, *J. Am. Chem. Soc.* 57: 1464 (1935); W. Luttke and R. Mecke, *Z. Elektrochem.* 53: 241 (1949); W. Luttke and R. Mecke, *J. Chem. Phys.* 21: 1606 (1953).
191. G. Habermehl, *Angew. Chem. Int. Ed.* 3: 309 (1964).
192. H.E. Simmons, E.P. Blanchard, and H.D. Hartzler, *J. Org. Chem.* 31: 295 (1966); P.G. Gassman and F.V. Zalar, *J. Org. Chem.* 31: 166 (1966); H.T. Tanida, Y. Hata, Y. Matsui, and I. Tanaka, *J. Org. Chem.* 30: 2260 (1965).
193. P. Yates, *Tetra. Letters* 4341 (1969).
194. P. Yates and L.L. Williams, *J. Am. Chem. Soc.* 80: 5896 (1958).
195. S. Masamuni and N.T. Castellucci, *J. Am. Chem. Soc.* 84: 2452 (1962).
196. G.L. Closs, *Adv. Alicyclic Chem.* 1: 53 (1966).
197. K. Nakamoto, P.J. McCarthy, A. Ruby, and A.E. Martell, *J. Am. Chem. Soc.* 83: 1272, 1066 (1961).
198. C.F. Wilcox, Jr., and R.R. Craig, *J. Am. Chem. Soc.* 83: 3866 (1961).

199. G. Geiseler, P. Richter, and K. Schmiedel, *Z. Elektrochem.* 65: 750 (1961).
200. D.G. Brown, R.S. Drago, and T.F. Bolles, *J. Am. Chem. Soc.* 90: 5708 (1968).
201. H. Yamada and K. Kozima, *J. Am. Chem. Soc.* 82: 1543 (1960).
202. S. Pinchas, *et al.*, *J. Chem. Soc.* 3063 (1961) and earlier papers.
203. G.J. Karabatsos, *J. Org. Chem.* 25: 315 (1960).
204. See M.F. El Bermani, A.J. Woodward, and N. Jonathan, *J. Am. Chem. Soc.* 92: 6750 (1970); N. Oi and J.F. Coetzee, *J. Am. Chem. Soc.* 91: 2473 (1969); C.D. Ritchie, B.A. Bierl, and R.J. Honour, *J. Am. Chem. Soc.* 84: 4687 (1962).
205. L.J. Bellamy and P.E. Rogasch, *J. Chem. Soc.* 2218 (1960).
*206. H.C. Smitherman, Jr., and L.N. Ferguson, *Tetrahedron* 24: 923 (1968).
207. H. Götz, E. Heilbronner, A.R. Katritzky, and R.A. Jones, *Helv. Chim. Acta* 44: 387 (1961).
*206a. F.R. Dollish, W.G. Fateley, and F.F. Bentley, *Characteristic Raman Frequencies of Organic Compounds*, New York: Wiley-Interscience, 1973.
208. H. Ziffer and I.W. Levin, *J. Org. Chem.* 34: 4056 (1969).
209. D.S. Bailey and J.B. Lambert, *J. Org. Chem.* 38: 134 (1973).
210. J.W. Linnett, *Quart. Revs.* (London) 1: 73 (1947).
211. R.M. Badger, *J. Chem. Phys.* 3: 710 (1935).
212. D.R. Herschback and V.W. Laurie, *J. Chem. Phys.* 35: 458 (1961).
213. For the theory of electronic absorption spectra, see: J.N. Murrell, *The Theory of the Electronic Spectra of Organic Molecules*, 2nd ed., New York: Wiley, 1971; H. Suzuki, *Electronic Absorption Spectra and Geometry of Organic Molecules*, New York: Academic, 1967; H.H. Jaffe and M. Orchin, *Theory and Applications of Ultraviolet Spectroscopy*, New York: Wiley, 1962.
214. D.E. Freeman, J.R. Lombardi, and W. Klemperer, *J. Chem. Phys.* 45: 52, 58 (1966).
215. See J.S. Swenton, *J. Chem. Educ.* 46: 217 (1969) for references on divergent views regarding assignment of the 150, 175, and 195 nm bands of formaldehyde.
216. H.H. Jaffe, D.L. Beveridge, and M. Orchin, *J. Chem. Educ.* 44: 383 (1967); H. Ley, *Z. Physik. Chem.* B17: 177 (1932).
217. F.S. Toby, S. Toby, and G.O. Pritchard, *J. Am. Chem. Soc.* 94: 4441 (1972).
218. L. Doub and J.M. Vandenbelt, *J. Am. Chem. Soc.* 71: 2414 (1949) and 69: 2714 (1947).
219. L.N. Ferguson, *Chem. Revs.* 43: 439 (1948).
220. M.J. Kamlet, J.C. Hoffsommer, and H.G. Adolph, *J. Am. Chem. Soc.* 84: 3925 (1962).
221. D.J. Cram, R.H. Bauer, N.L. Allinger, R.A. Reeves, W.J. Wechter, and E. Heilbronner, *J. Am. Chem. Soc.* 81: 5978 (1959).
222. See C. Reichardt, *Angew. Chem. Int. Ed.* 4: 29 (1965); M.R.J. Dack, *Chem. Tech.* Feb. 1971, p. 108; A.R. Katritzky, *J. Chem. Soc.* (B) 460 (1971).

223. (a) H.F. Herbrandson and F.R. Neufield, *J. Org. Chem.* 31: 1140 (1966).
     (b) A.H. Fainberg and S. Winstein, *J. Am. Chem. Soc.* 78: 2770 (1956).
     (c) S.G. Smith, A.H. Fainberg, and S.Winstein, *J. Am. Chem. Soc.* 83: 618 (1961). (d) M. Gielen and J. Nasielski, *Rec. Trav. Chim.* 82: 228 (1963). (e) A. Berson, Z Hamlet, and W.A. Mueller, *J. Am. Chem. Soc.* 84: 297 (1962). (f) E.M. Kosower, *J. Am. Chem. Soc.* 80: 3253 (1953). (g) K. Dimroth, C. Reichardt, T. Siepmann, and F. Bohlmann, *Ann.* 661: 1 (1963). (h) T. Kagiya, T. Sumida, and T. Inoue, *Bull. Chem. Soc. Japan* 41: 767 (1968). (i) S. Brownstein, *Canad. J. Chem.* 38: 1590 (1960)

224. C.S. Giam and J.L. Lyle, *J. Am. Chem. Soc.* 95: 3235 (1973).

225. J.D.C. Brand and D.G. Williamson, *Adv. Phys. Org. Chem.* 1: 365 (1963).

226. R.G. Warren, Y. Chow, and L.N. Ferguson, *Chem. Commun.* 1521 (1971); D.A. Nelson and J.J. Worman, *Tetra. Letters* (5) 507 (1966).

227. E.M. Kosower and M. Ito, *Proc. Chem. Soc.* 25 (1962).

228. W.D. Closson and P. Haug, *J. Am. Chem. Soc.* 86: 2384 (1964).

229. L.D. Weis, T.R. Evans, and P.A. Leermakers, *J. Am. Chem. Soc.* 90: 6109 (1968).

230. H. Hart, *et al.*, *J. Am. Chem. Soc.* 90: 5296 (1960) and 93: 720 (1971).

231. K.W. Hauser, R. Kuhn, and A. Smakula, *Z. Physik. Chem.* B29:403 (1935).

232. S.D. Worley, *Chem. Revs.* 71: 295 (1971); D.A. Sweigert and J. Daintith, *Science Progr.* (Oxford) 59: 325 (1971); M.J.S. Dewar, *J. Am. Chem. Soc.* 92: 19 (1970); D.W. Turner, A.D. Baker, C. Baker, and C.R. Brundle, *Molecular Photoelectron Spectroscopy*, London: Interscience, 1970.

233. See E. Haselbach and A. Schmelzer, *Helv. Chim. Acta* 54: 1575 (1971).

234. C. Batich, *et al.*, *J. Am. Chem. Soc.* 95: 930 (1973).

235. E. Haselbach, E. Heilbronner, and G. Schröder, *Helv. Chim. Acta* 54: 153 (1971).

236. P. Bischof and E. Heilbronner, *Helv. Chim. Acta* 53: 1677 (1970).

237. C.R. Brundle and M.B. Robin, *J. Am. Chem. Soc.* 92: 5550 (1970).

238. D. Chadwick, D.C. Frost, and L. Weiler, *J. Am. Chem. Soc.* 93: 4320, 4962 (1971).

239. P.G. Gassman and P.G. Pape, *J. Org. Chem.* 29: 160 (1964); L.D. Hess and J.N. Pitts, Jr., *J. Am. Chem. Soc.* 89: 1973 (1967).

240. A.D. Bain, J.C. Bünzli, D.C. Frost, and L. Weiler, *J. Am. Chem. Soc.* 95: 291 (1973).

241. O.S. Khalil, J.L. Meeks, and S.P. McGlynn, *J. Am. Chem. Soc.* 95: 5876 (1973).

242. The theory for the chemical shift is complex; see W.N. Lipscomb, *Adv. Mag. Resonance* 2: 137 (1966); J.I. Musher, *Adv. Mag. Resonance* 2: 177 (1966); B.V. Cheney and D.M. Grant, *J. Am. Chem. Soc.* 89: 5319 (1967).

243. H. Günther, H. Schmickler, H. Königshofen, K. Recker, and E. Vogel, *Ange Chem. Int. Ed.* 12: 241 (1973); R.H. Levin and J.D. Roberts, *Tetra. Lette* (2) 135 (1973); A.V. Kemp-Jones, A.J. Jones, M. Sakai, C.P. Beeman, and S. Masamune, *Canad. J. Chem.* 51: 767 (1973).

244. For discussions of shielding effects of cyclopropyl groups, see C.D. Poulter, R.S. Boikess, J.I. Brauman, and S. Winstein, *J. Am. Chem. Soc.* 94: 2291 (1972); R.C. Hahn and P.H. Howard, *J. Am. Chem. Soc.* 94: 3143 (1972).

245. R.W. Franck and M.A. Williamson, *J. Org. Chem.* 31: 2420 (1966); R.I. Herrmann and I.D. Rae, *Australian J. Chem.* 25: 811 (1972).

246. F.A. Bovey, *Chem. and Eng. News* Aug. 30, 1965, p. 98.

247. Spectrum No. 310, *NMR Spectra Catalogue*, Palo Alto, Calif.: Varian Associates.

248. R.M. Silverstein and R.G. Silberman, *J. Chem. Educ.* 50: 484 (1973).

249. J.R. Dyer, *Applications of Absorption Spectroscopy of Organic Compounds*, Englewood Cliffs, N.J.: Prentice-Hall, 1965, p. 100.

250. See E. Lustig, E.P. Ragelis, N. Day, and J.A. Ferretti, *J. Am. Chem. Soc.* 89: 3953 (1967) on the difficulty of analyzing spectra of AA'BB' systems.

251. D.F. DeTar, ed., *Computer Programs for Chemistry*, Vol. I, New York: Benjamin, 1968.

252. F.A.L. Anet and G.C. Levy, *Science* 180: 4082 (1973); J.B. Stothers, *Carbon-13 Spectroscopy*, New York: Academic, 1972; G.C. Levy and G.L. Nelson, *Carbon-13 Nuclear Magnetic Resonance for Organic Chemists*, New York: Wiley-Interscience, 1972.

252a. D.E. Dorman, M. Jautelat, and J.D. Roberts, *J. Org. Chem.* 36: 2757 (1971).

252b. G.B. Savitsky, P.D. Ellis, K. Namikawa, and G.E. Maciel, *J. Chem. Phys.* 49: 2395 (1968).

253. T.C. Farrar and E.D. Becker, *Pulse and Fourier Transform NMR*, New York: Academic, 1971.

254. R.H. Levin, J.-Y. Lallemand, and J.D. Roberts, *J. Org. Chem.* 38: 1983 (1973); G.A. Olah and P.W. Westerman, *J. Org. Chem.* 38: 1986 (1973); L.F. Johnson and W.C. Jankowski, *Carbon-13 NMR Spectra*, New York: Wiley-Interscience, 1972.

255. (a) For a brief reading, see N.L. Allinger, M.P. Cava, D. DeJongh, N.A. Lebel, and C.L. Stevens, *Organic Chemistry*, New York: Worth, 1971, Sec. 32.3; D.J. Pasto and C.R. Johnson, *Organic Structure Determination*, Englewood Cliffs, N.J.: Prentice-Hall, 1969, Chap. 8.
(b) For an introduction, see S.R. Shrader, *Introductory Mass Spectrometry*, Boston: Allyn and Bacon, 1971; R.M. Silverstein and G.C. Bassler, *Spectrometric Identification of Organic Compounds*, 2nd ed., New York: Wiley, 1967.
(c) For comprehensive or specialized treatments, see *Topics in Organic Mass Spectrometry*, ed. A.L. Burlingame, New York: Wiley-Interscience, 1970; H. Budzikiewicz, C. Djerassi, and D.H. Williams, *Mass Spectrometry of Organic Molecules*, San Francisco: Holden-Day, 1967; R.I. Reed, *Applications of Mass Spectrometry to Organic Chemistry*, New York: Academic, 1966: R.W. Kiser, *Introduction to Mass Spectrometry and Its Applications*, Englewood Cliffs, N.J.: Prentice-Hall, 1965; F.W. McLafferty, *Mass Spectrometry of Organic Ions*, New York: Academic, 1963; *Mass Spectrometry: Techniques and Applications*, ed. G.W.A. Milne, New York: Wiley-Interscience, 1971.

256. J. Lederberg, *J. Chem. Educ.* 49: 613 (1972).

257. J.H. Beynon and A.E. Williams, *Mass and Abundance Tables for Use in Mass Spectrometry*, Amsterdam: Elsevier, 1963. A reduced portion of these tables appears in Silverstein and Bassler (see ref. 255(b)).

258. F.W. McLafferty, "Mass Spectral Correlations," *Advances in Chemistry Series*, 40, Washington, D.C.: American Chemical Society, 1963.
259. M.M. Campbell and O. Runquist, *J. Chem. Educ.* 49: 104 (1972).
260. V. Wollrabe, *Phytochem.* 8: 623 (1969); P.E. Kolattukudy, *Lipids* 5: 398 (1969).
261. F.W. McLafferty and R.S. Gohlke, *Anal. Chem.* 31: 2076 (1959).
262. H. Budzikiewicz, C. Djerassi, and D.H. Williams, *Interpretations of Mass Spectra of Organic Compounds*, San Francisco: Holden-Day, 1964.
263. D.H. Williams, H. Budzikiewicz, Z. Pelah, and C. Djerassi, *Monatsh.* 95: 166 (1964).
264. J.H. Beynon, R.A. Saunders, and A.E. Williams, *The Mass Spectra of Organic Molecules*, New York: Elsevier, 1968, Chap. 11.
265. A.G. Sharkey, Jr., J.L. Shultz, and R.A. Friedel, *Anal. Chem.* 28: 934 (1956).
266. E. Pedersen and E. Larsen, *Org. Mass Spectrom.* 4: 249 (1970).

Part II: MODERN STRUCTURAL THEORY

The <u>structural</u> <u>theory</u> of organic chemistry was developed in the nineteenth century to facilitate the understanding and prediction of molecular properties. Organic molecules were regarded as an assemblage of functional groups, assigned on the basis of their chemical properties, and all compounds with a given functional group were expected to have the properties associated with that functional group. If a structural formula were written for a molecule, this immediately implied certain properties, physical as well as chemical. For example, one might infer from structure II.1 that the compound is basic to litmus, is moderately low boiling, reacts easily with aqueous $KMnO_4$, decolorized Br_2/CCl_4, and has the properties observed for amines such as reacting with alkyl and acyl halides and acid anhydrides.

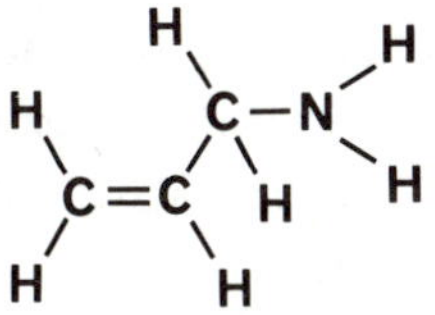

<u>II.1</u>

Slowly over the years it became evident that when certain functional groups are close together, the typical properties of each group are markedly altered. For illustration, <u>n</u>-pentyl chloride is easily hydrolyzed in hot aqueous alkali whereas neopentyl chloride (<u>II.2</u>) is not; and 2,6-dimethylbenzoic acid (<u>II.3</u>) cannot be esterified in the usual manner with alcoholic hydrogen chloride, like acids <u>II.4</u> and <u>II.5</u> (steric hindrance). Trichloroacetic acid is a strong acid (inductive effect) whereas trimethylacetic acid is a weak acid (inductive effect and hindrance to solvation). 1,3-Pentadiene gives predominantly a 1,4-addition

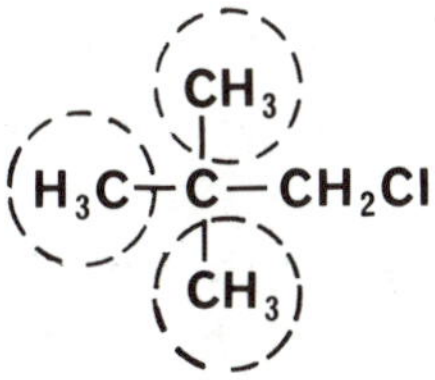

<u>II.2</u>

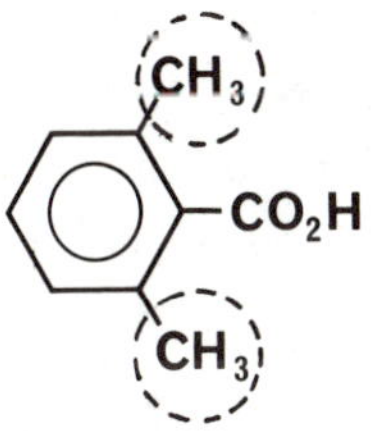

<u>II.3</u> 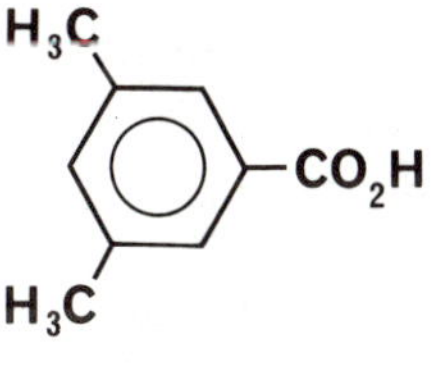   <u>II.4</u> 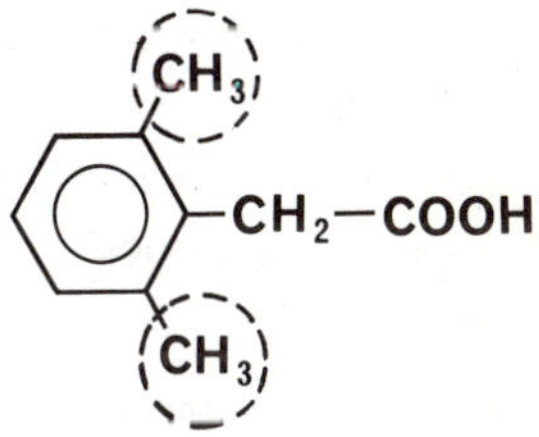   <u>II.5</u>

$$\underset{\text{Strong acid}}{\text{Cl}-\underset{\underset{\text{Cl}}{|}}{\overset{\overset{\text{Cl}}{|}}{\text{C}}}-\text{CO}_2\text{H}} \qquad \underset{\text{Weak acid}}{\text{H}_3\text{C}-\underset{\underset{\text{CH}_3}{|}}{\overset{\overset{\text{CH}_3}{|}}{\text{C}}}-\text{COOH}}$$

product with bromine rather than 1,2-addition products like the isomeric 1,4-pentadiene (conjugation). In order to explain these and other apparent failures of the earlier structural theory, the theory was modified to include the concepts of induction, steric hindrance, tautomerism, resonance, and orbital hybridization, among others. The combined multi-effect approach is now called the <u>modern structural theory</u>. As Erich Clar has said, this extends classical symbolism into modern structural theory.

If the structural formula written for a given compound is not consistent with all of its observed properties, then the structure must be revised. By the 1920s many compounds were known for each of which a single conventional structural formula could not be written that is symbolic of all the properties of a given compound. It was then proposed that more than one structure be written for such a compound but that it should be understood that the actual structure is a composite of the several structures written and not any single one. For instance, structure <u>II.6</u> is normally written for nitroethane. However, this would imply that one nitrogen-oxygen bond is different from the other. Since this is not the case, according to experimental measurement, two structures are written, <u>II.7a</u> and <u>II.7b</u>, and the actual structure is regarded as intermediate between the two. For simplicity, only one of the structures <u>II.7a</u> or <u>II.7b</u> is usually written with the tacit understanding that either could be used but that neither is correct and that the real structure is a composite of the two. This is resonance theory, which we see is merely an attempt to continue to use the classical structural formulas for representing molecules (1).

II.6   II.7 a   b

Chapter 5
RESONANCE THEORY

5.1 The Concept of Resonance

The concept of resonance was introduced through quantum mechanics by Heisenberg in the 1920s. The idea was then generalized by Pauling in the United States, by Robinson and Ingold in England (mesomerism), and by Arndt in Germany (Zwischenstufen). Each school had the common view that certain molecules cannot be adequately represented by a single structural formula but that the structure may be regarded as a composite of several conventional valence-bond structures. For example, neither Kekulé structure 5.1a nor 5.1b, nor any other single classical structural formula, is consistent with all the properties of o-diethylbenzene. Its chemical and physical properties are not characteristic of a conjugated triene (Table 5.1). Thus, o-diethylbenzene is considered to have a structure which is a composite of 5.1a and 5.1b. Of course, only one structure is generally written with the universal understanding that either could have been given and that neither is correct, that the actual structure is a hybrid of the two. This is now written as 5.2 in current symbolism.

There are several possible misconceptions about resonance to be avoided. One is that the actual structure assumes each written valence-bond structure part of the time, or that some of the molecules have one structure while the others have the other written structural formulas. Neither case is true. All molecules have the same fixed structure, which is closer to the most stable of the valence-bond structural formulas written. It is important to note that only electrons change positions in the structures contributing to a resonance hybrid and that all atoms have the same relative positions, within 0.3 Å.

Ch. 5 RESONANCE THEORY

Table 5.1. Comparison of olefinic with aromatic properties

Expected Chemical Properties for 5.1a or 5.1b	Observed Chemical Properties for o-diethylbenzene
Rapid decolorization of cold Br_2/CCl_4 sol'n.	Very slow decolorization of cold Br_2/CCl_4 sol'n.
Rapid decolorization of cold, dilute aq. $KMnO_4$	Inert toward cold, dilute aq. $KMnO_4$
Rapid polymerization in acid.	Not polymerized in acid.
Exothermic addition of H_2/cat.	Resists hydrogenation.
Diels-Alder with maleic anhydride.	No Diels-Alder reaction.

Expected Physical Properties for 5.1a or 5.1b	Observed Physical Properties for o-diethylbenzene
Alternating bond lengths, 1.48 Å and 1.34 Å, in ring.	Identical bond lengths, 1.39 Å, in ring.
Alternating force constants, 5 and 9 × 10^{-5} dynes/cm.	All identical force constants around ring, 7 × 10^{-5} dyne

Moreover, the structures written for a resonance hybrid are fictitious in the sense that they have no physical existence. That is, there are no molecules under normal conditions with either Kekulé structure. Such structures are unstable with respect to the actual structure of o-diethylbenzene and would spontaneously change into that of o-diethylbenzene. A resonance hybrid has properties generally resembling those expected for the various valence-bond structures written. Since 5.1a and 5.1b are equivalent, o-diethylbenzene gives ozonolysis products as if obtained equally from 5.1a and 5.1b. Both forms are needed, because only 5.1a yields Et-C-C-H ‖ ‖ O O whereas only 5.1b gives O O ‖ ‖ Et-C-C-Et , and both products are recovered.

Resonance differs from isomerism in the fact that structures contributing to a resonance hybrid cannot be isolated whereas isomers--

e.g., tautomers or conformers--may co-exist in solution and in some cases, two or more isomers of a molecule may be isolated.

Although we frequently use the expression "stabilized by resonance," we do not mean it in a literal sense. The inferred meaning is that the compound is a resonance hybrid and, therefore, is more stable than that implied by any written classical structural formula. Resonance energy is a correction which is made for the error in a calculation of the energy of a benzene ring, for instance, which results from the poor approximation of using a single Kekulé structure.

5.2 Writing Resonance Structures

An approximate structural formula can first be written for a compound based on a structure determination using chemical and physical methods. Then, when necessary, other structural formulas may be written as contributors to a resonance hybrid by a series of short electron-pair shifts. For example, valence-bond structures 5.3 and 5.4 may be written for ethyl and vinyl bromides.

	CH_3-CH_2-Br	$CH_2=CH-Br$
	5.3	5.4
C-Br bond length	1.91 Å	1.86 Å
Dipole moment	2.02 D	1.41 D

Formula 5.3 is consistent with all observed properties of ethyl bromide and will serve its purpose. However, in order to account for the shorter C-Br bond length in vinyl bromide, for the smaller dipole moment, for certain spectral properties, and for the resistance to nucleophilic displacements, the actual structure of vinyl bromide is regarded as a resonance hybrid of structures 5.5a and 5.5b.

5.5

This assigns some double-bondedness to the C-Br bond which accounts for its relative shortness and resistance to rupture for displacement reactions, and the charge distribution of 5.5b explains the reduction in dipole moment relative to that of ethyl bromide.

There are several rules commonly followed in assessing the relative importance or stability of structures comprising a resonance hybrid:

1. The more covalent bonds there are in a structure, the greater is its stability.
2. Resonance will be greater:
    a) for equivalent than for nonequivalent structures.
    b) when involving isovalent structures (structures containing the same number of covalent bonds).
    c) the larger is the number of nearly equivalent structures that can be written.
3. Separation of charge and close like charges in a structure reduce its contribution to the resonance hybrid.
4. Structures contributing to the same resonance hybrid must have the same number of unpaired electrons.
5. Structures that place more than eight electrons in the valence shell of the first row elements can be ignored.
6. Delocalization of π electrons in a structure is at a maximum when the σ system is planar.

The operation of these rules will be observed throughout this book and will not be demonstrated here. It should be noted, however, that sometimes two or more of these rules conflict and one must use experimental data or make a judgement on which is the prevailing rule. For example, 5.6a and 5.6b can be written for boron trifluoride.

$$\left[\begin{array}{c} F\diagdown_{B}\diagup F \\ | \\ F \end{array} , \begin{array}{c} F\diagdown_{B}={F}^{+} \\ | \\ F \end{array} \right] \text{(3 equivalent forms)}$$

5.6

Rule 1 would predict 5.6b to be more important, whereas 5.6a would be the major structure according to rule 3. The volatility, solubility, and other properties indicate that its structure is more like that of 5.6a, although 5.6b makes a substantial contribution to the resonance hybrid [2]. In the case of phenylisonitrile (5.7), again rules 1 and 3 are in conflict regarding the structure of the isonitrile group.

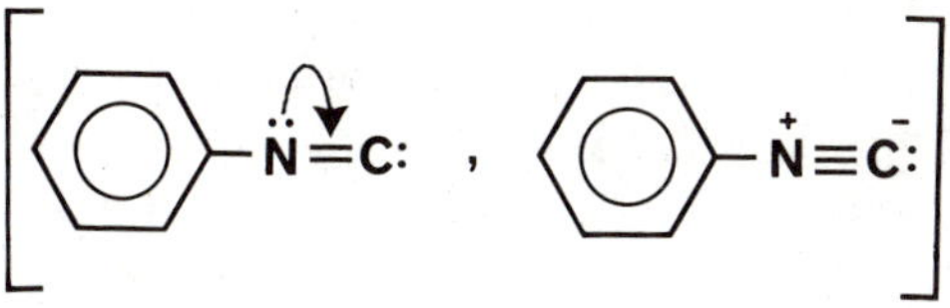

$$\left[\begin{array}{c} \underset{a}{\text{Ph-N=C:}} \quad , \quad \underset{b}{\text{Ph-}\overset{+}{\text{N}}\equiv\text{C:}^{-}} \end{array} \right]$$

5.7

This time, the properties of the molecule (dipole moment 3.5 D directed toward the carbon, ir absorption frequency in the triple bond range, etc.) indicate that the dipolar structure 5.7b makes the greater contribution.

5.3. Hyperconjugation [3]

In most cases, resonance will involve π and n (lone-pair) electrons but when there are C-C bonds with substantial p character (usually associated with strain) or C-H bonds α to a π bond, then σ electrons may also be involved in the resonance structures, e.g.,

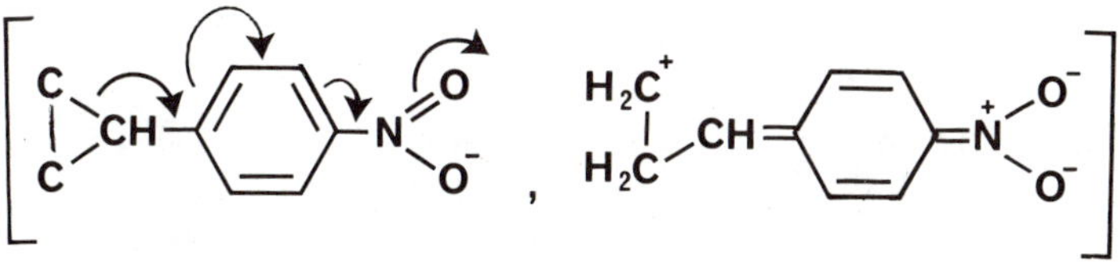

[structure showing cyclopropyl-substituted nitrobenzene resonance with 2 equivalent forms]

2 equivalent forms

[structure showing H-C(H)(H)-CH=CH-C(=O)-CH₃ ↔ H-C(H)=CH-CH=C(-O⁻)-CH₃ with H⁺]

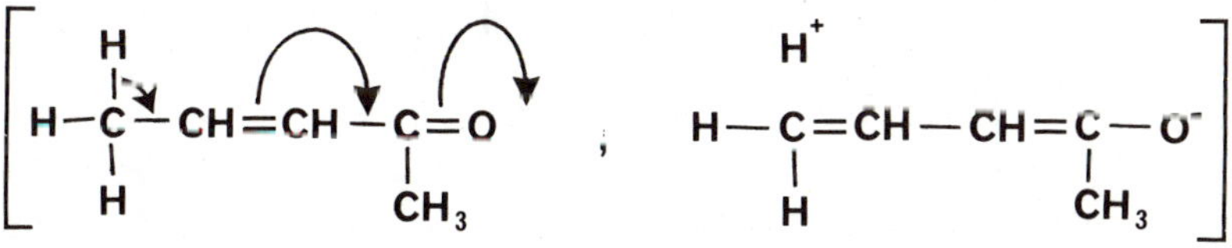

Such electron delocalization is called hyperconjugation, and other than the fact that it involves σ electrons, it is the same as resonance. It has been shown by photoelectron spectroscopy that hyperconjugation can be described by the model used for resonance [4]. Although MO calculations indicate that C-C hyperconjugation is more important than C-H hyperconjugation [5], the difference is certainly small.

The notion of C-H hyperconjugation was introduced to explain the apparent greater electron-releasing ability of methyl over t-butyl in certain cases, first reported by Baker and Nathan for the relative rates of reaction of p-alkylbenzyl halides with pyridine [6]. The relative rates of reaction are in the order Me > Et > i-Pr > t-Bu, which is opposite to that expected from their relative polarizabilities. It was proposed that Me has the greatest electron-releasing ability here since it has the largest number of C-H bonds for hyperconjugation, as it is now called. Pertinent to this point, read the discussion in Section 3.3c on the complex situation of electron releasing abilities of alkyl groups.

Nevertheless, hyperconjugation is valuable in accounting for structural features, physical properties, equilibria, and reaction kinetics in terms of stabilization of molecular species by electron delocalization. For example, ionization potentials (IP) and heats of hydrogenation (ΔH_{H_2}) of C=C bonds decrease with increasing alkylation or cyclopropylation (Table 5.2). This trend cannot be due to induction because cyclopropyl groups ($\underline{c}$-C_3H_5-) are certainly more electronegative than hydrogen atoms. The decrease in ionization potentials can be attributed to increased stabilities of the cation radicals produced upon loss of an electron from the olefins.

$$\left[\triangleright\text{-}\overset{+}{C}H\text{-}\dot{C}H\text{-}\triangleleft \quad , \quad \triangleright\text{-}\dot{C}H\text{-}CH=\overset{+}{\bigvee} \quad , \quad \triangleright\text{-}\dot{C}H\text{-}CH\underset{+}{=}\triangleleft \right]$$

Hyperconjugation is also felt in the neutral molecules, e.g.,

$$\left[\underset{H_3C}{\overset{H}{\diagdown}}C=C\underset{H}{\overset{CH_3}{\diagup}} \quad , \quad \underset{H_2C}{\overset{H}{\diagdown}}\underset{H^+}{C}\text{-}\dot{C}\underset{H}{\overset{CH_3}{\diagup}} \right]$$

Thus, alkylation decreases the heats of hydrogenation and the alkylated or cyclopropylated olefins are electron-rich relative to ethylene as shown by their relative ease of forming charge-transfer complexes (Section 6.1d) and undergoing cycloaddition reactions with TCNE [7a].

The ΔH_{H_2} of propene, 1-butene, isopropylethylene, and t-butyl-ethylene are virtually the same (8):

Table 5.2. Ionization potentials of some olefins

Compound	IP [7] (eV)	ΔH_{H_2} [8] (kcal/mol)
$H_2C=CH_2$	10.62	32.8
$CH_3CH_2CH=CH_2$	9.76	30.1
cis-CH_3-CH=CH-CH_3	9.34	28.6
trans-CH_3-CH=CH-CH_3	9.27	27.6
$(CH_3)_2C=CH_2$	9.26	28.4
$(CH_3)_2C=CH-CH_3$	8.89	26.9
$(CH_3)_2C=C(CH_3)_2$	8.53	26.6
$(c-C_3H_5)_2C=CH_2$	8.08	
$(c-C_3H_5)_2C=C(CH_3)_2$	7.82	
cis-c-C_3H_5-CH=CH-C_3H_5-c	7.70	
trans-c-C_3H_5-CH=CH-C_3H_5-c	7.72	
$(c-C_3H_5)_2C=CH-C_3H_5$-c	7.48	

Me-CH=CH_2	30.1 kcal/mol
Et-CH=CH_2	30.3
i-Pr-CH=CH_2	30.3
t-Bu-CH=CH_2	30.3

At first, this appears to be inconsistent with the notion that α C-H hyperconjugation lowers heats of hydrogenation. However, we saw in Figure 4.2 that ΔH_{H_2} depends upon the potential energies of the reactant and of the hydrogenation products. In this case, the potential energies of the olefins increase with decreasing hyperconjugation (methyl > t-butyl) but the energies of the respective alkanes increase in the same order due to increasing van der Waals strain. The net result appears to keep ΔH_{H_2} constant. Owing mostly to strain in the product (for instance, ΔH_{H_2} for cyclodecene (20.7 kcal/mol) is the lowest known for a simple olefin) and as a result of strain in the olefins, ΔH_{H_2} for barrelene (37.6 kcal/mol) and cis-di-t-butylethylene (36.2 kcal/mol) are among the highest.

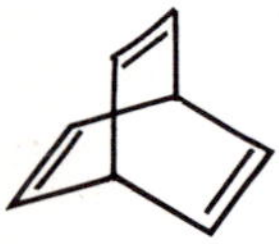

Ionization potentials and dipole moments of carbonyl compounds (Table 5.3) are also accounted for in terms of hyperconjugation.

Barrelene

$H_2C=O$ $CH_3-CH=O$ $CH_3CH_2-C=O$
                                      $|$
                                      CH_3

IP [9]   10.88 eV   10.26 eV   9.74 eV

The ionization potentials decrease and the dipole moments increase as hyperconjugation increases the electron density on the oxygen.

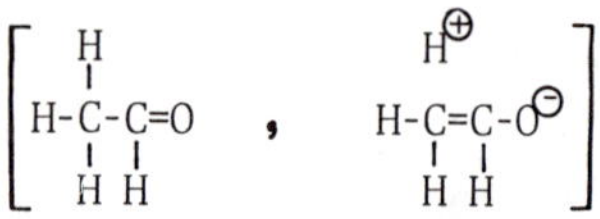

$$\left[\begin{array}{c} H \\ | \\ H-C-C=O \\ | \; | \\ H \; H \end{array} \; , \; \begin{array}{c} H^{\oplus} \\ \\ H-C=C-O^{\ominus} \\ | \; | \\ H \; H \end{array} \right]$$

Although the changes of ΔH_{H_2} and IP with increasing number of alkyl groups could be attributed to their inductive effects, this would not apply to the dipole moments of the aldehydes. There, we would expect a significant difference between the values for acetaldehyde and propionaldehyde, which is not observed. The marked difference between the moments of butylaldehyde and crotonaldehyde reflects the effect of resonance plus hyperconjugation in the latter.

Table 5.3. Dipole moments of some aldehydes

Compound	Dipole moment
H_2CO	2.27 D
CH_3-CHO	2.72
CH_3-CH_2-CHO	2.73
CH_3-CH_2-CH_2-CHO	2.72
CH_3-CH=CH-CHO	3.67

Sec. 5.3 HYPERCONJUGATION 189

$$\left[H_3C-CH=CH-\underset{H}{C}=O \;,\;\; CH_3-\overset{\oplus}{\underset{H}{C}H}-CH=C-O^{\ominus} \;,\;\; H_2C=CH-CH=\underset{H}{\overset{\overset{\oplus}{H}}{C}}-O^{\ominus} \right]$$

A good example of C-C hyperconjugation is found in the cyclopropyl group [10]. The effects on dipole moments, spectra and chemical reactivity are between those of vinyl and saturated alkyl groups, usually closer to the former. Some examples can be given here. In all cases, hyperconjugation structures can be written involving the cyclopropyl groups ($\underline{c}$-C_3H_5):

Dipole moments

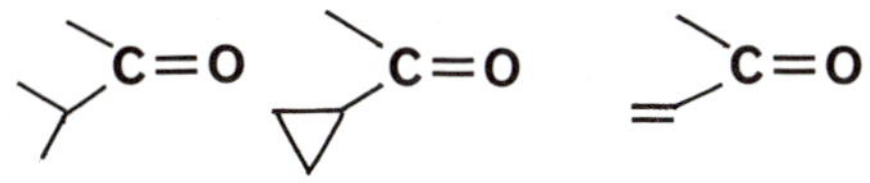

	>C=O	>C=O	>C=O
DM	2.7	2.84	2.98D

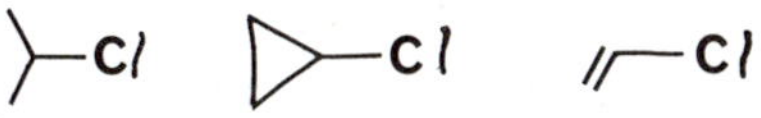

	>-Cl	>-Cl	//-Cl
DM	2.04	1.76	1.44D

$\underline{pK_a}$ values ($\underline{c}$-C_3H_5 makes the electron donor more basic than alkyl)

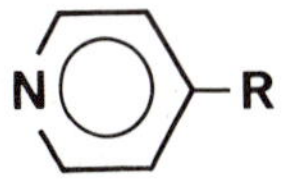

N⟨⟩-R

$R_2C=O$

R	pK_a	
$\underline{c}$-C_3H_5	6.44	
Me	6.07	
i-Pr	6.02	$\nu_{C=O}$
CH_3	-5.1	1718 cm^{-1}
$\underline{c}$-C_3H_5	-4.1	1694

Hammett-Brown substituent constants ($\underline{c}$-C_3H_5 more electron releasing than alkyl) [10a]

	σ_p^+
$\underline{c}$-C_3H_5	-0.44 to -0.48
Me	-0.311
i-Pr	-0.280

UV spectra (c-C$_3$H$_5$ increases λ_m 8-30 nm)

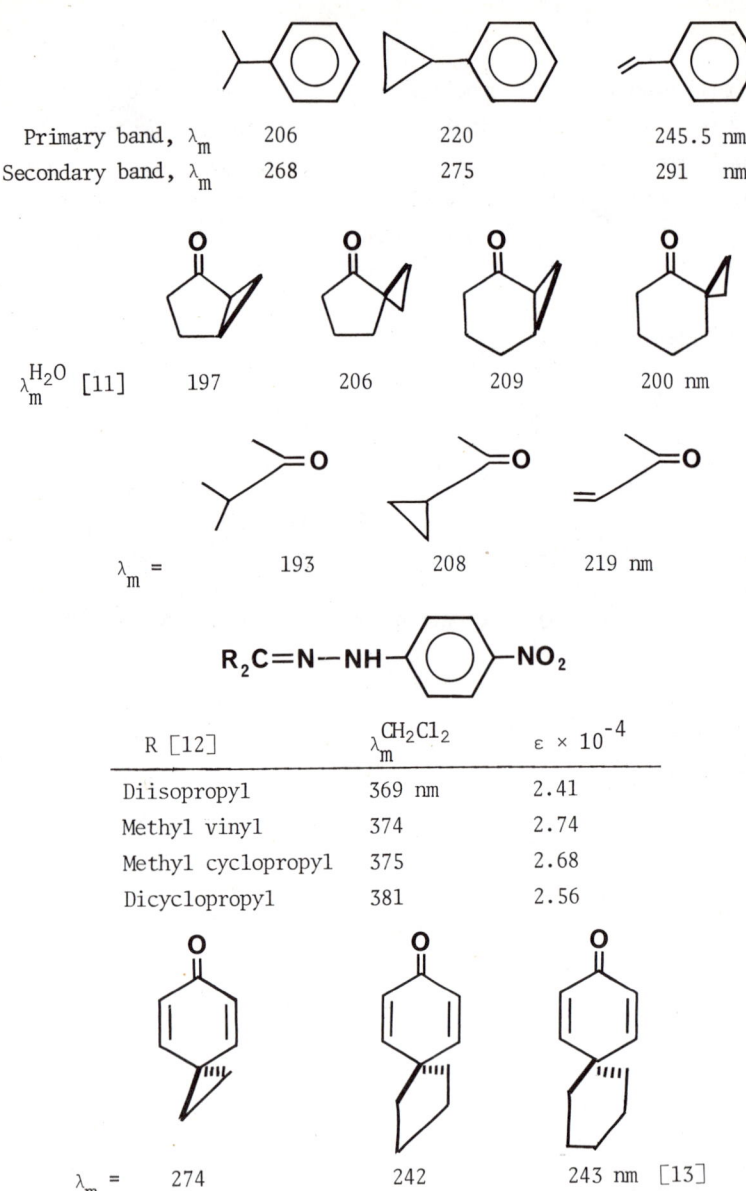

Primary band, λ_m	206	220		245.5 nm
Secondary band, λ_m	268	275		291 nm

$\lambda_m^{H_2O}$ [11]   197   206   209   200 nm

λ_m =   193   208   219 nm

R$_2$C=N—NH—⟨⟩—NO$_2$

R [12]	$\lambda_m^{CH_2Cl_2}$	$\varepsilon \times 10^{-4}$
Diisopropyl	369 nm	2.41
Methyl vinyl	374	2.74
Methyl cyclopropyl	375	2.68
Dicyclopropyl	381	2.56

λ_m =	274	242	243 nm [13]
ε	21,900	15,900	14,800

Sec. 5.3 HYPERCONJUGATION   191

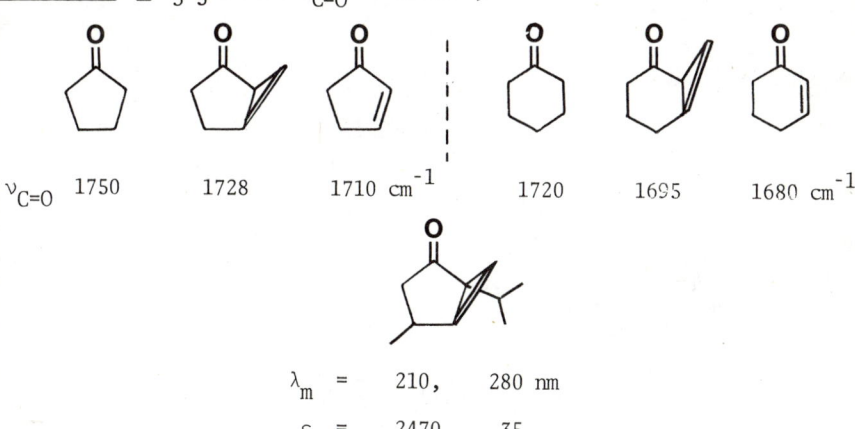

λ_m = 241   266 nm [13]
ε 16,000   13,700

IR spectra ($\underline{c}$-C_3H_5 lowers $\nu_{C=O}$ 10-30 cm^{-1})

$\nu_{C=O}$ 1750   1728   1710 cm^{-1} 1720   1695   1680 cm^{-1}

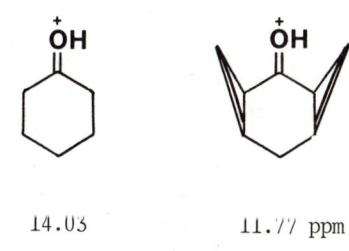

λ_m = 210,   280 nm
ε = 2470   35
$\nu_{C=O}$ = 1721 cm^{-1}

NMR spectra

$\delta\underline{\overset{\oplus}{O}H}$ =   14.03   11.77 ppm

The cyclopropyl group withdraws positive charge to move the $\overset{\oplus}{OH}$ proton chemical shift upfield of that for the protonated cyclohexanone [14].

Solvolysis rates [15]

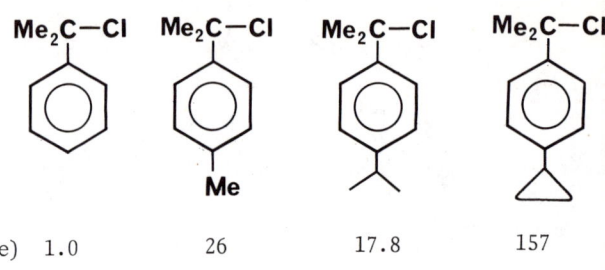

k_{rel} (aq. dioxane)   1.0   26   17.8   157

Aromatic nitration [16]

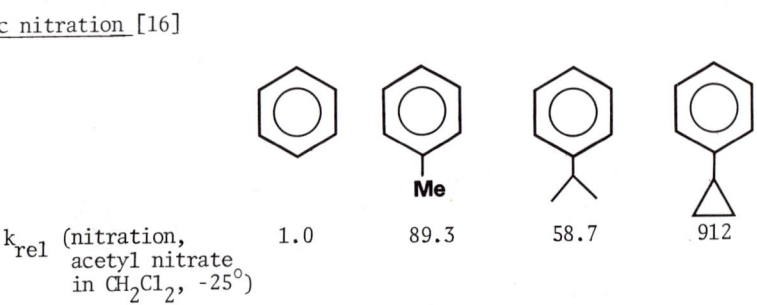

k_{rel} (nitration,   1.0   89.3   58.7   912
acetyl nitrate
in CH_2Cl_2, $-25°$)

Thus, the cyclopropyl group markedly enhances the rate of solvolysis of benzhydryl chlorides and the rate of electrophilic substitution by stabilizing the incipient benzyl or phenonium cations.

Nmr spectra reveal that positive charge actually increases in the cyclopropyl ring at the β-carbon. For example, the chemical shift of the α proton of 5.8 is larger than that of the β protons

5.8   5.9

since the former is closer to the cationic center. However, their positions are reversed in 5.9; the signal for the β proton is downfield from that for the α proton because hyperconjugation delocalizes positive charge to the β-carbon [17]. Similarly, the ^{1}H and ^{13}C chemical shifts for the cyclopropylcarbinyl cation are downfield from those of comparable models, reflecting an increase in positive charge on the respective atoms [18]. The ^{13}C chemical shifts (underlined in the drawing) are ppm from CS_2 and the ^{1}H chemical shifts are ppm from TMS. For comparison, $\delta(^{13}C^+)$ for Me_3C^+ is -135.4 ppm and $\delta(^{13}C)$ for $\underline{c}\text{-}C_3H_6$ is 196.3 ppm. Thus, the

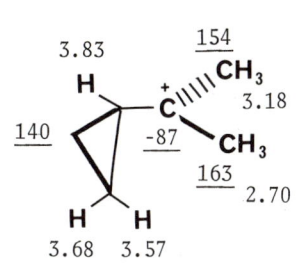

carbinyl carbon signal is upfield from that for Me_3C^+, indicating less positive charge on the former, and $\delta(^{13}C)$ for the β carbon of $\underline{c}\text{-}C_3H_5\overset{+}{C}Me_2$ is not far from that for the α carbon, even though the latter is closer to the C^+ atom. Other examples of the stabilizing ability of a cyclopropyl group on an α carbonium ion by delocalizing the positive charge are given in Section 6.1b and in Chapter 8.

It should be pointed out here that the foregoing comparisons provide evidence for a substantial ground-state conjugative interaction between a cyclopropyl group and an attached π system, which is greatest for the bisected conformation [19]. In this conformation, the planes of

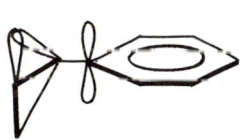

the phenyl and cyclopropyl groups, of cyclopropylbenzene for example, are perpendicular to each other, thereby allowing orbital overlap between the C-C $\underline{sp}^5$ hybridized orbitals of the cyclopropyl ring and the phenyl carbon $\underline{p}$ orbitals. This overlap between a cyclopropyl group and an adjacent π orbital decreases only slightly for deviations up to about 39° from the optimum geometry and thereafter decreases rapidly to a minimum at 90° [20]. Conjugation in the excited state is similarly sensitive to the dihedral angle but substantial only when there is a strong electron demand on the cyclopropyl group [21]. For example, there is only a modest red-shift for arylcyclopropanes compared to the corresponding arylalkyl compounds:

G	G-alkyl λ_m (nm)	G-C$_3$H$_5$-c λ_m (nm)	$\Delta\lambda$ (nm)
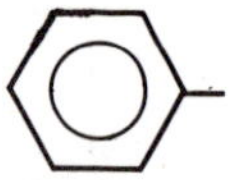 phenyl	268	275	7
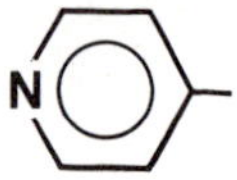 4-pyridyl	255	259	4
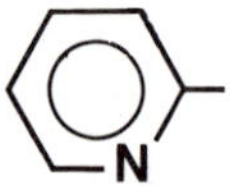 3-pyridyl	261.5	268.5	7
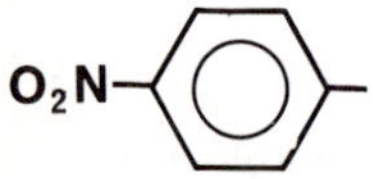 O_2N-phenyl	265.5	283	17.5
 2-thienyl	235	243	8

However, the bathochromic change is significantly larger for the p-nitrophenyl derivative, owing to the greater electron withdrawing ability of this group [22].

At one time the short C-C bond lengths α to a π bond, e.g.,

$H_3C\text{-}CH=CH_2$ and $H_3C\text{-}C\equiv CH$ compared to $H_3C\text{-}CH_2CH_3$
1.488 Å   1.459 Å   1.53 Å

were attributed to the double bond character which arises from hyperconjugation. However, it is generally accepted now that these are the normal bond lengths when one considers the natural covalent radii of the respective hybridized carbon atoms [23]. Actually, electronic effects of alkyl groups do not vary over a wide range and the modes of transmission via polarization, induction, field effects, and hyperconjugation are so interrelated that a firm assignment of a given property to any one of these effects would be tenuous. (See Section 3.3.)

Sec. 5.3 HYPERCONJUGATION   195

QUESTIONS

1. The dipole moments of the cyclopropyl compounds

▷=O	⟩=O	▷−	⟩=
DM 2.67	2.89	0.402	0.509 D

are smaller than those of the corresponding dimethyl analogues. How do you explain the consistent differences?

2. Predict the relative proton affinities of methanol, ethane, and cyclopropane and give the basis for your prediction.

ANSWERS

1. One explanation would be that the cyclopropyl carbons are more electronegative than methyl carbons. This is generally regarded to be so, although the differences are slight, e.g., σ* 0.017 and 0.00 for c-C$_3$H$_5$ and CH$_3$. [10a]. An alternate explanation is that the cyclopropyl ring delocalizes the positive charge on the C$_{sp^2}$ carbon less than do the CH$_3$s. The C$^+$ orbital is perpendicular to the molecular plane whereas the ring sp^5-hybridized C-C orbitals are in the molecular plane. Hence, there is less overlap in the cyclic compounds, whereas there is some hyperconjugation in the methyl compounds. (V. W. Laurie and W. M. Stigliani, J. Am. Chem. Soc. 92: 1485 (1970).)

2. Methanol, with its basic oxygen atom, should have the largest proton affinity. That of cyclopropane should be larger than that of ethane because of the p character of the C-C bond. Protonated cyclopropane is stabilized by nonclassical resonance. (For comments and references on the structure of protonated cyclopropane, see C. C. Lee, S. Vassie, and E. C. F. Ko, J. Org. Chem. 37: 8931 (1972).) The measured proton affinities in kcal/mol are methanol (179), cyclopropane (179), ethane (127), and the heats of formation of the corresponding protonated species are 139, 200, and 219 kcal/mol, respectively. (S.-L. Chong and J. L. Franklin, J. Am. Chem. Soc. 94: 6347 (1972); M. Saunders, P. Vogel, E. L. Hagen, and J. Rosenfeld, Accts. Chem. Res. 6: 53 (1973).)

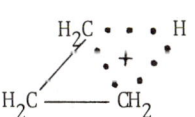

5.4 Resonance energies

Resonance energies are one of the few semiquantitative aspects of resonance theory (Taft σ_R constants are another, Section 7.1). The resonance energy (RE) of a compound is conventionally defined as the difference between its measured bonding energy and that calculated for its most stable valence-bond structure. The latter value can be taken from data in Table 1.5 and the measured value from thermochemical data such as heats of combustion or heats of hydrogenation. For example, the calculated heat of atomization of one of the Kekulé structures of benzene is

$$\Delta H_a \text{ (calc)} = 6(C-H) + 3(C-C) + 3(C=C)$$
$$= 6(99) + 3(83) + 3(147)$$
$$= 1284 \text{ kcal/mol.}$$

The experimental heat of combustion of benzene is 786 kcal/mol, so that from the equation for the combustion

$$C_6H_6 + 7.5\ O_2 \rightarrow 6\ CO_2 + 3\ H_2O + 786 \text{ kcal}$$

and the heats of atomization of O_2, CO_2, and H_2O (Table 1.5), the heat of atomization of benzene is

$$6(411) + 3(245) - 7.5(146) - 786 = 1320 \text{ kcal.}$$
$$\quad CO_2 \quad\quad H_2O \quad\quad\ O_2$$

The difference, 1320 - 1284 = 36 kcal/mol is the RE of benzene.

An alternative thermochemical method of determining resonance energies is from heats of hydrogenation (ΔH_{H_2}) data. As we saw in Section 5.3, ΔH_{H_2} for two olefins, especially if they are being hydrogenated to the same product, will reflect the relative stabilities of the two olefins. For example, ΔH_{H_2} is normally smaller for <u>trans</u> than for <u>cis</u> isomers because steric repulsion between <u>cis</u> groups either inhibits delocalization across the double bond or produces a destabilizing torsional strain.

Sec. 5.4 RESONANCE ENERGIES

ΔH_{H_2} (kcal/mol)

trans-2-butene	27.62
cis-2-butene	28.57
trans-stilbene	20.6
cis-stilbene	26.3
Methyl trans-cinnamate	24.7
Methyl cis-cinnamate	29.0
Diethyl fumarate	29.8
Diethyl maleate	34.0

The ΔH_{H_2} for 1-pentene is 30.3 kcal/mol and the total for both double bonds of 1,4-pentadiene is 60.6 kcal/mol, as expected. However, the value for 2-pentene is only 28 kcal/mol, and the difference between this and ΔH_{H_2} for 1-pentene, 2.3 kcal, can be attributed to greater hyperconjugation in 2-pentene. The value to expect for 1,3-pentadiene, in the absence of resonance, would be 30.3 kcal for the terminal bond and 28 for the inner double bond, or a total of 58.3 kcal/mol. The measured value is 54.1, giving the molecule a resonance energy of 4.2 kcal/mol. From the ΔH_{H_2} value of cyclohexene (28.6 kcal/mol, see Figure 5.1), one would expect a value of 85.8 kcal/mol for a Kekulé

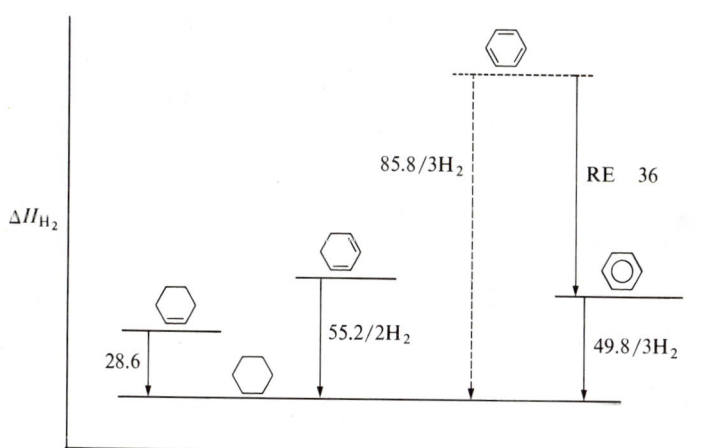

Figure 5.1. Chart of some heats of hydrogenation.

structure of benzene. The measured value is only 49.8 kcal for the addition of 3 H_2, giving a resonance energy of 85.8 - 49.8 = 36 kcal/mol. It is significant that hydrogenation of olefins is an exothermic reaction but, because of the RE of the benzene ring, the addition of the first H_2 to benzene to give cyclohexadiene is endothermic.

Resonance energies for various compounds are listed in Table 5.4. They have also been computed from constituent constants [24] and by quantum mechanical methods [25]. A recent proposal has been to take reference bond energy terms from acyclic conjugated systems [26]. Called the Dewar resonance energy (DRE), this method represents the conjugation in a ring system which is in excess of that in a similar acyclic system. Thus, DRE makes a comparison with "real" molecules having mostly localized π bonds whereas the conventional method uses hypothetical molecules for reference. Even recent MO calculations have taken this approach to compute resonance energies and to define aromaticity [27, 28]. The principal effect which the DRE method brings out is the aromaticity of rings: DRE is positive for aromatic, close to zero for non-aromatic, and negative for anti-aromatic rings [28].

Table 5.4. Resonance energies (kcal/mol)

Compound	RE	Compound	RE
Benzene	36	Phenol	43
Naphthalene	61	Aniline	42
Anthracene	84	Benzaldehyde	40
Phenanthrene	92	Benzoic acid	40
(18)Annulene	100	Furan	16
Biphenyl	71	Pyrrole	22
Triphenylene	119	Thiophene	29
Azulene	46	Propionic acid	24
Styrene	41	Ethyl propionate	24
trans-Stilbene	77	Acetamide	25
1,3,5-Cycloheptatriene	7	Carbon monoxide	58
Cyclooctatetraene	5	Carbon dioxide	33
Butadiene	4	Carbon disulfide	29
Tropolone	29	Urea	41
Pyridine	37		

Sec. 5.4   RESONANCE ENERGIES   199

Finally, a rarely used method of estimating resonance energies is one obtained from pK data. For example, it is shown in Section 6.2a that the difference in pK_b between an aliphatic amine and aniline, $\sim 10^6$, is almost entirely due to the phenyl-amino group resonance in the free base but which is absent in the cation. Accordingly, RT ln (ΔK) = 8.4 kcal/mol can be assigned to this resonance. This, added to the 36 kcal for the benzene ring, gives a value near the thermochemical value. Similarly, resonance energies of some heterocyclic compounds have been estimated from pK data [29].

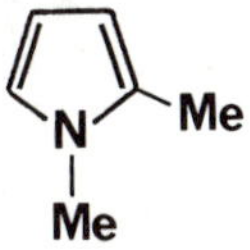

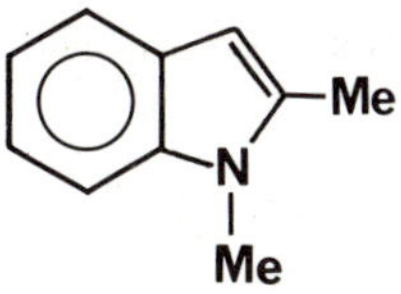

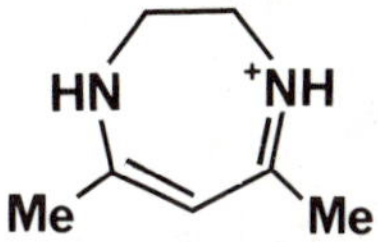

RE (kcal/mol) $\sim$22 $\sim$43.5 $\sim$19

Chapter 6

APPLICATIONS OF RESONANCE THEORY

The great value of resonance theory is the ability it gives chemists to understand and predict physical properties of molecules in terms of classical valence-bond structures. By and large, this includes the structures, spectra, and dynamic properties of substances. Some of these properties will be discussed in this chapter and the application of resonance to the interpretation of spectra will be taken up in Chapter 10.

6.1. Structure

6.1a. Molecular geometry

Bond lengths and planarity of conjugated systems are often explained with the aid of resonance theory. The observed bond order of a given bond in a molecule -- that is, its resemblance to a pure single, double, or triple bond with respect to length, strength, force constant, ir stretching frequency, etc. -- is accounted for in terms of a proposed resonance hybrid. The bond is drawn as a single (or double) bond in one or more valence-bond structures and double (or triple) in others. In this way, the resonance hybrid accounts for the observed intermediate bond characteristics. Some examples concerning bond lengths can be given here, with other properties cited in support of the proposed structures. The case of vinyl bromide was mentioned in Section 5.20

Urea $\quad \left[\begin{array}{c} H_2N \\ H_2N \end{array} \!\!\! C=O, \quad \begin{array}{c} H_2\overset{\oplus}{N} \\ H_2N \end{array} \!\!\! \overset{\ominus}{C-O}, \quad \begin{array}{c} H_2N \\ H_2\overset{\oplus}{N} \end{array} \!\!\! \overset{\ominus}{C-O} \right]$

	Observed	Normal
C=O bond length	1.26 Å	1.21 Å
C-N bond length	1.33 Å	1.47 Å

Therefore, the C-N bond has considerable double-bond character (30) and the C=O bond has some single-bond character. Urea has a large dipole moment, 4.56 D, and a large resonance energy, 41 kcal/mol; hence, the polar structure is almost as good a representation for the compound as the nonpolar structure.

Cyanogen chloride   [Cl-C≡N, $\overset{\oplus}{Cl}=C=N^{\ominus}$]

	Observed	Normal
C-Cl bond length	1.63	1.78 Å
C-Cl force constant	5.3	3.4×10^5 dynes/cm
C≡N force constant	16.7	18.1×10^5 dynes/cm
C≡N ir frequency	2201	2260 cm^{-1}

Thus, each bond has some double-bond character in which the C-Cl bond is shorter and stronger than usual and the C≡N bond is longer and weaker. In connection with the ir frequencies of some nitriles, note that resonance delocalization outweighs the inductive effect; the latter would raise rather than lower the frequencies in going from $CH_3C≡N$ to ICN to ClCN:

	ir frequency (31)
CH_3CN	2254 cm^{-1}
ClCN	2201
BrCN	2187
ICN	2158
$Cl-CH_2CH_2CN$	2260

We see then that we can use resonance to account for bond lengths but the effect is significant only when there is a substantial electron delocalization such as between equivalent or isovalent structures. For example, even the nitro-phenyl group resonance, which is not isovalent, has little effect on the C-N bond length:

Nonisovalent resonance   Isovalent resonance

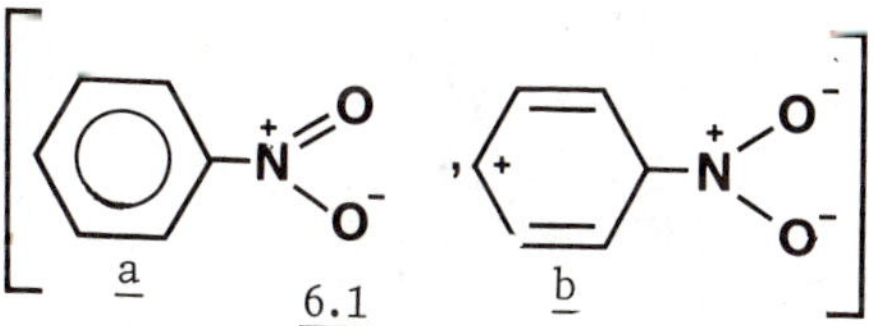

[structures 6.1 a, b]   [structures 6.2 a, b]

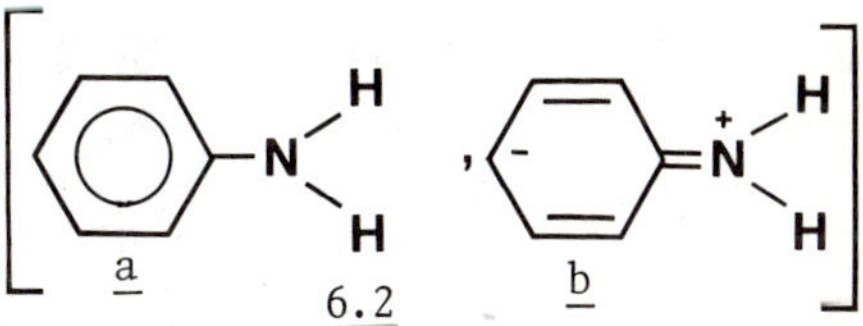

Observed C-N lengths:

Ph-NO$_2$	MeNO$_2$	Ph-NH$_2$	MeNH$_2$
1.486 Å	1.49 Å	1.431 Å	1.474 Å

Thus, the C-N distance in nitrobenzene is the same as in nitromethane, indicating that 6.1b makes little contribution to the resonance hybrid in the ground state. We saw in Section 4.2c that this same view is supported by uv data. The NH_2-phenyl group resonance is isovalent however, and we observe a definite bond contraction, 0.043 Å, which is larger than the 0.02 Å due to the change from C_{sp^3} to C_{sp^2} carbon atoms. Similarly, both bonds are substantially shorter in p-nitroaniline where there is an isovalent <u>interaction resonance</u> (see Section 6.1b).

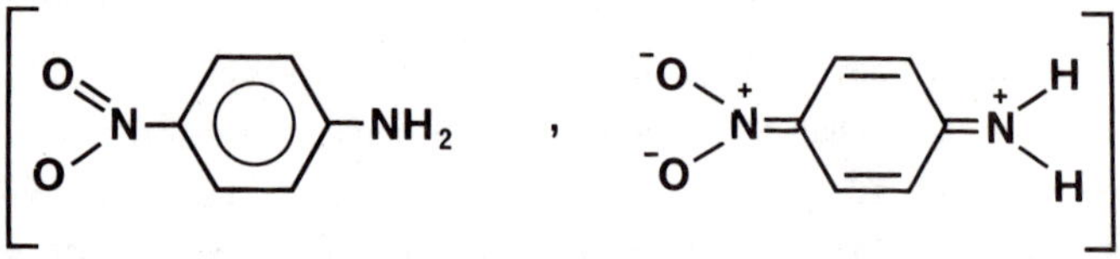

Among the conditions for substantial resonance listed in Section 5.2 is a planar σ system. Accordingly, a number of molecular fragments maintain a planar conformation, "locked" into this planar structure by the accrued resonance energy:

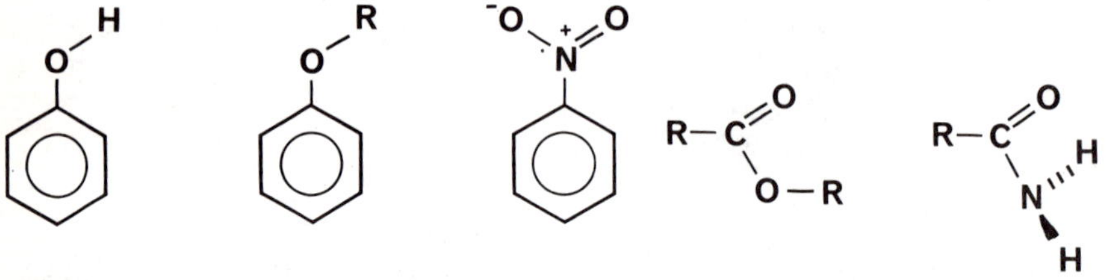

In fact, the barrier to rotation about the C_1-substituent bond in benzene derivatives is a crude measure of the resonance of the substituent with the ring or of interaction resonance if there is a <u>para</u> substituent. Scattered values for monosubstituted benzenes have been reported (32,33).

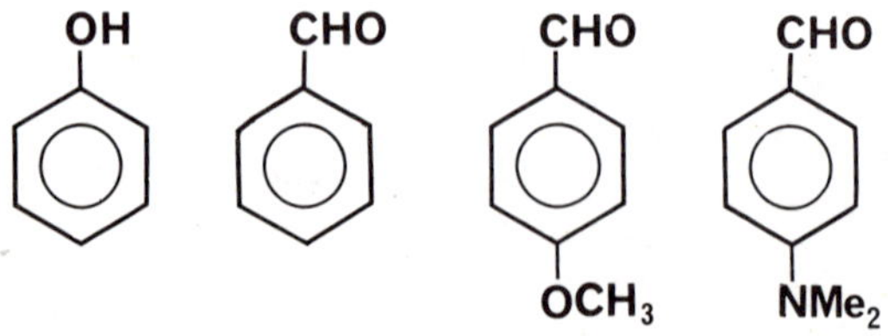

	OH	CHO	CHO	CHO
			OCH_3	NMe_2
Rotation barrier (kcal/mol)	3.56	7.9	9.2	10.8

The differences between the values for the disubstituted aldehydes and that for benzaldehyde can be taken as the interaction resonance energies between the CHO and substituents. A much larger value is found for the NHMe group of 6.3 because of an intramolecular H-bond (34). Surprisingly, the barrier to rotation of an NHR group about the C-N bond in N-alkyltrinitroanilines decreases in the order ethyl, isopropyl, t-butyl (35). The barrier to internal rotation of the NMe$_2$ group in dimethylanilines has been related to the nature of the para substituent (36).

The nitrogen atom of aromatic amines is pyramidal, although resonance flattens out the N to reduce ψ (ψ = 70.50 for a tetrahedral N). Thus, in p-F-C$_6$H$_4$NH$_2$, the competing fluorine-ring resonance dampens the ring-NH$_2$ resonance; hence ψ is enlarged and ϕ reduced (37). It appears that the only anilines with a virtually planar nitrogen atom are p-nitroaniline (38) and its N-alkyl derivatives (35).

	ψ	ϕ
C$_6$H$_5$NH$_2$	37°39'	113°16'
p-F-C$_6$H$_4$NH$_2$	46°22'	111°52'
NH$_3$		107°

With increasing NH$_2$-phenyl resonance, the lone-pair electrons become more p in character and hence, the N-H bonds have more s character. This increases ϕ and makes the N-H bond more polar and stronger. These effects are reflected in the N-H ir stretching frequencies (39). (see also the end of Chapter 13).

When strain or other steric conditions prevent a coplanarity of two attached π systems, the electron delocalization is cut off. For example, the C-C bond distances and resonance energy of butadiene and cyclooctatetraene indicate that they have about the same small degree of resonance.

| C-C length (40) | 1.467 | 1.476 Å |
| Resonance energy | 4 | 5 kcal/mol |

Note that $^{13}C-^{13}C$ coupling constants indicate a C_1-C_2 bond order less than 2 and a C_2-C_3 bond order greater than 1, as expected for the butadiene resonance hybrid (42).

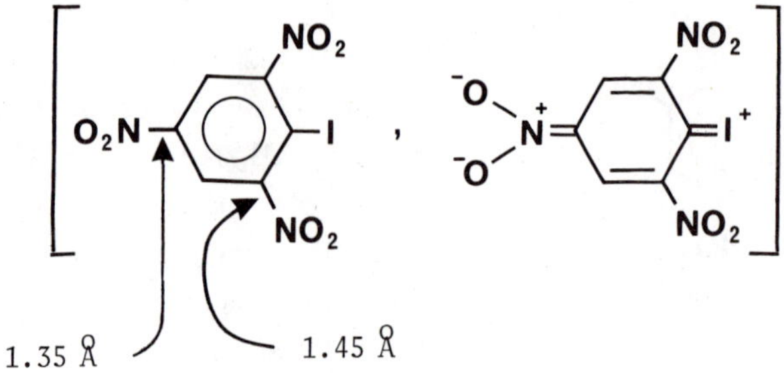

[picryl iodide resonance structures: left structure shows benzene ring with O₂N, I, NO₂ substituents with bond lengths 1.35 Å (C-N) and 1.45 Å (C-N); right structure shows quinoid form with N=, I+, NO₂]

Picryl iodide

In picryl iodide, the large iodine atom prevents the o-nitro groups from lying in the plane of the ring and thereby blocks the nitro-phenyl group resonance. The p-nitro, however, does not experience this restriction and exhibits an isovalent resonance. As a result, the C-N bond length is much shorter for the p-NO$_2$ group than for the o-NO$_2$ groups (41).

There are several ways to denote the resemblance of a given bond to a pure single, double, or triple bond. One is the <u>bond order</u>, which, based on valence-bond structures, is 1,2, and 3 for pure single, double, and triple bonds, respectively. Each MO method has its own set of bond orders which differ only slightly from each other (43). Alternately, the percentage double bond character of a bond might be given. This has been expressed by an empirical equation (44)

$$\underline{P}_d = \frac{100(\underline{r}_s - \underline{r})}{2\underline{r} + \underline{r}_s - 3\underline{r}_d} \qquad (6.1)$$

where $\underline{P}_d$ = percentage double bond character of a given bond

$\underline{r}$ = observed bond length

$\underline{r}_s$ = normal single bond length

$\underline{r}_d$ = normal double bond length.

For example, the percentage double bond characters of a few cases are:

C-N bond in (H$_2$N)$_2$CO   42%

C-Cl in Cl-CN   42

C-Cl in C_6H_5Cl 16

C-C in $H_2C=CH-CHO$ 18

Attempts to estimate the double bond character of bonds from other physical properties have been less satisfactory. One proposal is based on bond refractions, which gives several improbable values (45). For instance, that for the C-F bond (2%) in fluorobenzene is too low and the value for the C-N bond of nitrobenzene (40%) is far too high. Whereas the double bond character of the C-Cl bond of chlorobenzene is estimated as 16% from bond length data, it is calculated to be 2% from bond refractions (45) and 4% from dipole moments (46).

6.1b. Resonance and polar character

A resonance hybrid structure is written to explain the polar character and charge distribution in a molecule. For example, the resonance of monosubstituted benzenes accounts for the relative charge densitites at the ring positions, which is reflected in nmr chemical shifts (Chapter 10). Electron-withdrawing groups produce a downfield shift of $\delta^{15}N$ of the amino nitrogen atom in substituted anilines and the effect is greatest for those groups which exert an electron-accepting resonance interaction with the amino group (47).

R	$\delta^{15}N$	δ^1H
	(ppm from aniline)	
4-NO_2	19.7	1.73
3-NO_2	4.9	0.85
3-CF_3	2.7	0.62
4-I	1.4	0.30
4-Br	1.0	0.25
4-Cl	0.3	0.25
H	0.0	0.0
4-Me	-2.6	-0.2
4-F	-3.3	-0.02
4-MeO	-5.8	-0.42

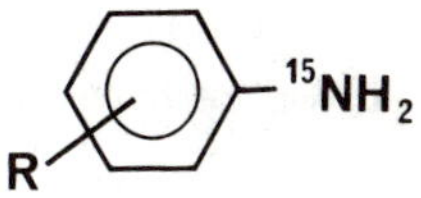

Similarly, polar resonance structures are written to explain various reactions. A simple example is one which reveals the anionic character of the para position to the phenoxide ion (48), as do all of the base-catalyzed condensations of phenols, e.g., the Reimer-Tiemann and the Kolbe-Schmitt reactions.

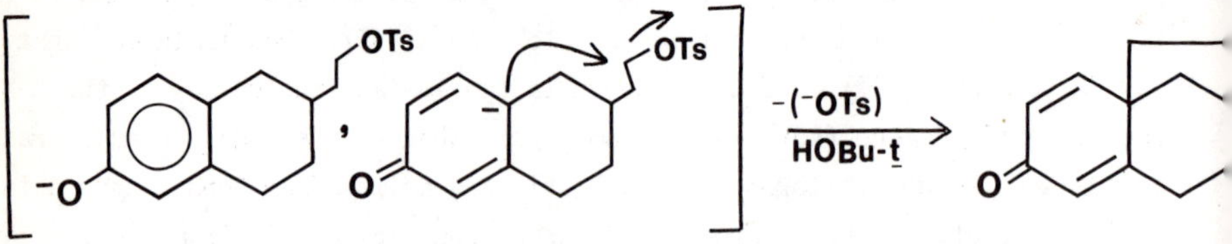

The more significant are the dipolar structures of a resonance hybrid, the more polar is the molecule. For example, the dipole moments of conjugated systems increase with length because more dipolar structures can be written.

	Dipole moment
CH_3CHO	2.72
$CH_3CH=CH-CHO$	3.67
p-$Me_2N-C_6H_4$-CHO	4.29
p-$Me_2N-C_6H_4$-CH=CH-CHO	5.4
p-$Me_2N-C_6H_4$-CH=CH-$COCH_3$	5.3
p-$Me_2N-C_6H_4$-(CH=CH)$_2$-$COCH_3$	6.7

The polarity of a molecule, of course, affects its transition temperatures, solubility, dipole moment, and other properties. For example, the dipole moment of formamide is larger rather than smaller (opposed bond dipoles would make formamide have the smaller moment) than that of formaldehyde as a result of the contribution

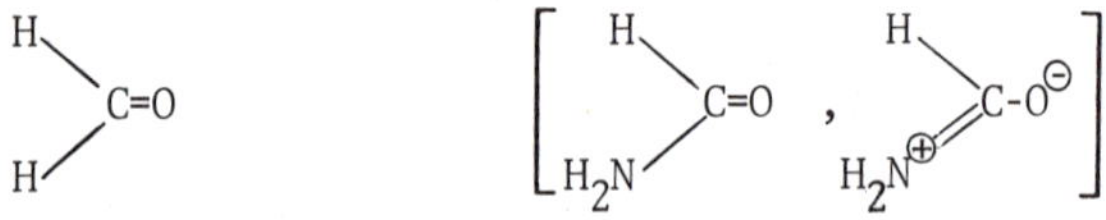

DM: 2.27   3.71 D

of the dipolar structure. Bond lengths also confirm this view, as the C-N bond is much shorter than normal (C_{sp^2}-N ≈ 1.45 Å), and so does the large barrier (18 kcal/mol) to rotation about the C-N bond (49).

The dipole moment, of course, depends upon the magnitude of the charges and the distance between opposite poles. For example, the moments of the members of the vinylogous series, $Me_2N\text{-}(CH=CH)_n\text{-}CHO$ increase with $\underline{n}$ (50). This occurs in spite of the fact that MO calculations indicate that the charges on the nitrogen and oxygen atoms decrease with chain length (51).

When the dipole moment for dipolar resonance structures augments the inductive moment, the observed moment for an unsaturated compound is larger than that for similar saturated compounds.

$Me_2N\text{-}(CH=CH)_{\underline{n}}\text{-}CHO$

$\underline{n}$	$\mu_{obs.}$	$\mu_{calc.}$
0	3.86 D	4.3 D
1	6.24	7.3
2	7.67	9.6
3	8.24	11.2
4	8.50	12.3

$CH_3\text{-}CH_2\text{-}CH=O$ $[H_2C=CH\text{-}CH=O,\ H_2\overset{\oplus}{C}\text{-}CH=CH\text{-}O^{\ominus}]$

DM   2.73 D   3.11 D

$CH_3CH_2CH_2CH=O$ $[H_3C\text{-}CH=CH\text{-}CH=O,\ H_3C\text{-}\overset{\oplus}{CH}\text{-}CH=CH\text{-}O^{\ominus},\ H_2C=CH\text{-}CH=CH\text{-}O^{\ominus}\ \overset{\overset{\oplus}{H}}{}]$

DM   2.72 D   3.67 D

When the resonance and induction dipoles are opposed, the observed moment of an unsaturated compound is smaller than that of the reference saturated compound. This is the case for aryl and vinyl halides (Table 6.1).

Table 6.1. Dipole moments of some halogen compounds

X	CH_3CH_2X	$CH_2=CH\text{-}X$	$HC\equiv CX$	C_6H_5X
Cl	2.05 D	1.44 D	0.44	1.60 D
Br	2.02	1.41	0	1.57
I	1.90	1.26		1.42

In practice, the difference between the moment of an aromatic compound and that of a suitable model is often taken as a measure of the resonance interaction between the substituent and the phenyl group. It is assumed that the moment of an aliphatic compound is due only to induction whereas the moment of the aromatic compound is the result of induction plus resonance.

Table 6.2. Resonance moments for some groups*

Group	μ_{arom}	μ_{ali}	μ_{res} (52)	μ_{res} (51)
N(CH$_3$)$_2$	1.61 D	0.86 D	1.66 D	1.66 D
NH$_2$	1.48	1.32	1.02	1.02
OCH$_3$	1.35	1.23	0.8	0.96
OH	-1.6	-1.7	0.6	
CH$_3$	0.35	0	0.35	0.35
F	-1.59	-1.89	0.30	0.41
Cl	-1.70	-2.01	0.30	0.41
Br	-1.73	-2.00	0.27	0.43
I	-1.7	-1.92	0.22	0.50
CF$_3$	-2.6	-2.4	-0.2	-0.2
CN	-4.39	-3.60	-0.79	-0.45
NO$_2$	-4.21	-3.25	-0.96	-0.76
COCH$_3$	-3.0	-1.57	-1.43	-0.46

* Positive values indicate negative charge into ring.

It is also assumed that the induction contributions to the aliphatic and aromatic systems are approximately equal. This latter assumption is very dubious; nevertheless, the results as determined in this fashion are given in Table 6.2. For atoms or linear groups, μ_{res} is simply $\mu_{arom} - \mu_{ali}$, whereas for bent groups, corrections were made for the angle between the resultant moment and the C_1-C_4 axis.

We saw in Section 3.2 that the dipole moments of disubstituted benzenes are close to the vector sum of the moments of the respective monosubstituted benzenes. When the substituents are the type which exhibits dissimilar or complementary resonance, there is an enhanced <u>resonance interaction</u> between them which gives the compound a larger dipole moment than expected on the basis of the corresponding monosubstituted benzenes. For example,

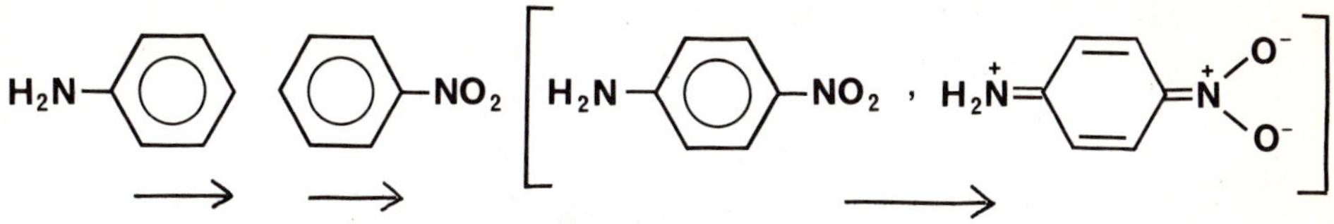

Sec. 6.1 STRUCTURE 209

$H_2N-C_6H_5 \rightarrow$ $C_6H_5-NO_2 \rightarrow$ $[H_2N-C_6H_4-NO_2 \rightarrow H_2\overset{+}{N}=C_6H_4=\overset{+}{N}(O^-)_2]$

DM: 1.53 D   3.95 D   6.1 D

Interaction moment = 6.1 − (1.53 + 3.95) = 0.62 D

The interaction moments for some other compounds are given in Table 6.3.

Table 6.3. Interaction resonance moments for some p-disubstituted benzenes

Compound	Interaction resonance moment
p-H_3C-C_6H_4-CHO	0.15 D
p-H_2N-C_6H_4-CF_3	0.15
p-HO-C_6H_4-COOH	0.27
p-CH_3O-C_6H_4-COOH	0.28
p-HO-C_6H_4-CO_2Et	0.35
p-Me_2N-C_6H_4-CN	0.35
p-Me_2N-C_6H_4-CF_3	0.44
p-H_2N-C_6H_4-CO_2Et	0.55
p-Me_2N-C_6H_4-NO_2	1.34
p-O=C=N-C_6H_4-NO_2	1.57
p-Me_2N-C_6H_4-NO	2.63

The increase in contribution from the quinoidal structure for disubstituted benzenes with dissimilar groups is reflected in other properties too, such as uv and nmr spectra (Chapter 10) and Hammett substituent constants (Chapter 7). For instance, nmr coupling increases with bond order. Hence, it is greater across a double bond than between two ortho protons (J_{ortho}) of a benzene ring, and J_{para} (coupling between para positions) is greater for a benzene ring than $J_{1,4}$ of a quinoidal ring (supposedly the bond order of three C_{ar}-C_{ar} bonds is greater than two single and one double bond). Since p-nitrophenol has a large interaction resonance, and thereby substantial quinoidal character, J_{ortho} ($J_{2,3}$) is larger than for most other phenols and J_{para} ($J_{2,5}$) is smaller. This effect is even more pronounced for the p-nitrophenoxide ion. Resonance interaction is, however, void in the cation of p-nitroaniline; consequently, J_{ortho} is larger and J_{para} smaller in the free base than the cation. Still, J_{ortho} and J_{para} undergo little change for the bromo- and cyanophenols. These rationalizations are demonstrated by the data below (53):

	J_{ortho}	J_{para}
p-O_2N-$C_6H_4O^-$	9.45 Hz	0.06 Hz
p-O_2N-$C_6H_4NH_2$	8.97	0.14
p-O_2N-C_6H_4OH	8.89	0.23
p-Br-$C_6H_4O^-$	8.77	0.25
p-Br-C_6H_4OH	8.71	0.26
p-O_2N-$C_6H_4\overset{+}{N}H_3$	8.59	0.31
p-NC-$C_6H_4O^-$	8.54	0.30
p-NC-C_6H_4OH	8.50	0.36

See Chapter 18 for a measurement of interaction resonance energy by nmr spectroscopy. These changes in dipole moments are in part due to alterations in geometry of the substituents accompanying changes in hybridization. This latter effect has not been assessed.

The fact that there is an interaction moment for the CF_3 compounds has been offered as evidence for the controversial issue of C-F hyperconjugation (54) (compare the view expressed in Section 7.1).

$$\left[F_3C-\langle\bigcirc\rangle-NMe_2 \;,\; F_2C=\langle\rangle=\overset{+}{N}Me_2 \;\overset{F^-}{} \right]$$

Interaction resonance or, as it is sometimes called, "push-pull" resonance has been proposed for the unusual stability of two cyclobutadiene derivatives (55). It is well known, of course, that the parent hydrocarbon is extremely reactive and barely isolable even at low temperatures (56).

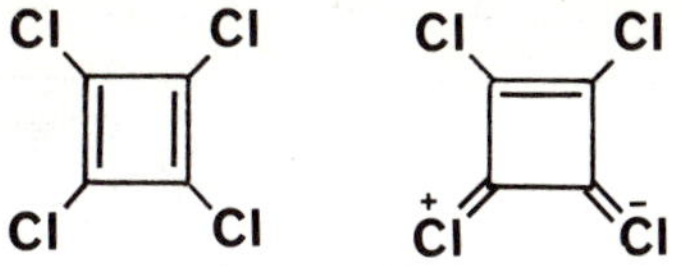

Sec. 6.1 STRUCTURE

Several equivalent structures

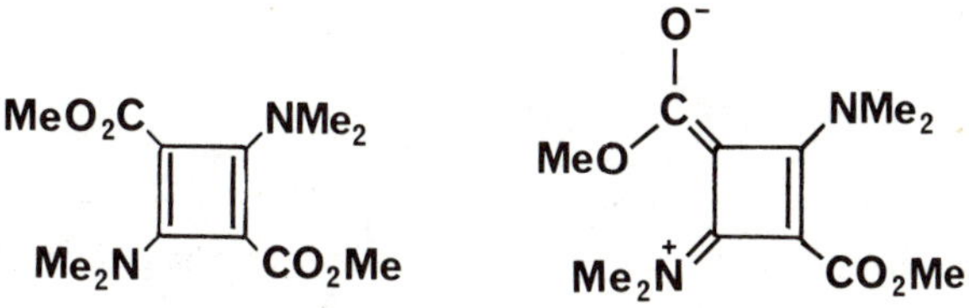

Several equivalent structures

When spatial requirements block coplanarity of two attached π systems, resonance delocalization is diminished and this will be reflected in the properties of the molecular aggregate (see also Chapters 11 and 20). For example, the ortho methyl groups in acetylmesitylene block the acetyl group from coplanarity

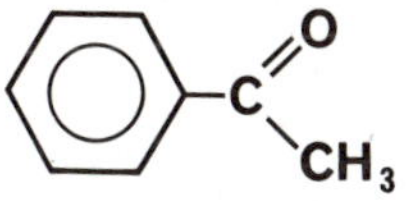

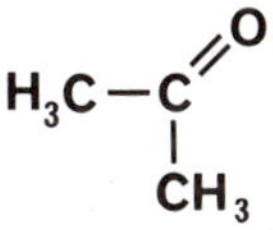

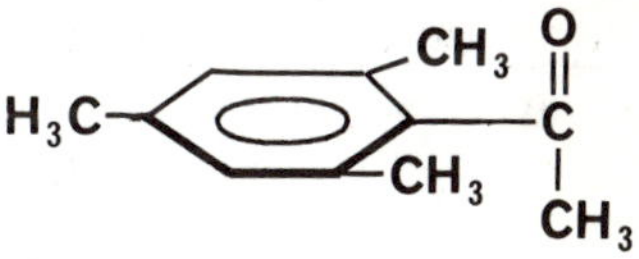

DM   2.88   2.72   2.71 D

with the ring and thereby prevent any resonance between the C=O and phenyl group. This eliminates the resonance moment found for acetophenone and reduces the moment of acetylmesitylene to that of the aliphatic acetone. For spectral effects of this steric hindrance see Chapter 11. Similarly, the moments of some disubstituted durenes are much smaller that those of the corresponding disubstituted benzenes (Table 6.4).

Table 6.4. Dipole moments of some disubstituted benzenes and durenes

		Dipole moment	
X	Y	p-X-C_6H_4-Y	X-durene-Y
NO_2	NMe_2	6.16 D	5.11 D
NO_2	NH_2	6.10	4.98 D
NO_2	OEt	4.74	3.68
NO_2	OH	5.04	4.08

The Me_2N-phenyl resonance is slightly larger than that of H_2N-phenyl, at least as reflected in the dipole moments of compounds 6.6 and 6.4 compared with those of 6.7 and 6.5, respectively. However, since it is larger in size, the NMe_2 group is more readily blocked from coplanarity with the phenyl group, in which case the respective dipole moment is decreased. This is observed for compounds 6.8 and 6.10, for which the moments are reduced below those of the corresponding NH_2 compounds (by the o-CH_3 in 6.8 and by the peri H in 6.10) (57). Similar changes are observed for the pair 6.12 and 6.13 and for the pair 6.14 and 6.15.

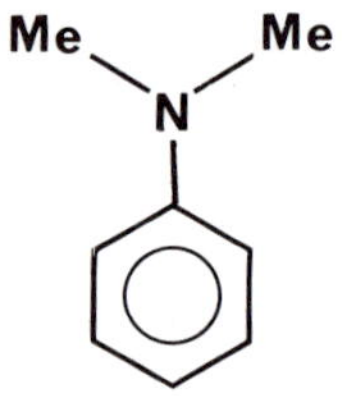

6.4
DM 1.61 D

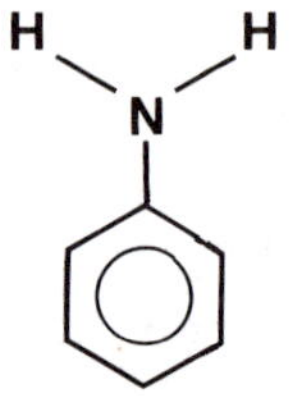

6.5
1.53 D

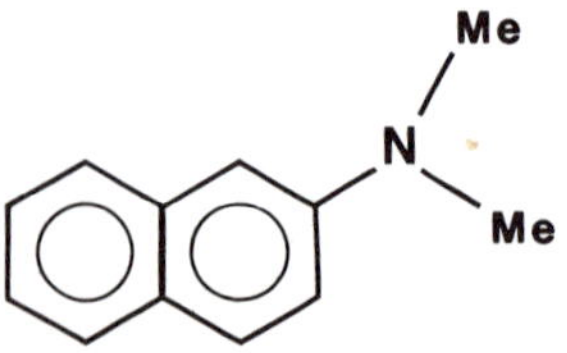

6.6
DM 1.90 D

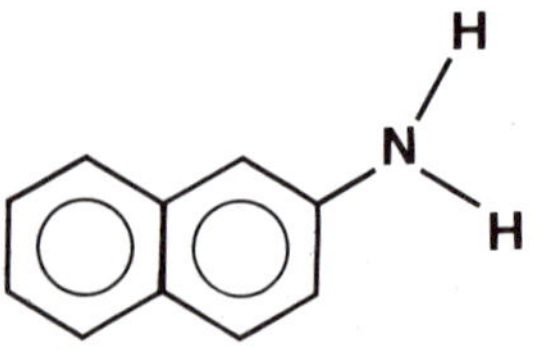

6.7
1.71 D

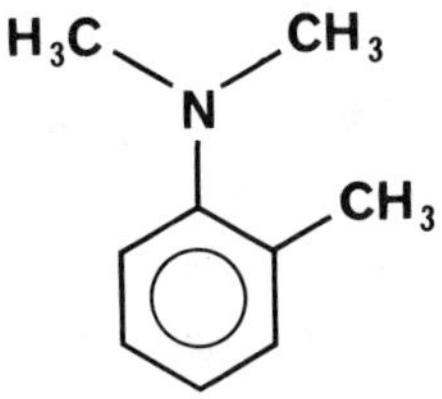

Sec. 6.1 STRUCTURE 213

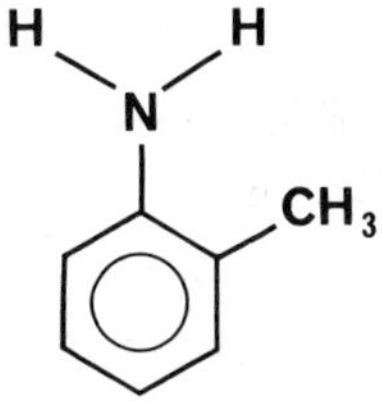

	6.8	6.9	6.10	6.11
DM	0.96 D	1.59 D	1.06 D	1.50 D

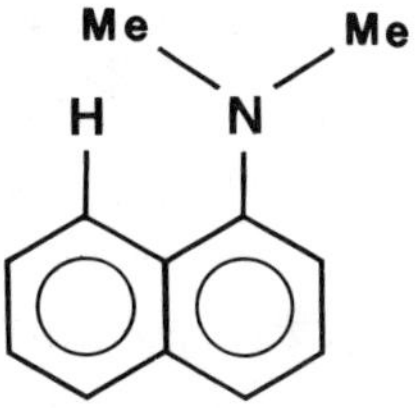

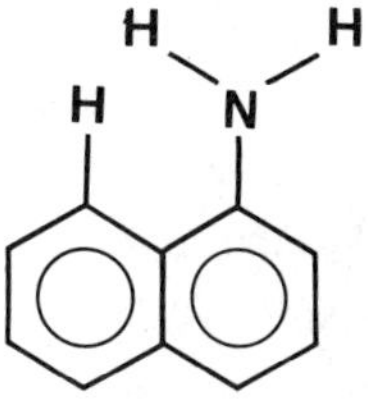

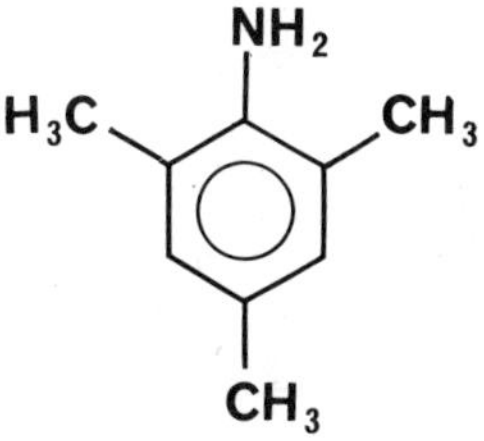	6.12	6.13
DM	1.45 D	1.03 D

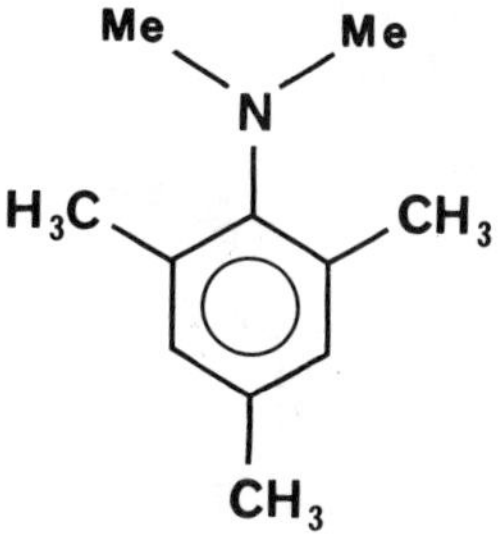

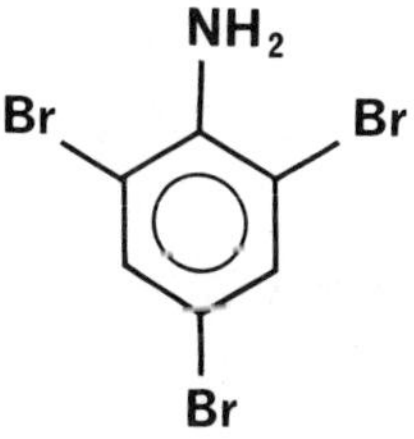	6.14	6.15
DM	1.73 D	1.02 D

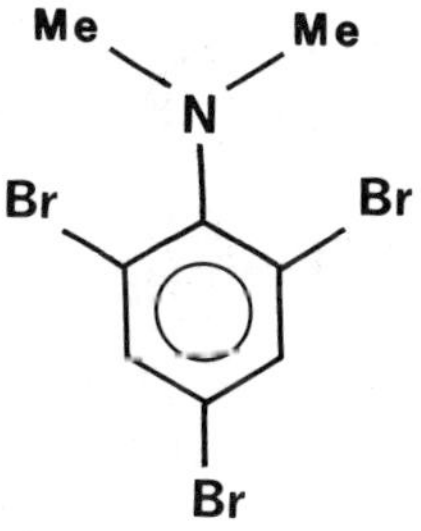

QUESTION

The observed dipole moment of p-nitrothioanisole (4.43 D; V. Baliah and M. Uma, Tetrahedron 19: 455 (1963)) is greater than that (3.99 D) calculated from the moments of the repective monosubstituted benzenes. However, the moment of dimethylaminothioanisole is _less_ than the calculated moment (2.24 D). This is not so for the p-dimethylaminoanisole. Offer a rationale for these opposite effects of the NO_2 and NMe_2 groups.

ANSWER

The nitro group on the one hand, is a strong electron-withdrawing group, which forces sulfur to exhibit π,p resonance.

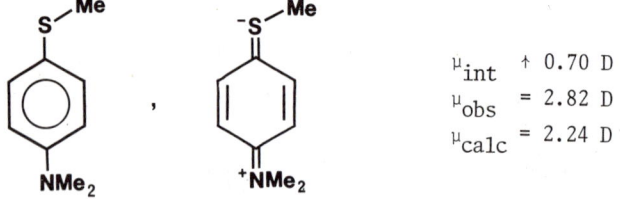

μ_{int} ↓ 0.46 D
μ_{obs} = 4.43 D
μ_{calc} = 3.99 D

On the other hand, the NMe_2 group is a strong electron donor and makes it energetically favored for sulfur to exhibit $\pi,\underline{d}$ resonance.

μ_{int} ↑ 0.70 D
μ_{obs} = 2.82 D
μ_{calc} = 2.24 D

The oxygen atom cannot undergo $p,\underline{d}$ resonance so the dipole moment of the oxygen analog is close to the additive moment from the two monosubstituted derivatives.

μ_{obs} = 1.76 D
μ_{calc} = 1.84 D

6.1c. Resonance and aromaticity

The classical definition of aromaticity is based on the chemical behavior of compounds, namely that an unsaturated compound, unlike typical olefins, is not easily oxidized, does not polymerize in acid, and undergoes <u>substitution</u> reactions in preference to <u>additions</u> (see Table 5.1). A modern definition is based on the presence of a ring current which is revealed by its magnetic properties (Table 6.5).

When a completely cyclic, conjugated, <u>planar</u> structure can be drawn for a substance, the substance can be expected to have aromatic character. Generally, two or more nearly equivalent valence-bond structures can be written, which implies a large resonance energy. For example, two Kekulé structures can be drawn for the hydrocarbons <u>6.16</u>, <u>6.17</u>, and <u>6.18</u>,

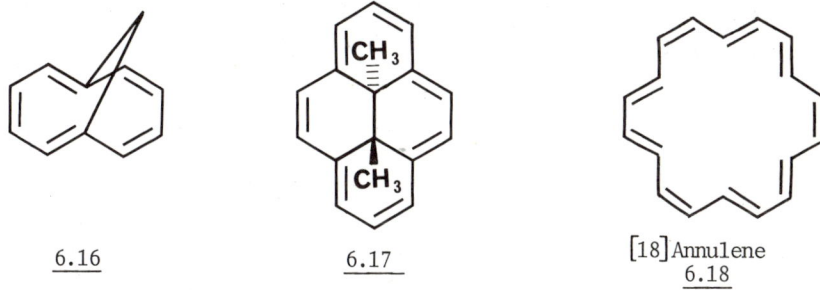

<u>6.16</u>   <u>6.17</u>   [18]Annulene
<u>6.18</u>

and they are found to be aromatic according to the classical as well as the modern definition. Thus, the compounds undergo electrophilic substitution but no Diels-Alder reaction, the rings are close to planar with C-C bond lengths (ca. 1.39 Å) and outer ring proton chemical shifts ($\delta \approx 8.8$) in the aromatic range, and the protons over the rings are greatly shielded ($\delta \approx -1$ to -4). The requirement for aromaticity of a nearly planar system is shown by the observation that both <u>cis</u>- and <u>trans</u>-(10)annulenes are nonplanar and are typical olefinic hydrocarbons, in contrast to the aromatic ring of <u>6.16</u> (58).

Hückel (59) offered the first theoretical basis for aromaticity. He showed that, among other things, in a conjugated monocyclic system the orbitals occur in degenerate pairs with a single orbital of lowest energy, illustrated by Figure 6.1. Accordingly, there is a closed shell only when there are $(4\underline{n} + 2)$ π electrons, $\underline{n} = 0,1,2,\ldots$. This is analogous to the "closed shell" arrangement of electrons about atoms of the inert elements ($2n^2$ electrons) for which there is marked chemical stability (chemical inertness). It has been predicted that the delocalization energy per double bond in these aro-

Table 6.5. Comparison of magnetic and spectral properties of aromatic and olefinic compounds

	Expected for a cyclohexatriene	Observed for a benzene ring
Magnetic susceptibility	$-36.9 \times 10^6 \chi$	$-55.6 \times 10^6 \chi$
NMR	$\delta \approx 5.5$ ppm Alternating vicinal coupling constants	$\delta \approx 7.2$ ppm Approximately constant vicinal coupling constants
Other spectral properties	IR: ν 1600-1620 cm^{-1} UV: λ_m with $\varepsilon > 10^4$	ν 1585 and 1450 cm^{-1} λ_m with $\varepsilon < 10^3$

matic systems will decrease with increasing ring size up to about (30)-annulene when the members will become simple polyolefins (60).

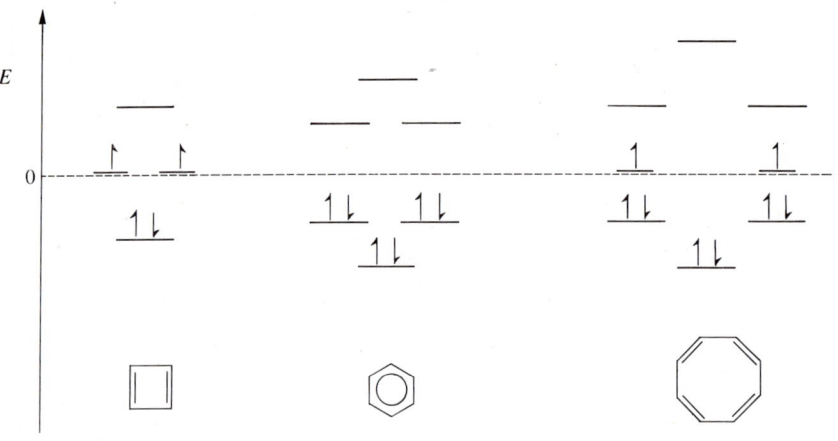

Figure 6.1. Molecular orbital energy diagram.

Although a molecular orbital basis for the Hückel rule of (4$\underline{n}$ + 2) π electrons for aromaticity was offered in 1931, it was noticed even before this that a sextet of π electrons promotes unusual stability. In 1922 it was suggested that a sextet of electrons confers reduced unsaturation and a tendency to be inert (61), and in 1928, the marked stability of the cyclopentadie ion was attributed to the presence of a sextet of electrons (62). It is like that even without the MO justification, an empirical rule equivalent to the (4$\underline{n}$ + 2) π electron rule would have been proposed just from the observation

Sec. 6.1 STRUCTURE 217

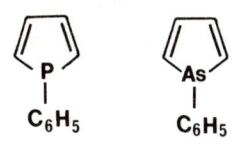

Phosphole   Arsole

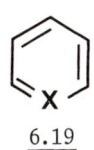

6.19

of aromatic character associated with 2, 6, 10, ..., π electrons.

The aromatic character of the six-π heterocyclic rings of pyrrole and thiophene, and possibly furan, are already well known. In contrast, the five-membered rings of phospholes and arsoles apparently are not aromatic, based on a variety of physical and reaction kinetic reaction measurements (63).

Some heterocyclic analogs of benzene, such as 6.19 have been synthesized and found to have aromatic character (64). Some of the 10-π heteronins (6.20) also have a certain aromaticity (see also later comments in this section).

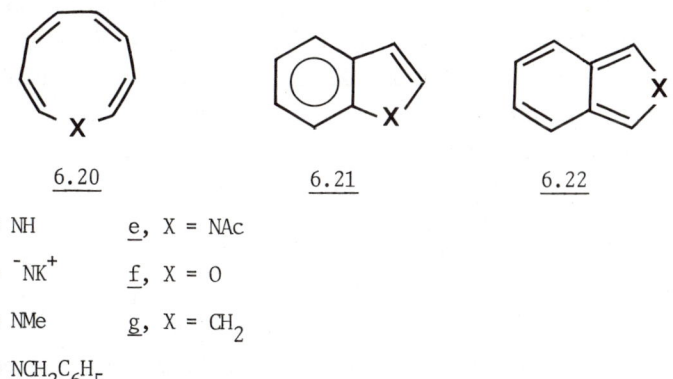

6.20   6.21   6.22

<u>a</u>, X = NH   <u>e</u>, X = NAc

<u>b</u>, X = $^-NK^+$ <u>f</u>, X = O

<u>c</u>, X = NMe   <u>g</u>, X = CH_2

<u>d</u>, X = $NCH_2C_6H_5$

Thus, 6.20a-d with a basic X are distinctly much more thermally stable than 6.20e-g and the proton chemical shifts for 6.20a are in the aromatic range (65). For comments and references on the aromaticity of heterocyclic compounds of type 6.21 and 6.22 see R. N. Warrener (66).

$B_{10}H_{10}$

6.23

Another interesting system is that of benzocarborane, 6.23. It is almost as inert as benzene toward sulfuric acid, bromine, and permanganate, and is thermally stable, but its nmr and uv spectra indicate that it has little aromatic character (67).

Lastly, it is interesting to note the effect of aromaticity upon carbenes. Aryl carbenes rearrange at elevated temperatures to aromatic carbenes, e.g.,

$$Ar-\ddot{C}-R \longrightarrow$$ [cycloheptatrienylidene with R]

The singlet aromatic carbenes, in which the vacant orbital is integrated into the conjugated two-, six-, or ten-π-electron carbocyclic ring system, behave differently from ordinary carbenes. Thus, the former do not insert into single bonds or add to electron-rich double bonds. On the contrary, they add as nucleophiles to styrenes and have a strong tendency to dimerize (

The general topic of aromaticity, covering a wide variety of systems, is well covered in recent surveys (69). Therefore, we will limit our discussion to experimental criteria for aromaticity. One of the earliest physical properties associated with aromatic character was thermodynamic stabilit as determined from heats of combustion or hydrogenation. We have seen that this led to the development of a table of resonance energies, and a proposed criterion for aromaticity was the assignment of a large resonance energy. It was also recognized prior to the adoption of the modern definition that aromatic rings have an unusually large diamagnetism (70) (Table 6.6). From time to time other physical properties, such as magnetic susceptibility anisotropies (74), equal adjacent bond lengths (75), and aromaticity index (76), have been proposed as criteria for aromaticity, and there are at least five parameters of nmr spectra which have been suggested as criteria.

1. $\underline{J}(^{13}C\text{-H}) = \underline{f}(\delta)$. A linear relationship was observed between the chemical shift and ^{13}C-H coupling (77). Aromatic compounds are found to fall below the line and the deviation increases with increasing aromatic char acter, e.g., -78 Hz for benzene and -40 Hz for thiophene. There is no obvious theoretical basis for the relationship.

Table 6.6 Diamagnetism of some compounds ($10^6 \chi$) (71)

Compound	Experimental	Calculated*	Exaltation (difference)
Benzene	54.8	41.1	13.7
Naphthalene	91.9	61.4	30.5
Azulene	91.0	61.4	29.6
Tropolone	64 (72)	52.6	8.4
Tropone	54 (72)	47.4	6.6
2,6-Dimethyl-4-pyrone	59.5 (73)	57.3	2.2
Cyclooctatetraene	53.9	54.8	-0.9

* See Chapter 4.

2. $J_{1,2}$ coupling constants (78). These are smaller across aromatic C-C bonds than for olefinic bonds because the latter are shorter and have a larger bond order.

benzene, 5-9 Hz
azulene, 3-5
(5 ring)

cyclohexene, 9-11 Hz
cyclopentenes, 5.1-5.4

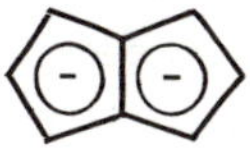

Pentalenyl dianion 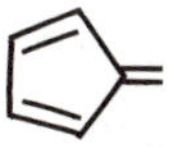   Fulvene

The value for the pentalenyl dianion (3.0 Hz) implies that it is aromatic whereas the values for fulvenes (5.17-5.45) indicate that they are olefinic. Fulvene itself adds halogens and undergoes a Diels-Alder reaction.

3. Constancy of J_{ortho} (79). One would expect all properties of the ring C-C bonds to be nearly the same for an aromatic ring. In this case, the J_{ortho} coupling constant, i.e., coupling across two adjacent carbons, is the proposed property.

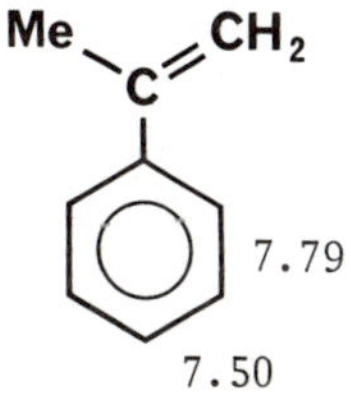

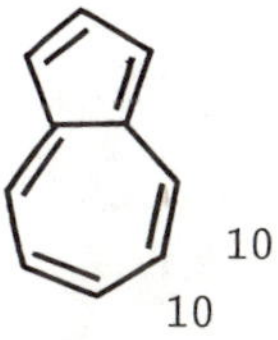

7.79
7.50

10
10

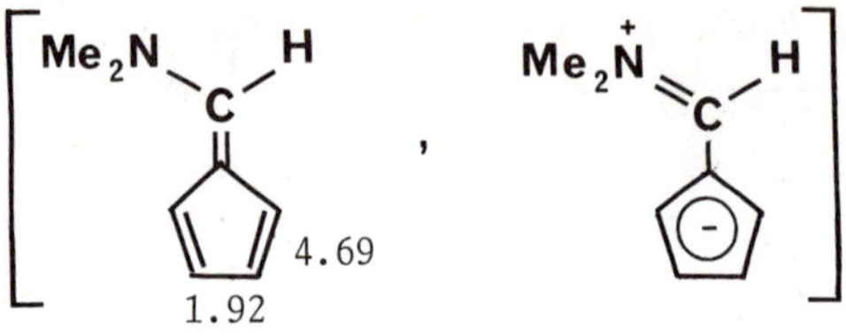

220   Ch. 6   APPLICATIONS OF RESONANCE THEORY

$$\left[\begin{array}{c} \text{Me}_2\text{N} \diagdown_{\text{C}} \diagup^{\text{H}} \\ \text{(cyclopentadienyl)} \\ 1.92 \quad 4.69 \end{array} , \begin{array}{c} \text{Me}_2\overset{+}{\text{N}} \diagdown_{\text{C}} \diagup^{\text{H}} \\ \text{(cyclopentadienyl anion)} \end{array} \right]$$

6.24

Thus, J_{ortho} values are fairly constant around the phenyl and azulene rings, implying aromatic character, but the values differ markedly around the fulvene ring, although 6.24 is one of the fulvenes with the greatest possibility of having some aromatic character. Surprisingly, the bond distances in 6.25 infer considerable aromaticity (80).

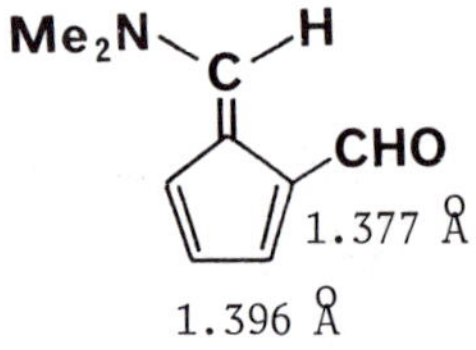

$$\text{Me}_2\text{N} \diagdown_{\text{C}} \diagup^{\text{H}} \text{, CHO}$$
1.396 Å   1.377 Å

6.25

It is interesting to apply this criterion to tropolone and tropone. Based on its dipole moment (4.3 D), resonance energy (11 kcal/mol), basicity, and other properties, varying degrees of aromatic character have been conferred on tropone on the assumption that the dipolar structure 6.26b makes a significant contribution to the resonance hybrid. However, from the J_{ortho} values around the tropone and tropolone rings (79), one can conclude that tropone is not aromatic and tropolone has only a moderate aromaticity. This view corroborated by recent magnetic susceptibility measurements (81).

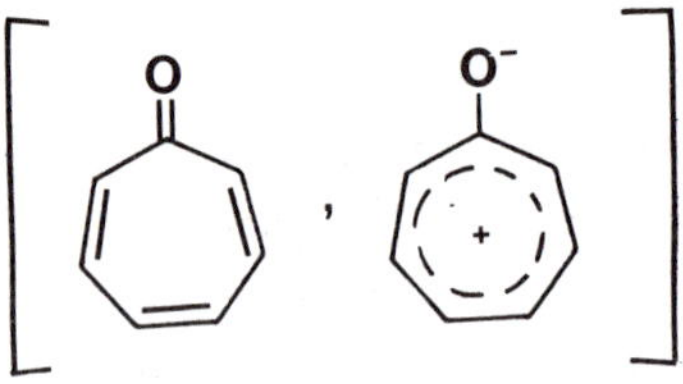

6.26 a b

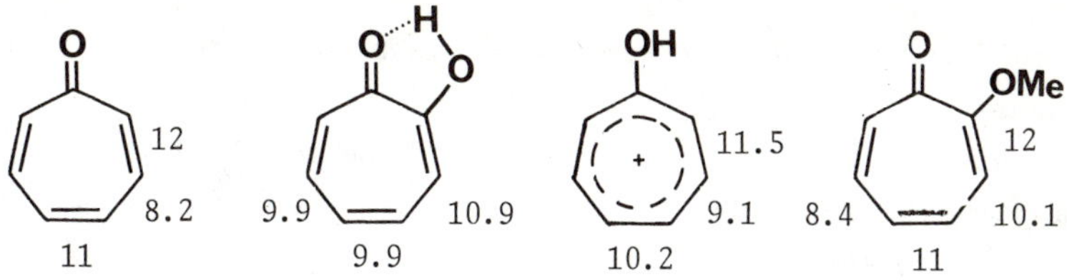

Tropone   Tropolone

Although ring carbon-carbon bond distances in tropone derivatives, e.g., 2-chlorotropone, and in tropolone and its salts and esters are in the "aromatic range," i.e., ~1.41 Å, there is a distinct bond-length alternation around the rings (82).

4. <u>Solvent shifts</u> (83). The effect of a solvent (a comparison method is also available for solid substrates) on the proton chemical shift of acetonitrile is compared with that of cyclohexane, expressed as the <u>S</u> factor:

$$S = \Delta\delta_x - \Delta\delta_{cyclohexane}$$

where $\Delta\delta_x$ is the difference in δ between cyclohexane and acetonitrile in solvent X and $\Delta\delta_{cyclohexane}$ is the analagous difference in cyclohexane solvent. The <u>S</u> values for several compounds, for example, are:

	S		S
Benzene	1.0 ppm	Tropilidene	0.33 ppm
Naphthalene	1.34	Cyclopentadiene	0.39
Pyrrole	0.90	1,4-Cyclohexadiene	0.01
Furan	0.42	1,3,5-Cyclooctatriene	0.04
Ferrocene	0.72	Cyclooctatetraene	-0.19

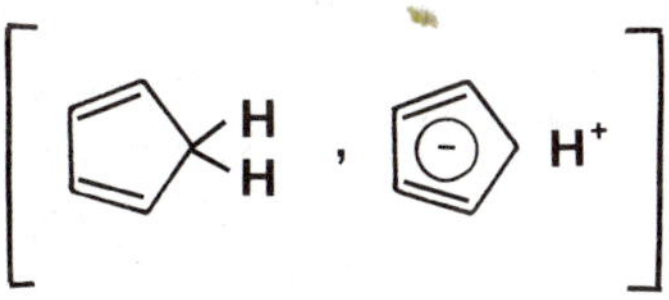

The moderate diamagnetic character of cyclopentadiene, based on its S value, can be attributed to hyperconjugation. This view also explains its relatively small λ_m in the uv spectrum compared to those of other cyclic dienes (Chapter 10).

The $\underline{S}$ values for the heteronins 6.20 support the deductions made earlier concerning their aromaticity from thermal stability studies (see earlier in this section) (84).

	$\underline{S}$		$\underline{S}$
6.20a	1.35	6.20f	-0.07
6.20c	0.34	6.20g	-0.05

Hence, 6.20a can be regarded as being strongly aromatic, 6.20c as mildly aromatic, and 6.20g as nonaromatic. Other solvents have been proposed as ring-current probes (85).

5. Deshielded peripheral H atoms. This is the most widely used and lea equivocal modern criterion for aromatic character. It is based on Pople's induced ring current concept. It is particularly good because it accounts for chemica shifts of ring-attached protons (deshielded as well as of protons lying over the ring (shielded) (Table 6.5). It applies to carb atoms too. Thus, the chemical shift of C15 in pyrene is upfield from that for the peri eral carbons. However, we saw in Section 4.2d that ring current effects on ^{13}C chemical shifts are small.

Pyrene

This criterion has been used to add nmr evidence to other evidence in the argument over the aromaticity of a metal chelate ring. Based on an abili to undergo electrophilic substitution react (86), on ir spectra (87), and bond lengths (88), it is proposed that the ring of metal chelates of 2,4-pentanedione is aromatic. instance, the C-C bond length is in the aromatic range and the C-O bonds are of identic lengths. However, divergent interpretations have been given to the nmr data (89). In o study, for example, (87), the shielding ($\Delta\delta = \delta_{\underline{p}} - \delta_{\underline{o}}$) of the ortho methyl protons

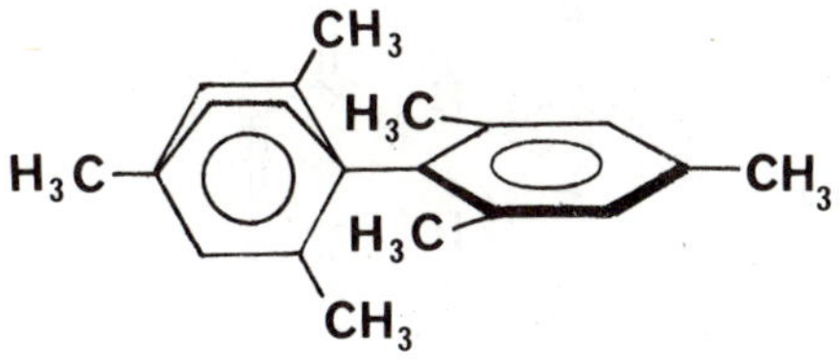

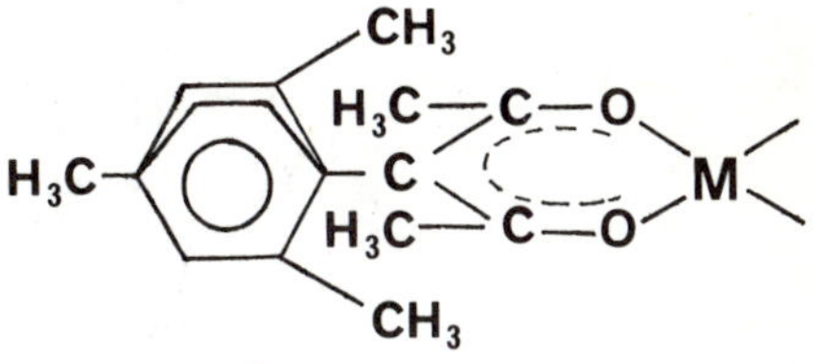

Sec. 6.1 STRUCTURE   223

$\Delta\delta = \delta_p - \delta_o = 0.45$ ppm

6.27

M	$\Delta\delta$ ppm
H	0.2
K	0.17
Co	0.16
Pd	0.22

bimesityl, which sit partially over the ring current of the other ring, is about twofold that of the 3-mesityl-chelate (6.27) and it might be argued that the chelate ring does have a moderate ring current. This deduction is weakened by the fact that $\Delta\delta$ is about the same for the H-chelate, which is not expected to have the aromaticity of the transition metal chelates. However, a recent electron diffraction study of the enol of acetylacetone in the gas phase indicates that the O-H'···O distance is only 2.381 Å and is symmetrical (90). Also, the C-O bonds are of equal distance and the C-C bonds of the enol ring are close to the aromatic length.

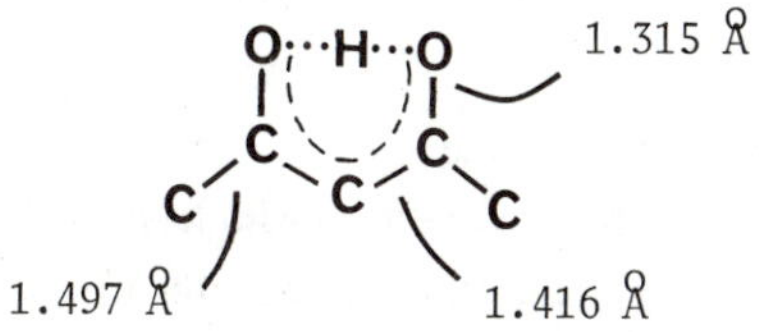

It is appropriate to insert here the fact that systems with $4n$ π electrons induce ring currents with the opposite effect on δ, i.e., they shift δ downfield for protons sitting over the ring and upfield for ring-attached protons. The nmr spectral data for (16)annulene, a $4n$ system, and (18)annulene, a $4n + 2$ system, provide a good illustration.

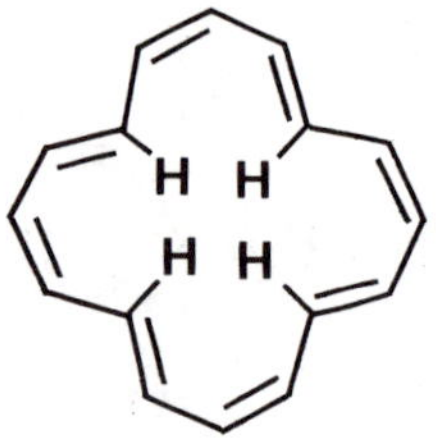

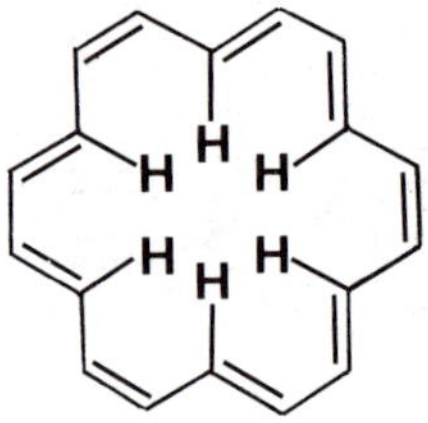

224   Ch. 6   APPLICATIONS OF RESONANCE THEORY

	(16)Annulene	(18)Annulene
Outer protons	δ 5.11-5.4 ppm	δ 9.25 ppm
Inner protons	δ 10.53	δ -2.88

X-ray crystallographic measurements on the solids show that (16)annulene is nonplanar with complete bond alternation of single and double bonds, in contrast to (14)- and (18)annulenes, which are close to planar with no bond alternation (91). The nonplanar $4n$ cyclooctatetraene ring also has a small paramagnetic ring current (92).

6.28   6.29

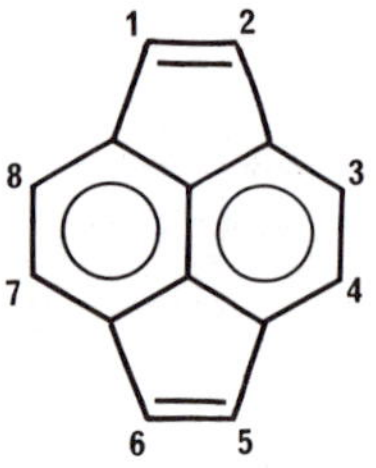

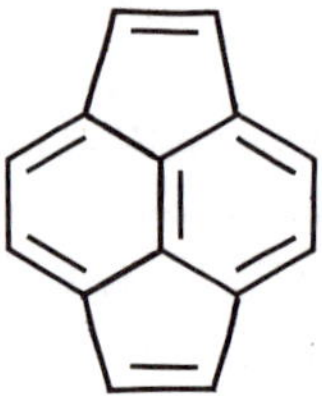

This difference in nmr characteristics of $4n$ and $4n + 2$ π rings can be used for structure elucidation. For example, pyracyclene could have a divinyl naphthalene structure (6.28), in which case the H3, H4, H7, and H8 protons would sit on the periphery of a diamagnetic aromatic ring and would have chemical shifts in the aromatic range. Pyracyclene, however, could have a fused (12)annulene system (6.29), and in this case the H3, H4, H7, and H8 protons would be attached to a paramagnetic ring system. The observed chemical shift for the H3 in pyracyclene (δ 6.52) is upfield of that in 5,6-dihydro-pyracyclene (δ 7.65) and the upfield shift implies that there is a weak paramagnetic ring current in pyracyclene and that it has structure 6.29 (93). Similarly, nmr data indicate that the homologous ketone 6.30 has a (13)annulenone structure rather than a naphthalene-based structure (94). Note also that a 2-electron reduction of a 4n + 2 system converts it to a 4n system and vice versa:

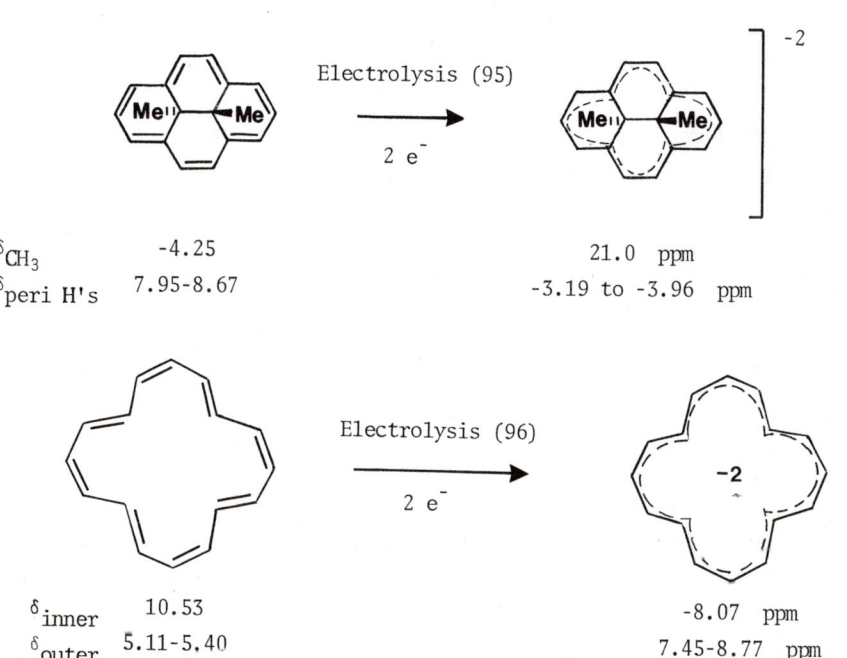

6.30

Electrolysis (95)
2 e⁻

δ$_{CH_3}$ −4.25   21.0 ppm
δ$_{peri\ H's}$ 7.95−8.67   −3.19 to −3.96 ppm

Electrolysis (96)
2 e⁻

δ$_{inner}$ 10.53   −8.07 ppm
δ$_{outer}$ 5.11−5.40   7.45−8.77 ppm

In addition to the 5-6-5-6 ring system of pyracyclene, other variations of (14)annulene are the 5-7-5-7 combinations of 6.31 (97) and 6.32 (98), both of which have aromatic ring currents.

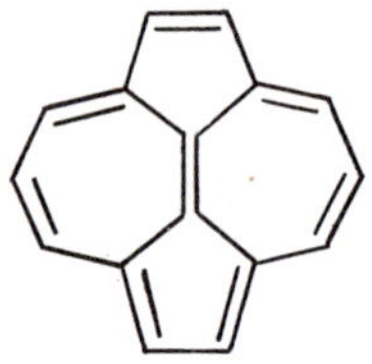

6.31

δ 7.78 - 8.4 ppm

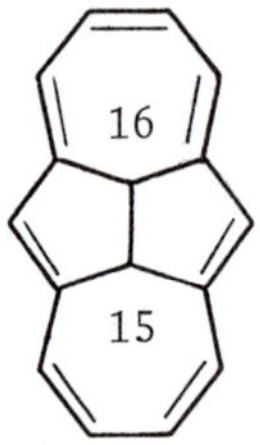

6.32

δ 1-14 8.00 - 8.17

δ 15, 16 -1.82

The recent synthesis of (20)annulene completes the series from (6) to (24) excepting (10)annulene (99). With the exception of (8)annulene, all of the 4n members have a paramagnetic ring current and the 4n + 2 rings have a diamagnetic ring current (100). The (8)annulene ring deviates too much from coplanarity to have a ring current.

It should also be pointed out here that antiaromaticity should be distinguished from nonaromatic character. The former is a ring with 4n π-electrons and in which electron delocalization produces destabilization. It is estimated, for instance, that the cyclobutadiene ring is destabilized by 12-16 kcal/mol relative to butadiene (101). Cyclopentyl iodide undergoes silver-assisted solvolysis over 10^5 times as fast as 6.33 even though the diene contains a doubly allylic halogen (102). This inertness of 6.33 can be attributed to the antiaromticity of the intermediate carbenium ion that is formed upon ionization. The pK_{R^+} of the cyclopentadienyl cation, -40, is very low compared to values for other conjugated systems (e.g., allyl cation

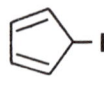
6.33

Sec. 6.1 STRUCTURE   227

$pK_{R^+} \sim -20$) and indicates that $C_5H_5^+$ is a destabilized antiaromatic species (103). Similarly, the fluorenone conjugate acid could have a benzyl cation resonance, however, the cation is antiaromatic (4n π electrons in center ring) making the ketone an unusually weak base (104).

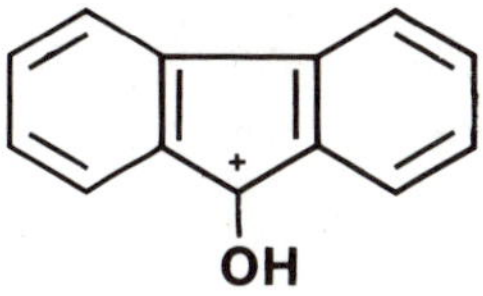

	pK_a
Fluorenone	-7.8
Benzophenone	-6.3
4,4'-Dimethoxybenzophenone	1.1
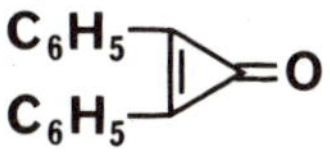 C_6H_5—△=O (C_6H_5)	-2.4

Fluorenone conjugate acid

In spite of greater classical resonance in the anion of 6.35, the rate of D for H exchange for 6.34 is over a thousand-fold the rate for 6.35. Thus, the anion of 6.35 has a 4n π ring and is destabilized relative to a simple cyclopropyl carbanion.

R = C_6H_5CO, $C_6H_5SO_2$, CN

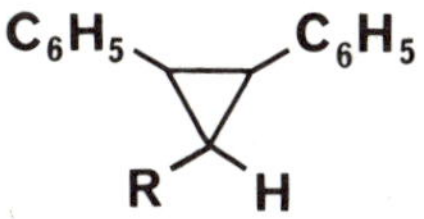

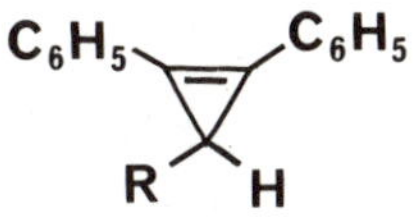

	6.34	6.35
k_{rel} (D/H exchange)	10^3-10^4	1.0

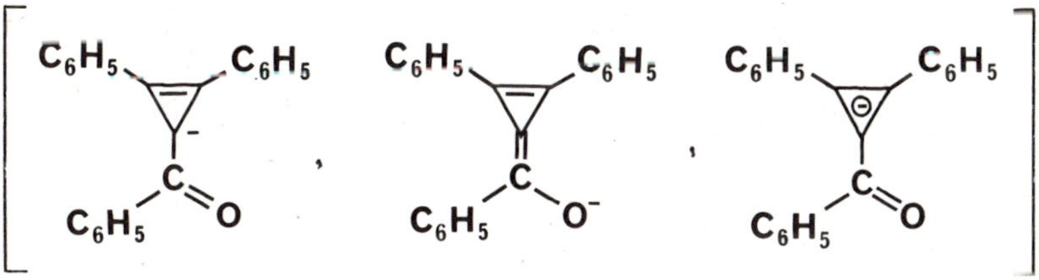

Anion of 6.35

228   Ch. 6   APPLICATIONS OF RESONANCE THEORY

An antiaromatic system is conjugated and delocalization is destabilizing only with respect to a corresponding conjugated linear system. The antiaromatic ring is still more stable than the same number of isolated multiple bonds. Thus, a planar cyclooctatetraene is antiaromatic but more stable than four isolated ethylene linkages.

Homoaromatic (105) and <u>bishomoaromatic</u> rings (106) have been produced (1 For example, the pmr chemical shifts (ppm) of cations <u>6.36</u> to <u>6.37</u> indicate that the seven-membered rings have an aromatic ring current which shields H_i

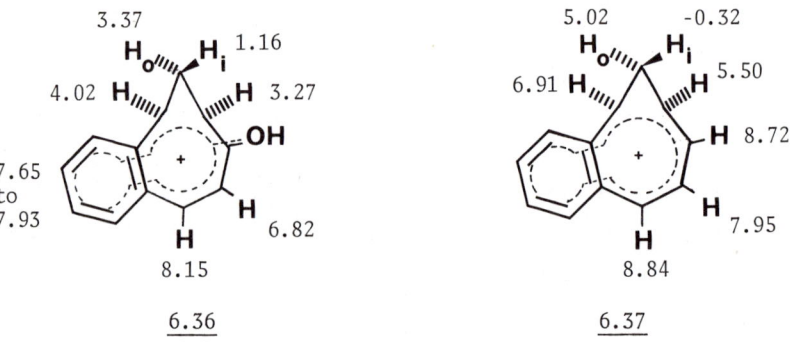

<u>6.36</u>   <u>6.37</u>

sitting over the ring, and it resonates significantly upfield of H_o (108). Comparison of $\Delta(\delta_{H_o} - \delta_{H_i})$ for <u>6.37</u> with that of <u>6.36</u> shows that the OH group markedly diminishes the ring current of the homoaromatic ring. Likewise, Δ is small for the bishomoaromatic ring of <u>6.38</u>. However, Δ for <u>6.39</u> is

<u>6.38</u>   <u>6.39</u>

too small to infer a ring current and this difference between <u>6.38</u> and <u>6.39</u> indicates that there is a stereochemical restriction for bishomoaromaticity in the benzobishomotropylium system (108).

The effects of ring strain in aromatic rings on their spectra are discussed in Chapter 11.

A theoretical parameter for aromatic character has been proposed, based on MO calculations. Since aromaticity is concerned only with π orbitals and depends on whether the π orbitals are stabilized relative to the σ orbitals, it is suggested that

$$\underline{A} = \left[\frac{E_\pi(M) - E_\pi(R)}{E_\pi(R)} \right] 100$$

where $\underline{A}$ = aromaticity and $E_\pi(M)$ and $E_\pi(R)$ are the sums of the occupied π-orbital energies for the molecule in question and for the open-chain reference molecule, respectively (109). Some pertinent calculated data for benzene, three of its isomers, and a reference triene are shown in Table 6.7.

Table 6.7. Aromaticity based on MO calculations

Molecule	Total Energy (eV)	ΣE_π	A
Benzene*	-197.3541	-0.82536	11.073
Fulvene*	-197.2362	-0.75075	1.032
2,3-Dimethylenecyclobutene*	-197.1555	-0.76913	3.505
Trimethylenecyclopropane	-197.0134	-0.76124	2.444
Hexatriene	-198.1446	-0.74308	0.000

* The successive π and σ ionization potentials have been measured by photoelectron spectroscopy (110).

Thus, only benzene has considerable aromaticity, and although fulvene is more stable than the 4- and 3-membered ring isomers, fulvene has less aromaticity than the other two isomers. Aromaticity need not parallel over-all stability of molecules. That is, aromaticity concerns the stability of a molecule relative to a hypothetical or open-chain structure, and is related to the heat of combustion. Chemical reactivity is a function of the difference in energy between the ground state and a transition state. If a transition state is low lying, then the compound would be reactive irrespective of its aromaticity.

QUESTIONS

1. Compound X is a relatively strong base, $pK_a \approx 3.8$. Write a resonance for X or its conjugate acid which would account for its unusual basicity.

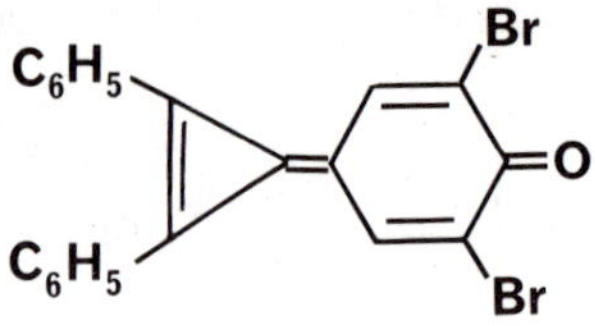

X

2. The dipole moment of diphenylcyclopropenone, 5.14 D, is larger than that of $Me_3\overset{+}{N}\text{-}O^-$ (5.03 D), tropone (4.3 D), or benzophenone (3.0 D). Propose a resonance for diphenylcyclopropenone which explains its relatively large polar character.

3. Explain why the protonation of compound Y occurs at the 5-membered ring carbon rather than a nitrogen atom.

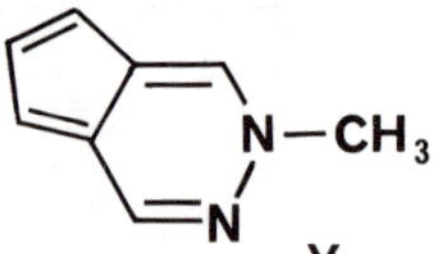

Y

4. Predict the more stable structure for Z and give the basis of your prediction.

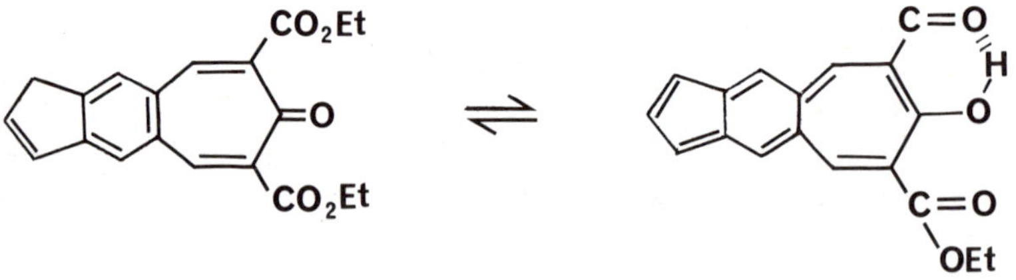

Za ⇌ Zb

5. Why should the diarylcyclopropenone be a weaker base than the dialkyl-cyclopropenone?

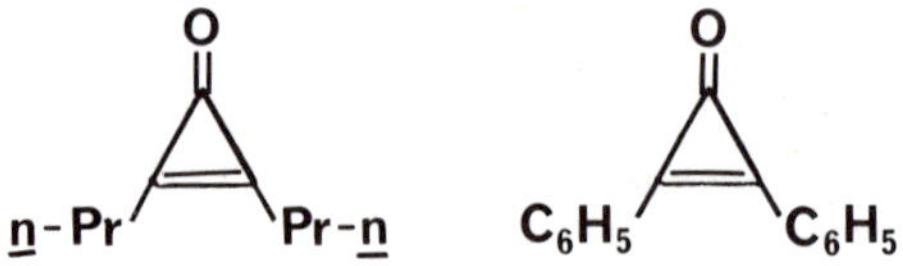

pK_a -1.9   -2.5

6. Predict the relative acidities of cyclopentadiene and indene.
7. Predict the relative proton chemical shifts for the two hydrocarbons M and N.

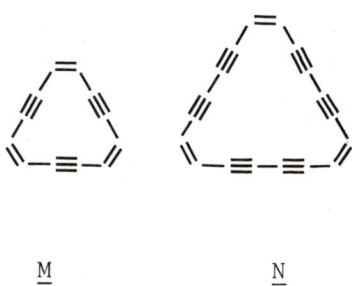

M   N

8. Predict the order of relative acid strengths of fluorene, indene, and cyclopentadiene and give the basis of your prediction.
9. What order of transmission would you expect for the intervening groups thiophene, furan, phenyl, in the three systems below ? That is, in which system does the substituent G have the greatest influence on $\nu_{C=O}$?

P   G–C₆H₄–C₄H₂S–CH=CH–CO–C₆H₅

Q   G–C₆H₄–C₄H₂O–CH=CH–CO–C₆H₅

R   G–C₆H₄–C₆H₄–CH=CH–CO–C₆H₅

10. The dipole moment of S is substantially greater than the additive value (K. Hafner, K. H. Vöpel, G. Ploss, and C. König, Ann. Chem. 661: 52 (1963). This is not the case for T. Offer an explanation.

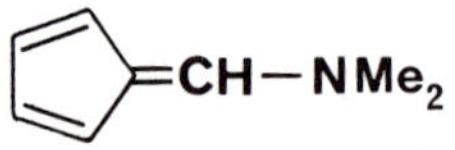

⌬=CH—NMe₂   ⌬=N—NMe₂

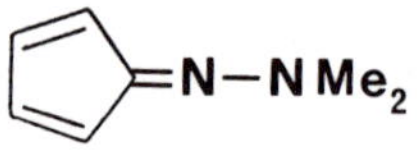

S   T
μ = 4.5 D   μ = 3.3 D

11. Write resonance structures which would account for the large dipole moments of U and V. Why is that of V the larger?

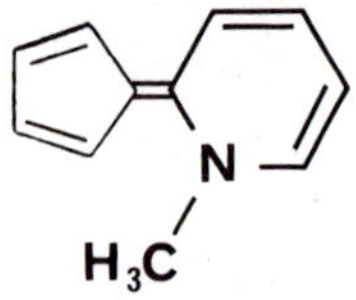

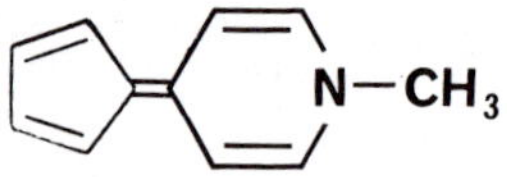

U   V
μ = 5.20 D   μ = 9.7 D

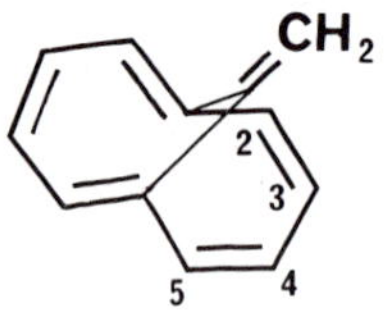

12. The following spectral and X-ray diffraction data are reported for W (H. Günther and H. H. Hinrichs, Tetrahedron 24: 7033 (1968)):

	J	Length	δ=CH_2
C_2-C_3	8.85 Hz	1.37 Å	3.19 ppm
C_3-C_4	9.43	1.41	

Comment on these data with respect to the aromaticity of the ring.

13. The pmr chemical shifts (ppm) of the three compounds shown reveal a paramagnetic ring current in the cyclobutadiene ring (S. Masamune, N. Nakamura, M. Suda, and H. Oma, J. Am. Chem. Soc. 95: 8481 (1973)).

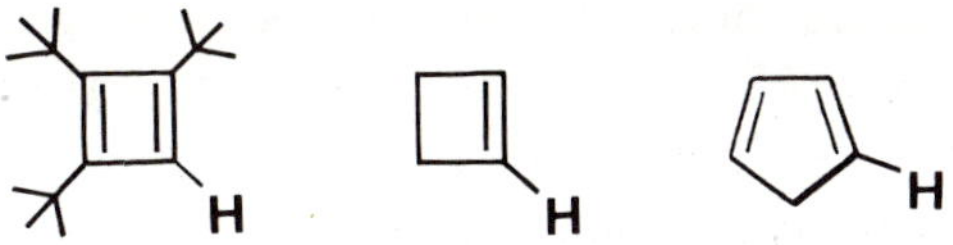

Sec. 6.1 STRUCTURE   233

| δ (ppm) | 5.58 | 5.97 | 6.42 |

The magnitude of the paramagnetic contribution by the induced ring current of the cyclobutadiene ring depends upon which compound is chosen as a reference. What factors favor cyclobutene or cyclopentadiene as the reference compound ?

ANSWERS

1.

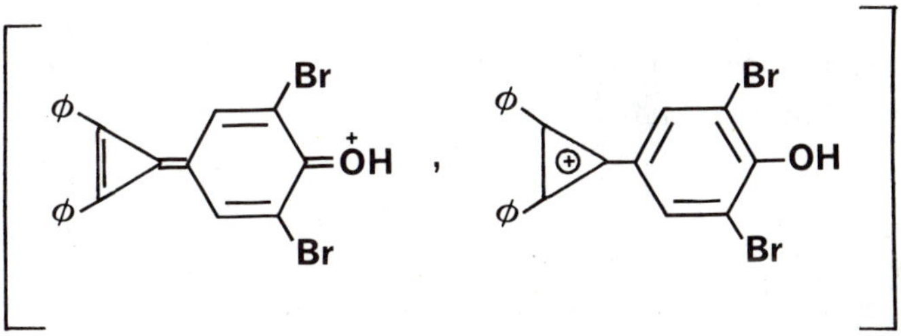

The conjugate acid is greatly stabilized by aromatic resonance for both rings.

2.

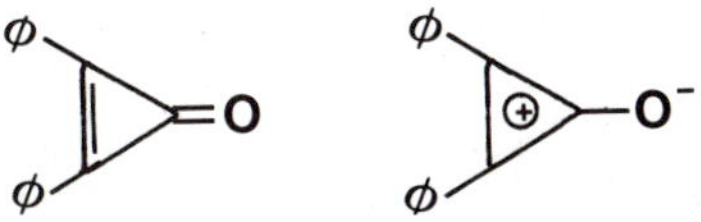

3.

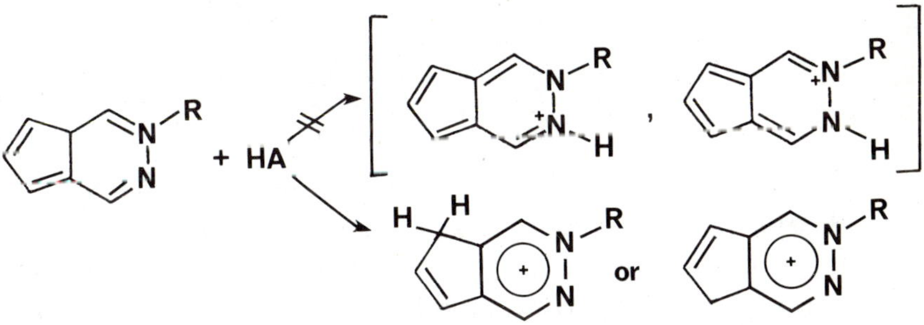

Protonation of a nitrogen does not give an aromatic ring whereas protonation of a carbon leads to an aromatic heterocyclic ring and hence to greater stability.

4. Za has in its favor a benzenoid aromatic ring plus a small stabilization from a tropone ring, whereas Zb has a 14-π electron ring plus an intramolecular H bond. This is a close one, and it might be expected that Zb would be the more stable, but ir, nmr, and lack of D/H exchange in D_2O indicates that all of the compound has structure Za (D. J. Bertelli, J. Org. Chem. 29: 3032 (1964)).

5. The aryl group can partially delocalize the positive charge of the C_3 ring there decreasing the aromatic character of the C_3 ring. This produces a slight destabilization of the conjugate acid of the aryl analog relative to that of the alkyl derivative.

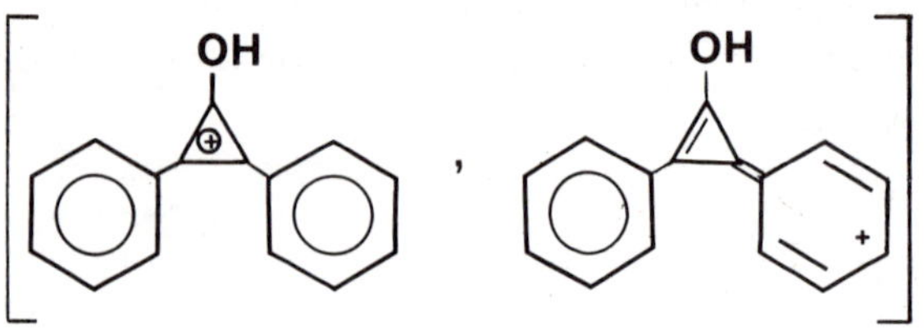

6. Five equivalent structures

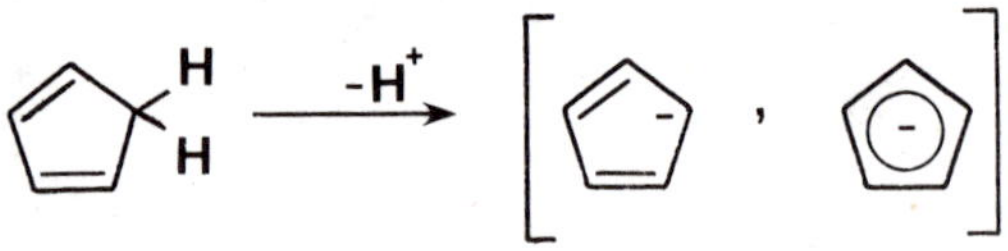

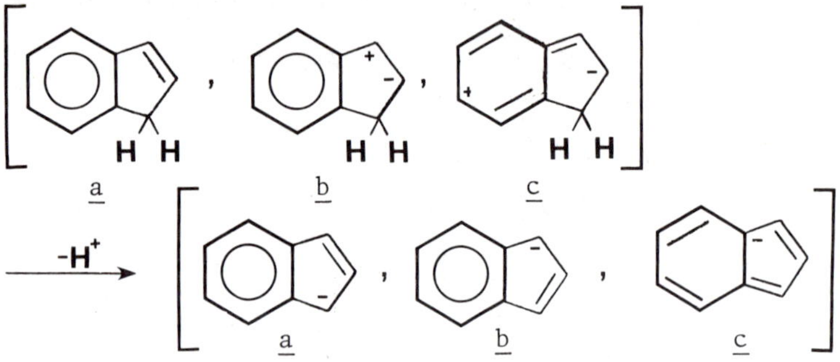

The cyclopentadienide anion is a resonance hybrid of five equivalent structures, hence it has a substantial resonance energy. Since there is only a small resonance in the acid form, there is a large net resonance energy of the anion, favoring dissociation. The resonance of indene and its anion is among the structures shown, that of c in each case making little contribution since it lacks a Kekulé ring. There is a small net resonance energy in favor of the anion because the resonance is among equivalent structures a and b whereas there is a separation of charge in b of the acid form. However, the net difference is not as large as it is for cyclopentadiene and the latter compound is the stronger acid.

7. Compound N has a (4n + 2) π electron ring so δ_H should be in the aromatic range, δ 7-9. Compound M has a 4nπ electron ring, producing a paramagnetic ring current so δ_H should be shielded relative to those of N, or δ 5-6. The actual values are for M, δ 4.45 (s) and for N, δ 7.02 (s). Thus, compared to the model compound O, δ 5.89, there is a downfield shift of 1.13 ppm for N and an upfield shift of 1.47 for M (W. H. Okamura and F. Sondheimer, J. Am. Chem. Soc. 89: 5991 (1967)).

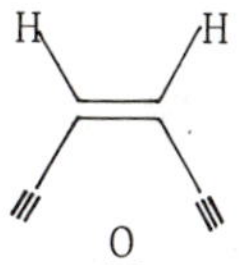

8. The anion of each is stabilized by an "aromatic" resonance of the five-membered ring. However, since the benzene rings can exhibit resonance in the undissociated acids as well as the anions, there is a decreasing importance of this resonance for the sequence fluorene, indene, and cyclopentadiene. Consequently, the cyclopentadienide type of resonance has a greater differential stabilization of the anion in the order fluorene, indene, cyclopentadiene, which should be the increasing order of acid strengths of the hydrocarbons. The observed approximate pK_a values are 25, 21, 17, respectively (B. L. McDowell and H. Rapoport, J. Org. Chem. 37: 3216 (1972)).

9. Probably, the greater the aromaticity of the intervening group, the less it transmits electrical effects of G, i.e., the more it resists forming a "quinoidal" structure. Hence, the decreasing order of transmission of electrical effects of G should be furan > thiophene > phenyl. (This is the observed order.)

10. The polar resonance forms of S and T are significant in the case of S.

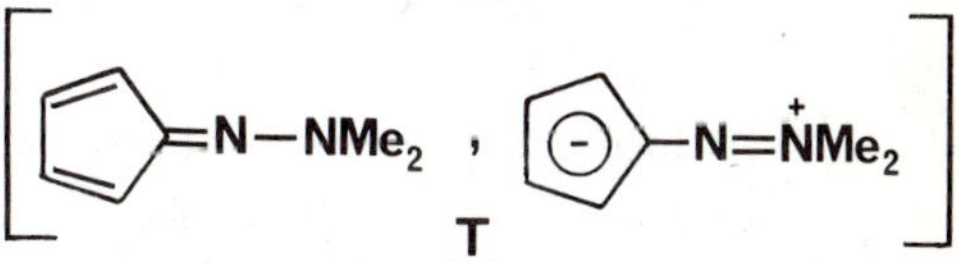

In T, even though the C_5 ring acquires an aromatic ring, the polar form makes less contribution to the ground state owing to N=N bond strain. The same situation accounts for the reduced resonance in azines

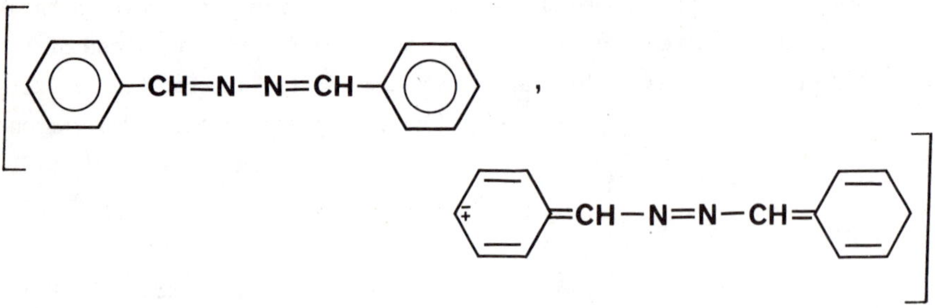

$$\left[\text{Ph-CH=N-N=CH-Ph} , \text{Ph}^+\text{=CH-N=N-CH=Ph} \right]$$

compared to diphenylbutadienes.

11. The polar forms of each resonance hybrid make a substantial contribution

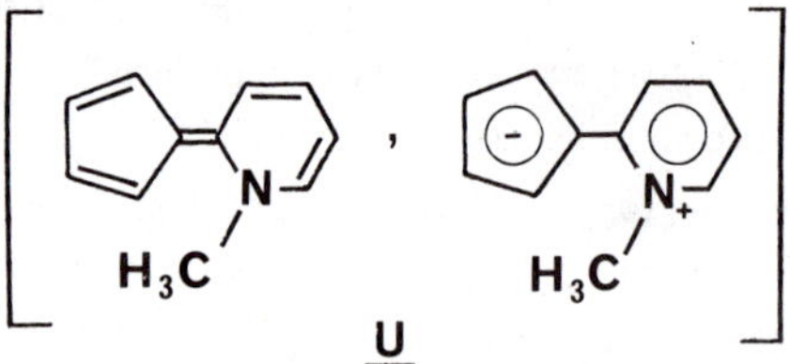

$$\left[\underset{U}{\text{structure with } H_3C-N} , \text{structure with } H_3C-N^+ \right]$$

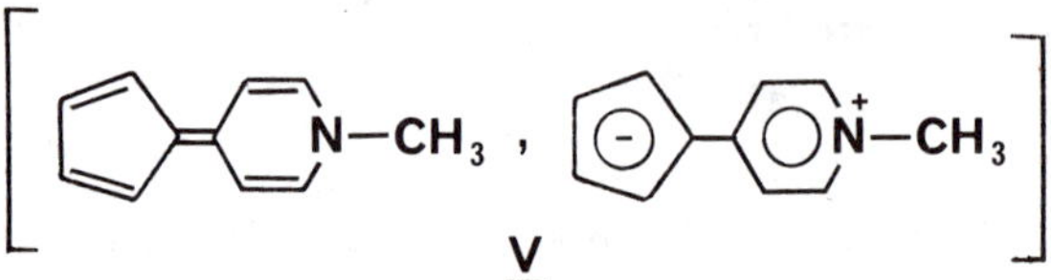

$$\left[\underset{V}{\text{structure with } N-CH_3} , \text{structure with } N^+-CH_3 \right]$$

to the ground states of the molecules because each ring acquires aromaticity. The moment of V is the larger because of a larger distance between the negative and positive centers of charges.

12. The two ortho-coupling constants are fairly close to each other, implying an aromatic ring. The bond lengths are in the aromatic range. The $\delta_{=CH_2}$ chemical shift is upfield of the usual 4.5 - 6 ppm olefinic range, indicating a diamagnetic ring current.

13. Cyclobutene should cancel any ring strain effect upon the chemical shift whereas cyclopentadiene should cancel the electronegative effect of the second olefinic bond upon the chemical shifts.

6.1d. Charge-transfer and metal π complexes

Charge-transfer (CT) complexes form another group of compounds for which resonance theory is helpful in understanding their properties. Again, we are not surveying the field (111) but merely examining general characteristics from the modern structural theory approach. There is more than academic interest in CT complexes (112). They have long been used for the isolation, purification, and identification of amines and polynuclear hydrocarbons. Racemic mixtures have been resolved by using optically active complexing agents, and cis-trans isomeric olefins are commonly separated by glc using silver nitrate in the stationary phase (113). CT complexes are found in biological systems (114), -- e.g., visual purple, the substance required for night vision -- and involvement of CT complexation has been proposed for several types of biological processes (115). Also, the storage and dispensing of drugs, explosives, fertilizers, etc., in the form of CT complexes have been explored.

CT complexes consist of a union of an electrophilic component E and an electron donor D held together by a mixture of forces and formed from a wide variety of E and D substance.

E	D
Polynitro, polycarbonyl, polynitroso, and polycyano compounds	Arenes
	Alkenes
Halogens	Amines
Transition cations	Alkyl halides
Acidic gases (116)	

The bonding in the CT compound can be described as a resonance hybrid, 6.40 (117).

$$E + D \rightleftharpoons [(E \cdots D) \;,\; (E^{-} - D^{+})]$$
$$\underline{a}\underline{b}$$
$$6.40$$

Form <u>a</u> symbolizes Coulombic forces such as dipole-dipole, H bonding, and van der Waals attraction, which contribute chiefly to the ground state. Form <u>b</u> represents an intermolecular delocalized coordinate covalent bond in which the covalent bond is not between two specific atoms in E and D. For example, for a xylene-bromine complex,

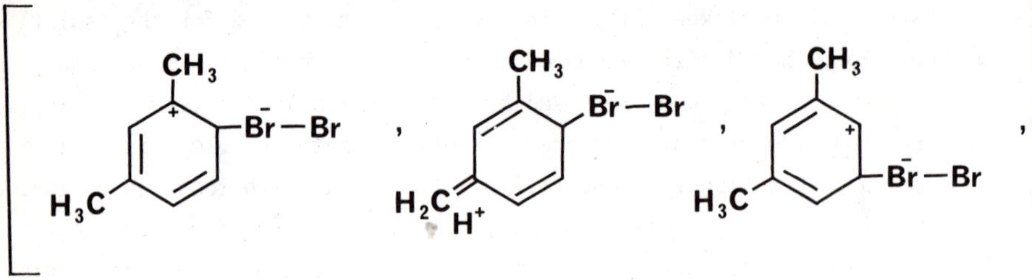

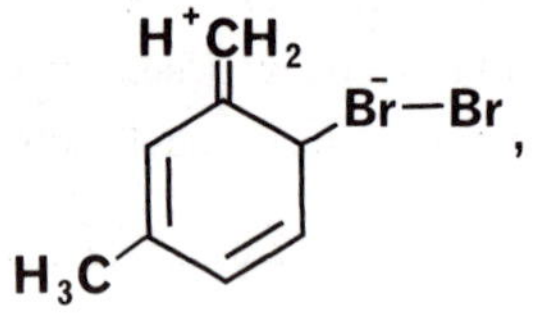

[structures of xylene-bromine resonance forms] and many others

for a Ag-Alkene complex

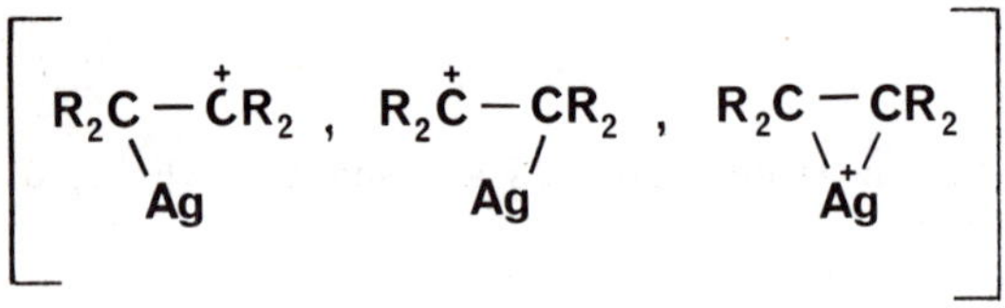

[resonance structures of Ag-alkene complex]

and for a TCNE-toluene complex

Sec. 6.1 STRUCTURE   239

[structure diagrams showing resonance forms of arene-TCNE complex with CH₃ substituent]

plus many structures in which the negative charge is on any one of the CN groups and the positive charge on any of the ring carbons or the CH_3 group.

For simplicity, these can be drawn as follows.

[simplified diagrams: toluene-Br₂ complex with δ-/δ+ charges; R₂C=CR₂···Ag⁺ complex; toluene-TCNE complex showing C(CN)₂=C(CN)₂ with δ- and δ+]

Some experimental observations in support of these structures are:

1. X-ray diffraction measurements of solid CT complexes show that they consist of parallel planes of the two components separated by 3-3.5 Å (118). This is too far apart to involve typical covalent bonds. In arene-Ag⁺ complexes, the silver ion is above a carbon-carbon bond of the ring, and in arene-halogen complexes, the axis of the halogen is perpendicular to the plane of the ring.

2. Even when E and D do not have dipole moments, the CT complexes do. For example, the dipole moments of some arene-TCNE complexes (in CCl_4) are listed in Table 6.7 along with the enthalpy of formation and the ionization potentials of the donors as computed from equation (6.2).

Since complex formation involves an electron withdrawal from the donor component, it is understood why the acidities of carboxylic acids and phenols increase upon forming a CT complex with an acceptor molecule. This partially explains the anomalous acidities of acids in aromatic solvents (119).

3. The ir and Raman spectra of CT complexes approximate the summation of spectra of its components E and D. Cerain bands are shifted, however, some up to 150 cm^{-1}, and the more stable the complex, the larger are the

frequency shifts (112). In arene-halogen complexes, for example, $^{\delta +}Ar \cdots ^{\delta -}X$ the X-Y frequencies are lowered as the formation constants increase (112, 120).

Table 6.7. Properties of some arene-TCNE CT complexes (121)

Donor	μ_D*	μ_g**	μ_{CT}†	$-\Delta H$ (kcal/mol)	IP (kcal/mol)
C_6H_6	0 D	0.75 D	0.75 D	2.7	216
C_6H_5Me	0.37	0.90	0.83	3.0	212.1
$1,2\text{-}Me_2C_6H_4$	0.52	1.06	0.94	3.4	207.6
$1,3\text{-}Me_2C_6H_4$	0.37	1.01	0.95	3.3	207.2
$1,4\text{-}Me_2C_6H_4$	0	0.96	0.96	3.6	206.4
$1,3,5\text{-}Me_3C_6H_3$	0	1.12	1.12	4.1	202.2
$1,2,4,5\text{-}Me_4C_6H_2$	0	1.32	1.32	4.8	196.9
Me_5C_6H	0.37	1.51	1.47	5.5	193.2
Me_6C_6	0	1.64	1.64	6.4	189.3

* μ_D = DM of donor.
** μ_g = DM of CT complex
† μ_{CT} = Portion of μ_g due to charge transfer, i.e., contribution of 6.40b.

4. Nmr and uv spectra, magnetic susceptibilities, and other properties support the model 6.40 for the CT bond. For example, the direction of greatest polarizability of CT complexes, based on polarized uv spectroscopy, is perpendicular to the planes of the components. The formation constant, K, and λ_m of cyclophane-TCNE complexes, for instance, are larger than for the single arenes (122).

Sec. 6.1 STRUCTURE 241

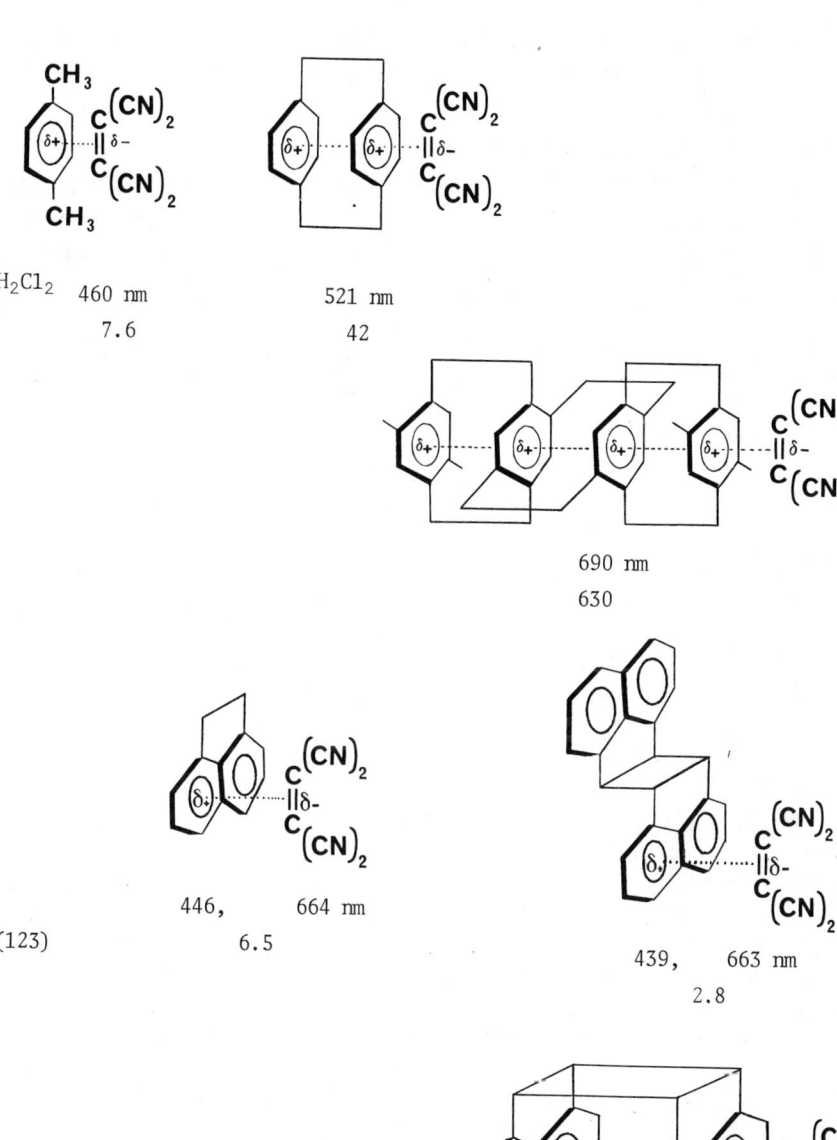

$\lambda_m^{CH_2Cl_2}$ 460 nm   521 nm
K   7.6   42

690 nm
630

λ_m 446,   664 nm
K (123)   6.5

439,   663 nm
2.8

459,   671 nm
8.2

5. The three major structural features which affect the stability of a CT complex are (i) the electron donor character of D, (ii) the polarizing power of E, and (iii) steric effects. In this respect, we make the following observations. (i) Substitution of electron-donating groups into D increases the stability of CT complexes. Hence, stabilities increase with greater number of alkyl groups on the benzene ring of D, and the decreasing stability of halobenzene complexes parallels the polarizabilities of the halogens (Table 6.8). Towards a given E, λ_m in the uv and formation constants

Table 6.8. Relative stabilities of some arene CT complexes

Electron donor (D)	Electron withdrawing agents				$-\Delta H^*$
	ICl	TCNE	TNB	$\overset{+}{\text{Ag}}$	
C_6H_5F			0.57	0.19	2.0 kcal/m
C_6H_5Cl		0.39	0.84	0.29	
C_6H_5Br	0.59	0.31	1.15	0.40	
C_6H_5I		0.62		2.07	
C_6H_6	1.00	1.00	1.00	1.00	
C_6H_5Me	1.61	1.85	1.96	1.22	2.3
$1,3-Me_2C_6H_4$	2.57	3.00	2.37	1.26	
$1,3,5-Me_3C_6H_3$	8.5	8.7	3.45	0.75	3.0
$1,2,4,5-Me_4C_6H_2$	7.88				3.9
C_6HMe_5	11.9	61.5			4.4
C_6Me_6	42	132			5.4

* Enthalpy of formation for fluoranil complex in CCl_4 (127).

are related to the ionization potential of D (112). Such a relationship has been used to determine ionization potentials of hydrocarbons (124) and of various bases (125). For alcohols, for example,

$$h\nu_{CT} = 0.197 \text{ IP} + 3.25 \text{ eV} \qquad (6.2)$$

where $h\nu_{CT} = 1/\lambda_m$ for the iodine CT complex and IP is the ionization potential of the alcohol (126). The IP calculated from (6.2) for four alcohols are

	$\lambda_m^{\underline{n}\text{-hexane}}$	IP
MeOH	231 nm	10.85 eV
EtOH	232.5	10.48
$\underline{i}$-PrOH	236	10.16
$\underline{t}$-BuOH	238	9.93

which are in the order of the polarizabilities of the alkyl groups. Equation (6.3) has been developed for arene-TCNE complexes (121),

$$h\nu_{CT} = 0.79\ IP - 96\ kcal/mol, \qquad (6.3)$$

and some ionization potentials calculated from it are listed in Table 6.7. In the case of ferrocene-TCNE complexes, we have

$$h\nu_{CT} = 0.92\ IP - 5.10\ eV.$$

The IP for ferrocene as determined in this fashion is 6.90 eV, which compares well with the value 7.05 eV from mass spectra and 7.57 eV by electron impact. The $E_{\frac{1}{4}}$ (quarter-wave polarographic oxidation potential) values for substituted ferrocenes are also related to λ_m, from which the $E_{\frac{1}{4}}$ could easily be determined for other ferrocenes (128).

Incidently, a simple, comprehensive correlation between polarographic oxidation potentials and ionization potentials which applies to virtually all classes of compounds, is

$$E_{\frac{1}{2}} = 0.89\ (IP) - 6.04\ eV$$

with a fair correlation coefficient (0.95) (129).

(ii) Many studies have been made of the effects of change in structure of the E component on the stabilities of CT complexes (130). Usually a number of electron-withdrawing groups such as NO_2, NO, CN, or CO, are attached to E to increase its polarizing power. Examples are

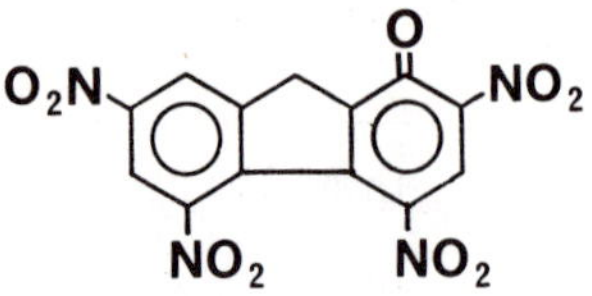

2,4,5,7-Tetranitrofluorenone (TENF)

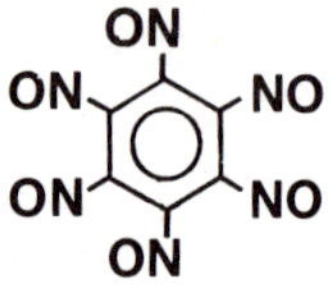

Hexanitrosobenzene

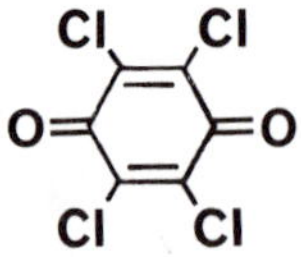

Chloranil

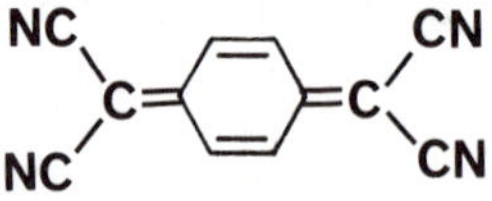

Tetracyanoquinomethane (TCNQ)

TCNE and TCNQ are about the strongest complexing agents found so far.

With strong acceptor and donor partners, complete electron transfer may occur in the CT complex to produce an ion radical (128, 131).

$$\text{Ferrocene} + \text{TCNE} \rightleftarrows \text{CT complex} \rightleftarrows \underset{\text{in solution}}{\text{TCNE}^{\ominus} + \text{Fe}(C_5H_5)_2^{\oplus}}$$

green in solid state, λ_m = 1150 nm

It can be shown by esr or magnetic susceptibility measurements that ion-radicals are formed. Such CT complexes are of great interest as superconducting solids. The most recent is a salt between tetrathiofulvalene (TTF) and TCNQ which has the highest conductivity reported for any organic compound, >10^6 ohm^{-1} at 58° K (132).

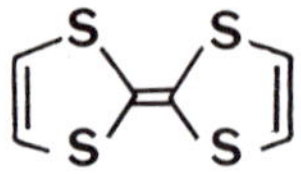

Tetrathiofulvalene (TTF)

(iii) Any steric conditions which tend to separate the components in a CT complex, reduce its stability. Thus, such nonplanar compounds as $(C_6H_5)_2CH_2$, $(C_6H_5)_3CH$, and $(C_6H_5CH_2)_2NH$ do not form solid CT compounds even with 2,4,7-trinitrofluorenone.

Another example of the steric effect is found in hexaethylbenzene. We saw in Table 6.7 that CT complex stability normally increases with the number of alkyl groups in the electron-donor component. This is true of methyl and ethyl groups up to hexaethylbenzene. Here, owing to the larger ethyl groups blocking close approach of the E component, hexaethylbenzene is an abnormally weak π-electron donor. This effect shows up also in its rate of reaction with ozone (133).

	k_{rel}	k_{rel}	
C_6H_5Me	1	12	C_6H_5Et
$1,3-Me_2C_6H_4$	28	40	$1,3-Et_2C_6H_4$
$1,3,5-Me_3C_6H_3$	150	143	$1,3,5-Et_3C_6H_3$
C_6Me_6	8750	120	C_6Et_6

QUESTION

At the same concentration of phenol and hydrocarbon, the shifts in the phenol ($\Delta\nu_{OH}$) band when mixed with some hydrocarbons are as listed below. Would you use the concept of hyperconjugation to explain the sequence?

Hydrocarbon	$\Delta\nu_{OH}$
Benzene	49 cm^{-1}
Toluene	58
o-, m-, or p-Xylene	68
Mesitylene	78
Durene	85
C_6Me_6	106

ANSWER

No. If the order were the result of hyperconjugation in the complex, then the value for m-xylene should be larger than that of the o- or p-xylenes and the value for mesitylene would be larger than it is. This sequence results from the inductive or polarizability effect of the alkyl groups increasing the electron-donor character of the ring.

6.1e. Metallocenes

An additional type of bond, largely covalent in character, which should be briefly discussed here is the metal π-complex bond.

One of the exciting rewards of laboratory work is that occasionally there will be entirely unexpected results which, if recognized, will open up a new area of science or will yield a product highly beneficial to mankind. Many chemical discoveries were made by accident (134). One of the most financially successful was Alfred Nobel's discovery of a method for the production of dynamite. Other familiar examples in organic chemistry are the vulcanization of rubber (Charles Goodyear); the plastic Bakelite (Leo Baekeland); synthetic dyes (William Perkins); the urea complexes (Friederich Bengen); Teflon (Roy Plunkett); the artificial sweetening agents saccharin (Remsen and Fahlberg), cyclamates (Sveda and Audrieth), and now neohesperidin dihydrochalcone (Horowitz and Gentilli); and LSD (Albert Hofmann). Another case occurred in 1951 when Kealy and Pauson (135) discovered ferrocene by accident.

The isolation and characterization of ferrocene opened up a new domain in chemistry bridging organic and inorganic chemistry. The quantity of recent literature on this topic makes it clear that metal π-complex chemistry fluorishes independently. Metal π complexes not only provide a means of learning more about metals and about chemical bonding, but also have several practical uses. Industrial processes such as the hydroformylation of olefins, the Ziegler-Natta catalysis of olefin polymerization (for which they received the 1963 Nobel Prize for chemistry), stereospecific and homogeneous hydrogenation, and others, have emerged. The metallocenes reached a pinnacle in the chemical world when the 1973 Nobel Prize for chemistry was awarded to Ernst Fischer and Geoffrey Wilkinson for their pioneering research in the field.

The literature abounds with very recent books, journals, and review series on metal π complexes. However, most of the attention is given to the metal component.* To stay within the scope of this book, then, only a brief description of the bonding in a few members of this class of compounds will be given.

Ferrocene is an orange, crystalline solid which melts at 173-4° and boils near 249°. It is soluble in common organic solvents, such as benzene, ether, and alcohol; it is insoluble in water; and it is unaffected by alkali

---

*Exceptions are the silver ion-catalyzed nonconformity of olefins to the orbital symmetry rules, and the isolation of highly reactive species such as cyclobutadiene in the form of their metal π complexes.

Sec. 6.1 STRUCTURE   247

and boiling concentrated hydrochloric acid. Diffraction studies show that many of these metal π complexes consist of the metal packed between two parallel cyclopentadienyl rings, which gave rise to the popular name, <u>sandwich compounds</u>. A more dignified term is <u>metallocene</u>.

The covalent character of the metallocenes is apparent from their solubilities, volatilities, and spectra. This indicates that each $C_5H_5^-$ ion, in the case of ferrocene for instance, donates an electron pair to the metal ion. Ferrocene is diamagnetic, hence its six $\underline{3d}$ electrons are paired to make available two open $\underline{3d}$ orbitals.

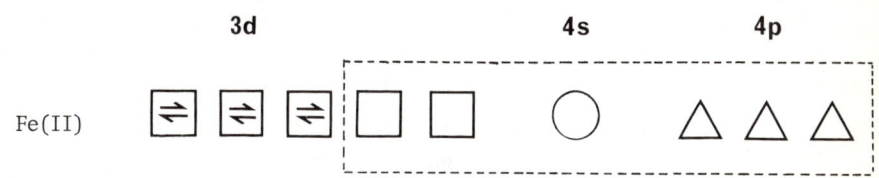

However, the electron pair does not come from any single carbon of the $C_5H_5^-$ ring. If it did, each ring would still have two typical double bonds and properties resembling those of cyclopentadiene. On the contrary, ferrocene resists addition reactions such as the Diels-Alder reaction, catalytic hydrogenation, polymerization, and it undergoes typical aromatic substitution reactions such as the Friedel-Crafts reaction. Furthermore, ir and nmr spectra and diffraction measurements show that the carbon atoms of each ring are equivalent with C-C bond lengths in the aromatic range, about 1.4 Å.

To rationalize the covalent and aromatic character of ferrocene, it is given the structure <u>6.41</u> or more conveniently, <u>6.42</u>.

6.41   6.42

The structure implies that the metal ion bonds equally with each carbon and that the bonding is covalent. Ferrocene exhibits exactly what is meant by the term covalent bond: one in which both components share electron pairs and act as if the electron pairs belonged solely to each. For instance, the iron in ferrocene is diamagnetic only because its 3d electrons are paired up to make available two d orbitals for an electron pair from each $C_5H_5^-$ ion. At the same time, each cyclopentadienyl group can have an aromatic character only if it has the three pairs of π electrons. Thus, the metal and the rings act as if each still has the bonding electron pairs.

The bonding in metallocenes is described better by MO theory. Bonding and antibonding molecular orbitals are formed by the addition or subtraction of atomic orbitals as was seen in elementary organic chemistry work. In the case of transition metals, 3d, 4s, and 4p (or 4d, 5s, and 5p) are the atomic orbitals. The ligands may have σ- and π-valence orbitals, and π-complex bonds are described in terms of overlap of appropriate valence atomic orbitals. Bond formation is expected only when the interacting orbital lobes have the same sign. Some examples can be given for illustration.

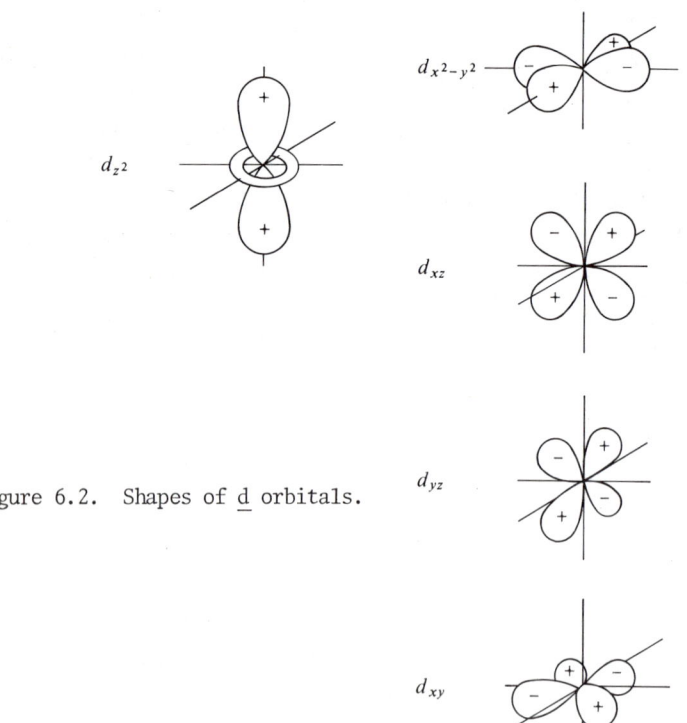

Figure 6.2. Shapes of d orbitals.

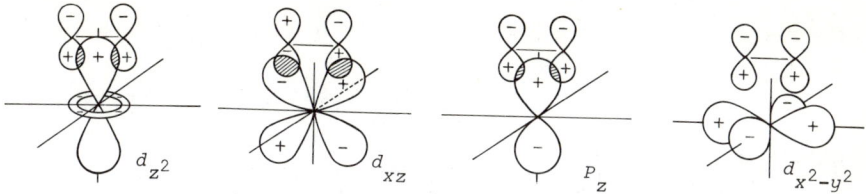

Figure 6.3. Diagramatic representation of orbital overlap between ethylene $2p_z$ atomic orbitals and orbitals of a transition element.

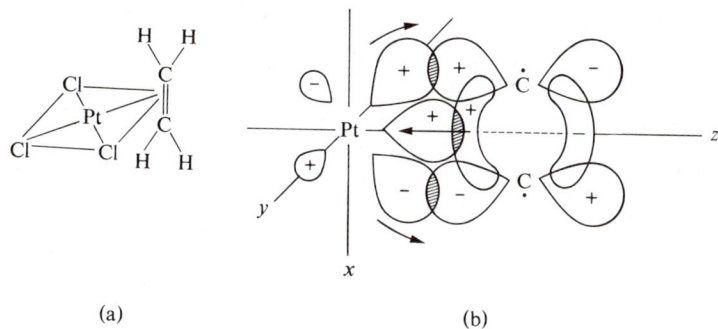

(a)   (b)

Figure 6.4. (a) Structure of the $PtCl_3:C_2H_4$ complex and (b) diagram of the π-complex bond with back-bonding from the metal ion.

Through overlap of the cyclopentadieneide orbitals with the $\underline{d}$ or hybridized $\underline{d^2sp^3}$ orbials of iron, the six π electrons of each ring fall into the valence shell of iron to give it a rare gas electron structure (18 electrons).

In order to account for the stability, bond lengths, ir spectra, dipole moments, and other properties of π complexes involving ligands such as CO, PF_3, C_6H_6, $C_5H_5^-$, etc., it has been proposed that the metals back-donate electrons to the ligand into vacant antibonding orbitals. This is diagrammed for a $PtCl_3$-ethylene complex in Figure 6.4. X-ray analysis establishes that the axis of the Pt-olefin bond is perpendicular to the plane of the ethylene moiety. For example carbon monoxide ($^-C\equiv O^+$) can react with a metal to produce a $^-M\text{-}C\equiv O^+$ bond, but the back-donation makes the M=C=O structure a major contributor to the resonance hybrid. This π bond between the metal and the carbon not only gives the bond a double-bond character, and therefore greater strength, but also prevents the accumulation of negative charge on the metal which would destabilize the system. An example is nickel tetracarbonyl, $Ni(CO)_4$, which is diamagnetic and tetra-

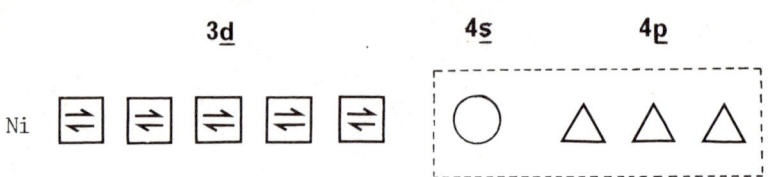

hedral (by X-ray diffraction) and is apparently using sp^3 hybridized orbitals. However, the Ni-C bond length is smaller than that expected for a single bond and the carbon-oxygen bond distance is shorter than that in carbon monoxide. Back-donation thus accounts for the electroneutrality of Ni(CO).

For olefin complexes, back-donation similarly increases stability of the π complex and lowers the bond order of the C=C bond. This is confirmed by the measured C-C bond lengths, which fall in the 1.40-1.47 Å region (the length is ~1.34 Å in olefins) and by the lower C-C ir stretching frequencies. The more stable the complex, generally the greater is the change in $\nu_{C=C}$. Metal-ligand bonds have ir bands in the 650-50 cm^{-1} region. Some illustrative data are given in Table 6.9.

Table 6.9. IR spectral data for some metal π complexes

Compound	$\nu_{C=C}$	$\nu_{metal-olefin}$
$K[Pt(C_2H_4)Cl_3]$	1516 cm^{-1}	407
trans$[Pt(C_2H_4)(NH_3)Cl_2]$	1521	383
$Fe(C_5H_5)_2$	1408	478
$Ru(C_5H_5)_2$	1410	397
$Ni(C_5H_5)_2$	1421	355

The entire corps of physical methods, such as uv, ir, nmr, mass and Mossbauer spectroscopy, X-ray diffraction, dipole moments, optical rotatory dispersion, have been used in the structure determination of metallocenes. is found, for instance, that the mass spectrum of cyclobutadieneiron tricarbonyl corresponds to the formation of a parent molecular ion and of derivative species having one, two, or three fewer CO groups. It is significant that peaks corresponding to the loss of C_2H_2 fragments are not observed, which

Sec. 6.1 STRUCTURE   251

implies that structure 6.43 is preferred over 6.44.

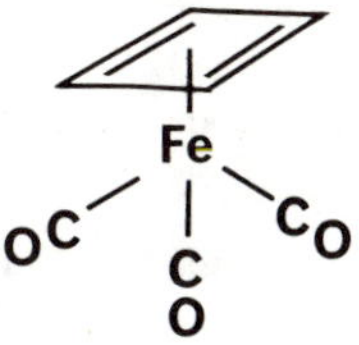

    6.43 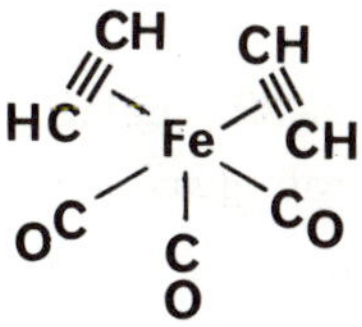   6.44

Generally, olefinic proton signals move upfield in metallocenes. An example is the Zeise salt:

	δ
C_2H_4	6.0 ppm
$[C_2H_4PtCl_3]$	4.7 $J(^{195}Pt-H)$ = 34 Hz

This upfield shift is the net result of shielding produced by the metal-olefin bond and diminished deshielding by the uncoordinated olefinic π electrons.

The pmr spectrum for ferrocene, singlet at δ 4.31 ppm. supports structure 6.45. In the solid state, the rings are fixed in the antiprismatic (staggered) conformation, but in solution or the vapor state the rings rotate freely. Ferrocene undergoes essentially all of the classical aromatic reactions. The cyclopentadienyl rings are electronically united through the iron atom. Thus, electron-donating groups in one ring enhance electrophilic substitution in the other ring.

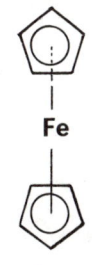

6.45

As stated earlier, the topic of metal π complexes can fill several books (136, 137). Hence, these few comments barely introduce the subject and are included merely to draw attention to this area of organic molecular structure.

252   Ch. 6 APPLICATIONS OF RESONANCE THEORY

6.2. Dynamics -- Eqilibrium and Kinetics

So far, we have used resonance theory to account for certain physical properties, such as
1. thermochemial properties, e.g., heats of combustion or hydrogenation, and aromaticity.
2. molecular geometry, such as bond lengths, planarity, and steric hindrance to planarity.
3. polar character, as indicated by dipole moments; steric effects on polarity.

6.2a. Resonance and equilibria.

With the concept of resonance stabilization, we shall adopt resonance theory to interpret several types of equilibria. When comparing two similar systems with approximately equal entropy changes, the equilibrium will be shifted toward the side favored by resonance. For example, consider the classical case of the dissociation of OH groups:

Alcohols   ROH $\rightleftarrows$ RO$^-$ + H$^+$ $\underline{K_a} \approx 10^{-16}$

Phenols

$$\left[\text{C}_6\text{H}_5\text{-OH} \, , \, \text{C}_6\text{H}_5\text{=}\overset{+}{\text{OH}} \right]$$
      $\underline{a}$ $\underline{b}$
      6.46
             6.46

$\rightleftharpoons \left[\text{C}_6\text{H}_5\text{-}\bar{\text{O}} \, , \, \text{C}_6\text{H}_5\text{=O} \right]$ $\underline{K_a} \approx 1$
                    $\underline{a}$ $\underline{b}$
                         6.47

Resonance has no appreciable effect on the dissociation of alcohols. In the case of phenols, resonance is greater in the ion (isovalent resonance) than in the acid (separation of charges diminishes the resonance effect) with the result that the equilibrium shifts toward the right relative to that of alcohols. This increases $\underline{K_a}$ by a factor of 10^6.

Carboxylic acids

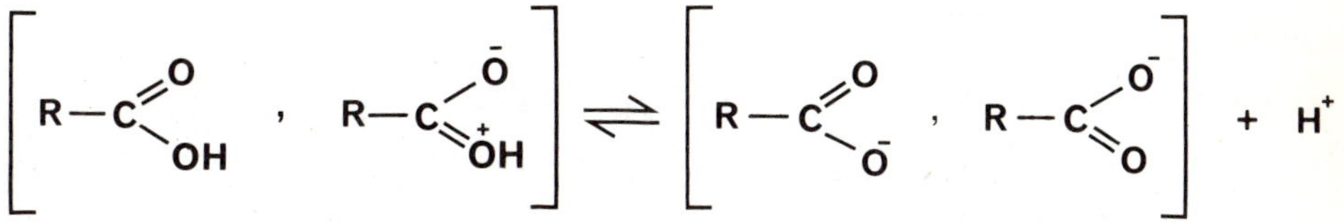

$$\left[R-C\overset{O}{\underset{OH}{\diagdown}} \;,\; R-C\overset{\bar{O}}{\underset{\overset{+}{O}H}{\diagdown}} \right] \rightleftharpoons \left[R-C\overset{O}{\underset{\bar{O}}{\diagdown}} \;,\; R-C\overset{\bar{O}}{\underset{O}{\diagdown}} \right] + H^+$$

$$\underline{K_a} \approx 10^{-5}$$

Here again, resonance is greater in the ion (equivalent structures) than the acid and $\underline{K_a}$ is increased by an additional factor of 10^5.

The same principles apply to bases:

Aliphatic amines

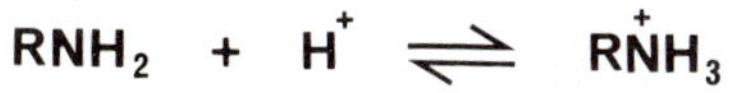

$$RNH_2 + H^+ \rightleftharpoons R\overset{+}{N}H_3 \qquad \underline{K_b} \approx 10^{-5}$$

Anilines

$$\left[\text{Ph}-NH_2 \;,\; \text{Ph}=\overset{+}{N}H_2 \right] + H^+ \rightleftharpoons \text{Ph}-\overset{+}{N}H_3$$

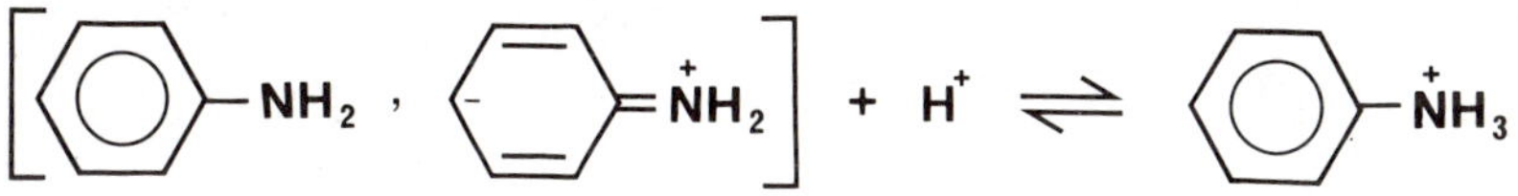

6.48 $\qquad\qquad \underline{K_b} \approx 10^{-9}$

Resonance has no significant effect on the equilibrium of the aliphatic amine. In the case of the anilines, no resonance occurs between the $\overset{+}{N}H_3$ group and the ring but there is resonance in the free base. Resonance of anilines, then, is base-weakening and $\underline{K_b}$ is smaller by a factor of 10^4. However, in addition to the resonance effect, the phenyl group exerts a base-weakening inductive effect, and differential solvation is a dominant factor. For example, the gas phase basicities of aniline and methylamine differ by a much smaller factor (138):

	Proton affinities relative to ammonia (ΔH, kcal/mol)	
	Gas phase	Aqueous phase
Ammonia	0	0
Aniline	-8.9	+6.3
Methylamine	-10.8	-1.9
N-Methylaniline	-15.1	+6.1
Dimethylamine	-18.3	-1.9
Trimethylamine	-23.3	-0.8

Groups which participate in the resonance of phenols or anilines enhance the resonance effect. Thus, electron-withdrawing groups, when <u>ortho</u> or <u>para</u> to the OH or NH_2, increase the acidity of phenols and diminish further the basicity of anilines, and the more such substituents there are, the larger is the effect.

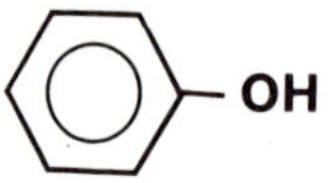

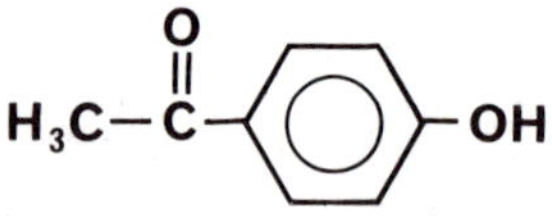

$10^9 K_a$ 0.13   3.7

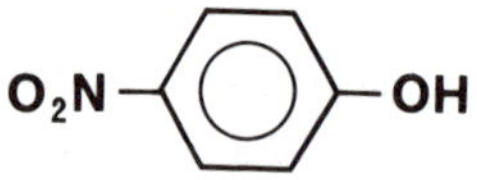

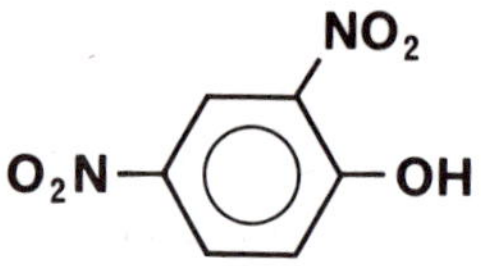

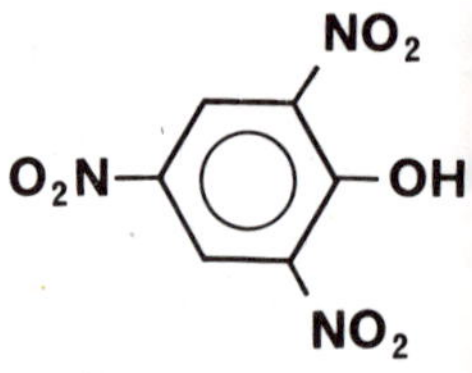

6.9 10^4 4.2×10^7

$10^8 K_a$

Thus, picric acid is a strong acid.

Whereas a 4-nitro group increases the acidity of phenol and of naphthol by a factor of about 10^3, a nitro group produces only a small inductive increase (0.4 pK_a units) from the 5 position of naphthol or the 4' position of 4-hydroxybiphenyl.

Sec. 6.2 DYNAMICS   255

	pK_a		pK_a
Phenol	10.92	4-Hydroxybiphenyl	9.40
4-Nitrophenol	7.81	4'-Nitro-4-hydroxybiphenyl (6.51)	8.95
3-Nitrophenol	9.25		
1-Naphthol	10.65		
4-Nitro-1-naphthol (6.49)	6.36		
5-Nitro-1-naphthol (6.50)	9.25		

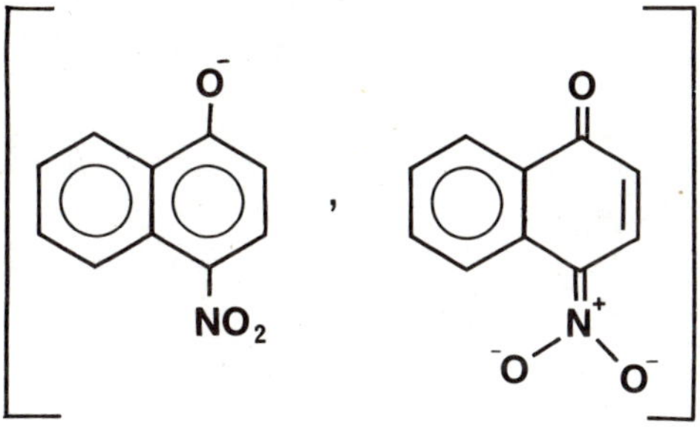

6.49 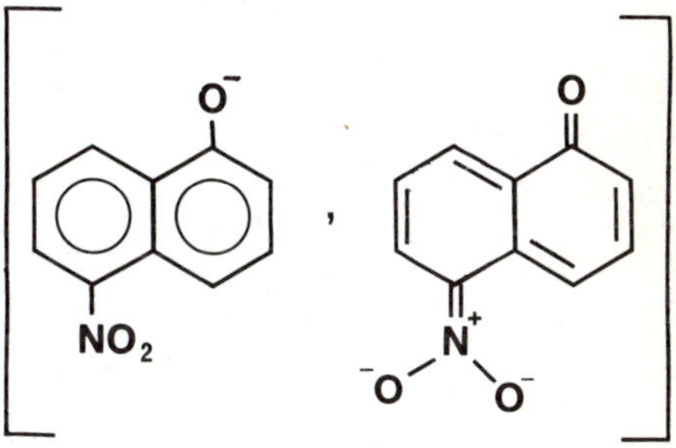
a, b

6.50 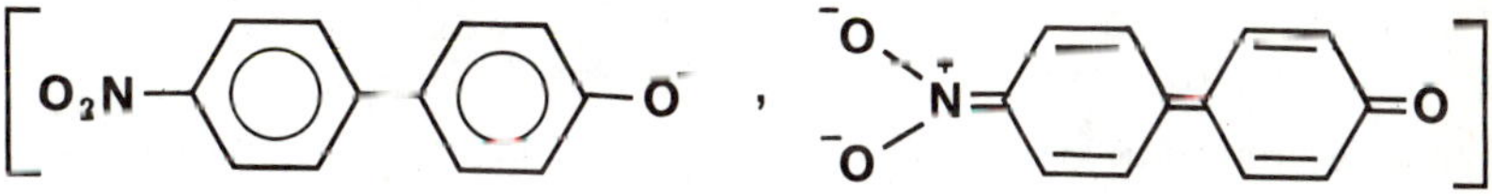
a, b

6.51
a, b

Ch. 6 APPLICATIONS OF RESONANCE THEORY

In polycyclic structures, the more Kekulé rings there are, the more stable is the structure. This explains the difference between 6.49 and 6.50. There is a large acid strengthening effect of the 4-nitro group for the anion of 6.49. Resonance of anions 6.50 and 6.51 does not involve Kekulé-type rings however, and accordingly, resonance involving the nitro group does not make any significant change in pK_a, only a small inductive effect.

There are many structures for anion 6.53 with the negative charge located in one of the four aromatic rings of each of the three large groups. Consequently, resonance stabilization of the anion of 6.52 makes it the most acidic hydrocarbon reported (139).

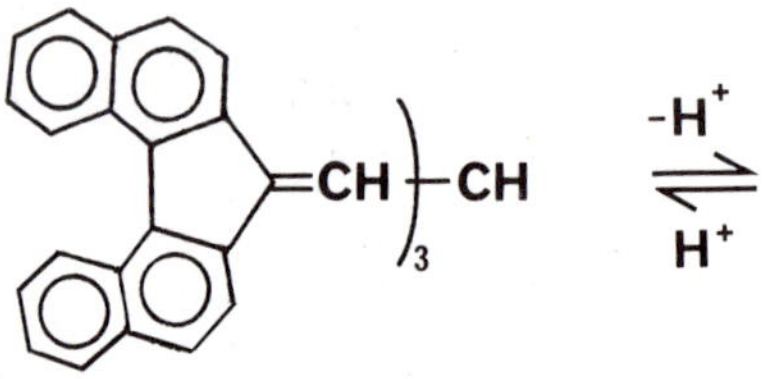

6.52

pK_a = 5.9

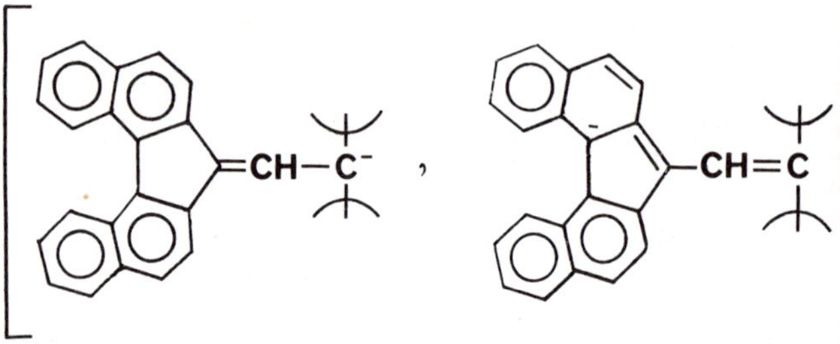

6.53

Another highly acidic hydrocarbon is 6.54, pK_a ≈ 11 (140). The marked

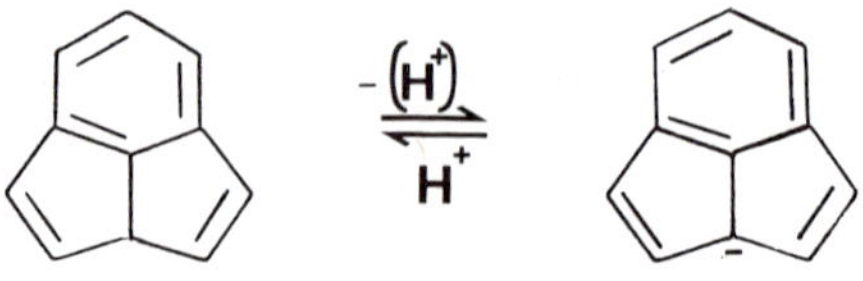

6.54

acidity can be attributed to a large net resonance stabilization of its anion, which has an aromatic character.

The acidities of many perfluorocarbon acids have been measured, and here too resonance stabilization of the corresponding carbanions is a major factor in determining the relative dissociation constants (141). For example, the relative acid strengths of the acids 6.55 decrease in the order

$(CF_3)_2CH-X$

6.55

$X = CF_3 > Br > Cl > I > F > OCH_3$. Except for $X = I$, this order agrees with that predicted on the basis of a destabilizing +R effect of X, which is greatest for OCH_3 and nil for CF_3. The exception $X = I$ indicates that other factors are operative. One important effect is obviously induction. For instance, separation of X from the C in 6.55, as in 6.56,

$(CF_3)_2CH-\langle\bigcirc\rangle-X$

6.56

markedly reduces the acidity, even though X could exert its +R effect through the phenyl group. As with other acids and bases, consideration must be given to the effects of solvation, steric requirements, and polarization, as well as resonance and induction (chain and field) in interpreting the relative acidities of the large variety of polyhalogenated carbon acids studied (141).

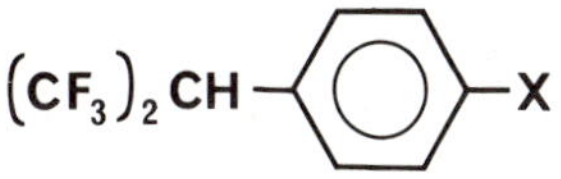

6.57

6.58

Another illustration of the resonance enhancement of Kekulé rings is found in the relative resonance effects in benzene and anthracene analogs. Structure 6.58 with its two Kekulé rings makes a larger contribution to its resonance hybrid than does 6.57 to the corresponding benzene resonance hybrid (142). This is shown by ^{19}F SCS values, which are greater for the anthracene compounds (143).

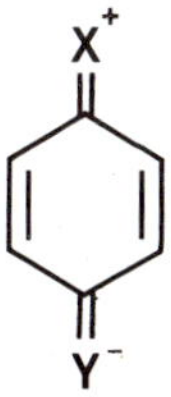

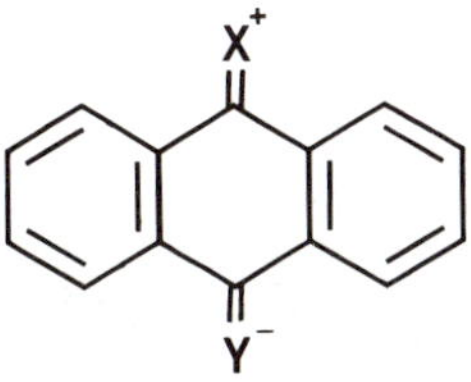

R			
CN	-9.80	-9.67	-13.51
O⁻	19.50		22.42
NO₂	-10.30	-5.51	
CO₂Me	-6.70	-2.80	
COOH	-6.05	-1.73	

The -R effect of CN and the +R effect of O⁻ are shown to be larger for the anthracene ring than the benzene ring (-13.51 compared to -9.80 for CN and 22.42 vs 19.5 for O⁻). The NO_2, COOH, and CO_2Me groups are sterically hindered from exerting any appreciable resonance effect in the 3,5-dimethyl compounds and the ^{19}F SCS are smaller than in the monosubstituted flourobenzenes. The difference gives an indication of the resonance effect. The CN and O⁻ groups are sterically unencumbered and CN is shown to exert about the same effect in both benzene rings.

Guanidine is a classic example in which resonance theory can account for the basicity of a substance (144).

$$\left[\begin{matrix} H_2N \\ H_2N \end{matrix} C=NH, \quad \begin{matrix} H_2N^{\oplus} \\ H_2N \end{matrix} C-\overset{\ominus}{N}H \right] \quad \overset{H^{\oplus}}{\underset{\leftarrow}{\rightarrow}} \quad \begin{matrix} H_2N \\ H_2N \end{matrix} C=\overset{\oplus}{N}H_2$$

pK_a = 13.6   Three equivalent
                                                            structures

The ion is a resonance hybrid of three equivalent structures whereas that for the free base involves separated charges. Hence, the greater resonance energy of the ion shifts the equilibrium toward the right to make guanidine about the strongest organic neutral base known. The basicities of ketones reflect the stabilizing effects of resonance in cations too. The relative basicities of some ketones, in terms of pK as measured by H_0 at half protonation in aqueous H_2SO_4 by nmr, are as follows (145):

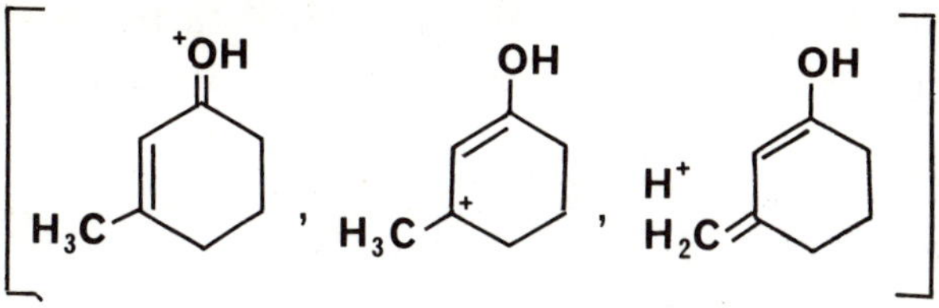

$$\left[\begin{array}{ccc} \text{[3-methylcyclohex-2-enone protonated on O]} & \text{[enol cation]} & \text{[exocyclic methylene enol]} \end{array} \right]$$

pK$_a$ (of the ketone)   −3.8

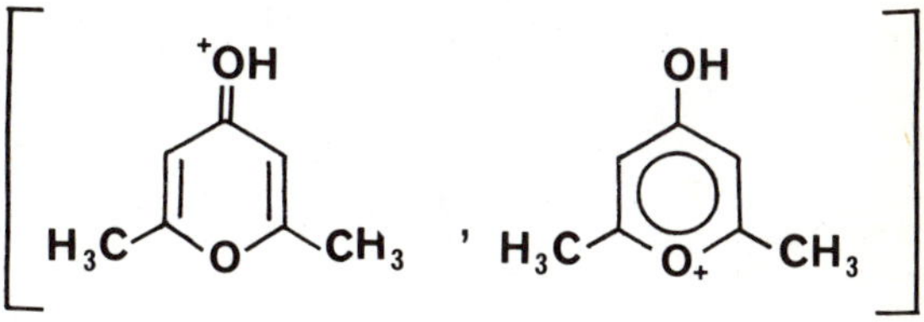

$$\left[\begin{array}{cc} \text{[2,6-dimethyl-4-pyranone protonated on carbonyl O]} & \text{[protonated on ring O]} \end{array} \right]$$

0.3

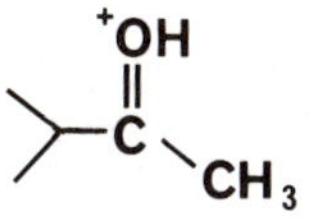

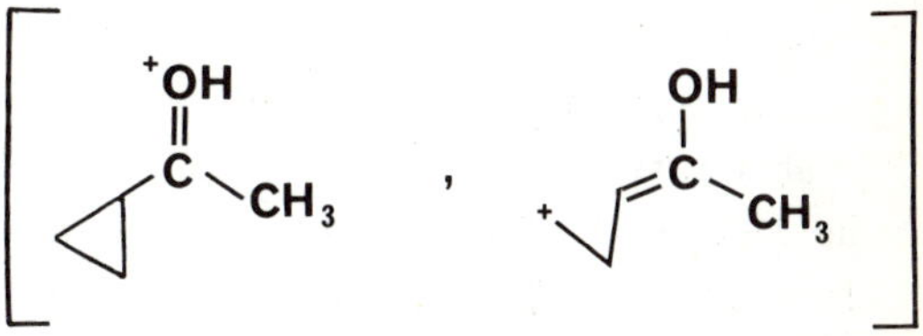

[isopropyl methyl ketone protonated]

$$\left[\begin{array}{cc} \text{[cyclopropyl methyl ketone protonated]} & \text{[enol cation]} \end{array} \right]$$

pK$_a$ (of the ketone) −7.4   −5.9

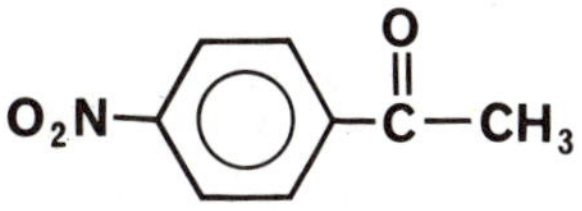

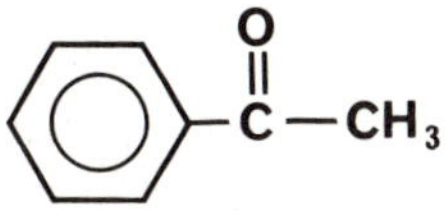

[p-nitroacetophenone]   [acetophenone]

pK$_a$ −8.5   −6.3

[p-methoxyacetophenone]

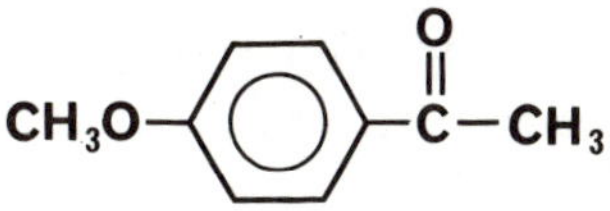

−4.9

The perchloroarylmethanes are interesting compounds (146). The triaryl analogs (6.59) react with alkali in ethyl ether-DMSO solution to give the corresponding carbanions 6.60 and the ether-DMSO solution can be

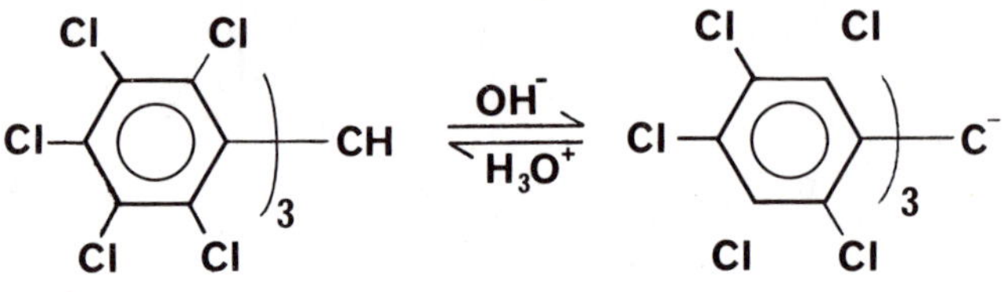

$$\text{6.59} \quad \xrightleftharpoons[H_3O^+]{OH^-} \quad \text{6.60}$$

diluted with water without complete hydrolysis of the carbanion. This unusual strong acidity of the perchloroarylmethanes can be attributed to two factors: (i) accumulated inductive effect of the chlorines and (ii) resonance stabilization of the carbanion. The first effect is evident from the greater acid strength of the perchloro compounds over that of the corresponding hydrocarbons. The second effect is revealed by the failure of the perchlorotriptycene (6.61) to form a carbanion under these conditions. There is little classical conjugation in the latter compound.* Although the perchlorophenyl groups in 6.60 radiate out from the central carbon and are too crowded to be coplanar, the dihedral angles are still small enough to permit some resonance involving the three rings.

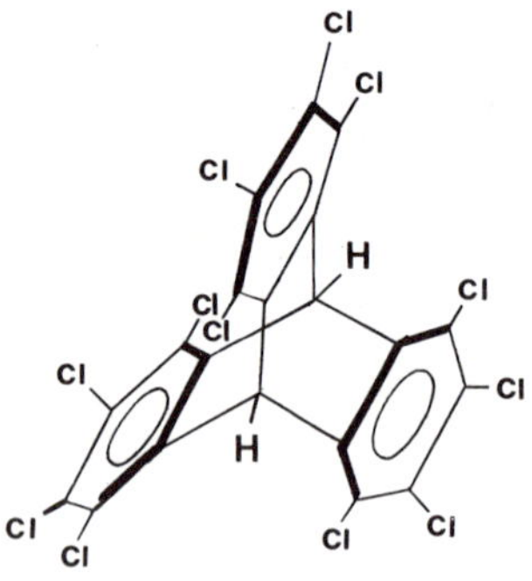

As to be expected, steric effects can strongly affect basicities of aromatic amines (148). Two ways in particular are by (i) internal strain and (ii) steric inhibition of resonance. Protonation of the nitrogen of aniline causes the tetrahedral N to be more

---

* The hydrocarbon triptycene does have a small homoconjugation which is revealed by its uv spectrum (147).

crowded than in the base, which is base-weakening. We saw above that resonance in the free base is base-weakening so that blocking this resonance increases the basicity of the amine. Alkyl groups ortho to the amino group act as in (i) and when attached to the nitrogen, they act as in (ii). For example,

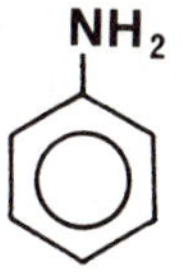

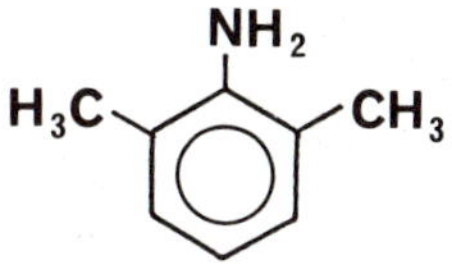

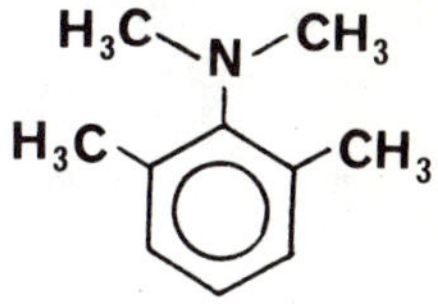

pK_a   4.25   3.42   4.69

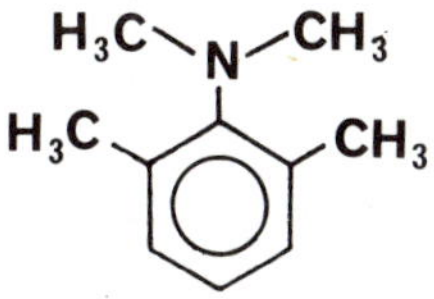

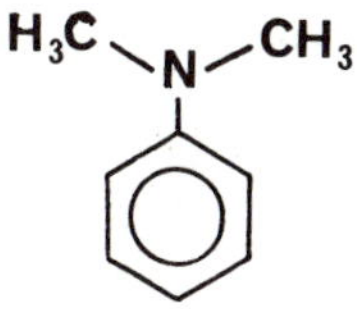

	6.62	6.63
pK_a	4.69	4.26
DM	0.94 D	1.55 D

In fact, the resonance in 6.62 is virtually completely blocked and its dipole moment is about that of aliphatic tertiary amines. Similarly, pK_a increases with increasing size of the alkyl group on N, and pK_a decreases with growing size of the ortho groups, as shown in Table 6.10.

To summarize, the relative basicities of substituted anilines are controlled by a variety of factors which are interrelated in a complex fashion. These include inductive and resonance effects of N and ring substituents, steric effects as just described, differential solvation of free base and conjugate acid, and hybridization changes of the amine nitrogen atom. Experimental attempts to delineate these factors have used nmr and uv spectral data, molecular polarizabilities, H bonding, and other solvation parameters, in addition to basicity measurements (150). In general, the configurations

Table 6.10. Base strengths of some hindered anilines (149)

Substituent	pK_a	Substituent	pK_a
2-Me	4.09	N-Me	4.29
2-Et	4.04	N-Et	4.71
2-i-Pr	4.06	N-i-Pr	5.14
2-t-Bu	3.38	N-t-Bu	6.51
		N,N-di-Me	5.18
		N,N-di-Et	6.65

about nitrogen in aniline, its N,N-dialkyl derivatives, and the N-(4-nitrophenyl)-polymethyleneimines, $4-O_2NC_6H_4N-(CH_2)_n$, n = 3-6, is pyramidal, i.e., near sp^3 hybridization. The configuration about the nitrogen in 4-nitroaniline, its N,N-dimethyl derivative, and the N-(4-nitrophenyl)-polymethyleneimines with n = 2, however, are essentially coplanar, i.e., near sp^2 hybridized nitrogen (150).

Another equilibrium that is partly accounted for by resonance is keto-enol tautomerism (151).

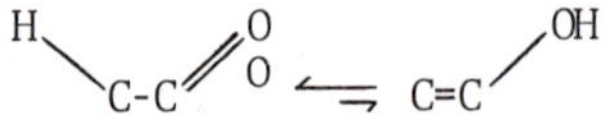

$$\underset{C-C}{\overset{H}{\diagup}}\overset{O}{\underset{}{\diagdown}} \rightleftharpoons \underset{C=C}{\overset{}{\diagup}}\overset{OH}{\underset{}{\diagdown}}$$

The sum of bond energies in the keto form outweighs that of the enol by 15 to 20 kcal/mol (in the gas phase) and indeed, the presence of the enol form is not measurable by simple physical or chemical methods. At the other extreme, the Kekulé resonance of the phenol ring, which can be regarded as an enol, outweighs the bond energy of the keto form to shift the equilibrium

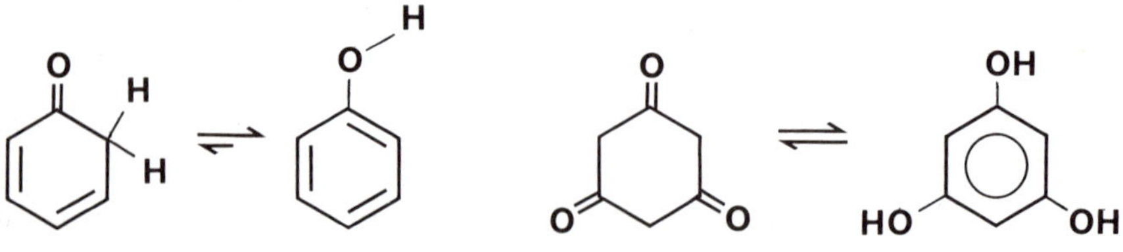

virtually all the way toward the enol. However, the bond energies of several keto groups may outweigh the Kekulé resonance energy. Thus, 1,3,5-cyclohexanetrione has ketonic properties (forms a trioxime) as well as phenolic properties.

When a keto group is introduced β to a carbonyl group, the corresponding enol is partially stabilized by an intramolecular H bond.

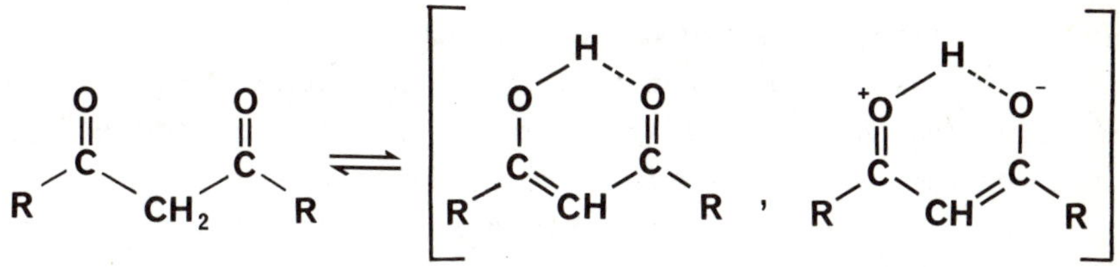

The H bond plus the resonance of the enol ring shifts the equilibrium toward the enol. Resonance also accounts for the direction of enolization. For example, benzoylacetone exists over 90 per cent in the enol structure 6.64 (152).

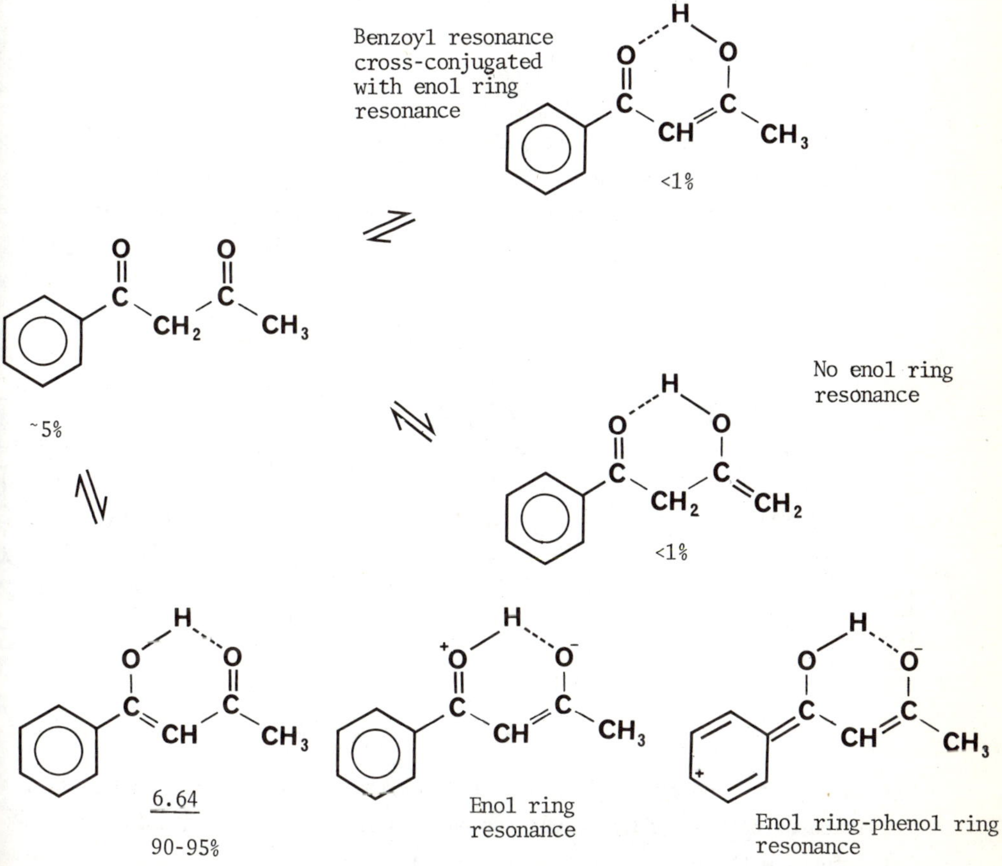

Benzoyl resonance cross-conjugated with enol ring resonance

<1%

~5%

No enol ring resonance

<1%

6.64

90-95%

Enol ring resonance

Enol ring-phenol ring resonance

Thus, the phenyl group augments rather than competes with the enol ring resonance (153).

It is now recognized that the enol-keto equilibrium is controlled by a variety of factors. In addition to resonance and H bonding, induction, steric conditions, and solvation have pronounced effects on the equilibrium, so that one must be cautious in rationalizing structural effects (153, 154). Numerous techniques have been applied to the study of the keto-enol equilibrium, particularly spectroscopic methods (155).

Another type of equilibrium that is often explained in terms of resonance is the dissociation of compounds into free radicals. One well-studied group is the hexaarylethanes.

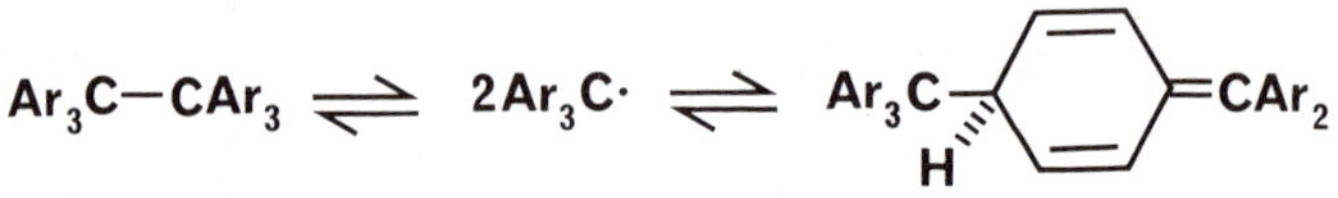

$$Ar_3C-CAr_3 \rightleftharpoons 2Ar_3C\cdot \rightleftharpoons Ar_3C-\underset{H}{\bigcirc}=CAr_2$$

$$\underline{6.65} \qquad \underline{6.66} \qquad \underline{6.67}$$

It has been shown that the free radicals $\underline{6.66}$ usually do not recombine to give the hexaarylethane but yield a dimer such as $\underline{6.67}$ (156). Nevertheless, the amount of free radical present in the equilibrium is greatly dependent upon its resonance stabilization. For example, resonance structures for the triphenylmethyl radical may be written with the odd electron in the ortho or para position of either ring. Since each Kekulé ring is independent

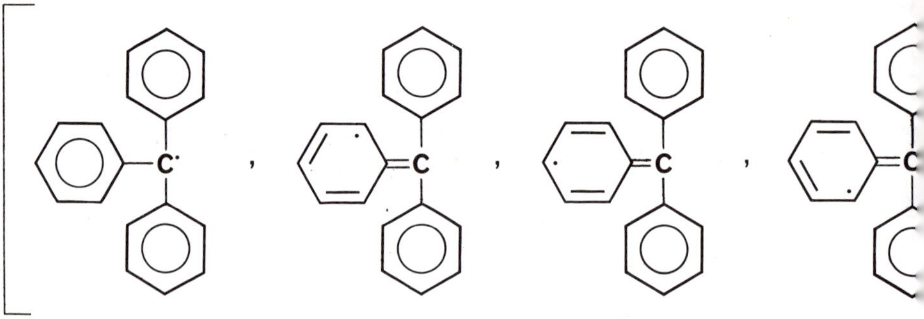

$$\underline{6.68} \qquad \underline{6.69} \qquad \underline{6.70} \qquad \underline{6.71}$$

of the others, there are 2 × 2 structures with the odd electron in the ortho position of $\underline{6.69}$ and four structures each as in $\underline{6.70}$ and $\underline{6.71}$, making twelve structures with the odd electron in any one ring. For the three rings, then there are 36 more structures to be written for the free radical than for the undissociated hexaphenylethane. For a biphenyldiphenylmethyl radical, there are 2 × 2 × 2 = 8 structures with the odd electron as in 6.72 and with three

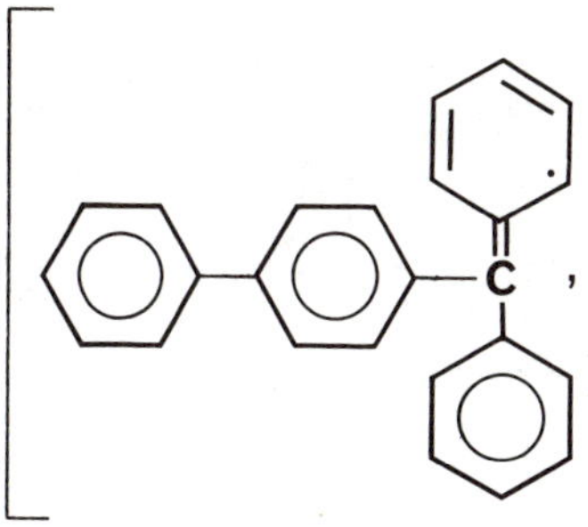

<u>6.72</u> 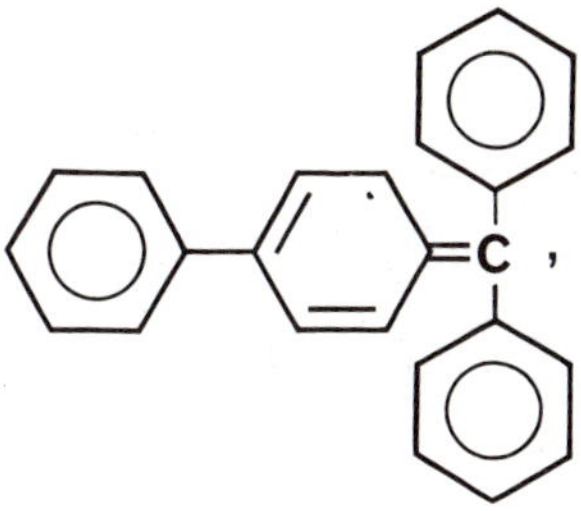   <u>6.73</u> 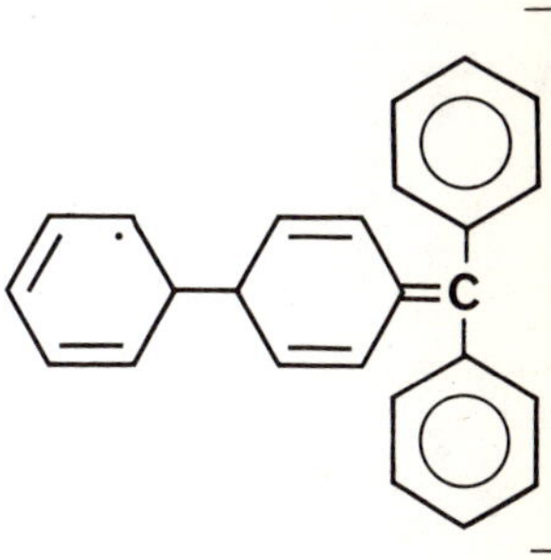   <u>6.74</u>

positions in that ring, there are 24 structures with the odd electron in that ring. There is an equal number for the bottom ring and 24 more for the inner ring as in <u>6.73</u>. There are 2 × 2 × 3 for the odd electron in the outer ring as in <u>6.74</u>, making a total of 84 structures. Thus, the replacement of one phenyl group by a biphenyl group increases the number of resonance structures from 36 to 84 which is reflected in a slightly larger dissociation. With more biphenyl or other aryl groups and with substituents present, the degree of dissociation may be as high as 100 per cent.

	Degree of dissociation
$(C_6H_5)_3C{-}]_2$	1-3%
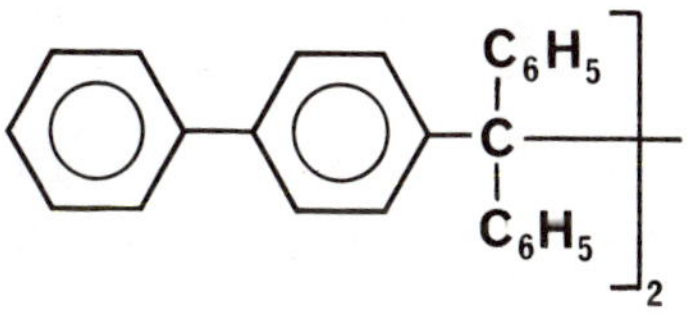 $[C_6H_5{-}C_6H_4{-}C(C_6H_5)_2{-}]_2$	5%
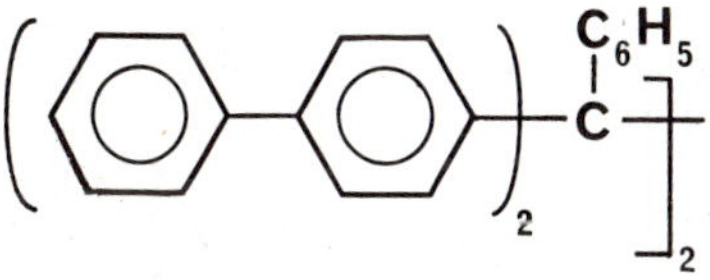 $[(C_6H_5{-}C_6H_4{-})_2C(C_6H_5){-}]_2$	16%

	Degree of dissociation
$\left(\left(C_6H_5-C_6H_4\right)_3 C-\right)_2$	37%
$\left(\left(MeO-C_6H_4\right)_3 C-\right)_2$	100%

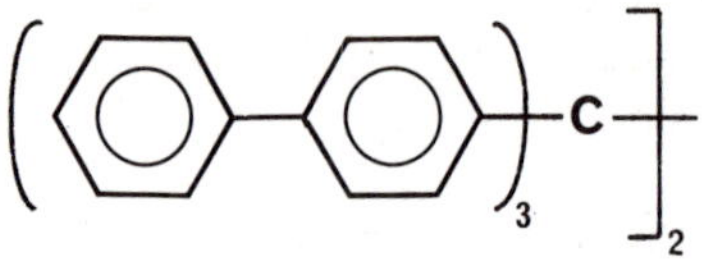

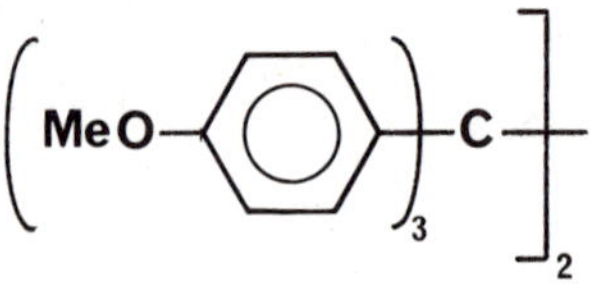

In addition to structures with the odd electron in the rings, structures for the anisyl compound above include 6.75.

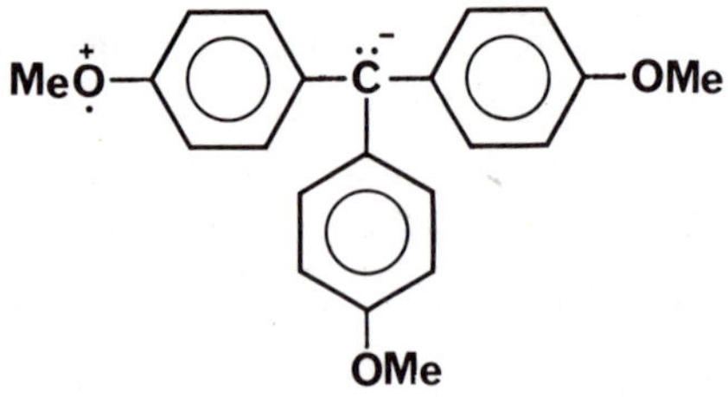

$$\text{MeO}^+{=}C_6H_4{-}\ddot{C}{-}C_6H_4{-}OMe \text{ with third ring bearing } OMe$$

6.75

The combined effects of resonance stabilization, steric hindrance to dimerization, and inductive effects of the chlorines make perchlorotriarylmethyl radicals even more stable and chemically inert (see Section 20.5).

Resonance stabilization of free radicals is also reflected in bond dissociation constants. One might expect, for example, the C-C bond dissociation energy in ethylbenzene to be larger than that of ethane because phenyl is more electronegative than H. The opposite is observed, however, because the arising benzyl radical is unusually stable.

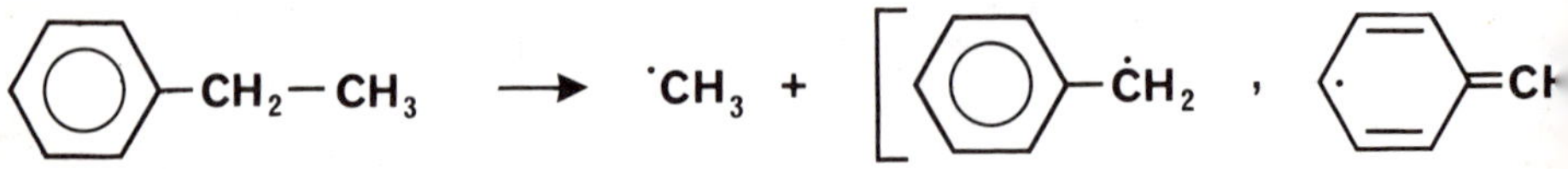

$$C_6H_5-CH_2-CH_3 \longrightarrow \cdot CH_3 + [C_6H_5-\dot{C}H_2 \;,\; \text{(ring radical form)}]$$

Accordingly, the C-C and C-H dissociation energies in some related compounds are found to decrease with increasing resonance in the product radical.

	D_{C-C} (kcal/mol)			D_{C-H} (kcal/mol)
H_3C-CH_3	83	CH_3CH_2-H		96
$C_6H_5CH_2-CH_3$	63	$C_6H_5CH_2-H$		77
$(C_6H_5CH_2-)_2$	47	$H_2C=CHCH_2-H$		76.5
$(H_2C=CHCH_2-)_2$	38			
$(C_6H_5)_3C-)_2$	11			

However, the C-X dissociation energy differences between CH_3X and $H_2C=CHCH_2X$ vary with the nature of X (157):

	X = H	OH	I	Me	Et	i-Pr	t-Bu	Phenyl
$X-CH_3$	104	91	56	88	85	84	80	93
$X-CH_2CH=CH_2$	85	77	41	72	69	67	63	77

QUESTIONS

1. Predict the order of relative C_3-H and OH proton chemical shifts for 1,3-diketones A, B, C, and D and give the basis of your reply.

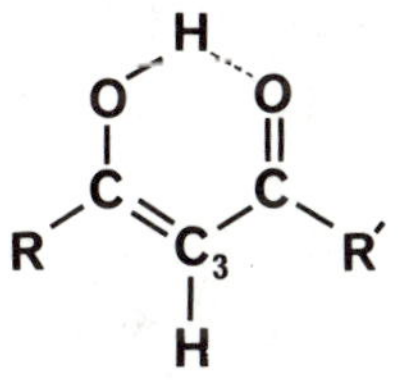

Compound	R	R'
A	CH_3	CH_3
B	CH_3	CF_3
C	CF_3	CF_3
D	C_6H_5	C_6H_5

268   Ch. 6 APPLICATIONS OF RESONANCE THEORY

2. Should E have a high enol content, and why or why not ?

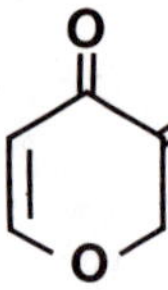

E 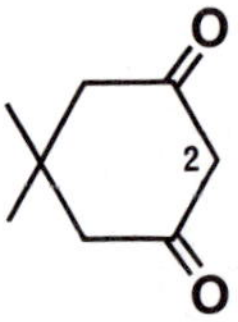   F

3. In $CHCl_3$ solution, F has carbonyl bands in the 1735-1745, 1708-1716, and 1607-1628 cm^{-1} regions. The first two bands are regarded as a Fermi resonance doublet for the keto tautomer. As the concentration is increased the intensities of bands 1 and 2 decrease and that of band 3 increases. Also, upon replacing the 2-H by R, as R changes from Me, Et, to i-Pr, the intensity of band 3 decreases and even disappears for the i-Pr derivative. Offer an explanation for the changes in intensity of the bands for the two cases: concentration change and structural change.

4. Benzoylacetone, C_6H_5-$COCH_2COCH_3$, has uv spectra in ether and water as shown. Which curve represents the spectrum in ether and to what chromophores do you assign the bands ?

247   310
λ (nm)

5. The pK_a values in water for the bases G, H, I, and J are given below.

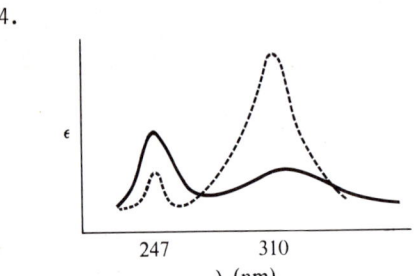

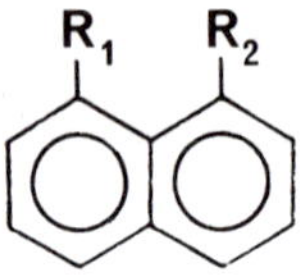

Compound	R_1	R_2	pK_a
G	NH_2	NH_2	4.61
H	NHMe	NHMe	5.61
I	NMe_2	NHMe	6.43
J	NMe_2	NMe_2	12.34

How do you account for the trend in basicities and the unusual value for $\underline{J}$ as an aromatic amine?

6. 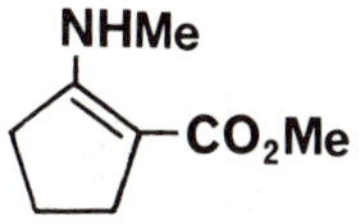 [Structure K: cyclopentene with NHMe and CO₂Me substituents on the double bond carbons]

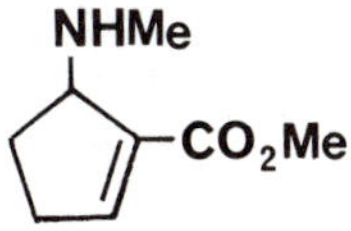

   $\underline{K}$

   Predict which would be the more stable, $\underline{K}$ or $\underline{L}$. The actual compound has ir bands at 3300 and 1650 cm^{-1}. How would you assign these bands?

   [Structure L: cyclopentadiene with NHMe and CO₂Me substituents]

   $\underline{L}$

7. The lowest ionization potentials of the NO_2 group (as measured by pes) in nitromethane and nitrobenzene are 11.32 and 10.90 eV, respectively (O. S. Khalil, J. L. Meeks, and S. P. McGlynn, J. Am. Chem. Soc. 95: 5876 (1973)). How would you explain the smaller value for nitrobenzene if the phenyl group is more electronegative than methyl?

8. Offer an explanation for the relative magnitudes of $\Delta G^{\ddagger}$ for rotations about the amide C-N bond in the following groups of amides. (a) $\underline{M}$, $\underline{N}$, and $\underline{O}$; (b) $\underline{N}$, $\underline{P}$, and $\underline{Q}$; (c) $\underline{N}$ and $\underline{S}$. (M. B. Shambhu, G. A. Digenus, and R. J. Moser, J. Org. Chem. 38: 1227 (1973).)

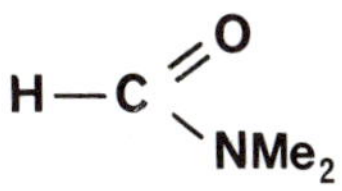

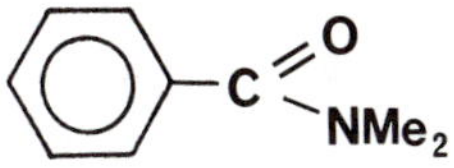

   [Structure M: H-C(=O)-NMe₂]   [Structure N: Ph-C(=O)-NMe₂]

   $\underline{M}$ $\underline{N}$

   $\Delta G^{\ddagger}$ =   21   15.5

   [Structure O: 4-pyridyl-C(=O)-NMe₂]   [Structure P: 4-O₂N-C₆H₄-C(=O)-NMe₂]

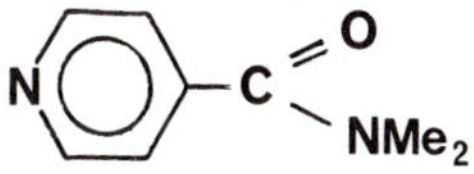

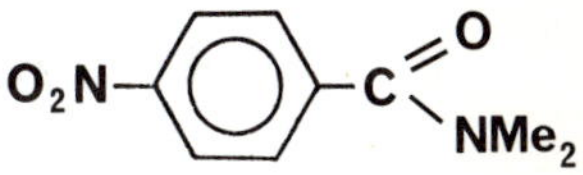

   $\underline{O}$ $\underline{P}$

   16.0   16.4 (kcal/mol)

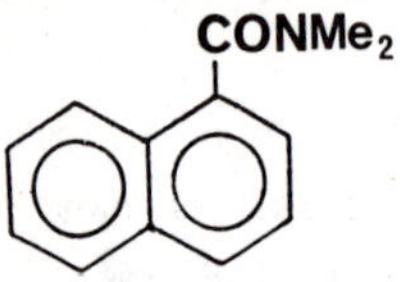

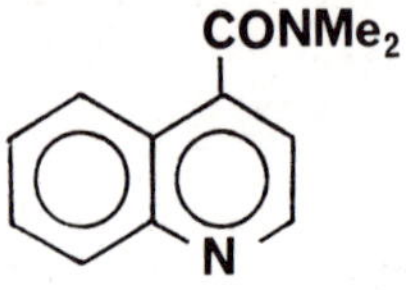

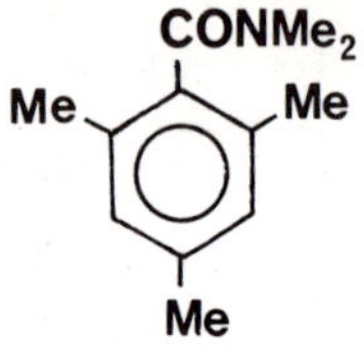

Ch. 6 APPLICATIONS OF RESONANCE THEORY

<u>Q</u> (naphthalene with CONMe₂ at 1-position)

<u>R</u> (quinoline with CONMe₂ at 4-position)

<u>S</u> (2,4,6-trimethylphenyl with CONMe₂)

$\Delta G^{\ddagger}$ = 16.6   17.2   22.5 (kcalmol)

ANSWERS

1. Predictions: δ_{C_3-H} $\underline{C} > \underline{D} \simeq \underline{B} > \underline{A}$

   δ_{O-H} $\underline{D} > \underline{A} > \underline{B} > \underline{C}$

With increasing electronegativity of R or R', C_3 becomes more electronegative and the C_3-H proton signal should move downfield in the order $\underline{C} > \underline{B} > \underline{A}$. The position of $\underline{D}$ is doubtful, perhaps between $\underline{C}$ and $\underline{B}$. As we saw in Section 1.2, α electronegative groups reduce the basicity of a carbonyl group, decrease the corresponding H-bond strengths, and δ_{OH} moves upfield. Hence, the order for $\underline{A}$-$\underline{C}$ is as given. The phenyl-enol ring resonance, which is H-bond strengthening, outweighs the inductive effect of the phenyl to give it the strongest H bond and move δ_{OH} farthest downfield. The actual values are:

C_3-H:	<u>D</u>	<u>C</u>	<u>B</u>	<u>A</u>	OH:	<u>D</u>	<u>A</u>	<u>B</u>	<u>C</u>
δ	6.80	6.43	5.90	5.44	δ	17.13	15.40	14.24	13.00

2.

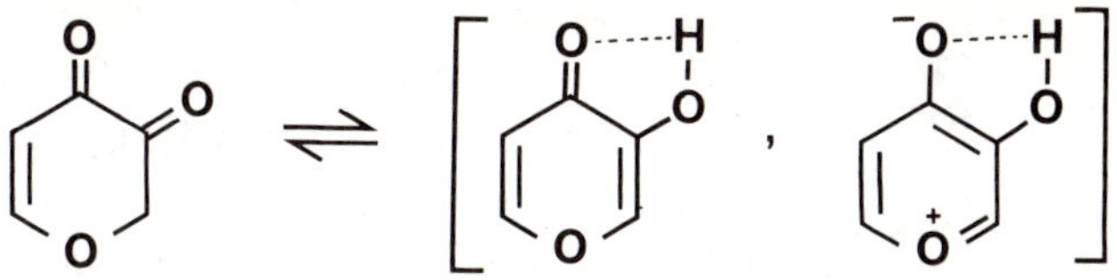

[diketone E ⇌ enol M (structures a and b, with aromatic dipolar form)]

As a result of the aromaticity of the dipolar structure $\underline{Mb}$ and the intramolecular H bond, the compound is virtually 100 per cent enolic.

3.

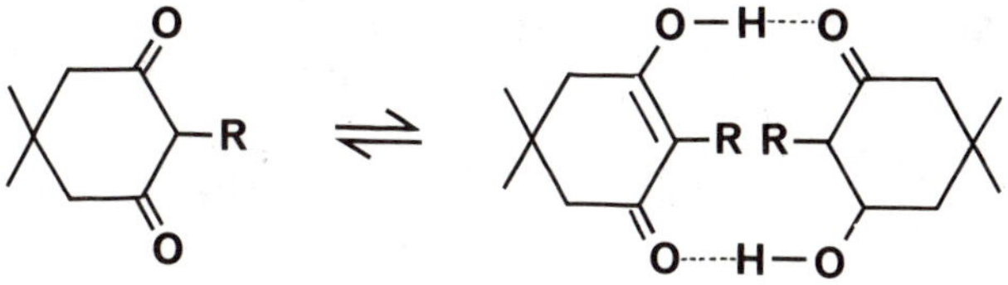

[dimedone-type diketone with R ⇌ enol dimer with intermolecular H bonds]

$\nu_{C=O} 1735, 1708 \text{ cm}^{-1}$ $\nu_{C=O} 1607 \text{ cm}^{-1}$

As the concentration increases, there is more dimerization through H bonding which shifts the equilibrium toward the enol to increase its band intensity. As R gets larger, it hinders dimerization, so there is a shift toward the diketone.

4. Since the two bands are at the same position in each solvent, there must be an equilibrium affected by the solvent:

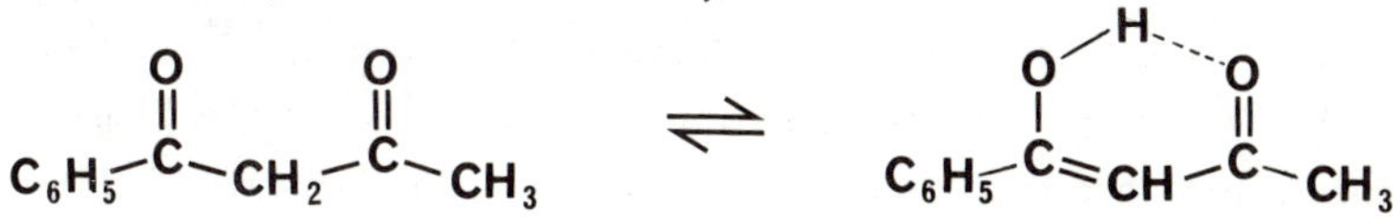

$$C_6H_5-\overset{O}{\overset{\|}{C}}-CH_2-\overset{O}{\overset{\|}{C}}-CH_3 \quad \rightleftharpoons \quad C_6H_5-\overset{O\cdots H\cdots O}{\overset{|}{C}}=CH-\overset{\|}{C}-CH_3$$

The ketone is favored by H-bonding solvents, H_2O in this case, and the enol is favored by nonpolar solvents where it can form an intramolecular H bond not in competition with the solvent. The enol has a cinnamoyl chromophore $C_6H_5-C=C-C=O$ with the larger λ_m (310 nm) and the ketone the smaller λ_m (247 nm) for the benzoyl chromophore. Hence, the dotted line is for ether solvent.

5. As ϕNR_2 resonance decreases, base strength increases. As we go from G to J, steric hindrance increasingly blocks resonance resulting in a stronger base. This steric effect causes NR_2 groups to rotate to an orientation for which the H bond between Ns is more effective. Since G-I can form the H bond in the free base as well, the H bond produces the greatest net stabilization of the ion for J.

6.

K

Structure K is the more stable owing to the aminoketone resonance. The 3300 and 1650 cm^{-1} bands are for the N-H and C=O bonds in the fragment N-H···O=C.

7. Apparently, electron delocalization outweighs the inductive effect to increase the electron density on the NO_2 group in nitrobenzene.

8. (a) The amide group resonance, [O=C-NMe$_2$, $^-$O-C=$\overset{+}{N}$Me$_2$], increases $\Delta G^{\ddagger}$, but any cross resonance of the C=O bond with an attached group competes with this resonance to decrease $\Delta G^{\ddagger}$. Hence, $\Delta G^{\ddagger}$ is much smaller for N than for M. The pyridine ring exhibits greater electron-withdrawing resonance than electron-donating resonance. This results in less aroyl resonance and more amide group resonance to increase $\Delta G^{\ddagger}$ for O relative to N.

(b) Similarly, the nitrophenyl group involves electron-withdrawing resonance. As in O, this diminishes $\Delta G^{\ddagger}$ for P relative to N. In Q, steric hindrance with the peri H hinders coplanarity of the amide group with the ring to diminish the aroyl resonance and increase $\Delta G^{\ddagger}$ relative to N.

(c) The 2,6-methyl groups of S completely prevent coplanarity of the amide and phenyl groups, which is required for the benzoyl resonance. As a result, amide group resonance is substantial as in M and $\Delta G^{\ddagger}$ is relatively large.

6.2b. Resonance and reaction mechanisms

Resonance stabilization of reaction transition states or of reactive intermediates has a profound effect on rates of reaction. It is well known, for instance, that benzyl and allyl substrates undergo S_N1 reactions much faster than simple alkyl halides. This is attributed to the transient formation of resonance-stabilized benzyl or allyl cations. Reaction rates are affected as well as the composition of mixtures. Thus, an allylic cation, anion, or radical exists as a resonance hybrid and generally yields a single product mixture irrespective of the structure of its precursor. For example,

$H_2C=CH-CH_2-Cl$

or

$CH=CH-CH_3$
$|$
Cl

$\xrightarrow{Cl\cdot}$

$[H_2C=CH-\overset{\cdot}{C}HCl, \quad H_2\overset{\cdot}{C}-CH=CHCl]$

resonance hybrid

$\downarrow Cl_2$

$H_2C=CH-CHCl_2 \;+\; CH_2-CH=CHCl$
$ |$
$ Cl$

10%   90%

$CH_3-CH=CH-CH_2Cl$

or

$CH_3-CH-CH=CH_2$
$|$
Cl

$\xrightarrow{Mg}$

$[CH_3-CH=CH-\overset{\ominus}{C}H_2, \quad CH_3-\overset{\ominus}{C}H-CH=CH_2]$

resonance hybrid

$\downarrow H_3O^{\oplus}$

$CH_3-CH=CH-CH_3 \;+\; CH_3-CH_2-CH=CH_2$

43%   57%

Thus, $CH_2=CH-CH_2-Cl$ gives the same proportion of the two products as does $ClCH=CH-CH_3$. Other examples of radical reactions are:

$CH_3-CH-CH=CH_2$
$|$
Cl

or

$CH_3-CH=CH-CH_2Cl$

$\xrightarrow{Ni(CO)_4}$ $CH_3CH=CH-CH_2 \;\; CH_2-CH-CHCH_3 \;+\; CH_3CH=CHCH_2-CH-CH=CH_2$
$|$
CH_3

80%   20%

CH$_3$CH$_2$CH=CHCH$_2$-COOH

or

CH$_3$CH$_2$CHCH=CH$_2$
　　　|
　　　COOH

$\xrightarrow{\text{Kolbe electrolysis}}$ same mixture of 6.76, 6.77, and 6.78

CH$_3$CH$_2$CH=CHCH$_2$$\mathrm{-\!\!\!\!|}_2$

6.76

CH$_3$CH$_2$-CH-CH=CH$_2$
　　　　　|
CH$_3$CH$_2$-CH-CH=CH$_2$

6.77

CH$_3$CH$_2$-CH=CH-CH$_2$
　　　　　　　　　|
CH$_3$-CH$_2$-CH-CH=CH$_2$

6.78

A good estimate of the allyl radical resonance energy is between 9 and 12 kcal/mol (158), although estimates range up to 25 kcal/mol.

Resonance stabilization of the reactive intermediates also accounts for the relative reactivities of vinyl substrates. For example, the vinyl halide 6.79 reacts quickly at room temperature with aqueous AgNO$_3{}_{\text{alc}}$ to yield cyclo-

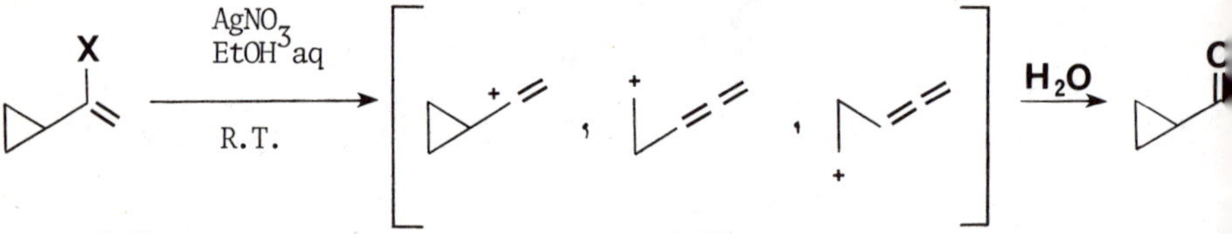

6.79 $\xrightarrow[\text{R.T.}]{\text{AgNO}_3 \\ \text{EtOH aq}}$ [6.80] $\xrightarrow{\text{H}_2\text{O}}$

propyl methyl ketone whereas the isopropyl analogue 6.81 is inert at 25° and reacts only very slowly at 150° (159). Apparently the vinyl cation intermedi is stabilized by resonance with the cyclopropyl group to promote its forma tion. Only a trace of rearrangement t the cyclobutyl or homoallyl products occurs. Only recently have vinyl cati been seriously proposed as reaction intermediates and some reviews of their role in solvolysis reactions have appeared (160). Vinyl cations prefer a linear structure, and the reactivity of vinyl substrates via vinyl cations decreases as the respective vinyl catio deviate from linearity (161).

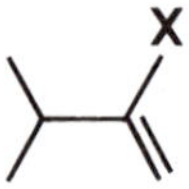

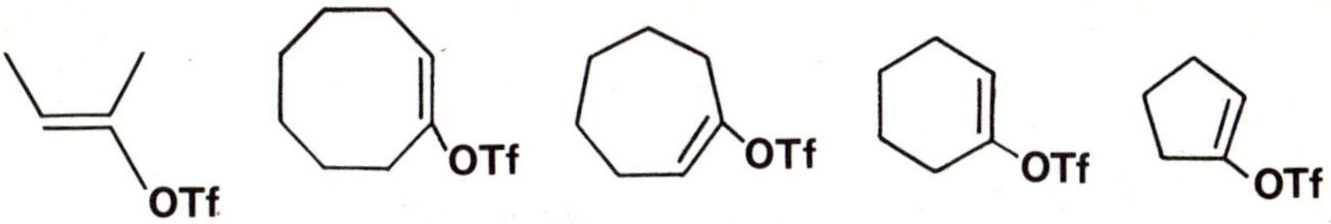

Sec. 6.2 DYNAMICS 275

structure	CH₂=C(CH₃)OTf	cyclooctenyl-OTf	cycloheptenyl-OTf	cyclohexenyl-OTf	cyclopentenyl-OTf
k_{rel}	1.0	3.4	0.32	10^{-4}	10^{-5}

(50% EtOH, 100°)

$Tf = CF_3SO_2-$

In contrast to its alicyclic counterparts which are quite reactive owing to resonance stabilization of the vinyl cation, bromocyclohexadienes (6.82) also do not react with silver ion or undergo solvolysis up to 180° (162). Here again, the inability to form a linear, resonance stabilized vinyl cation prohibits reaction. Other examples of the influence of nonclassical resonance on reaction kinetics will be discussed in Chapter 9.

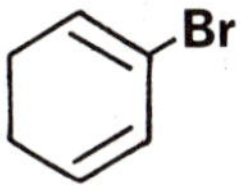

6.82

Chapter 7

HAMMETT-TYPE TREATMENTS (163)

Chemists are always seeking empirical relationships for the prediction of chemical, physical, or biochemical properties of a molecule on the basis of its structure, or conversely, chemists look for a guide for the design in structure of a molecule which would have any desired property. The two most widely used correlations of this type are the periodic table and relative electronegativities. Another extensively used relationship is that developed by L. P. Hammett (164). He reasoned that substituents in the para or meta position of the benzene ring should exert the same relative effects on various reactions. Hence, a straight line should result when the effects on two reactions are plotted against each other. For instance, if the log of the dissociation constants of substituted benzoic acids is plotted against the log of the dissociation constants of the similarly substituted phenylboric acids, the result is a reasonably straight line which can be reproduced by equation (7.1)

$$\log \underline{K} = \rho \log \underline{K}' + \underline{C} \qquad (7.1)$$

where $\underline{K}$ and $\underline{K}'$ are the two series of dissociation constants, ρ is the slope of the line, and $\underline{C}$ is the intercept. If the equilibrium constants for the unsubstituted compounds are denoted $\underline{K}_0$ and $\underline{K}'_0$ in equation (7.1) we have

$$\log \underline{K}_0 = \rho \log \underline{K}'/\underline{K}'_0. \qquad (7.2)$$

Subtracting equation (7.2) from (7.1), we get (7.3).

$$\log \underline{K}/\underline{K}_0 = \rho \log \underline{K}'/\underline{K}'_0. \qquad (7.3)$$

Equation (7.3) is applicable to any two pairs of equilibria, and one equilibrium may be chosen as a reference for comparing all the others. Since the dissociation constants of substituted benzoic acids were known very accurately for a large number of substituents, this reaction was selected as the standard. Accordingly, a new constant may be chosen equal to $\log \underline{K}'/\underline{K}'_0$, which would be characteristic of each substituent. This reduces equation (7.3) to (7.4)

Ch. 7 HAMMETT-TYPE TREATMENTS

$$\log \underline{K}/\underline{K}_0 = \sigma\rho \qquad (7.4)$$

in which σ is called the <u>substituent constant</u> and is a measure of the effect which the substituent has upon the equilibrium resulting from the electron density change at the reaction center, and ρ is a <u>reaction constant</u> which reflects the sensitivity of the equilibrium in question to a change in electron density at the reaction center.

Since thermodynamically $\Delta \underline{G}^{\circ}$ = -$\underline{RT}$ ln $\underline{K}$, equation (7.4) is really a <u>linear free-energy</u> relationship. It also applies to rates of reactions, in which case rate constants are used in place of equilibrium constants.

Hammett carried out this procedure for a large number of equilibria and reaction rates. Each series of reactions yields a specific value for σ for a given substituent. These values of σ for a given substituent vary over a small range and a mean value must be chosen. Hence, Hammett selected a set of "best value" <u>meta</u> and <u>para</u> substituent constants for a fairly large number of groups. His original σ values have been revised from new and better data (163), and some representative values are listed in columns 2 and 3 in Table 7.1.

Now, an established set of σ values enables us to do the following:
(1) If we know ρ for a reaction, we can estimate the rate of reaction for a substituted derivative from the σ constant. Hence, a vast amount of information is stored in the Hammett equation. (2) ρ for a new reaction is obtained from the slope of the line which best fits the data when values of log $\underline{K}/\underline{K}_0$ or log $\underline{k}/\underline{k}_0$ are plotted against a few σs. (3) For a new group, σ is determined from equation (7.4) using the log $\underline{K}$ or log $\underline{k}$ data and ρ.

Hammett and others soon noted that dual σ constants for some groups seemed to be needed because of interaction resonance between the substituent and the side-chain reaction center. Gradually it was recognized that there are three major ways in which the substituent transmits its electrical effect to the reaction center: (1) induction, (2) polar effect of resonance, and (3) interaction resonance with the reaction center.

      (1)   (2)   (3)

The σ constants set up by Hammett involved only modes (1) and (2), except

278   Ch. 7 HAMMETT-TYPE TREATMENTS

Table 7.1. Substituent constants for some common groups

Group	Hammett σ		Taft		Electrophilic σ^+_{para}	Nucleophilic σ^-_{para}	σ_I^p	Hansch partition π (166a)
	para	meta	σ_R	σ_I				
NH_2	-0.66	-0.16	-0.76	0.12	-1.3			-1.23
OH	-0.37	0.12	-0.60	0.25	-0.92		-0.39	-0.67
OCH_3	-0.27	0.12	-0.51	0.25	-0.76		-0.12	-0.02
CH_3	-0.17	-0.07	-0.11	-0.05	-0.31		-0.96	0.50
$C(CH_3)_3$	-0.20	-0.10	-0.09	-0.07	-0.25		-1.55	1.68
$CH=CH_2$	-0.08	0.08	-0.23	0.15				
H	0.0	0.0	0.0	0.0	0.0	0.0	0.0	0.0
F	0.06	0.34	-0.44	0.52	-0.07		0.56	0.14
Cl	0.23	0.37	-0.24	0.47	0.11	0.27	0.93	0.71
Br	0.23	0.39	-0.22	0.45	0.15	0.29		0.86
I	0.18	0.35	-0.11	0.38	0.13	0.32		1.12
COOH	0.41	0.37				0.73		-0.3
CO_2Et	0.45	0.37	0.11	0.32		0.68		0
COMe	0.50	0.38	0.15	0.28		0.87		-0.45
CF_3	0.54	0.43	0.09	0.41		0.74	0.70	0.88
CN	0.66	0.56	0.10	0.58		1.0		-0.57
NO_2	0.78	0.71	0.16	0.63	0.78	1.27		-0.28

for his few dual values. Later, others developed constants which include mode (3). These are called electrophilic (σ^+) (165) or nucleophilic (σ^-) (166) constants, depending upon whether the reaction center increases in positive or negative charge in the transition state. Some values are listed in Table 7.1.

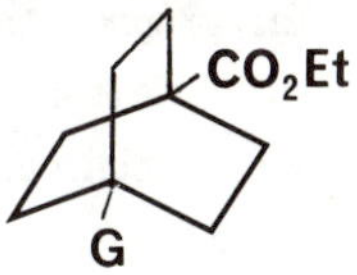

In an attempt to evaluate the inductive effect alone, Roberts measured the rates of hydrolysis of several esters of type 7.1. Here, the effect is transmitted across the six-membered ring by induction (chain and field) only. The σ constants so obtained were called inductive constants. Later, Taft observed a fixed relationship between these inductive constants and his polar substituent constants σ^* (see Section 3.1b), $\sigma^* = 0.45\ \sigma_I$, which enabled him to compute σ_I values for many more groups. Some values are listed in Table 7.1. Three series where they apply, for example, are:

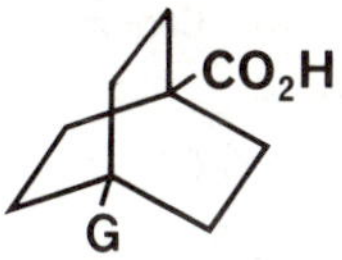

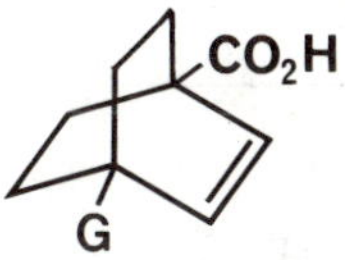

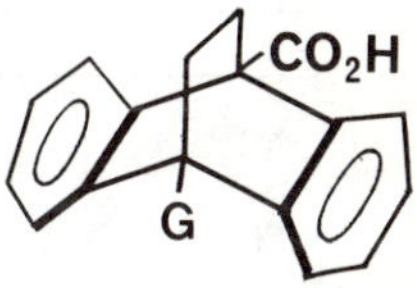

$\log \underline{K_a}$ (167) = $1.63\underline{\sigma_I}$ -6.88 $1.75\underline{\sigma_I}$ -6.49 $1.15\underline{\sigma_I}$ -5.75

Taft also showed that Hammett's σ values could be expressed as the sum of the inductive constant and a resonance constant: $\sigma = \sigma_R + \sigma_I$. Hence, $\underline{\sigma_R}$ values could be determined indirectly from σ and $\underline{\sigma_I}$ values. Later, it was found that σ_R and σ_I could be determined independently from physical measurements:

$$\underline{\sigma_R} = 0.0074\ \sqrt{\underline{A}} \quad \text{and} \quad \underline{\sigma_I} = \frac{\delta_m^F + 0.05}{0.61}$$

where $\underline{A}$ represents the integrated intensities of the 1600- and 1585-cm^{-1} bands in the ir spectra of the respective monosubstituted benzenes (168), and δ_m^F is the ^{19}F nmr chemical shift for the respective meta-substituted fluorobenzenes (169). However, some chemists have charged that "the effects of substituents on chemical properties and on ^{19}F chemical shifts present entirely different problems and that attempts to combine the two will prove fruitless" (170).

Since there is no direct resonance between the substituent and the meta carbon, it might be thought that σ_m would equal σ_I for meta substituents. This is approximately true for groups which exhibit little ground-state resonance. However, for groups which have substantial ground-state resonance, like the NH_2 group, the polar effect of resonance can be felt at the meta position. This has been called a relayed resonance effect, a $\sigma_R^{\underline{m}}$ for meta groups (169), a mesomeric-field effect (170), and a secondary resonance effect (171). A measure of this mesomeric-field effect, in terms of $\sigma_R^{\underline{m}}$ is given in Table 7.2.

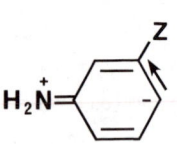

Table 7.2. <u>Meta</u> resonance or mesomeric-field effects (172)

Group	$\sigma_R^{\underline{m}} = \sigma_{\underline{m}} - \sigma_{\underline{I}}$	Group	$\sigma_R^{\underline{m}} = \sigma_{\underline{m}} - \sigma_{\underline{I}}$
NH_2	-0.25	COOEt	0.04
OH	-0.17	$COCH_3$	0.07
OCH_3	-0.17	CN	0.04
F	-0.17	CF_3	0.01
Cl	-0.10	NO_2	0.08
Br	-0.06		
I	-0.04		
CH_3	-0.02		

Again, we have an indication that ground-state resonance for NR_2, OR (R = alkyl or H), and F is substantial, and that F is distinctly different from the other halogens. That is, the resonance effect of F is comparable to its large inductive effect and often is dominant. For instance, a p-F atom has a <u>shielding</u> effect upon $\delta^{15}N$ and δ^1H of the amino group in p-fluoroaniline (Section 6.1b) and the p-F atom has a partial rate factor slightly larger than unity for the bromination of fluorobenzene (173).

Similarly, a p-fluorine atom stabilizes a phenylmethyl cation whereas meta and ortho fluorine atoms are destabilizing (174). That is, the electron-donating resonance of the para fluorine outweighs its inductive effect but the opposite is true for the ortho fluorine atom. Also, nmr chemical shift differences between fluorine and chlorine compounds are much larger than between chlorine and bromine analogs. Thus, owing to the polar form of the resonance hybrids

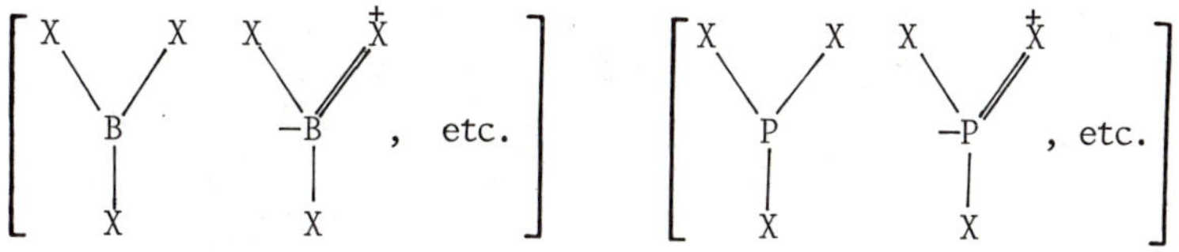

$$\left[\begin{array}{c} X \\ \diagdown \\ B \\ | \\ X \end{array} \diagup X \quad , \quad \begin{array}{c} X \\ \diagdown \\ -B \\ | \\ X \end{array} = \overset{\pm}{X} \quad , \quad \text{etc.} \right] \quad \left[\begin{array}{c} X \\ \diagdown \\ P \\ | \\ X \end{array} \diagup X \quad , \quad \begin{array}{c} X \\ \diagdown \\ -P \\ | \\ X \end{array} = \overset{\pm}{X} \quad , \quad \text{etc.} \right]$$

and

$$\left[\begin{array}{c} CH_3 \\ \diagdown \\ \diagup \\ CH_3 \end{array} \overset{\pm}{C} - X \quad , \quad \begin{array}{c} CH_3 \\ \diagdown \\ \diagup \\ CH_3 \end{array} C = \overset{\pm}{X} \right]$$

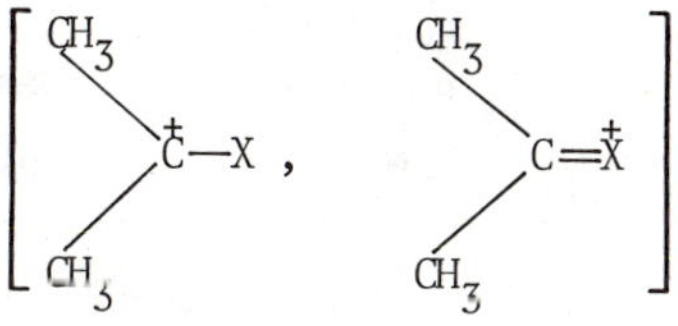

the chemical shifts of the central boron, phosphorus, and carbon atoms are shielded and the effect is greatest for X = F. This also accounts for the greater acidity of BCl_3 over that of BF_3, which is opposite to what is expected from relative electronegativities of the halogens.

It is not clear why the resonance effect of fluorine is greater than that of the other halogens. One view has been that with carbon and fluorine each using 2p orbitals, the smaller disparity between orbital sizes allows greater orbital overlap and provides greater feedback of electrons from the halogen. Some MO calculations show that there is no correlation between orb overlap and resonance substituent constants σ_R for the halogens, and that the ring π-electron distribution results from repulsion between the halogen vale electrons and the p_π electrons on the attached ring carbon which sets up a π charge distribution as follows (175):

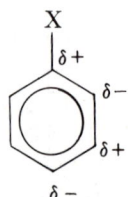

The table of σ_I and σ_R values provides semiquantitative support for two common notions: (1) the resonance and inductive effects of OH, OR, NH_2, and the halogens are opposed in each case; and (2) the resonance effect of the halogens decreases in the order F >> Cl > Br > I.

Observe from Table 7.1 that σ is positive for electron-withdrawing groups, and that $\sigma_m > \sigma_p$ for o,p-directing groups, including the halogens. Always keep in mind that σ and ρ are not fixed constants but are mean values

Substituent constants have also been developed for homolytic reactions (173a). Hammett did not determine ortho substituent constants because steri effects are often considerable. However, constants have been proposed for assessing steric effects (176), and ortho substituent constants were later determined by a variety of ways (177). Additionally, sets of substituent constants have been established for other nuclei (178) including heterocycli rings (179), for substituted phenols (180), for aliphatic compounds (181), and for organophosphorus compounds with substituents attached directly to phosphorus (182). The latter substituent constants, σ^P, do not correlate well with σ_I or σ_R constants although there is a limited correlation with polar substituent constants, σ*. By and large, Hammett σ_m and σ_p constants can apply to pyridine rings, sometimes treating nitrogen as a substituent in the benzene ring (183). For the alpha position of pyridine, the inductive effect usually predominates and the results are best correlated with σ_I (184).

Approaches which are analogous but not equivalent to the Hammett treatment of substituent effects have been proposed by several others. These are attempts to separate field and inductive effects more succinctly from resonance effects (185). Supportive (186a) and adverse criticisms of the validity of substituent constants continue to appear (186b).

Taft polar substituent constants for a few groups are listed in Table 3.2.

7.1. Applications of Hammett-type constants

Among the many applications of Hammett-type constants, two are very common. One provides information about reaction mechanisms and the other correlates properties to probe the mode of transmission of electrical effects of substituents. Information about reaction mechanisms is obtained by noting whether the data are best correlated with σ, σ^+, or σ^-, or by considering the sign and magnitude of ρ. Positive ρ constants imply an increase in negative charge at the reaction site in the transition state. Although there is a continuous spectrum of ρ constants from highly negative values for Friedel-Crafts reactions to highly positive constants for the hydrolysis of aryl halides, it is generally expected that a large negative ρ value indicates an electrophilic mechanism, a large positive value a nucleophilic mechanism, and a free radical reaction will have a ρ value fairly close to zero. For example, the ρ value for some well-established electrophilic re-reactions with benzene and a couple of solvolysis reactions are:

	ρ		ρ
Br_2/AcOH-H_2O	-12.1	$\phi CH_2 OTs$, Aq. acetone	2.20
$CH_3 COCl/AlCl_3$, $ClCH_2 CH_2 Cl$	-9.1	$\phi CH_2 OTs$, AcOH	-4.75
HOBr/$HClO_4$	-6.2		
HNO_3/various solvents	-6.0		
EtBr/$GaBr_3$	-2.4		

In the reaction

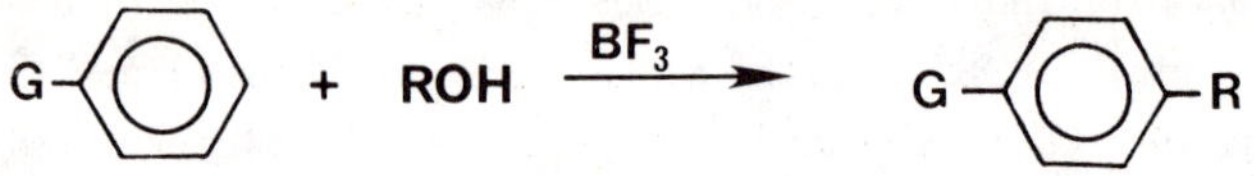

$$G\text{-}C_6H_4\text{-}H + ROH \xrightarrow{BF_3} G\text{-}C_6H_4\text{-}R$$

a ρ value of -5.8 is obtained and the rate of reaction correlates well with σ^+ (187). These two observations indicate that a substantial positive charge is generated in the transition state.

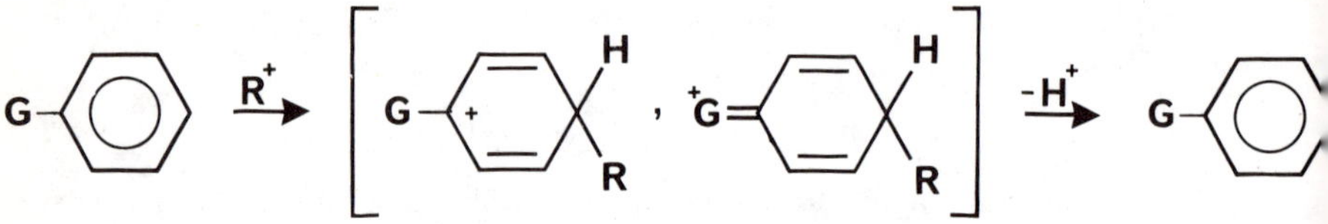

$$G\text{-}C_6H_5 \xrightarrow{R^+} \left[G\text{-}(+)C_6H_5(H)(R) , \overset{+}{G}=C_6H_4(H)(R) \right] \xrightarrow{-H^+} G\text{-}C_6H_4\text{-}R$$

The fact that optically active alcohols racemize in the reaction with little inversion supports this mechanism.

In the reaction of phenoxide ions with ethyl iodide in EtOH, ρ = -0.99. Since ρ is negative and small, a transition state only slightly less negative (i.e., more positive) than the reactant would be proposed. For example,

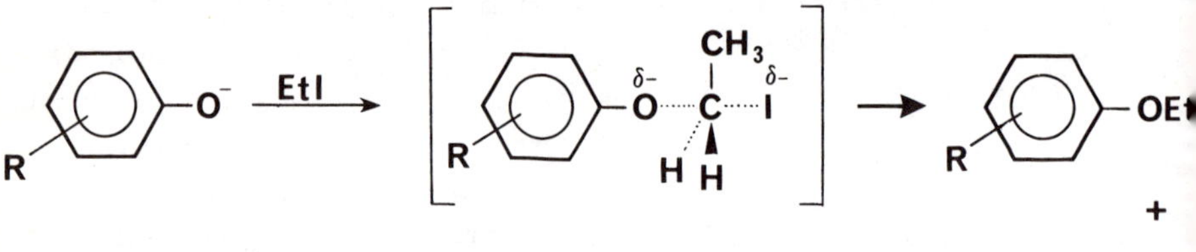

$$R\text{-}C_6H_4\text{-}O^- \xrightarrow{EtI} \left[R\text{-}C_6H_4\text{-}O \overset{\delta-}{\cdots} \underset{H\ H}{\overset{CH_3}{C}} \overset{\delta-}{\cdots} I \right] \rightarrow R\text{-}C_6H_4\text{-}OEt + $$

In the reaction

Sec. 7.1 APPLICATIONS OF HAMMETT-TYPE CONSTANTS

$$Ar-CH(O-CO-CH_3)-CH_3 \longrightarrow \left[Ar-CH\cdots CH_2 \cdots O\cdots H \cdots O=C-CH_3 \right] \longrightarrow Ar-CH=CH_2 + O=C(O-H)-CH_3$$

$\rho = -0.66$, hence there is little charge developed in the transition state. Since the rate correlates with σ^+, groups stabilize the transition state through a mild resonance interaction with the α carbon. A large negative entropy factor supports the proposed ring structure in the transition state.

For the reaction

$$ArSO_2-CH_2-CH_2-CH_2-Cl \xrightarrow{\text{base}} ArSO_2-\overset{\ominus}{CH}-CH_2-CH_2Cl \longrightarrow ArSO_2-\triangleleft$$

$\rho = +2.32$, which supports the carbanion intermediate.

In the polarographic reduction of halobenzenes, σ^- gives a better correlation than σ. Hence, the process must involve a carbanion mechanism (188).

There have been controversial views in the literature on the relative significance of fluorine hyperconjugation vs. the inductive effects of CF_3 groups (189). For example, the presence of CF_3 groups greatly increases the acidity of α C-H bonds, which conceivably could be attributed to resonance stabilization of the anions through F hyperconjugation.

$$F-\underset{\underset{F}{|}}{\overset{\overset{F}{|}}{C}}-\underset{\underset{R}{|}}{\overset{\overset{R}{|}}{C}}-H \xrightarrow{-H^{\oplus}} F-\underset{\underset{F}{|}}{\overset{\overset{F}{|}}{C}}-\underset{\underset{R}{|}}{\overset{\overset{R}{|}}{\overset{\ominus}{C}}}, \quad F-\underset{\underset{F}{|}}{\overset{\overset{F^{\ominus}}{|}}{C}}=C\underset{R}{\overset{R}{\diagup}}$$

3 equivalent forms

The relative rates of D/H exchange in CH_3O^-, CH_3OD for some perfluoro compounds are:

Compound	k_{rel} D/H exchange (190)
F_3CH	1
$F_3C-(CF_2)_5-CF_2H$	6
$(F_3C)_2CFH$	10^5
$(F_3C)_3CH$	10^9
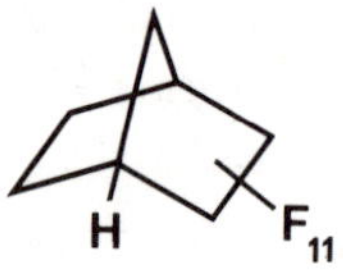 norbornyl H/F$_{11}$	5×10^9

Since hyperconjugation cannot occur for the perfluoronorbornyl ion,

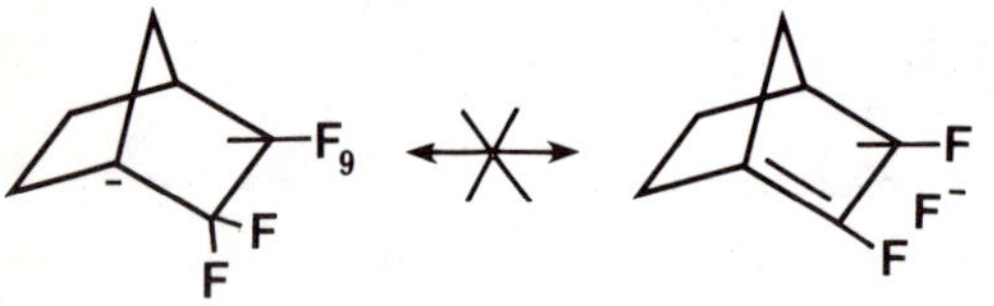

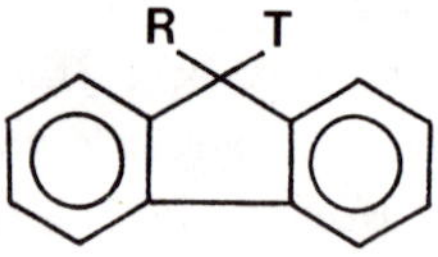

7.2

following Bredt's rule of no double bonds to the bridgehead carbon (Section 1 the acid strengthening effect of the fluorines must be inductive (field and σ induction). Moreover, the perfluoronorbornyl acid is five times <u>more</u> acidi than $(CF_3)_3CH$, which should exhibit the maximum fluorine hyperconjugation, indicating that the acid strengthening factor in these perfluoro hydrocarbon acids must not be hyperconjugation.

A straight line results when the relative rate of H/T exchange in CH_3O^-, CH_3OH for 7.2 is plotted against the ionization constants of RCOOH acids for various R groups, including CF_3. The effect of CF_3 must be operati similarly in both series, and since it is by σ or field induction or both in the RCOOH series, it must be the same in the 7.2 series. Also, a straight line results when σ values for the fluorinated hydrocarbon acid series 7.3 are plotted against the same values for the benzoic acid series (191), and the $\sigma_{p-CF_3}/\sigma_{m-CF_3}$ ratio for the phenol series (1.25) is the same for the benzoic acid series (54).

R—⌬—OH   R—⌬—CH(CF$_3$)$_2$ R—⌬—COOH

7.3   7.4

Both comparisons indicate that there is no significant effect of fluorine hyperconjugation in these compounds. The possible interaction resonance via F hyperconjugation inferred from dipole moments (p. 210) can be attributed to polarization by the CF$_3$ group in p-trifluoromethylaniline beyond that expected from the respective monosubstituted benzenes. The Hammett ρ values for ionization of the acids 7.4 is 4, a value rarely exceeded (191). The fact that the data are correlated by σ rather than σ⁻ implies that there is no significant resonance interaction between the substituents and the carbanion center, including the CF$_3$ substituent.

Deviations from linearity of the Hammett equation can be due to several causes such as an unknown simultaneous reaction, or catalytic impurities, or change in reaction mechanism as the reaction proceeds. Some of these cases have been reviewed recently (192).

Not only can Hammett-type substituent constants serve as diagnostic probes for the study of reaction mechanisms but they can also be used to correlate most physical properties of aromatics. Some examples can be given for illustration.

Infrared spectra. $\log A/A_0 = k\sigma$

where A and A_0 are the intensities of a given band for the substituted and unsubstituted benzene analog and k is the slope of the line when $\log A/A_0$ is plotted against the respective σ constants. Fairly good agreement has been observed for acetophenones ($\nu_{C=O}$), aldehydes ($\nu_{C=O}$), benzonitriles ($\nu_{C=N}$), benzoate esters ($\nu_{C=O}$), phenols (ν_{OH}), anilines (ν_{N-H}), and others (193). For molecules in which there is no resonance effect, such as alcohols, nitroalkanes, alkyl cyanides, and phosphones, R$_3$P=O, σ* is used in place of σ (194).

Infrared frequencies also conform fairly well to a Hammett-type equation, $\log \nu/\nu_0 = k\sigma$, where ν and ν_0 are the band frequencies of the substituted and parent compounds. Such correlations have been found for nitriles, phenols, acetophenones, anilines, nitrobenzenes, and the para C-H of monosubstituted benzenes, among others (193). For instance, the relationship $\nu_{OH}^{CH_3CN} =$

$-66.6\sigma + 3402\ \text{cm}^{-1}$ is applicable to a series of *para*- and *meta*-substituted phenols (195). Correlations of $\nu_{C=O}$ of chalcones are generally better with σ^+ than with σ (196). Again, σ^* is used in place of σ for aliphatic series (

$$\nu_{C=O} = 1690 + 52.56\ \sigma^* \text{ (for R-COCH}_3\text{)}$$

$$\nu_{N=O} = 1500 + 110.1\ \sigma^* \text{ (for R-NO)}$$

Ultraviolet spectra. Since λ_m is determined by excited and ground state energies, λ_m values are not well correlated by Hammett σ constants. However, for *p*-disubstituted benzenes containing dissimilar groups, in which there is substantial interaction resonance in the ground state (Section 6.1), there is a fair linear relationship between λ_m and σ, $\Delta\lambda = \underline{k}\sigma + \underline{C}$, in which $\Delta\lambda$ is the λ_m -203 nm, $\underline{k}$ is the slope of the line, and $\underline{C}$ the intercept when $\Delta\lambda$ is plotted against σ values (198).

Nmr spectra. There are numerous cases of correlation of Hammett constants with nmr chemical shifts, including ^{1}H, ^{13}C, ^{19}F, and other nuclei. Some systems can be given for illustration (193, 199, 200):

^{1}H

^{1}H

^{1}H

^{1}H

^{1}H

^{13}C

^{19}F

^{1}H and ^{13}C

^{1}H

^{1}H

^{1}H

^{1}H

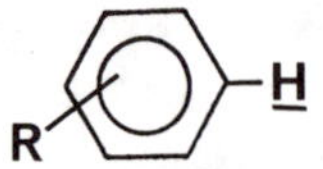

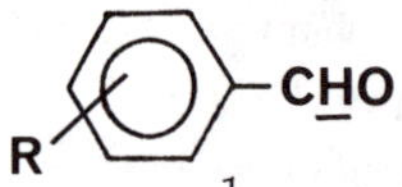

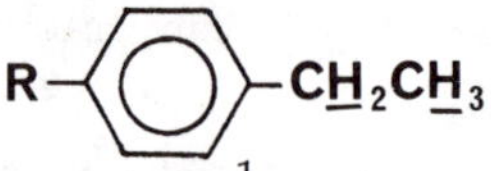

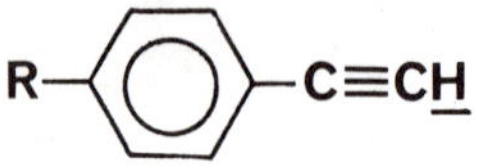

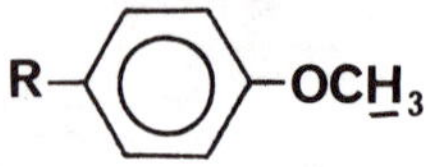

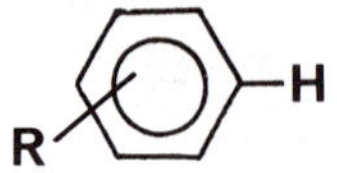

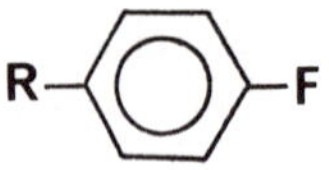

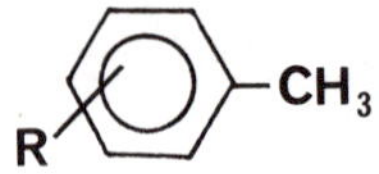

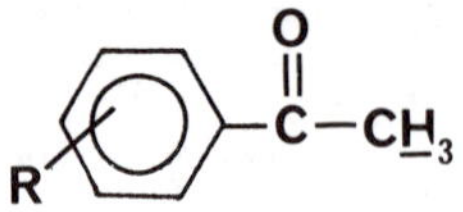

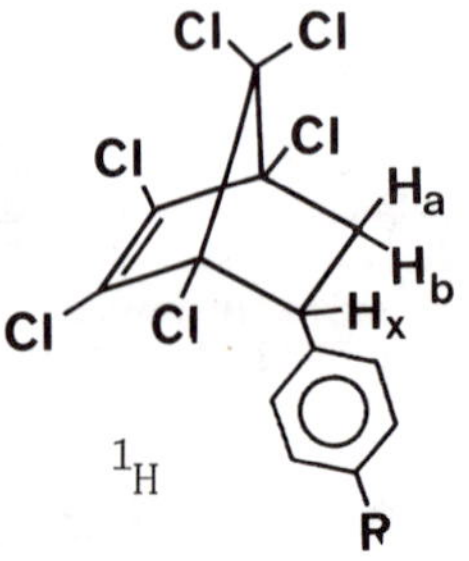

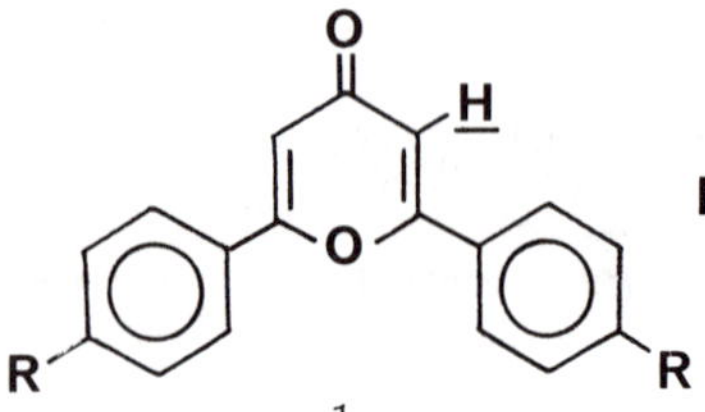

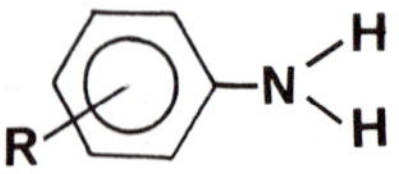

Thus, the electrical influence of a para substituent on protons as far away as attached to a β carbon can be measured by nmr chemical shifts. In some cases, there is better correlation if $\alpha\sigma_I + \beta\sigma_R$ is used in place of σ (201). Fluorine nmr chemical shifts of p-FC_6H_4OH have been used to determine linear free energy relationships in H-bonded complexes (202).

Resonance of the carbonyl group which increases the contribution from the polar structure is reflected by a variety of properites. Not only is the ir stretching frequency altered, but also several properties which depend on the electron density on the oxygen atom such as pK_a, half-wave reduction potential ($E_{1/2}$), and H bonding ability. The latter, in turn, affects the ir, uv, and nmr spectra (203). Finally, several of these properties can be correlated by Hammett-type equations (204). For example, in the series 7.5, with increasing electron-donor ability of R, the C=O acquires greater single-bond character, the carbonyl oxygen becomes more basic to shift ν_{OH} of phenol when mixed, and the C=O becomes more difficult to reduce. Each of these properties can be linearly related to Hammett substituent constants and all are therefore indirectly related (200):

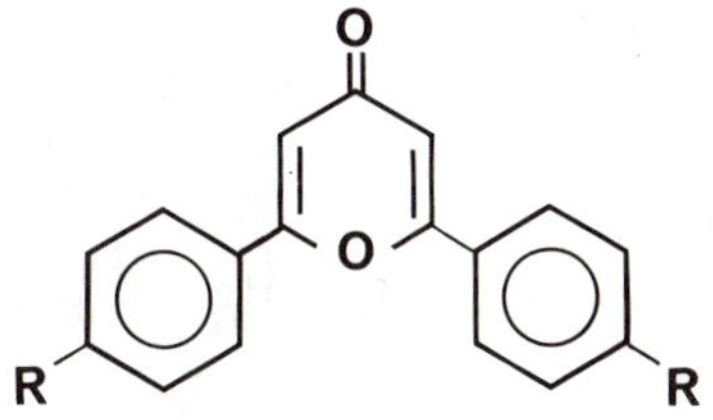

$$\nu_{C=O} = 1649 + 15\sigma \text{ cm}^{-1}$$

$$\Delta\nu_{OH} \text{ (phenol)} = 330 - 185\sigma \text{ cm}^{-1}$$

$$E_{\frac{1}{2}} = -1.896 - 0.44\sigma \text{ eV}$$

Similarly for acetophenones, $\nu_{C=O}$, $\Delta\nu_{OH}$ (for phenol), and ionization potentials are interrelated (205).

$$IP = 1.27\sigma + 9.34 \text{ eV}$$

$$\nu_{C=O} = 1543 + 15.4 \text{ IP (cm}^{-1})$$

whereas, the pK_a, δ^+_{OH}, and ΔH of protonation of protonated acetophenones are correlated best with σ^+ (206). Likewise, $\nu_{N\equiv N}$ for 7.6 and $\delta^{13}C$ for 7.7 correl well with σ^+.

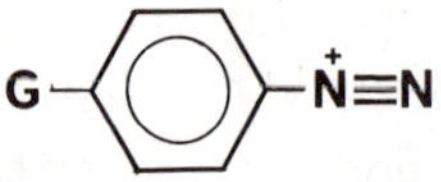

G—⟨benzene⟩—$\overset{+}{N}\equiv N$ G—⟨benzene⟩—$\overset{+}{C}(CH_3)_2$

7.6 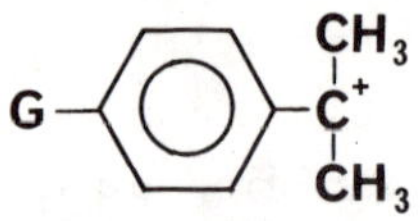   7.7 (207)

For an o-hydroxyacetophenone, the major dipolar resonance structures are of the type

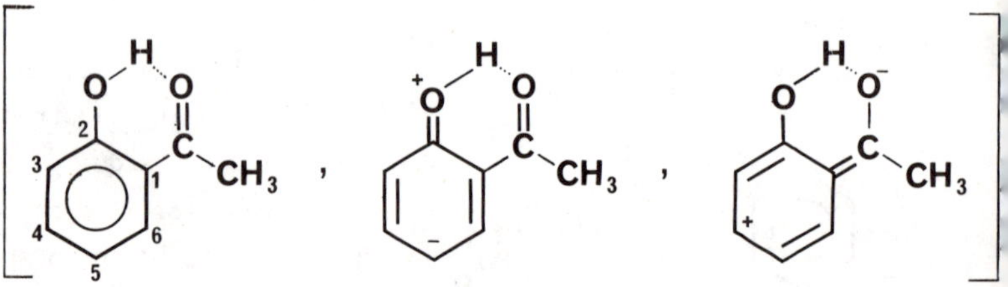

[structure a, structure b, structure c]

7.8

The charge distributions in both dipolar structures, 7.8b and 7.8c, increase the H-bond strength. Hence, when groups in the 5 position delocalize the negative charge of 7.8b, i.e., electron-withdrawing groups, or when there are electron-donating groups in the 4 position to delocalizae the positive charge in 7.8c, then the H-bond strength increases and spectral properties are related to this change. Such changes are observed for $\nu_{C=O}$, γ_{OH} (out-of-plane O-H bending), ν_{OH}, and δ_{OH} (see also Chapter 12). Recall that δ_{OH} moves downfield with increasing H-bond strength. In this case, ν_{OH} and γ_{OH} are linearly correlated by the function, $(\sigma_p - \sigma_I)$ and $(\sigma_m - \sigma_I)$ (208).

The ionization potentials of some monosubstituted benzenes and the λ_m of their TCNE CT complexes are linearly related to σ^+ substituent constants (2

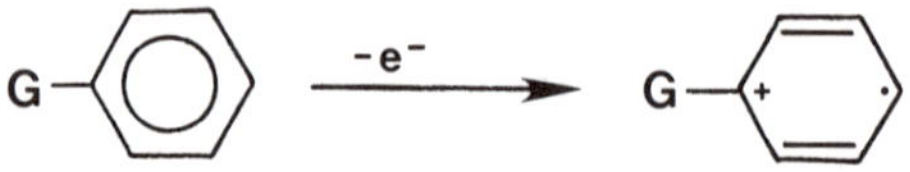

G—⟨benzene⟩ $\xrightarrow{-e^-}$ G—⟨benzene$^+$⟩

7.9

Sec. 7.1 APPLICATIONS OF HAMMETT-TYPE CONSTANTS

$$IP = -17.4\sigma^+ + \underline{C}$$

$$\underline{\nu_{CT}} = 9300\sigma^+ + 26{,}200 \text{ cm}^{-1}$$

This is to be compared with the relationship found for some monosubstituted benzenes of the type 7.10 in which polar substituent constants are used (210).

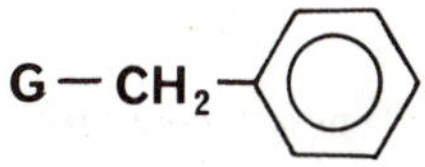

G—CH$_2$—⟨phenyl⟩ $\underline{E_{CT}}$ (eV) = $0.254\sigma^* + 2.999$

7.10

As will be discussed in Chapter 12, $\Delta\nu_{OH}$ for a given proton donor, when mixed with a series of proton acceptors, is sometimes used as a measure of their relative basicities. For instance, $\Delta\nu_{OH}$ for phenol when mixed with a series of monosubstituted benzenes has been used to determine their π-donor basicities (see also Table 16.2), and the $\Delta\nu_{OH}$ values are correlated with the respective Hammett substituent constants (211).

<u>Miscellaneous properties.</u> Dipole moments (207a) and redox potentials of various π systems usually correlate well with Hammett σ constants. Illustrative systems studied are (212, 213):

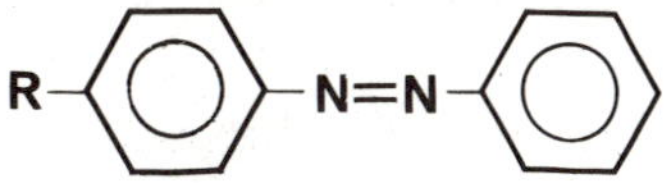

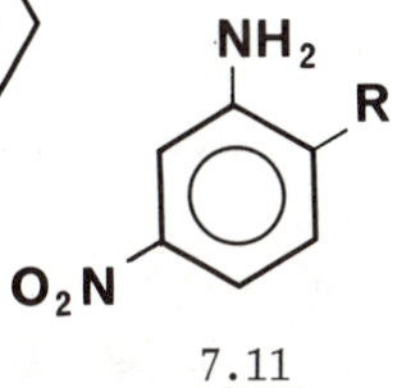

R—⟨phenyl⟩—N=N—⟨phenyl⟩   ⟨phenyl with NH$_2$, R, O$_2$N substituents⟩   R—⟨phenyl⟩—CH=CH—C=O—⟨phenyl-R′⟩

7.11

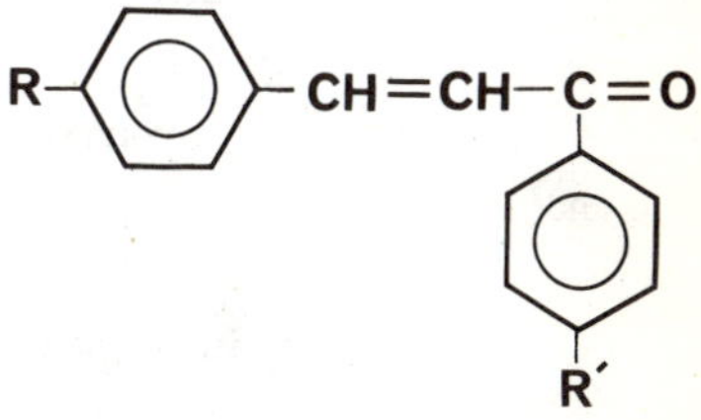

$$E_{\frac{1}{2}} = 0.185\sigma - 0.38$$

Interestingly, for series 7.11, certain taste qualities (relative sweetness, RS) can also be correlated with $E_{\frac{1}{2}}$ and Hammett σ constants:

$$\text{Log RS} = 12.28\ E_{\frac{1}{2}} - 1.19\sigma + 7.18.$$

Again, for aliphatic systems, one uses σ^*, e.g.,

$$R_2\underset{NO_2}{C}-Cl \xrightarrow{red.} R_2\underset{NO_2}{C}-H \qquad E_{\frac{1}{2}} = 0.19\sigma^* - 0.61 \text{ (eV)}$$

$$\nu_{NO_2} = 12\sigma^* + 1562 \text{ (cm}^{-1})$$

The ionization potentials of alkyl free radicals are correlated with σ^* (214) and ionization potentials of alcohols, ethers, amines, and mercaptans correlate well with σ^* and with σ_I (215). Hammett correlations have even been observed in mass spectrometry, although they are not necessarily related to reaction mechanisms (216).

Extensive use has been made of linear free energy equations in recent years to correlate the bioactivity of chemicals (217). The general procedure is to take a biologically active compound and study the effects of substituents on its activity, directed toward finding the substitution pattern of the derivative expected to have the most potent activity. An equation is used of the type

$$\log \frac{1}{\underline{C}_{\underline{s}}} = \underline{a}\pi + \underline{b}\sigma + \underline{c}\underline{E}_{\underline{s}} + \underline{d}$$

where $\underline{C}_{\underline{s}}$ = concentration of member $\underline{s}$ which gives a standard biological response e.g., $\underline{LD}_{50}$, I_{50}, EC_{50}. π = hydrophobic substituent constant (determined from distribution coefficients of model compounds between n-octanol and water) $\underline{E}_{\underline{s}}$ = Taft steric constant (172). $\underline{a}$, $\underline{b}$, $\underline{c}$, $\underline{d}$ = constants determined by multiple regression analysis. The rationale is that the bioactivity of a drug is affected by its hydrophilic-lipophilic character, its charge distribution, and its molecular shape. Thus, parameters are used to reflect the hydrophobic, electronic, and steric effects of substituents or molecular fragments. Moreover, by ascertaining how each of the physicochemical properties is concerned, it is sometimes possible to gain an insight into the mechanism of the drug-receptor site interaction or which process (diffusion, membrane transport, lipoid or aqueous solution, etc.) is the critical step in getting the drug to the site of action.

QUESTION

In correlating $\nu_{C=O}$ with σ or σ^+ of substituted acetophenones, what slope would you expect, positive or negative?

ANSWER

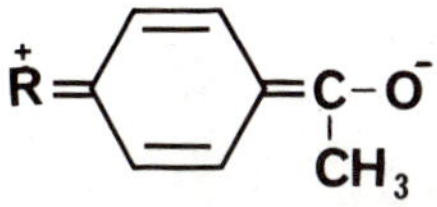

$$\nu = k\sigma + \nu_0$$

Since electron-donating groups (negative σ) decrease $\nu_{C=O}$, $\underline{k}$ (the slope) would have to be positive.

7.2. The Q parameter

In this chapter we have shown that the electrical effects of substituents on positions <u>meta</u> or <u>para</u> to the substituent can be correlated well by Hammett-type constants and in some cases similar correlations are possible for <u>ortho</u> positions. However, the best general correlation of carbon, proton, and fluorine nmr chemical shifts for ortho positions are in terms of the Q parameter (218, 219). This is a semiempirical quantity, defined by the term $Q = P/I\underline{r}^3$, where P = polarizability of the C-X bond (X is the substituent), I is the first ionization potential of X, and $\underline{r}$ the C-X bond length. Q was initially calculated for X = halogen (Table 7.3) but later an experimental method was developed (220) for determining Q and a table of Q values was presented (Table 7.4) (221).

Table 7.3. Calculated Q values for the halogens

X	r_{C-X} (Å)	P_{C-X} ($\times 10^{24}$ cc)	I ($\times 10^{12}$ ergs)	$Q = P/Ir^3$ ($\times 10^{-14}$ ergs)
H	1.09	0.645	21.8	2.28
F	1.30	0.633	27.8	1.04
Cl	1.70	2.604	20.8	2.55
Br	1.85	3.754	18.8	3.16
I	2.05	5.752	16.7	3.98

Table 7.4. Q values for various substituents

Substituent	Q(1)	Q(2)	Substituent	Q(1)	Q(2)
NH_2	-0.67	-0.03	CN	*	3.43
OH	0.30	1.20	NO_2	6.33	4.00
OCH_3	0.54	0.4-0.8	CF_3	3.71	*
CH_3	*	1.77	CCl_3	*	5.15
t-Bu	*	2.50	CHO	4.90	4.22
H	2.28†	*	$COCH_3$	5.20	**
F	1.50	*	COCl	6.00	**
Cl	2.55†	*			
Br	3.16†	*			
I	3.98†	*			

* One Q value suffices. † Calculated Q (218).
** Should have a different Q(2) value.

Two values of Q are required for groups which are not symmetrical about the C-X axis. When X is flanked by two Hs, Q(1) is used, and when X is sterically distorted from its normal position, i.e., rotated out of the benzene ring plane or forms an H bond with an adjacent group, then the Q(2) value will be required.

Accordingly, Q will correlate chemical shift data for ortho protons or fluorine atoms very well (222) (H_3 with Y or H_6 with X in disubstituted compounds, o-C_6H_4XY). The failure of the nitro group to fall on the lines, when δ_{H_3} of o-substituted benzaldehydes or $\delta^{19}F$ of o-substituted fluorobenzenes are plotted against Q values, denotes that the nitro group is rotated out of the molecular plane in both cases. This deduction is supported by other data (see also Chapter 11).

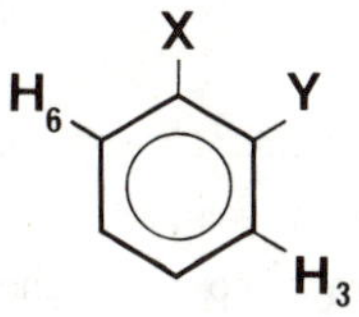

X, Y = a variety of substituents

A theoretical basis for the applicability of the Q parameter has not yet been formed. It is believed that resonance, field, inductive, and paramagnetic anisotropic effects contribute to Q but a model to assess and separate these factors is not available. Q can be applied to H_3 of 2-substituted pyridines and to the cis and trans protons of substituted ethylenes and propenes but poor correlations are obtained for meta and para protons in monosubstituted benzenes.

Chapter 8

NONCLASSICAL RESONANCE

In order to account for the solvolytic behavior and the spectral (see Chapter 10) or photochemical properties of certain systems, it has been necessary for classical resonance theory to incorporate some of the geometric concepts of molecular orbital theory. For instance, we have seen that for maximum conjugation or π overlap, the atomic orbitals of conjugated systems must be coparallel. However, when the atomic orbitals are at an angle with respect to each other and can overlap in σ fashion, even though they are separated by one or more saturated atoms, there still can be some conjugation (223). This is described by resonance structures but is called <u>nonclassical</u> <u>resonance</u>. Some illustrations are as follows.

σ overlap or homoconjugation   Nonclassical resonance
―――――――――――――――――――――   ―――――――――――――――――

 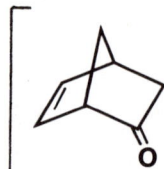 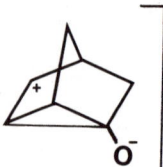

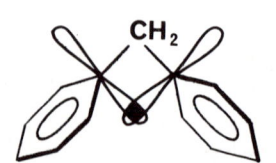 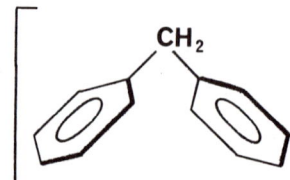 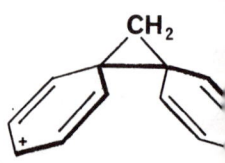

One area which draws heavily upon the concept of nonclassical resonance is the structural description of carbocations (224). Carbocations is the proposed name for all cationic carbon compounds. Carbenium ions, the conventional or classical carbocations, are tricovalent and either contain a linear sp-hybridized or near planar trigonal sp^2-hybridized carbon center. Typical examples of the latter are the methyl $^+CH_3$ and t-butyl $(CH_3)_3C^+$ cations, and of the former, vinyl $R_2C=\overset{+}{C}H$ and oxocarbenium acyl cations $R-\overset{+}{C}=O$. Such ions, of course, are stabilized by electron delocalization with neighboring groups, such as the resonance of benzyl cations and hyperconjugation of t-alkyl cations.

The other limiting class of carbocations is the nonclassical ions which have penta- or tetracoordinated carbon centers. A frequently used definition is that a nonclassical carbocation ion is one which in its ground state has delocalized σ-bonding electrons (225, 226). Two good examples are the 2-norbornyl cation 8.1 and the cyclopropylcarbinyl cation 8.2.

These two classes of carbocations are characterized by distinctly different nmr chemical shifts and coupling constants (J_{CH}) (Table 8.1) (227). The classical carbenium ions, with more electron-deficient carbon centers, have greater deshielded chemical shifts. Carbocations exhibit varying degrees of delocalization such that there is a spectrum of carbocation structures from extreme classical ($^+CH_3$ is the parent) to extreme nonclassical ($^+CH_5$ is the parent).

Table 8.1.

Nmr parameters of representative classical and nonclassical carbocations (227)

	Classical				Nonclassical		
Ion	$\delta^{13}C$	$\delta_{\alpha H}$	J_{CH} (Hz)	Ion	$\delta^{13}C$	δ_H	J_{CH} (Hz)
$(CH_3)_2\overset{+}{C}H$	-125	13	169				
$(C_6H_5)_2\overset{+}{C}H$	-5.6	9.7	164				
				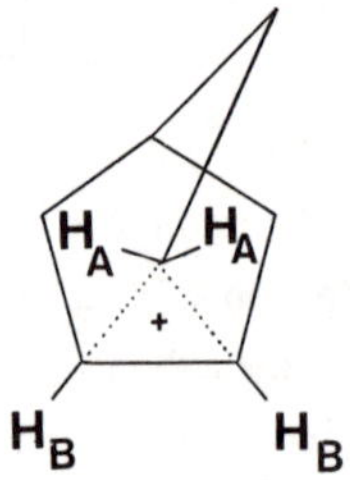 2-Norbornyl	C_A 173 C_B 70	H_A 3.05 H_B 6.59	H_{A+B} = 320
				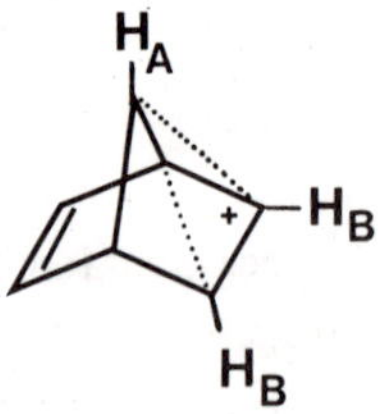 7-Norbornyl	C_A 152.7 C_B 75.8	H_A 3.24 H_B 7.48	H_A 220 H_B 200

δ_H ppm rel to TMS δ_C ppm rel to $^{13}CS_2$

The use of nonclassical resonance or homoconjugation for interpreting spectra will be taken up in Chapter 10. Here we shall discuss the solvolytic behavior of a few well-studied systems. One of the most exhaustively examined systems is that related to the norbornyl cation 8.5 (226).

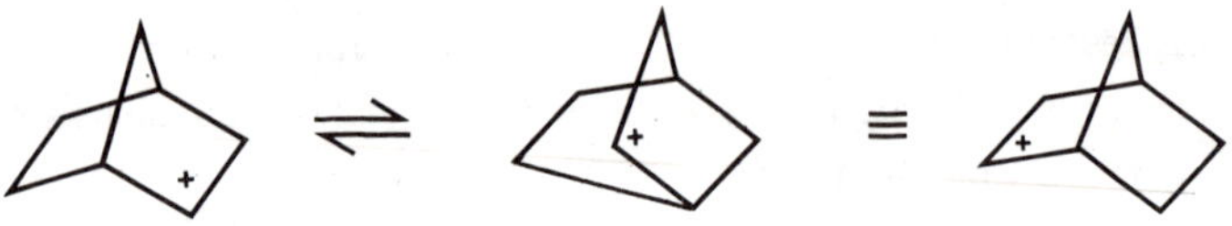

8.3 ⇌   ≡ 8.4

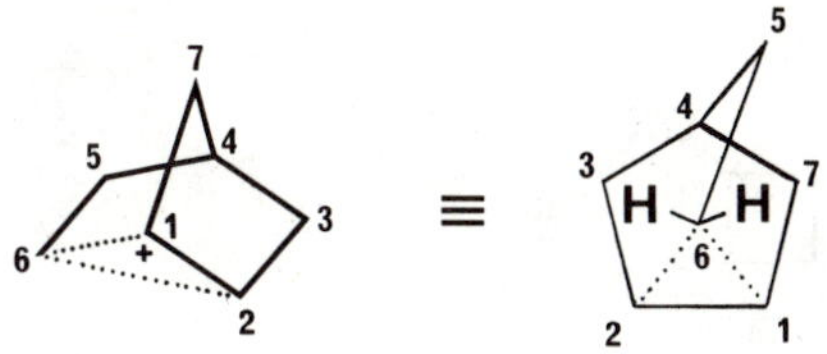

8.5

There is a great deal of literature on this ion and its derivatives, because of interest in the question of its structure: is it a rapidly equilibrating pair of ions, 8.3 ⇌ 8.4, or does it have a nonclassical structure, 8.5 ? Without summing up the arguments, it appears that in an extremely acidic solution, the 2-norbornyl cation has the protonated nortricycline structure 8.5 (228). In more nucleophilic solvents or with substituents, particularly aromatic groups at the 2-position (229), the ions are more classical, i.e., there is less σ delocalization and classical resonance structures may be written:

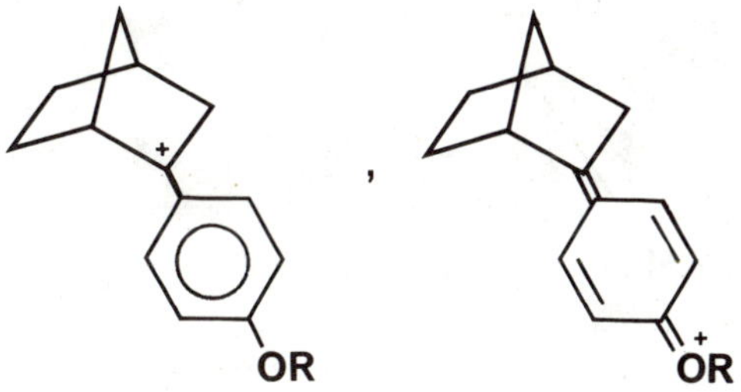

Whereas simple secondary carbocations generally prefer classical structures, the 2-norbornyl cation has the corner protonated nonclassical structure. Calculations intimate that structure 8.5 is stabilized by relief of strain through a flattening of the 5-membered rings (230).

The point to be made is that nonclassical resonance is used to rationalize the reactivity and product mixtures from various reactions just as classical resonance has been used. That is, stabilization of transition states or reactive intermediates by nonclassical resonance enhances rates of reactions. A good illustration is cyclopropylcarbinyl derivatives (226). They solvolyze tens of times faster than alkyl analogues to give a mixture of cyclopropylcarbinyl, homoallyl, and cyclobutyl products. Moreover, the same mixture of products can be obtained from the respective cyclobutyl compounds.

This can be explained by the formation of a common transition state visualized as in Figure 8.1.

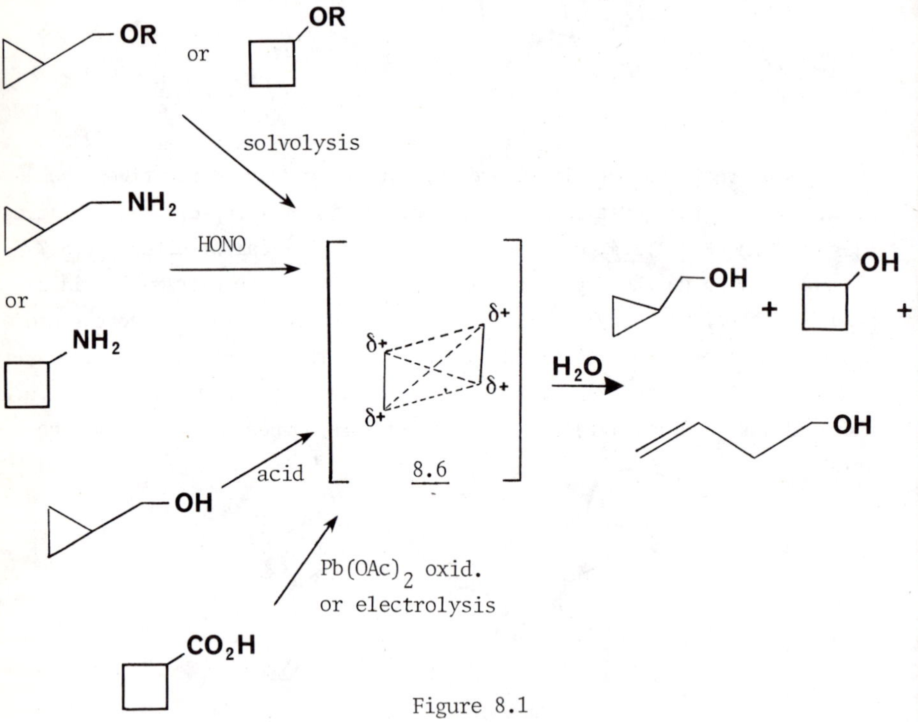

Figure 8.1

Thus, the cyclobutyl or cyclopropylcarbinyl substrates yield a carbocation which is stabilized by nonclassical resonance, diagrammed as 8.6, which may react at the various carbons to yield the mixture of products. The unusual rate of reaction of cyclopropylcarbinyl derivatives, the S_N1 character of the reaction, the large salt effect, the formation of ethers from solvolyses or deaminations in alcohol solvents, and the rate enhancement by methyl substitution at all positions, are all indicative of a highly stable carbocation intermediate.

Other examples of the extraordinary ability of cyclopropyl groups to stabilize an α carbocation can be given.

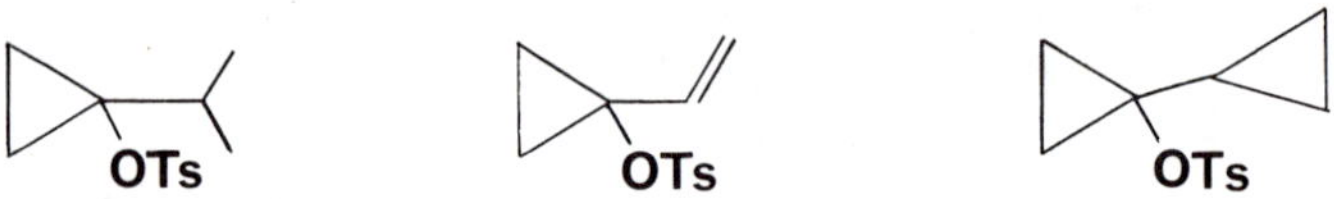

k_{rel} (EtOH$_{50\%\ aq}$)   1   10.3   16,000   (231)

Ch. 8 NONCLASSICAL RESONANCE   301

k_{rel} 1   246   23,500 10^7 (232)
(acetolysis)

The cyclopropyl group is even more effective as a participating group in solvolysis than is phenyl. For instance, cyclopropylcarbinyl esters solvolyze faster than the corresponding phenyl esters, and dicyclopropylcyclopropenone is slightly more basic than diphenylcyclopropenone (233a). Olah (233b) has pointed out, however, that phenyl delocalizes charge from neighboring carbenium centers more than cyclopropyl. That is, benzyl resonance is greater than cyclopropylcarbinyl hyperconjugation, which in turn is more significant than methyl hyperconjugation. Cyclopropyl, on the other hand, is very ineffective in <u>transmitting</u> conjugation. This has been observed in a variety of studies (233c).

The structures of protonated cyclopropanes have been studied extensively, particularly concerning the relative stability of corner-, edge-, or face-protonated structures. The nonclassical resonance hybrids are important because of their probable role as intermediates in many rearrangements of secondary and tertiary alkyl cations (234).

The combined effects of anchimeric assistance, nonclassical resonance-stabilized transition states, and relief of ring strain produce a range in reactivities of norbornyl derivatives of over 10^{23}. For example, the transition state <u>8.8</u> can result from π delocalization or from σ delocalization in the respective cations:

    8.7   8.8   8.9

As a result, <u>8.7</u> solvolyzes 10^{11} times faster than the saturated analogue and <u>8.9</u> about 10^{23} times faster (235). The relative reactivities of other related norbornyl compounds are given in Table 8.2. When a chloro derivative of <u>8.7</u> is reacted with a cyanide ion, for instance, a mixture of products <u>8.7</u>-CN and <u>8.9</u>-CN is obtained. Similarly (236),

Table 8.2 Relative solvolytic rates of some norbornyl-type substrates

X = OTs k_{rel} (AcOH) (237)		X = OPNB k_{rel} (aq. acetone) (238)	
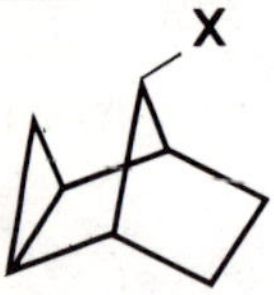	0.4	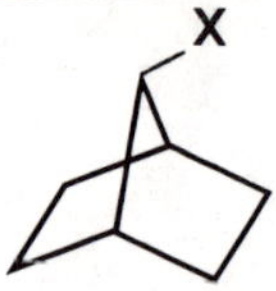	1.0
	1.0	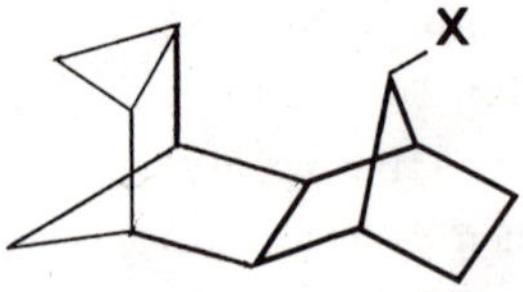	10^8
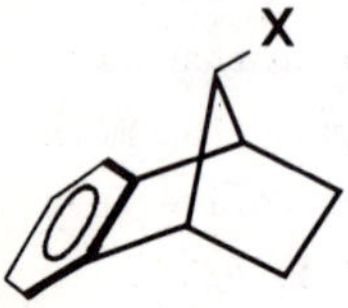	10^5	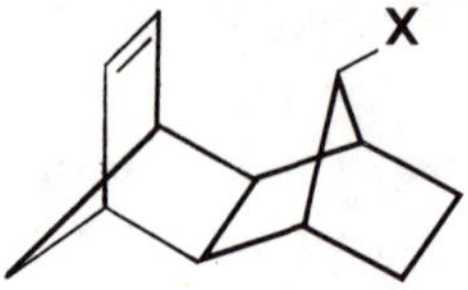	10^{11}
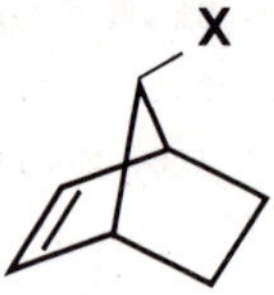	10^{11}		10^{12}
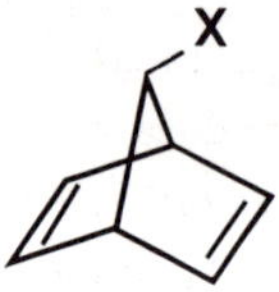	10^{14}	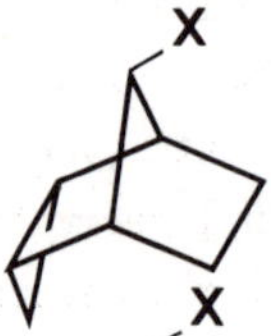	10^{13}
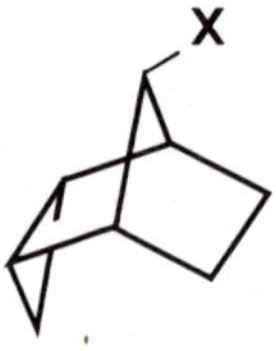	10^{14}	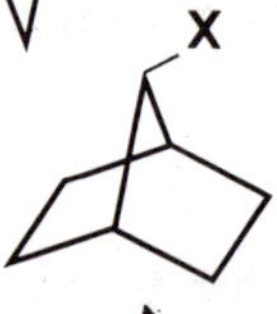	10^{13}
			10^{23}

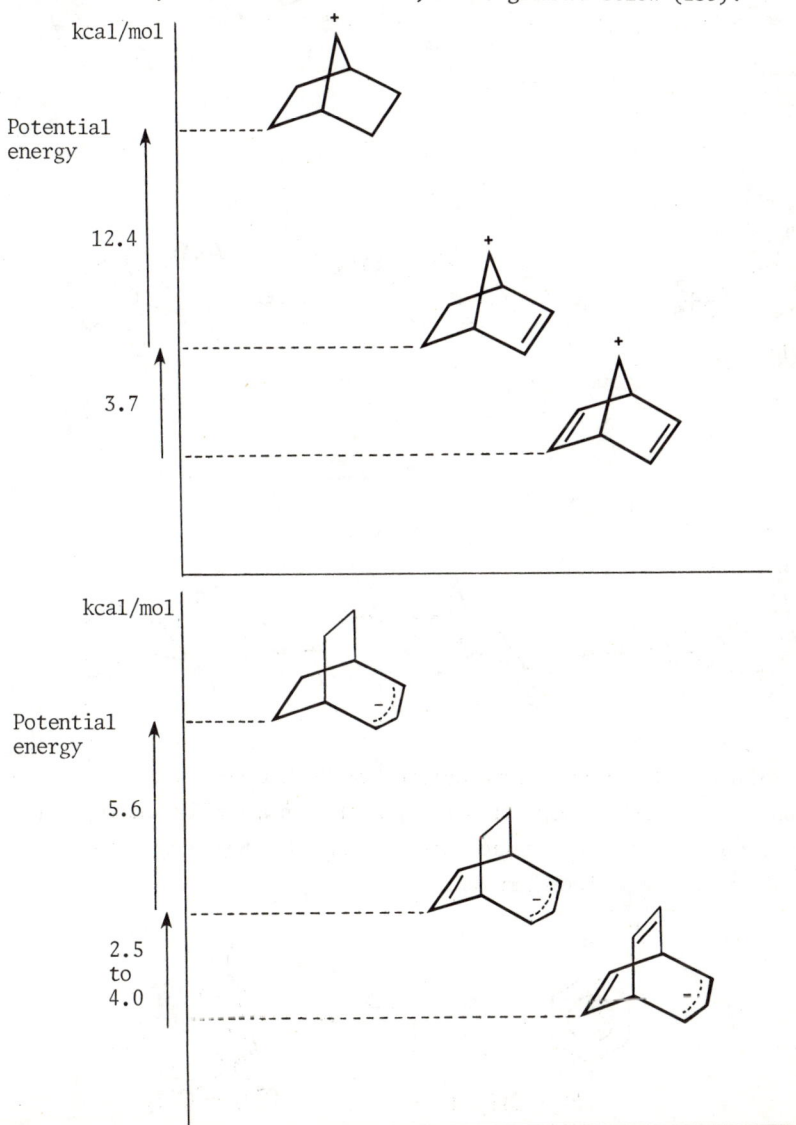

The homoconjugation (nonclassical resonance) stabilization of some such ions, as determined by kinetic measurements, is diagrammed below (239):

The relative solvolytic rates of 8.11/8.10 = 1.6 and 8.13/8.12 = 3, indicate that σ delocalization is as effective as π delocalization in these systems (240).

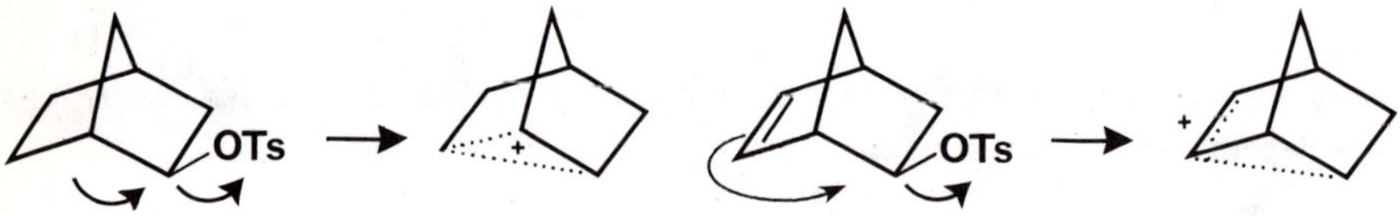

Solvent and the aryl group compete in the nucleophilic assisted ionization of β-arylethyl substrates (226, 241). In a weakly nucleophilic solvent, such as trifluoroacetic acid or SbF_5-SO_2, even a p-nitrophenyl group produces primarily a phenonium ion.

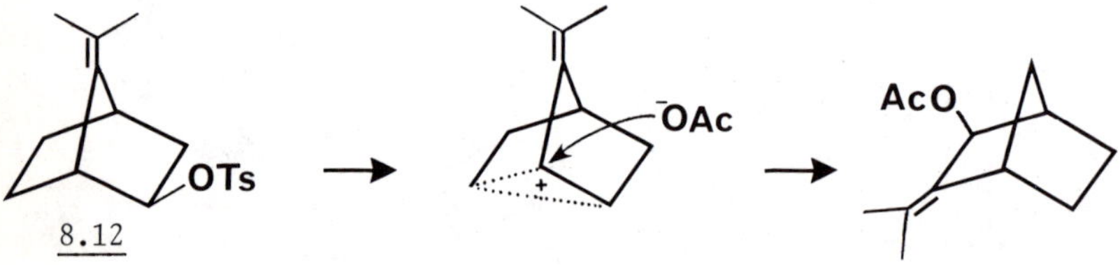

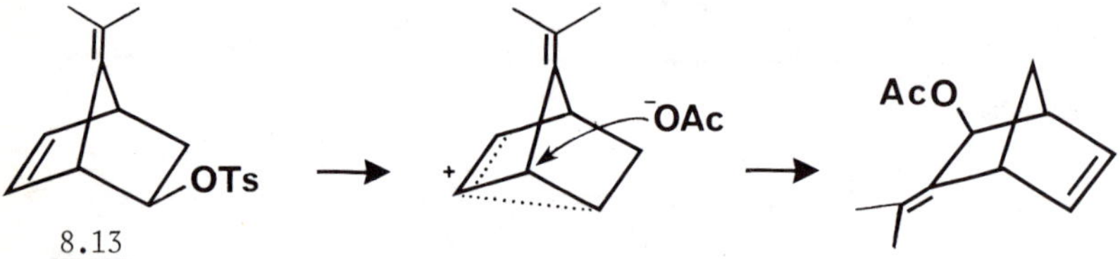

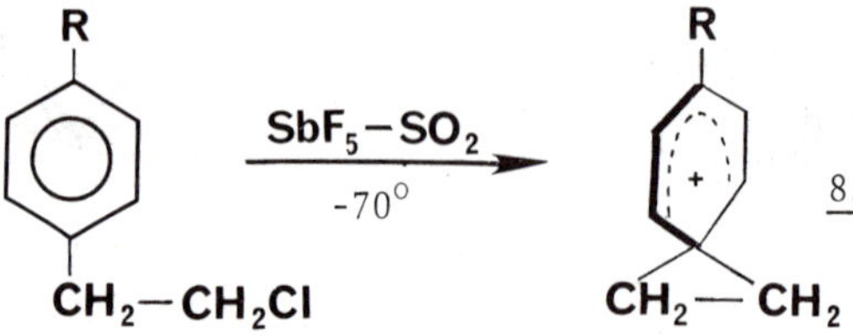

The phenonium ion accounts for the enhanced rate of reaction as well as the product mixture.

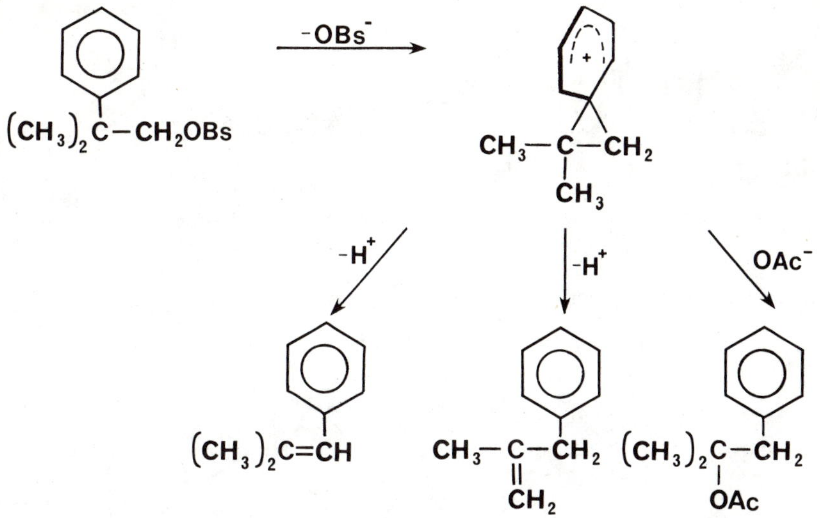

Nmr spectra (242) reveal that the phenyl ring in phenonium ions is no longer benzenonium, as it would be for a pair of rapidly equilibrating classical ions 8.15 ⇌ 8.16.

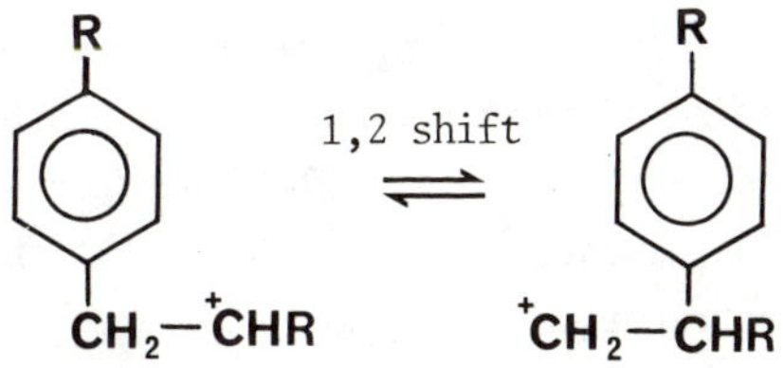

4-Chloro-3,4'-dinitrodiphenylmethane reacts with methoxide ion in methanol more than four times as fast as the 4-chloro-3,3'-dinitro isomer, which can be attributed to homoconjugation in the reaction intermediate (8.17) from the 4'-isomer (243).

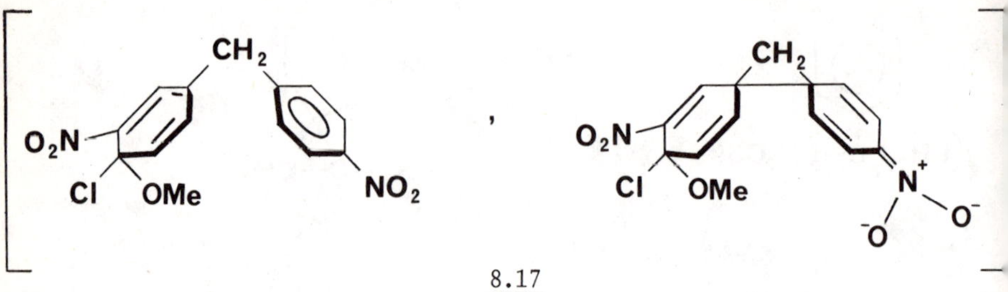

$$[\text{structure with } O_2N, Cl, OMe, CH_2, NO_2 \quad , \quad \text{dipolar structure}]$$

8.17

A similar resonance occurs in properly substituted N,N'-diarylureas (244).

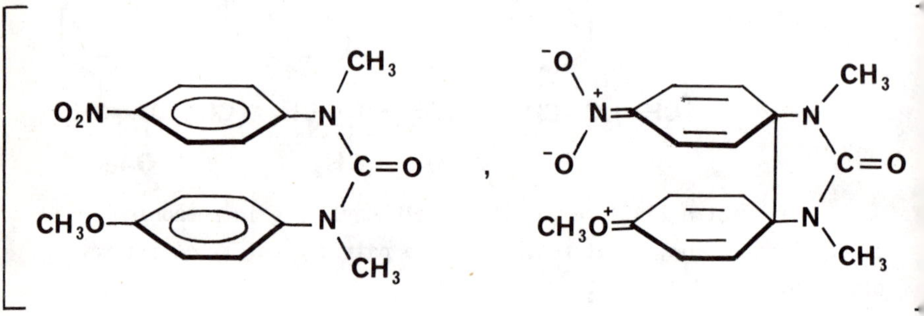

$$[\text{diarylurea structure} \quad , \quad \text{dipolar diarylurea structure}]$$

Evidence for the approximately face-to-face orientation of the phenyl rings is found in the shielding of the aromatic protons in the nmr compared with the values for the mono arylureas. The chemical shifts are not appreciably affected by changes in solvent, which indicates that the dipolar structure makes only a small contribution to the ground state. The uv absorption bands are at longer wavelengths for the 4-NO_2-4'-methoxy derivative than the 4,4'-dinitro analog and λ_m is solvent dependent.

Solvent	4,4'-dinitro	4-NO_2-4'-methoxy
chloroform	325 nm	347 nm
acetonitrile	333	354
acetone	337	355
DMSO	350	365

The solvent effect substantiates the view that the homoconjugation is primarily in the excited state.

QUESTION

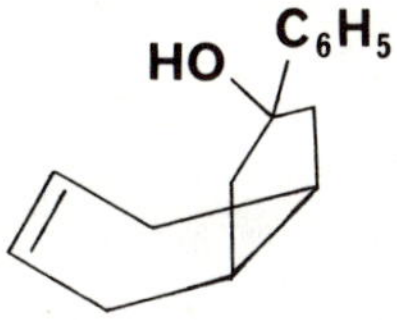

I

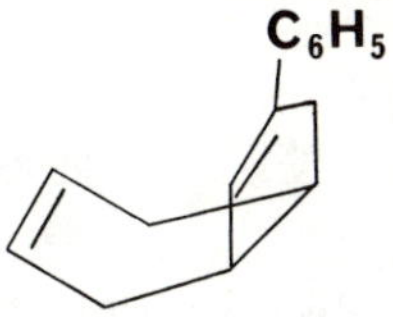

II

How would you rationalize the observations that I and II dissolve in strong acids to give a deep yellow solution (λ_m 386, 427 nm) whereas the trans-fused ring isomers show no strong absorption above 300 nm under the same conditions ? (G. Leal and R. Pettit, J. Am. Chem. Soc. 81: 3160 (1959).)

ANSWER

Both compounds produce the nonclassical ion III which is not geometrically feasible for the trans-fused isomers.

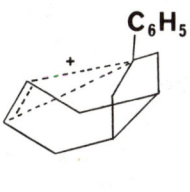

III

Chapter 9

RESONANCE VS. MOLECULAR ORBITAL THEORY

Molecular orbital (MO) theory is without doubt the most promising method available to organic chemists for describing molecules in terms of a variety of molecular parameters. However, it is beyond the scope of this book to attempt to provide an introduction to MO methods, as this alone would comprise a book itself (245). Nevertheless, any modern coverage of organic chemistry would be incomplete if it did not adopt the fruits of MO theory. For instance, we have adopted some of the energy and geometric MO concepts in discussing spectroscopy, homoconjugation, and aromaticity. Numerous papers have pointed out ways in which resonance theory and MO theory are interwoven (246). The major disadvantage of resonance theory relative to MO theory is the limited quantitative information which resonance theory provides. Other than resonance energies and, indirectly, Hammett substituent constants, resonance theory is virtually completely qualitative. MO calculations, however, can provide a host of semiquantitative data concerning bond characteristics, thermodynamic stabilities, spectra, dipole moments, and chemical reactivities. There are so many excellent examples that it is an injustice to cite just a few. Nevertheless, references can be given as a lead to the literature (247). It is also important to note that in some cases, the simple MO methods are ineffective (248).

The most prominent achievement of MO theory, which has no counterpart in resonance theory, was the recognition of the importance of orbital symmetry as defined by the Woodward-Hoffmann rules (249). This has been regarded as one of the most important new theories in organic chemistry in decades because it explains many reactions as well as predicting many unstudied reactions. Alternate but related approaches have been proposed by others (250).

Another outstanding case in which MO theory is far superior to resonance theory for the elucidation and description of molecular structure is the relatively recent area of π organometallic chemistry (Section 1.1). Although valence bond theory (VBT) and crystal field theory (CFT) have been used, MO theory offers the most perceptive view. In a sense, VBT accounts for the geometry and magnetism of metal π complexes but not their spectra (uv, nmr,

Ch. 9 RESONANCE VS. MOLECULAR ORBITAL THEORY

esr), whereas CFT explains their spectra, especially electronic, but is inadequate for rationalyzing the stereochemical and certain physical properties of these complexes. Ligand field theory (251) is a modification of CFT which makes use of molecular orbitals and is often used interchangeably with MO theory.

There are a number of other isolated examples where MO theory is better than resonance theory in explaining or predicting a given property. For instance, resonance theory is inadequate for describing the benzene anion radical (252), and resonance theory is silent regarding the interaction along σ bonds between suitably oriented γ and δ substituted ketones, e.g., 9.1 (see Chapter 10) (253).

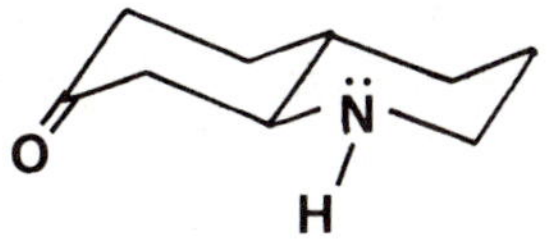

9.1

Photoelectron spectroscopy is almost uniquely versatile for the measurement of first and higher ionization potentials. However, interpretation of the data for determining energies of π and σ molecular orbitals or for detecting conjugation and interactions between nonbonded orbitals cannot be done without MO calculations and the develpment of correlation diagrams based on a MO model (254).

REFERENCES
Part II

1. G.W. Wheland, *Resonance in Organic Chemistry*, New York: Wiley, 1955.

2. B.A. Olah, Y.K. Mo, and Y. Halpern, *J. Am. Chem. Soc.* 94: 3551 (1972).

3. M.J.S. Dewar, *Hyperconjugation*, New York: Ronald Press, 1962; J.W. Baker *Hyperconjugation*, New York: Oxford U. Press, 1952.

4. H. Schmidt and A. Schweig, *Angew. Chem. Int. Ed.* 12: 307 (1973).

5. L. Radom, J.A. Pople, and P.v.R. Schleyer, *J. Am. Chem. Soc.* 94: 5935 (1972).

6. J.W. Baker and W.S. Nathan, *J. Chem. Soc.* 1844 (1935).

7. (a) S. Nishida, I. Moritani, and T. Teraji, *Chem. Comm.* 1114 (1972); (b) J. Collin and F.P. Lossing, *J. Am. Chem. Soc.* 81: 2064 (1959).

8. R.B. Turner, *Conference on Hyperconjugation*, New York: Pergamon, 1959, p. 1; G.B. Kistiakowsky, *J. Am. Chem. Soc.* 60: 440 (1938); J.B. Conant and G.B. Kistiakowsky, *Chem. Revs.* 20: 181 (1937).

9. K. Higasi, L Omura, and H. Baba, *Nature* 178: 652 (1956).

10. L.N. Ferguson, *Highlights of Alicyclic Chemistry*, Palisades, N.J.: Franklin, 1973, Chap. 3.

10a. The various substituent constants for cyclopropyl groups have been summarized in Y.E. Rhodes and L. Vargas, *J. Org. Chem.* 38: 4077 (1973).

11. F.A. Van-Catledge, *J. Am. Chem. Soc.* 95: 1173 (1973).

12. M.F. Hawthorne, *J. Org. Chem.* 21: 1523 (1956).

13. See F.J. McQuillin, *Alicyclic Chemistry*, New York: Cambridge U. Press, 1972, p. 17.

14. M. Brookhart, G.C. Levy, and S. Winstein, *J. Am. Chem. Soc.* 89: 1735 (1967).

15. H. Shechter, R.C. Hahn, and T.F. Corbin, *J. Am. Chem. Soc.* 90: 3404 (1968); H.C. Brown and J.D. Cleveland, *J. Am. Chem. Soc.* 88: 2051 (1966).

16. L.M. Stock and P.E. Young, *J. Am. Chem. Soc.* 94: 4247 (1972).

17. T. Sharpe and J.C. Martin, *J. Am. Chem. Soc.* 88: 1815 (1966).

18. G.A. Olah, D.P. Kelly, D.L. Jeuell, and R.D. Porter, *J. Am. Chem. Soc.* 92: 2544 (1970).

19. D.S. Kabakoff and E. Namanworth, *J. Am. Chem. Soc.* 92: 3234 (1970); L. Radom, J.A. Pople, V. Buss, and P.v.R. Schleyer, *J. Am. Chem. Soc.* 92: 6380 (1970); B.R. Ree and J.C. Martin, *J. Am. Chem. Soc.* 92: 1660 (197

20. W.A. Bernett, *J. Chem. Educ.* 44: 17 (1967); L. Martinelli, S.C. Mutha, R. Ketcham, L.A. Strait, and R. Cavestri, *J. Org. Chem.* 37: 2278 (1972).

21. R.M. Kellogg and J. Buter, *J. Org. Chem.* 36: 2236 (1971); S.S. Hixson, *J. Am. Chem. Soc.* 93: 5293 (1971).

22. R.C. Hahn, P.H. Howard, and G.A. Lorenzo, *J. Am. Chem. Soc.* 93: 5816 (1971).

23. M.J.S. Dewar and H.N. Schmeising, *Tetrahedron* 5: 166 (1959).

24. J.L. Franklin, *J. Am. Chem. Soc.* 72: 4278 (1950); F. Klages, *Chem. Ber.* 82: 358 (1949).
25. See B.A. Hess and L.J. Schaad, *J. Am. Chem. Soc.* 93: 2413 (1971).
26. M.J.S. Dewar, A.J. Harget, and N. Trinajstic, *J. Am. Chem. Soc.* 91: 6321 (1969); N.C. Baird, *J. Chem. Educ.* 48: 509 (1971).
27. B.A. Hess, Jr., and L.J. Schaad, *J. Org. Chem.* 37: 4179 (1972); M. Milun, Z. Sobotka, and N. Trinajstic, *J. Org. Chem.* 37: 139 (1972); W.D. Hobey, *J. Org. Chem.* 37: 1137 (1972).
28. R. Breslow, J. Brown, and J.H. Gajewski, *J. Am. Chem. Soc.* 89: 4383 (1967).
29. D. Lloyd and D.R. Marshall, *Chem. and Ind.* (London) 335 (1972).
30. Bond force constants also support this view; see L. Keldner, *Proc. Roy. Soc.* A177: 456 (1941).
31. M.F.A. El-Sayed and R.K. Sheline, *J. Inorg. Nucl. Chem.* 6: 187 (1958).
32. L. Radom, *et al.*, *Chem. Commun.* 308 (1972).
33. F.A.L. Anet and M. Ahmad, *J. Am. Chem. Soc.* 86: 119 (1964).
34. J. Heidberg, J.A. Weil, G.A. Janusonis, and J.K. Anderson, *J. Chem. Phys.* 41: 1033 (1964).
35. J.v. Jouanne and J. Heidberg, *J. Am. Chem. Soc.* 95: 487 (1973).
36. R.K. Mackenzie and D.D. MacNicol, *Chem. Comm.* 1299 (1970).
37. A. Hastie, D.G. Lister, R.L. McNeil, and J.K. Tyler, *Chem. Commun.* 108 (1970).
38. K.N. Trueblood, E. Goldfish, and J. Donohue, *Acta Crystalogr.* 14: 1009 (1961).
39. S.F. Mason, *J. Chem. Soc.* 1275 (1959).
40. W. Haugen and M. Traetteberg, *Acta Chem. Scand.* 20: 1724, 1726 (1966).
41. G. Huse and H.M. Powell, *J. Chem. Soc.* 1398 (1940).
42. G. Becher, W. Luttke, and G. Schrumpf, *Angew. Chem. Int. Ed.* 12: 339 (1973).
43. C.A. Coulson, *Proc. Roy. Soc.* A169: 413 (1939); N.F. Phelan and M. Orchin, *J. Chem. Educ.* 45: 633 (1968).
44. See G.W. Wheland, *Resonance in Organic Chemistry*, New York: Wiley, 1955, p. 185.
45. W.T. Cresswell, G.H. Jeffery, J. Leicester, and A.I. Vogel, *J. Chem. Soc.* 514 (1952).
46. C.P. Smyth, *J. Am. Chem. Soc.* 63: 57 (1941).
47. T. Axenrod, *et al.*, *J. Am. Chem. Soc.* 93: 6536 (1971).
48. J.A. Marshall and S.F. Brady, *J. Org. Chem.* 35: 4068 (1970).
49. B. Sunners, L.H. Piette, and W.G. Schneider, *Canad. J. Chem.* 38: 681 (1960).
50. See reference 51, p. 236.
51. V.I. Minkin, O.A. Osipov, and Y.A. Zhdanov, *Dipole Moments in Organic Chemistry*, New York: Plenum, 1970.

52. L.E. Sutton, *Determination of Organic Structures by Physical Methods*, ed. E.A. Braude and F.C. Nachod, New York: Academic, 1955, p. 395.
53. W.B. Smith and T.J. Kmet, *J. Phys. Chem.* 70: 4084 (1966).
54. C.L. Liotta and D.F. Smith, Jr., *Chem. Commun.* 416 (1968).
55. K.V. Scherer, Jr., and T.J. Meyers, *J. Am. Chem. Soc.* 90: 6253 (1968); M. Neuenschwander and A. Niederhauser, *Chimia* 22: 491 (1968). However, for an opposing view, see R. Gompper, F. Holsboer, W. Schmidt, and G. Seybold, *J. Am. Chem. Soc.* 95: 8479 (1973).
56. L. Watts, J.D. Fitzpatrick, and R. Pettit, *J. Am. Chem. Soc.* 87: 3253 (1965); S. Masamune, N. Nakamura, M. Suda, and H. Ona, *J. Am. Chem. Soc.* 95: 8481 (1973).
57. J.W. Smith, *J. Chem. Soc.* 81 (1961).
58. A.V. Kemp-Jones, *et al.*, *Canad. J. Chem.* 51: 767 (1973).
59. E. Hückel, *Z. physik.* 72: 310 (1931).
60. See C.A. Coulson and W.T. Dixon, *Tetrahedron* 17: 215 (1962).
61. W.D. Karmack and R. Robinson, *J. Chem. Soc.* 121: 427 (1922).
62. F.R. Goss and C.K. Ingold, *J. Chem. Soc.* 1268 (1928).
63. W. Schäfer, *et al.*, *Angew. Chem. Int. Ed.* 22: 145 (1973).
64. A.J. Ashe, III, *J. Am. Chem. Soc.* 93: 3293, 6690 (1970).
65. A.G. Anastassiou, *Accts. Chem. Res.* 5: 281 (1972).
66. R.N. Warrener, *J. Am. Chem. Soc.* 93: 2346 (1971).
67. D.S. Matteson and N.K. Hota, *J. Am. Chem. Soc.* 93: 2893 (1971).
68. R.A. LaBar and W.M. Jones, *J. Am. Chem. Soc.* 95: 2359 (1973).
69. P.J. Garratt, *Aromaticity*, London: McGraw-Hill, 1971; *Nonbenzenoid Aromatics*, ed. J.P. Snyder, Vol. 1, 1970, Vol. 2, 1971, New York: Academic; G.M. Badger, *Aromatic Character and Aromaticity*, New York: Cambridge U. Press, 1969.
70. H.J. Dauben, Jr., J.D. Wilson, and J.L. Laity, *Nonbenzenoid Aromatics*, ed. J.P. Snyder, Vol 2, New York: Academic, 1971; A.J. Jones, *Rev. Pure Appl. Chem.* 18: 253 (1968).
71. H.J. Dauben, Jr., J.D. Wilson, and J.L. Laity, *J. Am. Chem. Soc.* 91: 1991 (1969).
72. J.D. Bertelli and T.G. Andrews, Jr., *J. Am. Chem. Soc.* 91: 5280 (1969).
73. H.C. Smitherman, Jr., and L.N. Ferguson, *Tetrahedron* 24: 923 (1968).
74. J.M. Pochan and W.H. Flygare, *J. Am. Chem. Soc.* 91: 5928 (1969).
75. J.F. Chiang and S.H. Bauer, *J. Am. Chem. Soc.* 92: 261 (1970).
76. J. Kruszewski and T.M. Krygowski, *Tetra. Letters* (4) 319 (1970).
77. J.H. Goldstein and G.S. Reddy, *J. Chem. Phys.* 36: 2644 (1962).
78. W.B. Smith and T.J. Kmet, *J. Phys. Chem.* 70: 4084 (1966).
79. D.J. Bertelli, T.G. Andrews, Jr., and P.O. Crews, *J. Am. Chem. Soc.* 91: 5286 (1969).
80. H.L. Ammon and G.L. Wheeler, *Chem. Commun.* 1032 (1971).

81. C.L. Norris, R.C. Benson, P. Beak, and W.H. Flygare, *J. Am. Chem. Soc.* 95: 2766 (1973).
82. J.P. Schaefer and L.L. Reed, *J. Am. Chem. Soc.* 93: 3902 (1971); C.A. Veracini and F. Pietra, *Chem. Commun.* 262 (1972).
83. F.A.L. Anet and G.E. Schenck, *J. Am. Chem. Soc.* 93: 556 (1971).
84. A.G. Anastassiou and H. Yamamoto, *Chem. Commun.* 286 (1972).
85. T.J. Barton, R.W. Roth, and J.G. Verkade, *Chem. Commun.* 1101 (1972).
86. J.P. Collman, *Angew. Chem. Int. Ed.* 4: 132 (1965).
87. B. Bock, *et al.*, *Angew. Chem. Int. Ed.* 10: 225 (1971).
88. E.C. Lingefelter and R.L. Braun, *J. Am. Chem. Soc.* 88: 2953 (1966).
89. R.C. Fay and N. Serpone, *J. Am. Chem. Soc.* 90: 5701 (1968).
90. A.H. Lowrey, G.P. D'Antonio, and J. Karle, *J. Am. Chem. Soc.* 93: 6399 (1971).
91. S.M. Johnson, I.C. Paul, and G.S.D. King, *J. Chem. Soc.* (B) 643 (1970).
92. G.W. Buchanan and A.R. McCarville, *Canad. J. Chem.* 51: 177 (1973).
93. F. Sondheimer, *Accts. Chem. Res.* 5: 81 (1972).
94. I. Murata, K. Yamamoto, T. Hirotsu, and M. Moricka, *Tetra. Letters* (5) 331 (1972).
95. R.H. Mitchell, C.E. Klopfenstein, and V. Boekelheide, *J. Am. Chem. Soc.* 91: 4931 (1969).
96. J.F.M. Oth, H. Baumann, J.-M. Gilles, and G. Schröder, *J. Am. Chem. Soc.* 94: 3498 (1972).
97. A.G. Anderson, Jr., A.A. MacDonald, and A.F. Montana, *J. Am. Chem. Soc.* 90: 2993 (1968).
98. E. Vogel and H. Reel, *J. Am. Chem. Soc.* 94: 4388 (1972).
99. B.W. Metcalf and F. Sondheimer, *J. Am. Chem. Soc.* 93: 6675 (1971); S. Masamune and N. Darby, *Accts. Chem. Res.* 5: 272 (1972).
100. F. Sondheimer, *Accts. Chem. Res.* 5: 81 (1972).
101. R. Breslow, *Accts. Chem. Res.* 6: 393 (1973); R. Breslow, D.R. Murayama, S.-I. Murahashi, and R. Grubbs, *J. Am. Chem. Soc.* 95: 6688 (1973); see also C.F. Wilcox, Jr., J.P. Uetrecht, and K.K. Grohman, *J. Am. Chem. Soc.* 94: 2532 (1972); compare W.D. Hobey, *Tetra. Letters* 23: 2967 (1973).
102. R. Breslow and J.M. Hoffmann, Jr., *J. Am. Chem. Soc.* 94: 2110 (1972).
103. R. Breslow and S. Mazur, *J. Am. Chem. Soc.* 95: 585 (1973).
104. R. Breslow, *J. Am. Chem. Soc.* 92: 4139 (1970) and earlier papers.
105. S. Winstein in *Carbonium Ions*, Vol. III, ed. G.A. Olah, New York: Wiley-Interscience, 1972; W.J. Hehre, *J. Am. Chem. Soc.* 95: 5807 (1973).
106. S. Winstein, *et al.*, *J. Am. Chem. Soc.* 93: 1284 (1971).
107. For an additional variation, bicycloaromaticity, see J.B. Grutzner and S. Winstein, *J. Am. Chem. Soc.* 94: 2200 (1972).
108. H.A. Corver and R.F. Childs, *J. Am. Chem. Soc.* 94: 6201 (1972).

109. R.E. Christoffersen, *J. Am. Chem. Soc.* 93: 4104 (1971).
110. E. Heilbronner, *et al.*, *Helv. Chim. Acta* 54: 783 (1971).
111. There are many volumes on CT complexes. Some leading references are: R.S. Mulliken and W. Person, *Molecular Complexes*, New York: Wiley, 1969; R. Foster, *Organic Charge-Transfer Complexes*, New York: Academic 1969; G.R. Anderson, *J. Am. Chem. Soc.* 92: 3553 (1970). A concise summary of the important characteristics of CT complexes is given by A.I. Scott and A.D. Wrixton, *Tetrahedron* 29: 933 (1972).
112. L.N. Ferguson, *Modern Structural Theory of Organic Chemistry*, Englewood Cliffs, N.J.: Prentice-Hall, 1963, p. 103 ff.
113. S. Moon and C.R. Ganz, *J. Org. Chem.* 34: 465 (1969).
114. R.G. Jesaitis and A. Krantz, *J. Chem. Educ.* 48: 137 (1971).
115. M.A. Slifkin, *Charge Transfer Interactions of Biomolecules*, New York: Academic, 1971.
116. H. Ishida, H. Takahashi, H. Soto, and H. Tsubomura, *J. Am. Chem. Soc.* 92: 275 (1970).
117. See R.S. Mulliken and W.B. Person, *J. Am. Chem. Soc.* 91: 3409 (1969); J.L. Lippert, M.W. Hanna, and P.J. Trotter, *J. Am. Chem. Soc.* 92: 3553 (1970).
118. I. Ikemoto and H. Kuroda, *Bull. Chem. Soc. Japan* 40: 2009 (1967).
119. A.K. Colter, S.S. Wang, G.H. Megerle, and P.S. Ossip, *J. Am. Chem. Soc.* 86: 3106 (1964).
120. P. Klaboe, *J. Am. Chem. Soc.* 89: 3667 (1967).
121. R.K. Chan and S.C. Lao, *Canad. J. Chem.* 48: 299 (1970).
122. A. Bier, *Rec. Trav. Chim.* 75: 866 (1956); R.E. Merrifield and W.D. Phillips, *J. Am. Chem. Soc.* 80: 2778 (1958); D.T. Longone and H.S. Chow, *J. Am. Chem. Soc.* 92: 994 (1970).
123. T.D. Walsh, *J. Am. Chem. Soc.* 90: 6390 (1968).
124. P. Machmer and J. Duchesne, *Nature* 206: 618 (1965).
125. J.B. Birks and M.A. Slifkin, *Nature* 191: 761 (1960); A. Streitwieser, Jr., *J. Am. Chem. Soc.* 82: 4123 (1960).
126. M.J. Kurylo and N.B. Jurinski, *Tetra. Letters* (12) 1083 (1967).
127. M.I. Foreman, R. Foster, and C.A. Fyfe, *J. Chem. Soc.* (B) 528 (1970).
128. M. Rosenblum, R.W. Fish, and C. Bennett, *J. Am. Chem. Soc.* 86: 5166 (1964).
129. L.L. Miller, G.D. Nordblom, and E.A. Mayeda, *J. Org. Chem.* 37: 917 (1972).
130. T. Sulzberg and R.J. Cotter, *J. Org. Chem.* 35: 2762 (1970); T.K. Mukherjee and L.A. Levasseur, *J. Org. Chem.* 30: 644 (1965); A.K. Colter and L.M. Clemens, *J. Am. Chem. Soc.* 87: 849 (1965).
131. J.C. Goan, E. Berg, and H.E. Podall, *J. Org. Chem.* 29: 975 (1964).
132. W.D. Metz, *Science* 180: 1041 (1973).
133. T.W. Nakagawa, L.J. Andrews, and R.M. Keefer, *J. Am. Chem. Soc.* 82: 269 (1960).

134. "History of Inventions" weekly series, *Sunday Times Magazine* (London) June 21-August 23, 1970; "Chance Favors the Prepared Mind," B.E. Schaar, Schaar and Co., Chicago, Ill., 1968.
135. T.J. Kealy and P.L. Pauson, *Nature* 168: 1039 (1951).
136. Some leading references are: D.F. Shriver, *Chemistry in Britain* 8: 419 (1972); G.E. Coates, M.L.H. Green, and K. Wade, *Organometallic Compounds*, Vol. 2 (3rd ed.), London: Methuen, 1968. Pertinent journals are: *J. Organometallic Chemistry* and *Organometallic Chemistry Reviews*.
137. M. Tsutsui, M.N. Levy, A. Nakamura, M. Ichikawa, and K. Mori, *Introduction to Metal π-Complex Chemistry*, New York: Plenum, 1970.
138. J.P. Briggs, R. Yamdagni, and P. Kebarle, *J. Am. Chem. Soc.* 94: 5128 (1972).
139. R. Kuhn and D. Rewicki, *Angew. Chem. Int. Ed.* 6: 635 (1967).
140. B.L. McDowell and H. Rapoport, *J. Org. Chem.* 37: 3261 (1972).
141. K.J. Klabunds and D.J. Burton, *J. Am. Chem. Soc.* 94: 5985 (1972).
142. R.O.C. Norman and P.D. Ralph, *J. Chem. Soc.* 222 (1961).
143. G.L. Anderson, R.C. Parish, and L.M. Stock, *J. Am. Chem. Soc.* 93: 6984 (1971).
144. P. Gund, *J. Chem. Educ.* 49: 100 (1972).
145. G.C. Levy, J.D. Cargioli, and W. Racela, *J. Am. Chem. Soc.* 92: 6240 (1970); E.M. Arnett, R.P. Quirk, and J.W. Larsen, *J. Am. Chem. Soc.* 92: 3977 (1970).
146. M. Ballester, *et al.*, *J. Am. Chem. Soc.* 93: 2215 (1971).
147. L.N. Ferguson, *Highlights of Alicyclic Chemistry*, Palisades, N.J.: Franklin, 1973, p. 64.
148. J. Clark and D.D. Perrin, *Quart. Revs.* (London) 18: 295 (1964); D.D. Perrin, *IUPAC Dissociation Constants of Organic Bases in Aqueous Solution: Supplement 1972*, London: Butterworth, 1972.
149. B.M. Wepster, *Progr. in Stereochemistry*, 2, ed. W. Klyne and P.B.D. de la Mare, New York: Academic, 1958, Chap. 4.
150. See M.J. Kamlet, *et al.*, *J. Org. Chem.* 36: 3847, 3852 (1971) and references cited there.
151. L.N. Ferguson, *Modern Structural Theory of Organic Chemistry*, Englewood Cliffs, N.J.: Prentice-Hall, 1963, p. 368 ff.
152. J.U. Lowe, Jr., and L.N. Ferguson, *J. Org. Chem.* 30: 3000 (1965).
153. R.L. Lintvedt and H.F. Holtzclaw, Jr., *J. Am. Chem. Soc.* 88: 2713 (1966); G. Allen and R.A. Dwek, *J. Chem. Soc.* (B) 161 (1966); A.I. Kol'tsov and G.M. Kheifets, *Russ. Chem. Revs.* 40: 773 (1971).
154. M. Gorodetsky, Z. Luz, and Y. Mazur, *J. Am. Chem. Soc.* 89: 1183 (1967); N. Allinger, L.W. Chow, and R.A. Ford, *J. Org. Chem.* 32: 1994 (1967); J.A. Hirsch and F.J. Cross, *J. Org. Chem.* 36: 955 (1971); R. Filler, *et al.*, *J. Org. Chem.* 35: 930 (1970).
155. J.S. Chickos and R.E. Winter, *J. Am. Chem. Soc.* 95: 506 (1973); R. Herscovitch, J.J. Charette, and E. de Hoffmann, *J. Am. Chem. Soc.* 95: 5135 (1973).

156. W.B. Smith, *J. Chem. Educ.* 47: 535 (1970); H. Lankamp, W. Th. Nauta, and C. MacLean, *Tetra Letters* (2) 249 (1968).
157. C. Ruchardt, *Angew. Chem. Int. Ed.* 9: 830 (1970).
158. D.M. Golden, N.A. Gac, and S.W. Benson, *J. Am. Chem. Soc.* 91: 2136 (1969); W. Tsang, *Internat. J. Chem. Kinetics* 1: 245 (1969); A.B. Trentwith, *Trans. Farad. Soc.* 66: 2805 (1970).
159. S.A. Sherrod, D.R. Kelsey, and R.G. Bergman, *J. Am. Chem. Soc.* 93: 1925, 1941, 1953 (1971).
160. M. Hanack, *Accts. Chem. Res.* 3: 209 (1970); C.A. Grob, *Chimia* 25: 87 (1971); G. Modena and L. Tonellato, *Adv. Phys. Org. Chem.* 9: 185 (1971).
161. P.v.R. Schleyer, *J. Am. Chem. Soc.* 93: 1513 (1971).
162. C.A. Grob, *et al.*, *Helv. Chim. Acta* 53: 2119, 2130 (1970).
163. Recent reviews: C.D. Johnson, *The Hammett Equation*, New York: Cambridge U. Press, 1973; P.R. Wells, *Linear Free Energy Relationships*, New York: Academic, 1968; J. Shorter, *Chemistry in Britain* 5: 269 (1969).
164. L.P. Hammett, *Physical Organic Chemistry*, 2nd ed., New York: McGraw-Hill, 1970.
165. H.C. Brown and Y. Okamoto, *J. Am. Chem. Soc.* 79: 1913 (1957); *J. Org. Chem.* 32: 485 (1957).
166. J. Miller, *Australian J. Chem.* 9: 61 (1956).
166a. T. Fujita, J. Iwasa, and C. Hansch, *J. Am. Chem. Soc.* 86: 5175 (1964).
167. F.W. Baker, R.C. Parish, and L.M. Stock, *J. Am. Chem. Soc.* 89: 5677 (1967).
168. A.R. Katritzky, R.F. Pinzelli, M.V. Sinnott, and R.D. Topson, *J. Am. Chem. Soc.* 92: 6861 (1970) and earlier papers.
169. R.W. Taft, Jr., and I.C. Lewis, *J. Am. Chem. Soc.* 81: 5343 (1959).
170. M.J.S. Dewar, R. Golden, and J.M. Harris, *J. Am. Chem. Soc.* 93: 4187 (1971).
171. P.R. Wells and W. Adcock, *Australian J. Chem.* 18: 1365 (1965).
172. R.W. Taft, Jr., N.C. Deno, and P.S. Skell, *Ann. Rev. Phys. Chem.* 9: 287 (1958).
173. L.N. Ferguson, A.Y. Garner, and J.L. Mack, *J. Am. Chem. Soc.* 76: 1250 (1954).
173a. T. Yamamoto and T. Otsu, *Chem. and Ind.* (London) 787 (1967).
174. S.V. Kulkarni, R. Schure, and R. Filler, *J. Am. Chem. Soc.* 95: 1859 (1973).
175. P. Politzer and J.W. Timberlake, *J. Org. Chem.* 37: 3557 (1972).
176. R. Fellows and R. Luft, *J. Am. Chem. Soc.* 95: 5593 (1973); C.K. Hancock, E.A. Meyers, and B.J. Yager, *J. Am. Chem. Soc.* 83: 4211 (1961); R.W. Taft, Jr., *Steric Effects in Organic Chemistry*, ed. M.S. Newman, New York: Wiley, 1956, Chap. 13.

177. M. Charton, *Progr. Phys. Org. Chem.* 8: 235 (1971); I.M. Koltoff and M.K. Chantooni, Jr., *J. Am. Chem. Soc.* 93: 3843 (1971); M.T. Tribble and J.G. Traynham, *J. Am. Chem. Soc.* 91: 379 (1969); G.G. Smith and K.K. Lum, *Chem. Commun.* 1208 (1968).

178. B.M. Wepster, *J. Am. Chem. Soc.* 95: 102 (1973); C.C. Price and R.H. Michel, *J. Am. Chem. Soc.* 74: 3652 (1952); H.H. Jaffe and G.O. Doak, *J. Am. Chem. Soc.* 77: 4441 (1955); E.A. Hill, M.L. Gross, M. Stasiewicz, and M. Manion, *J. Am. Chem. Soc.* 91: 7381 (1969); M.S. Melzer, *J. Org. Chem.* 27: 496 (1962); P.R. Wells, S. Ehreson, and R.W. Taft, *Progr. Phys. Org. Chem.* 6: 147 (1968).

179. D.S. Noyce, J.A. Virgilio, and B. Bartman, *J. Org. Chem.* 38: 2657, 2660 (1973); G.S. Marx and P.E. Spoerri, *J. Org. Chem.* 37: 111 (1972); D.S. Noyce, C.A. Lipinski, R.W. Nichols, and H.J. Pavez, *J. Org. Chem.* 37: 2620 (1972); J.M. Angelelle, A.R. Katritzky, R.F. Pinzelli, and R.D. Topson, *Tetrahedron* 28: 2037 (1972); Review: H.H. Jaffe and H.L. Jones, *Adv. Heterocyclic Chem.* 3: 209 (1964).

180. L.A. Cohen and W.M. Jones, *J. Am. Chem. Soc.* 85: 3397, 3402 (1963).

181. M. Charton, *Progr. Phys. Org. Chem.* 10 (1973).

182. T.A. Mastryukova and M.I. Kabachnik, *J. Org. Chem.* 36: 1201 (1971).

183. A.D. Campbell, *et al.*, *J. Chem. Soc.* (B) 1068 (1970).

184. M. Charton, *J. Am. Chem. Soc.* 86: 2033 (1964).

185. K. Sekigawa, *Tetrahedron* 28: 505, 515 (1972); B. Kamienski and T.M. Krygowski, *Tetra. Letters* (8) 681 (1972); M. Godfrey, *Tetra. Letters* (9) 753 (1972); C.G. Swain and E.C. Lupton, *J. Am. Chem. Soc.* 90: 4328 (1968).

186a. S. Ehrenson, R.T.C. Brownlee, and R.W. Taft, *Prog. Phys. Org. Chem.* 10 (1973).

186b. C.D. Johnson and K. Schofield, *J. Am. Chem. Soc.* 95: 270 (1973); S.K. Dayal and R.W. Taft, *J. Am. Chem. Soc.* 95: 5595 (1973).

187. A. Streitwieser, *Chem. and Eng. News*, Sept. 20, 1965, p. 58.

188. J.W. Sease, F.G. Burton, and S.L. Nickol, *J. Am. Chem. Soc.* 90: 2595 (1968).

189. D. Holtz, *Chem. Revs.* 71: 139 (1971).

190. A. Streitwieser, Jr., *et al.*, *J. Am. Chem. Soc.* 89: 691, 693 (1967); S. Andreades, *J. Am. Chem. Soc.* 86: 2003 (1964).

191. K.J. Klabunde and D.J. Burton, *J. Am. Chem. Soc.* 94: 820 (1972).

192. J.O. Schreck, *J. Chem. Educ.* 48: 103 (1971).

193. L.N. Ferguson, *The Modern Structural Theory of Organic Chemistry*, Englewood Cliffs, N.J.: Prentice-Hall, 1963, Sec. 4.11.

194. W.F. Baitinger, P.v.R. Schleyer, T.S.S.R. Murty, and L. Robinson, *Tetrahedron* 20: 1635 (1964); T.L. Brown and M.T. Rogers, *J. Am. Chem. Soc.* 79: 577 (1957).

195. V. Bekárek, *Chem. Commun.* 1565 (1971).

196. A. Perjéssy, D.W. Boykin, Jr., L. Fišera, A. Krutošikova, and J. Kováč, *J. Org. Chem.* 38: 1807 (1973).

197. D.G. O'Sullivan and P.S. Sadler, *J. Chem. Soc.* 4144 (1957).

198. L.A. Cohen and W.M. Jones, *J. Am. Chem. Soc.* 85: 3402 (1963); C.N.R. Rao, *Chem. and Ind.* (London) 1239 (1957) and 666 (1956).

199. C.H. Yoder, R.H. Tucke, and R.E. Hess, *J. Am. Chem. Soc.* 91: 539 (1969); K.L. Williamson, N.C. Jacobus, and K.T. Soucy, *J. Am. Chem. Soc.* 86: 4021 (1964); S.H. Marcus, W.F. Reynolds, and S.I. Miller, *J. Org. Chem.* 31: 1872 (1966); M. Kasai, M. Hirota, Y. Hamada, and H. Matsuoka, *Tetrahedron* 29: 267 (1973).

200. H.C. Smitherman, Jr., and L.N. Ferguson, *Tetrahedron* 24: 923 (1968).

201. R.C. Lauterbur, *J. Am. Chem. Soc.* 83: 1846 (1961); P. Diehl, *Helv. Chim. Acta* 45: 568 (1962).

202. L. Joris, J. Mitsky, and R.W. Taft, *J. Am. Chem. Soc.* 94: 3438, 3442 (1972).

203. J.W. Baker, M.M. Davies, and J. Gaunt, *J. Chem. Soc.* 25 (1949).

204. L.A. Yanovskaya, *et al.*, *Tetrahedron* 29: 2053 (1973).

205. A. Buchs, G.P. Rossetti, and B.P. Susz, *Helv. Chim. Acta* 47: 1563 (1964); D. Cook, *J. Am. Chem. Soc.* 80: 49 (1958).

206. E.M. Arnett, J.V. Carter, and R.P. Quirk, *J. Am. Chem. Soc.* 92: 1770 (1970); T.G. Bonner and J. Phillips, *J. Chem. Soc.* (B) 650 (1966).

207. G.A. Olah, C.L. Jeuell, and A.M. White, *J. Am. Chem. Soc.* 91: 3962 (1969).

207a. M. Charton, *J. Org. Chem.* 30: 552 (1965).

208. Z. yoshida and M. Haruta, *Tetra. Letters* (42) 3745 (1965); R.A. Nyquist, *Spectrochim. Acta* 19: 1655 (1963).

209. W. Hanstein, H.J. Berwin, and T.G. Traylor, *J. Am. Chem. Soc.* 92: 829 (1970).

210. H. Sakurai, *J. Org. Chem.* 35: 2807 (1970).

211. B.B. Wayland and R.S. Drago, *J. Am. Chem. Soc.* 86: 5240 (1964).

212. T.M. Florence, *Australian J. Chem.* 18: 609 (1965); P.T. Lansbury and R.E. MacLeay, *J. Am. Chem. Soc.* 87: 831 (1965).

213. M.L. Ash, F.L. O'Brien, and D.W. Boykin, Jr., *J. Org. Chem.* 37: 106 (1972).

214. A. Streitwieser, Jr., *Progr. Phys. Org. Chem.* 1: 1 (1963).

215. L.S. Levitt and B.W. Levitt, *J. Org. Chem.* 37: 332 (1972).

216. T.W. Bentley, R.A.W. Johnstone, and D.W. Payling, *J. Am. Chem. Soc.* 91: 3978 (1969); R.W. McLafferty and M.M. Bursey, *J. Am. Chem. Soc.* 90: 5299 (1968).

217. Two recent reviews: A. Verloop, in *Drug Design*, Vol. III, ed. E.J. Ariens, New York: Academic, 1972, Chap. 2; *Biological Correlations -- The Hansch Approach*, Advances in Chemistry Series, Vol. 114, ed. R.F. Gould, Washington, D.C.: American Chemical Society, 1972.

218. F. Hruska, H.M. Hutton, and T. Schaeffer, *Canad. J. Chem.* 43: 2392 (1965).

219. For comments on the correlation coefficients of SCS-pmr correlations using Hammett, Swain-Lupton, or other modifications, see G.R. Wiley and S.I. Miller, *J. Org. Chem.* 37: 767 (1972).

220. W.B. Smith and J.L. Roark, *J. Am. Chem. Soc.* 89: 5018 (1967).
221. W.B. Smith, A.M. Ihrig, and J.L. Roark, *J. Phys. Chem.* 74: 812 (1970).
222. W.B. Smith, D.L. Deavenport, and A.M. Ihrig, *J. Am. Chem. Soc.* 94: 1959 (1972).
223. K.W. Shen, *J. Chem. Educ.* 50: 239 (1973); R. Hoffmann, *Accts. Chem. Res.* 4: 1 (1971).
224. G.A. Olah, *J. Am. Chem. Soc.* 94: 808 (1972); O.C. Dermer and J.C. Traynham, *J. Chem. Educ.* 50: 545 (1973).
225. P.D. Bartlett, *Nonclassical Ions*, New York: Benjamin, 1965.
226. *Carbonium Ions*, Vol. III, ed. G.A. Olah, New York: Wiley-Interscience, 1972.
227. G.A. Olah, *Angew. Chem. Int. Ed.* 12: 173 (1973).
228. G.A. Olah, G. Liang, G.D. Mateescu, and J.L. Riemenschneider, *J. Am. Chem. Soc.* 95: 8698 (1973).
229. H.L. Goering and J.V. Clevenger, *J. Am. Chem. Soc.* 94: 1010 (1972).
230. L. Radom, J.A. Pople, V. Buss, and P.v.R. Schleyer, *J. Am. Chem. Soc.* 94: 311 (1972).
231. B.A. Howell and J.G. Jewett, *J. Am. Chem. Soc.* 93: 799 (1971).
232. H. Hart and P.A. Law, *J. Am. Chem. Soc.* 86: 1957 (1964).
233a. R.C. Kerber and C.-M. Hsu, *J. Am. Chem. Soc.* 95: 3239 (1973).
233b. G.A. Olah and P.W. Westerman, *J. Am. Chem. Soc.* 95: 7530 (1973).
233c. R.S. Brown and T.G. Traylor, *J. Am. Chem. Soc.* 95: 8025 (1973); C.F. Wilcox, L.M. Loew, and R. Hoffmann, *J. Am. Chem. Soc.* 95: 8192 (1973); L.N. Ferguson, *Highlights of Alicyclic Chemistry*, Palisades, N.J.: Franklin, 1973, p. 217.
234. M. Saunders, P. Vogel, E.L. Hagden, and J. Rosenfeld, *Accts. Chem. Res.* 6: 53 (1973).
235. J.J. Tufariello and R.J. Lorence, *J. Am. Chem. Soc.* 91: 1546 (1969); J. Lhomme, A. Diaz, and S. Winstein, *J. Am. Chem. Soc.* 91: 1548 (1969).
236. P. Story, *J. Am. Chem. Soc.* 83: 3347 (1961).
237. H. Tanida, *Accts. Chem. Res.* 1: 239 (1968); J.W. Wilt and P.J. Chenier, *J. Org. Chem.* 35: 1571 (1970).
238. R.M. Coates and J.L. Kirkpatrick, *J. Am. Chem. Soc.* 90: 4162 (1968); Y.E. Rhodes and T. Takino, *J. Am. Chem. Soc.* 92: 5270 (1970).
239. M.V. Moncur and J.B. Grutzner, *J. Am. Chem. Soc.* 95: 6449 (1973); M.J. Goldstein and S. Natowsky, *J. Am. Chem. Soc.* 95: 6451 (1973).
240. C.H. DePuy, I.A. Ogawa, and J.C. McDaniel, *J. Am. Chem. Soc.* 83: 1668 (1961).
241. J.A. Thompson and D.J. Cram, *J. Am. Chem. Soc.* 91: 1778 (1969).
242. G.A. Olah and R.D. Porter, *J. Am. Chem. Soc.* 93: 6877 (1971).
243. S. Clementi, V. Mancini, and G. Marion, *Chem. Commun.* 1457 (1970).
244. G. Lepore, S. Migdal, D.E. Blagdon, and M. Goodman, *J. Org. Chem.* 38: 2590 (1973).

245. See reference 3 in Part I.
246. See W.C. Herndon, *Tetrahedron* 29: 3 (1973) and references cited there.
247. W.B. Smith, *J. Chem. Educ.* 48: 749 (1971); W.C. Herndon and E. Silber, *J. Chem. Educ.* 48: 503 (1971); N.F. Phelan and M. Orchin, *J. Chem. Educ.* 45: 633 (1968); J.K. Elwood, *J. Org. Chem.* 38: 2430 (1973); V.M. Csizmadia, et al., *J. Org. Chem.* 38: 2281 (1973); B. Halton, *Chem. Revs.* 73: 113 (1973); M.J.S. Dewar and T. Morita, *J. Am. Chem. Soc.* 91: 796 (1967); J.A. Pople and M. Gordon, *J. Am. Chem. Soc.* 89: 4253 (1967); L. Radom, P.C. Hariharan, J.A. Pople, and P.v.R. Schleyer, *J. Am. Chem. Soc.* 95: 6531 (1973); W.G. Richards and J.A. Horsley, *Molecular Orbital Calculations for Chemists*, New York: Oxford U. Press, 1971.
248. See Y. Tanaka and S.I. Miller, *J. Org. Chem.* 38: 2708 (1973).
249. R.B. Woodward and R. Hoffmann, *The Conservation of Orbital Symmetry*, New York: Academic, 1970.
250. See K.-W. Shen, *J. Chem. Educ.* 50: 238 (1973).
251. M. Gerloch and R.C. Slade, *Ligand-Field Parameters*, New York: Cambridge U. Press, 1973.
252. L.L. Miller, *J. Chem. Educ.* 48: 168 (1971).
253. J. Hudec, et al., *Chem. Commun.* 805, 807 (1971).
254. See, for example, P. Bischoff, E. Heilbronner, H. Prinzbach, and H.D. Martin, *Helv. Chim. Acta* 54: 1072 (1971).

Part III: SPECTRA-STRUCTURE CORRELATIONS

Organic chemical spectroscopy has developed in an empirical fashion. That is, spectra of compounds of known structure have been studied in detail and from these studies many relationships between structure and spectra have been recorded. Generalizations based on these relationships then may be applied to newly encountered compounds for their characterization (1). To cite just a few, many characteristics of the C=C bond may be quickly learned from spectra: the number of conjugated double bonds by uv; whether they are mono-, di-, tri-, or tetrasubstituted by ir; the number of olefinic and allylic protons by nmr; whether they are cis or trans substituted by ir, uv, or nmr; C-C=C bond angles and ring size by ir (2); ring size of cyclenes and dihedral angles by nmr (3).

This section will discuss some of these correlations and, in particular, will note the effects of certain structural parameters, e.g., electronegativity, resonance, H bonding, etc. In a sense, this will provide a rationale for some of the generalizations mentioned above.

Chapter 10
SPECTRA AND ELECTRON DELOCALIZATION

Resonance is merely a device for interpreting and predicting molecular properties in terms of structural formulas. Accordingly, a great effort has been made to develop the theory to account for absorption spectra. Recall that ir spectra result from the presence of specific <u>bonds</u> in a molecule, uv spectra relate to <u>chromophoric systems</u>, and nmr spectra reflect electron densities at a given <u>atom</u> in a molecule. The common simple chromophores were given in Section 4.2c. A molecule may contain many such groups and if they are not conjugated, then the molecule will absorb light characteristic of the isolated chromophores. For instance, the natural rubber molecule contains hundreds of C=C bonds but the hydrocarbon is colorless; and proteins, which consist of long chains of amino acids, have absorption bands much like the constituent monomer amino acids (4).

Resonance theory accounts for spectra in terms of proposed resonance hybrid structures. A simple example is the case of isolated and conjugated C=C and C=O bonds.

	$\nu_{C=O}$ (cm^{-1})	$\nu_{C=C}$ (cm^{-1})	λ_m (nm)	ϵ
$(CH_3)_2C=O$	1720		195	900
			275	22
$CH_3CH_2CH=CH_2$		1647	185	10^3
$CH_3-\overset{O}{\underset{\|\|}{C}}-CH=CH_2$	1685	1623	212	6500
10.1			324	24
10.2 (O=⟨⟩=O)	1690	1600	219	15700
			365	58

Thus, $\nu_{C=O}$ and $\nu_{C=C}$ are smaller for the conjugated systems, reflecting the resonance hybrid <u>10.3</u>

$$\underset{a}{C=C-C=O}, \quad \underset{b}{\overset{(+)}{C}-C=C-O^-}$$

10.3

in which the C=C and C=O bonds have some single-bond character.

QUESTION

Why is it not surprising that $\nu_{C=O}$ is <u>larger</u> for <u>10.2</u> than for <u>10.1</u>, and $\nu_{C=C}$ for <u>10.2</u> is <u>smaller</u> than that for <u>10.1</u> ?

ANSWER

There are two opposed α,β-enone systems in <u>10.2</u> and resonance of type <u>10.3</u> for each comes at the expense of the other. Hence, dipolar resonance from <u>10.3b</u> makes a little less contribution to each enone fragment than in <u>10.1</u>, giving each C=O bond a little less single bond character. However, resonance in either direction of <u>10.2</u> gives the central C=C bond a single-bond structure, so on the whole it has a greater single bond character than the C=C bond of <u>10.1</u>. Similarly, in aromatic systems, the greater the resonance contribution from dipolar structures

$$R-\bigcirc-\underset{|}{C}=O \quad , \quad R-\overset{+}{\bigcirc}=\underset{|}{C}-O^- \quad , \quad R^+=\bigcirc=\underset{|}{C}-O^-$$

the lower is the $\nu_{C=O}$ frequency. Semiquantitative relationships between $\nu_{C=O}$ and the resonance effect of R are expressed in terms of Hammett substituent constants (Section 7.1).

...

The uv bands for both groups of the conjugated enone move to longer wavelengths, i.e., smaller transition energies, as diagrammed in Figure 10.1. Although resonance among dipolar structures (such as <u>10.3b</u>) is greater in the electronic excited state, it does occur somewhat in the ground state and is revealed by ground state properties such as dissociation constants, dipole moments, and ir and nmr spectra. For instance, the protons β to a C=O group, because they are farther away from the oxygen atom than the α protons, give an nmr signal upfield from that of the α protons in a saturated system. However, the reverse is true for an α,β-enone system.

δ_H 1.7   2.3   7.0   5.85 ppm

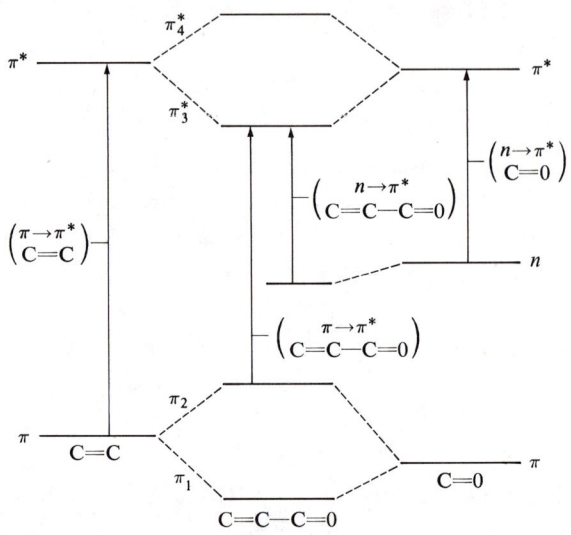

Figure 10.1 Energy level diagram for isolated and conjugated C=C and C=O bond

This too is a reflection of the contribution of the dipolar resonance hybrid. Furthermore, ^{13}C chemical shifts reveal the contribution of 10.3b to the resonance hybrid. For example, $\delta^{13}C$ for β carbons are downfield from those of α carbons (5):

	$\delta^{13}C$ (ppm from TMS)	
	C_α	C_β
H_2C=CHCOOH	129.2	130.8
H_2C=CHCO$_2$Me	128.7	129.9
CH$_3$CH=C(CH$_3$)CO$_2$Me (Z)	128.3	137.9
CH$_3$CH=C(CH$_3$)CO$_2$Me (E)	128.6	136.4
CH$_3$CH=C(CH$_3$)CO$_2$H (Z)	127.4	136.6
CH$_3$CH=C(CH$_3$)CO$_2$H (E)	128.2	137.9

Likewise, $\delta^{13}C$ values for cyclohexanone and cyclohexenone indicate that the

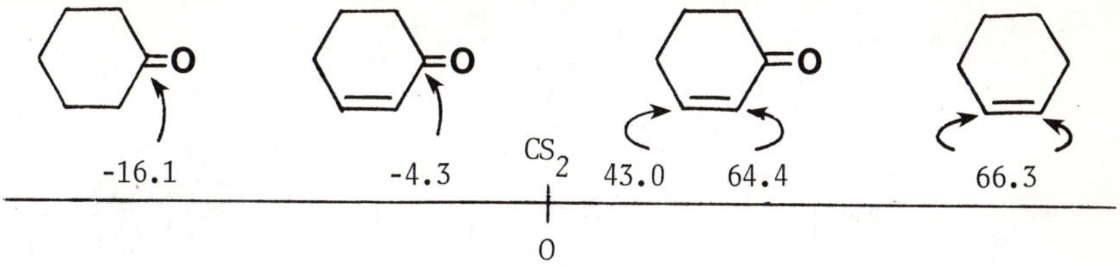

Ch. 10 SPECTRA AND ELECTRON DELOCALIZATION   325

$\delta^{13}C$ values (ppm rel to CS_2): cyclohexanone C=O at −16.1; cyclohexenone C=O at −4.3; CS_2 at 0; cyclohexenone α-C at 43.0, β-C at 64.4; cyclohexene =C at 66.3.

positive charge on the carbonyl carbon is reduced by delocalization ($\delta^{13}C$ of the unsaturated ketone moves upfield) and although the olefinic $\delta^{13}C$ values are downfield relative to cyclohexene, $\delta^{13}C$ for the β carbon is again farthest downfield. A similar charge distribution occurs in the protonated α,β-enone system.

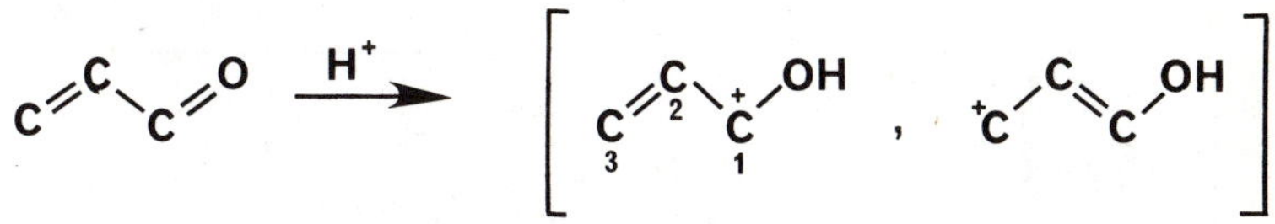

$$C=C-C=O \xrightarrow{H^+} \left[\overset{3}{C}=\overset{2}{C}-\overset{+1}{C}-OH \;,\; \overset{+}{C}-C=C-OH \right]$$

Carbon-13 nmr measurements indicate that the charge density at C-3 is greater than at C-1 and that there is practically no positive charge at C-2 (6). The spectral changes of vinyl ethers are accounted for by the resonance hybrid **10.4**, in which the C=C bond has some single-bond character and the C-O bond

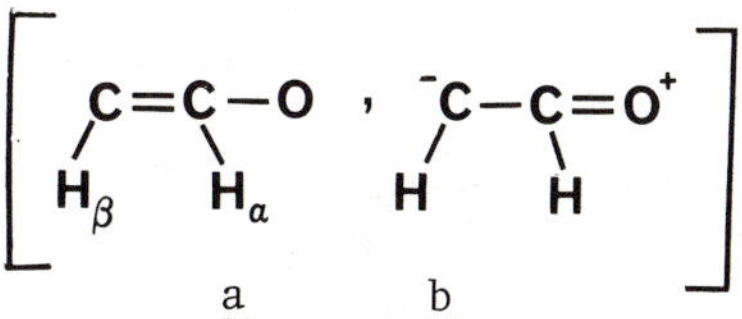

$$\left[\underset{H_\beta \;\; H_\alpha}{C=C-O} \;,\; \underset{H \;\;\; H}{{}^-C-C=O^+} \right]$$

              a   b
            10.4

some double-bond character. Thus, the ir C-O frequencies are higher than for alkyl ethers.

ν_{C-O}: R-O-R 1050; furan 1170; diphenyl ether 1250 cm^{-1}

In this case, the nmr chemical shifts for the β protons, δH_β are moved upfield, thereby widening the difference between δH_α and δH_β for protons α and β to oxygen atoms.

$$\delta_H \text{ (ppm)} \begin{bmatrix} \underset{8.64}{\overset{4.96}{}} & \text{[pyranone]} , & \text{[pyranone]}^- , & \text{[pyranone]}^- \end{bmatrix}$$

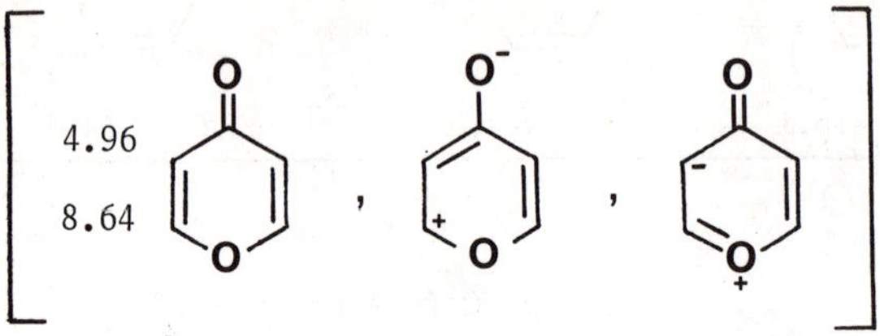

We can estimate the double-bond character of carbon-oxygen bonds, in terms of π-bond order, from carbonyl ir frequencies:

Structure	$\nu_{C=O}$	C=O π-bond order
$[(CH_3)_3C-\overset{O}{\overset{\|}{C}}-CH_3 \;,\; (CH_3)_3C-\overset{O^{\ominus}}{\overset{\|}{C}{}^{\oplus}}=CH_3]$	1710 cm^{-1}	0.90
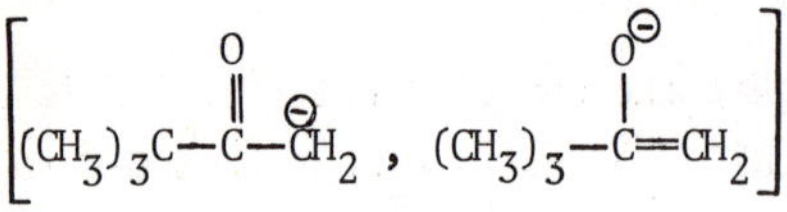 $[(CH_3)_3C-\overset{O}{\overset{\|}{C}}-\overset{\ominus}{C}H_2 \;,\; (CH_3)_3-\overset{O^{\ominus}}{\overset{\|}{C}}=CH_2]$ 10.5	1575	0.73 (7)
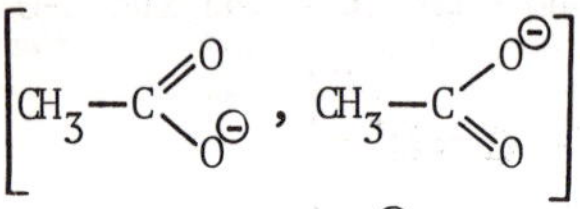 $[CH_3-C\overset{O}{\underset{O^{\ominus}}{\diagdown}} \;,\; CH_3-C\overset{O^{\ominus}}{\underset{O}{\diagdown}}]$	1560	0.5
$[\overset{O}{\underset{\ominus O}{\overset{\|}{C}}}\diagdown_{O^{\ominus}} \;,\; \cdots \;,\; \cdots]$	1445	0.33
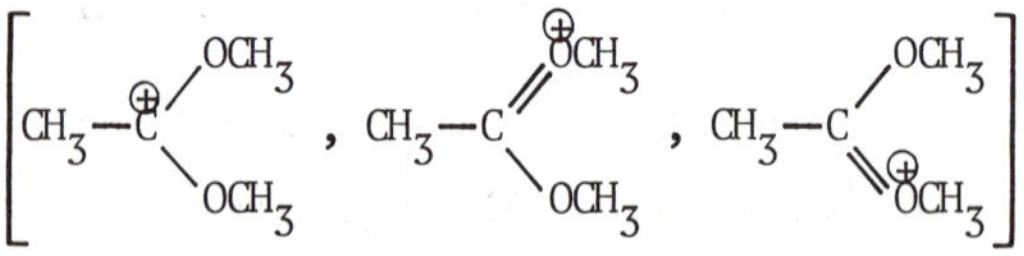 $[CH_3-\overset{\oplus}{C}\overset{OCH_3}{\diagdown_{OCH_3}} \;,\; CH_3-C\overset{\overset{\oplus}{O}CH_3}{\diagdown_{OCH_3}} \;,\; CH_3-C\overset{OCH_3}{\diagdown_{\overset{\oplus}{O}CH_3}}]$ 10.6	1440	0.2-0.3 (8)

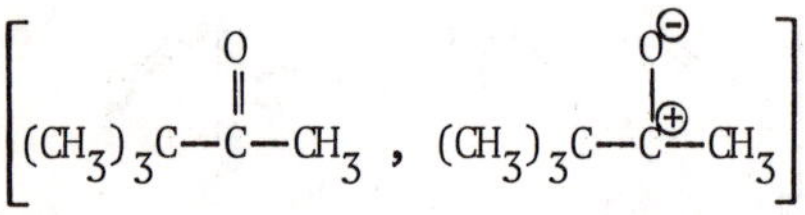

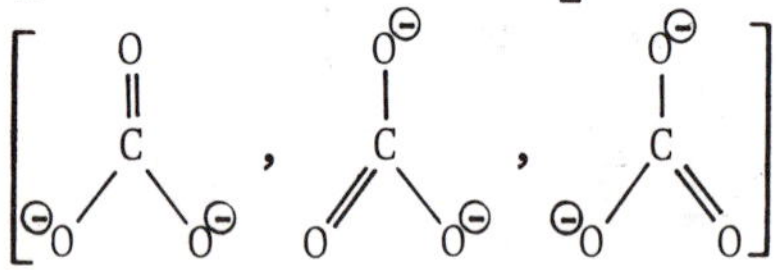

Thus, the C=O bond in 10.5 has lost about 25 per cent of its double-bond character and the C-O bond in 10.6 has acquired about 25 per cent double-bond character. Each case is accounted for by the resonance shown. Likewise, the low C-H frequency (ν_{C-H} 2830 cm^{-1}) for the <u>t</u>-butyl cation Me$_3$C$^\oplus$ can be attributed to hyperconjugation

$$\left[\begin{array}{c} \text{CH}_3 \\ | \\ \text{H}_3\text{C} - \overset{\oplus}{\text{C}} \\ | \\ \text{CH}_3 \end{array} , \begin{array}{c} \text{H}^\oplus \;\; \text{CH}_2 \\ \| \\ \text{H}_3\text{C} - \text{C} \\ | \\ \text{CH}_3 \end{array} \right] \text{ 3 equivalent structures}$$

in which there is "no bond" resonance.

We see, therefore, that any structural feature which affects the carbonyl hybrid $\left[\text{C=O}, \overset{\oplus\ominus}{\text{C-O}} \right]$ alters the carbon-oxygen bonding and the electron density on the oxygen atom. This in turn will be reflected in such properties as ir frequencies, <u>pK$_a$</u>, ionization potentials, $\Delta\nu_{OH}$ of added proton donors, etc. (see also Section 3.1 and Chapters 7 and 12). Thus, resonance and hyperconjugation delocalization lower $\nu_{C=O}$ whereas induction and ring strain (which alters the electronegativity of the carbonyl carbon) produce the opposite effect. This is demonstrated by some properties of the following ketones (9):

Ketone	$\nu_{C=O}$ (cm^{-1})	pK$_a$	IP (eV)	δ^+_{OH} (in FSO$_3$H)
CH$_3$-CO-CH$_3$	1718	-5.1	9.69	14.24 ppm
<u>c</u>-C$_3$H$_5$-CO-C$_3$H$_5$-<u>c</u>	1694	-4.1		12.08
CH$_3$-CH=CH-CO-CH$_3$	1697	-2.4	9.05	
Cl$_3$C-CO-CCl$_3$		-14.7		
Cyclohexanone	1716	-5.6		14.03
Cyclopentanone	1740	-5.9		14.17
Cyclobutanone	1791	-7.7		14.73

That is, with increasing contribution of the polar form ($^\oplus$C-O$^\ominus$), there is a decrease in $\nu_{C=O}$ and IP, an increase in <u>pK$_a$</u>, and the nmr chemical shift of the protonated ketone moves upfield.

We saw in Section 4.2 that $\Delta\lambda$ for the primary band of monosubstituted benzenes ($\Delta\lambda = \lambda_m - 203.5$ nm) is correlated with the amount of resonance interaction between the substituent and the ring regardless of direction with respect to the ring. For disubstituted benzenes with similar resonance groups, or <u>meta</u>-oriented if dissimilar, $\Delta\lambda_{1,2}$ rarely exceeds the larger of $\Delta\lambda_1$ and $\Delta\lambda_2$ ($\Delta\lambda$s for the respective monosubstituted benzenes). However, when two dissimilar groups are <u>para</u> to each other, $\Delta\lambda_{1,2}$ is greater than $\Delta\lambda_1 + \Delta\lambda_2$, and the difference is often taken as a measure of the resonance interaction between the two groups (see Table 10.1). This generalization is sometimes used to ascertain the direction of electron delocalization by a group. For example when X in <u>10.7</u> is NO_2, the $\overset{C}{\underset{C}{>}}C$ group as Y produces an enhancement in $\Delta\lambda_{x,y}$

$$X-\bigcirc-Y$$

10.7

whereas there is no enhancement for Y = $\overset{C}{\underset{C}{>}}O$ or $\overset{C}{\underset{C}{>}}S$. When X is OCH_3 however, the latter two groups as Y give an enhancement in $\Delta\lambda_{x,y}$ but the $\overset{C}{\underset{C}{>}}C$ group does not. We deduce from this that cyclopropyl and OCH_3 groups are of similar resonance type, i.e., electron-donating, and that the NO_2, $\overset{C}{\underset{C}{>}}O$, and $\overset{C}{\underset{C}{>}}S$ groups are of similar resonance types, i.e., electron-withdrawing.

We have seen in this section that certain uv, ir, and nmr spectral effects may be rationalized in terms of structures contributing to a resonance hybrid. Having established such relationships, we may reverse the process and use spectra to provide evidence for a proposed resonance. For example, λ_m for cyclopentadiene is unusually low in comparison with the values for other cyclic dienes.

λ_m = 238.5   250   257   254   256 nm

This can be attributed to ground state stabilization of cyclopentadiene through hyperconjugation, giving the ring partial aromatic character. This view is corroborated by nmr solvent shift data (10).

Ch. 10 SPECTRA AND ELECTRON DELOCALIZATION   329

Table 10.1. Substituent effects on the uv first primary band of benzene

Group 1	Group 2	$\Delta\lambda_1 + \Delta\lambda_2$	$\Delta\lambda$ observed	Difference
CH_3	CN	23.5 nm	30.5 nm	7 nm
Cl	CN	26.5	34	7.5
CH_3	NO_2	68	81.5	14
CH_3O	$COCH_3$	55.5	73	18
NH_2	CN	47	66.5	20
OH	CHO	53	80	27
NH_2	$COCH_3$	68.5	108	40
OH	NO_2	72	114	42
O^-	$COCH_3$	71.3	121	48
O^-	CHO	77.5	126.5	49
NH_2	NO_2	91.5	177.5	86
O^-	NO_2	96.5	199	103

Comparison of the short-wavelength bands of these dienes, however, indicates that there is some electronic interaction between the spirocyclopropyl and vinyl groups with the cyclopentadienyl ring:

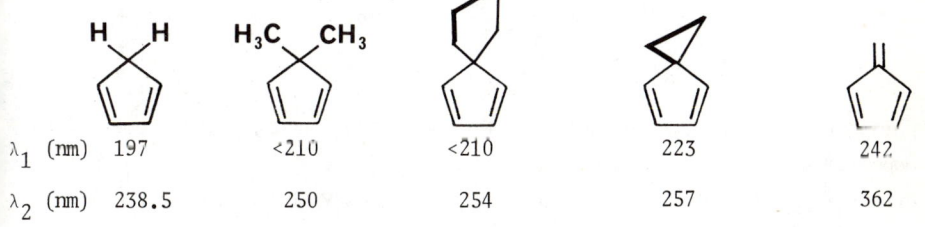

λ_1 (nm)	197	<210	<210	223	242
λ_2 (nm)	238.5	250	254	257	362

This delocalization in the spirocyclopropylcyclopentadiene is corroborated by its nmr spectrum which shows the cyclopropyl hydrogen absorption to be about 1.0 ppm downfield from that in model compounds (11). The small delocalization in the ground state of the molecule has an insignificant effect upon its bond lengths and angles (12).

A good example of the use of spectroscopy to provide evidence for a given resonance hybrid is that involving phospholenes. Two types of polar resonance forms might be written:

p_π - p_π $\left[>\!\!\underset{..}{P}\!\!-\!\!C\!=\!\overset{\frown}{C} \, , \, >\!\!\overset{+}{P}\!=\!C\!-\!\overset{-}{C}\!: \right]$

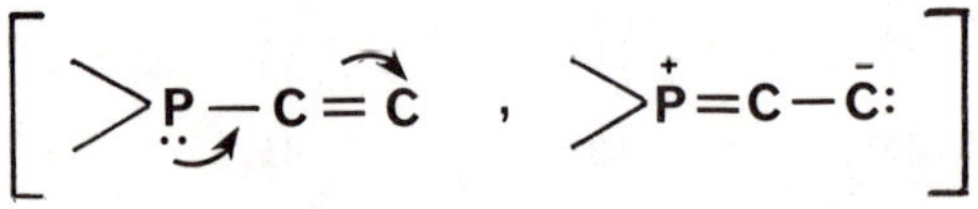

p_π - d_π $\left[>\!\!\underset{..}{P}\!\!-\!\!C\!=\!C \, , \, >\!\!\overset{-}{P}\!=\!C\!-\!\overset{+}{C} \right]$

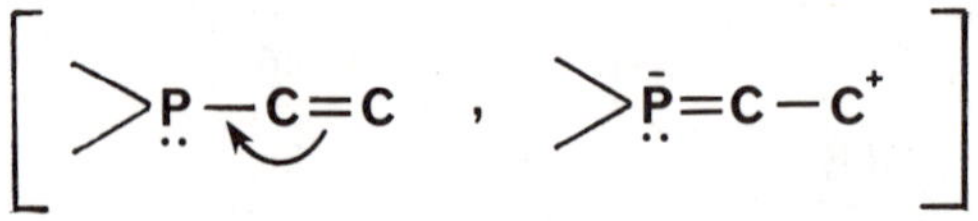

p_π-p_π resonance increases the positive charge on phosphorus whereas p_π-d_π resonance has the opposite effect. A choice can be made from nmr spectra, which show that the chemical shifts for phosphorus and groups attached to

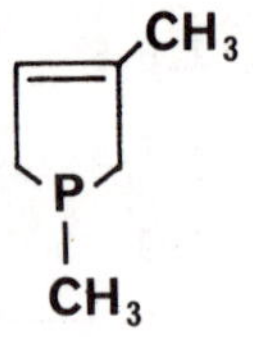

                CH₃   CH₃
              (phospholane)   (phospholene)
                P   P
                | |
                CH₃   CH₃

                 10.8 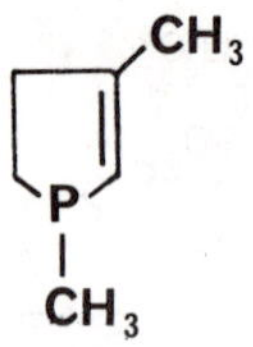   10.9

Compound	δ_P (rel. to H_3PO_4)	$\delta H(P\text{-}CH_3)$ (rel. to TMS)	$\nu_{C=C}$
10.8	+ 32.6 ppm	1.14 ppm	1658 cm^{-1}
10.9	+ 15.2	1.23	1613

phosphorus in 10.9 are deshielded relative to the nonconjugated isomer 10.8. Opposite nmr effects would be observed for p_π-d_π resonance. The conjugative effect in 10.9 is confirmed by the C=C ir frequency which is much smaller in 10.9 than in 10.8.

The extent of p_π-d_π bonding in the C-P bond of phosphorus substituents attached to aryl and heteroaryl rings has been studied by the usual physical methods (14). For example, dipole moment data on phosphinobenzenes reveal a p_π-d_π resonance effect.

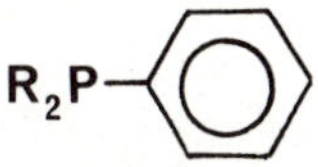

R_2P—⟨phenyl⟩

10.10

R	μ
H	1.11 D
C_4H_9	1.45

R_2P—⟨phenyl⟩—CH_3

10.11

R	μ
H	1.41 D
C_4H_9	1.55

From the moments of 10.10 and 10.11 (15), it follows that the resonance moment of the R_2P group coincides with the moment of the CH_3 group; that is it is directed toward the phosphorus atom. For comparison, CH_3 substitution diminishes the dipole moments of anilines because the amino group can only exhibit p_π-p_π resonance.

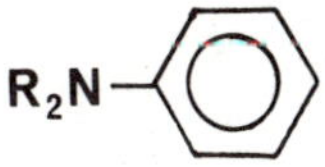

R_2N—⟨phenyl⟩

R	μ
H	1.53 D
C_4H_9	1.91

R_2N—⟨phenyl⟩—CH_3

R	μ
H	1.32 D
C_4H_9	1.57

Apparently, p_π-d_π bonding occurs in the C-P bonds of furan, thiophene, and pyrrole derivatives but not always in phenyl or pyridine derivatives (14).

Consequently, nmr chemical shifts are commonly used to infer a certain resonance, as we have just done for the enone, vinyl ether, and phospholene systems. Likewise, the ^{13}C and ^1H chemical shifts of substituted benzenes reflect the resonance of the substituents, i.e., move downfield for +R groups (electron-withdrawing by resonance) and upfield for -R groups (16):

Ch. 10 SPECTRA AND ELECTRON DELOCALIZATION

Substituent*	$\Delta\delta^C_1$	$\Delta\delta^C_o$	$\Delta\delta^C_m$	$\Delta\delta^C_p$
CO_2H	1.9	1.3	0.2	4.8
CO_2Me	2.1	1.1	0.1	4.5
OH	28.7	-12.8	1.2	-8.5
OMe	31.6	-14.3	1.2	-7.7

$\Delta\delta = \delta_x - \delta_{benzene}$ $\delta^C = \delta^{13}C$ $\delta^H = \delta^1H$

* Subscripts 1, o, m, p, are for the substituted, ortho, meta, and para carbons, respectively. Positive values indicate a shift downfield.

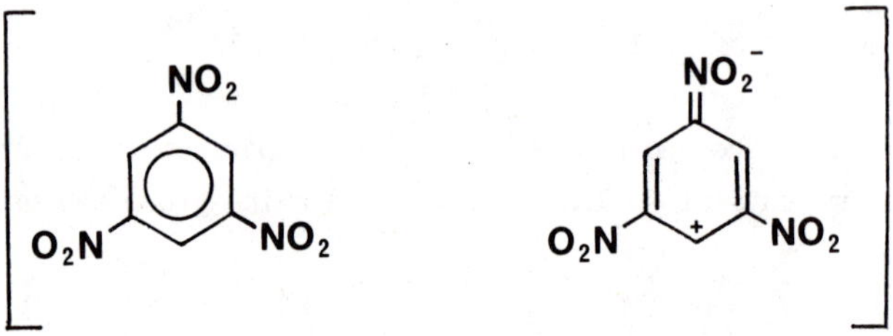

$\Delta\delta = 2.2$ ppm

$\Delta\delta = 0.0$ ppm

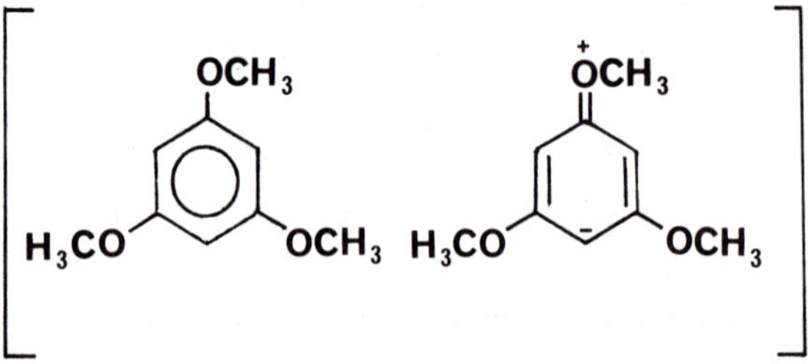

$\Delta\delta = -1.32$ ppm

Ch. 10 SPECTRA AND ELECTRON DELOCALIZATION

Another example of the confirmation of resonance hybrids from nmr spectra is found in the data for some carbanions (17):

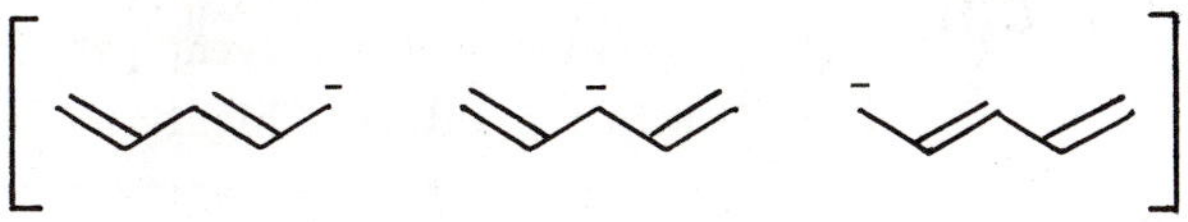

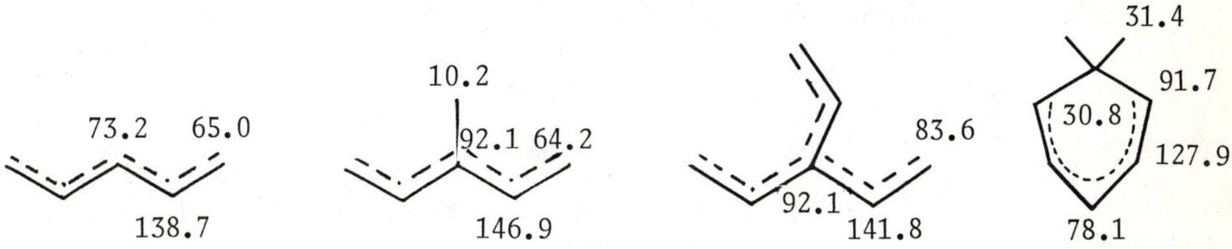

[structures of three resonance forms of pentadienyl anion]

[structures with chemical shift values:
73.2, 65.0, 138.7;
10.2, 92.1, 64.2, 146.9;
83.6, 92.1, 141.8;
31.4, 91.7, 30.8, 127.9, 78.1]

Observe that the odd-numbered carbons resonate (cmr shifts given in ppm from TMS) at higher field (65 to 92) than the even-numbered carbons (127 to 147) in accord with the resonance hybrid written for these carbanions. Conversely, resonance forms are written to account for observed chemical shift data. Pi electron densities may be computed from the relationships

$$\delta_H = 10.6\rho + \underline{c} \text{ and } \delta_C = 160\rho + \underline{k}$$

where ρ are π electron densities, $\underline{c}$ and $\underline{k}$ are constants, and δ_H and δ_C are ^{1}H and ^{13}C chemical shifts (18).

Solvent effects on spectra are often interpreted to infer large or small contributions of polar structures to a resonance hybrid. For example, the nmr chemical shift values for cyclopropanones differ very little in neutral and acidic solvents, which implies that the compounds have a very polar character.

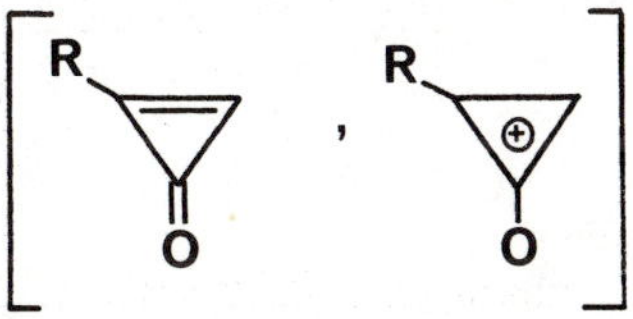

[structures of cyclopropanone resonance forms with R groups]

R = n-C_5H_{11}

$\delta_H^{CDCl_3}$ 1.53 ppm $\delta_H^{CF_3COOH}$ 1.32 ppm

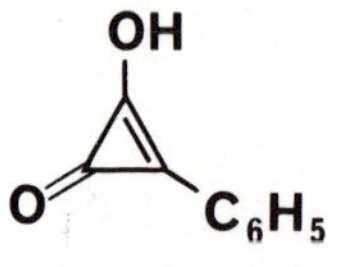

Likewise, the uv spectra of 10.12 and its anion are very similar. This indicates that 10.12 is largely ionized in solution, and indeed, its pK_a ≈ 2. Similarly, the uv λ_m for the derivative 10.13 shifts to the blue as the solvent gets more polar. This too infers a highly polar ground state, which is confirmed by its very large dipole moment (7.9 D). This is even larger than that of the salt-like trimethylamine oxide, $Me_3\overset{+}{N}-\overset{-}{O}$, DM = 5.03 D.

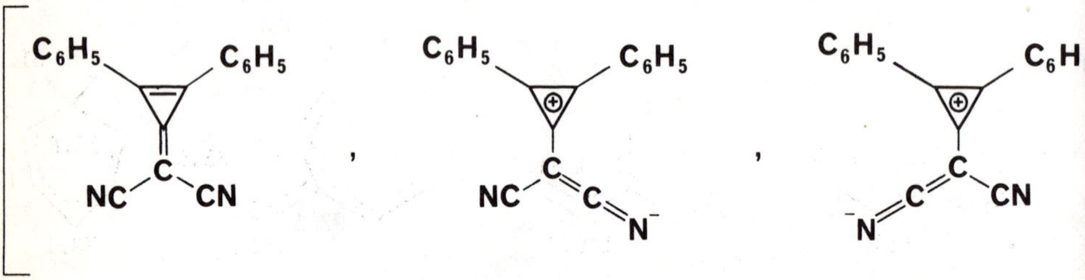

10.12

10.13

QUESTIONS

1. When the ^{19}F chemical shifts for pentafluorobenzyl fluoride and its carbocation are compared

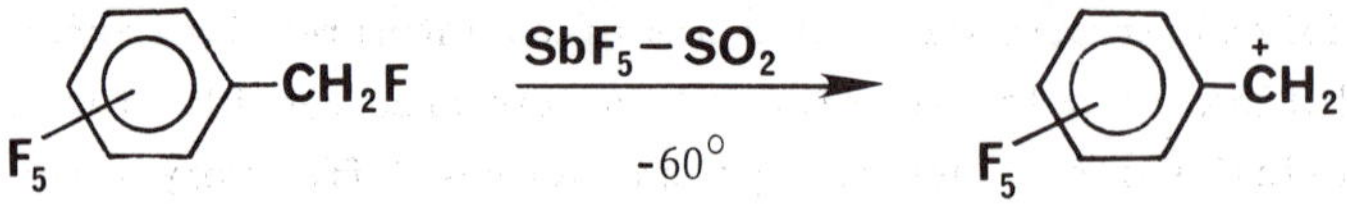

$F_5C_6H_4-CH_2F \xrightarrow[-60°]{SbF_5-SO_2} F_5C_6H_4-\overset{+}{C}H_2$

there are downfield changes in the ortho, meta, and para fluorine chemical shifts as follows:

$\Delta\delta_o$ = 40.4 $\Delta\delta_m$ = 11.0 $\Delta\delta_p$ = 77.4 ppm

How do you account for these changes? (G. A. Olah and M. B. Comisarow, J. Am. Chem. Soc. 89: 1027 (1967).)

2. Write resonance structures to account for the following spectral changes in the uv:

	λ_m
3-Hydroxybiphenyl	250 nm
4-Hydroxybiphenyl	260
3'-Nitro-4-hydroxybiphenyl	263
4'-Nitro-4-hydroxybiphenyl	340

3. Some spectral data for $\underline{I}$ are as shown (R. C. DeSelms and F. Delay, J. Am. Chem. Soc. 95: 274 (1973).)

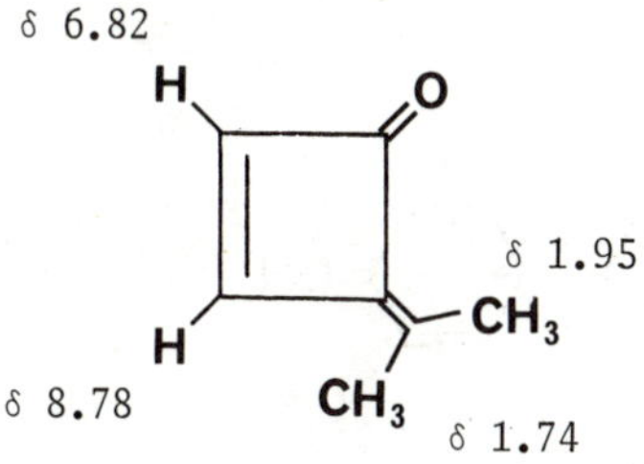

δ 6.82

δ 8.78

δ 1.95 CH₃

CH₃ δ 1.74

$\underline{I}$

ir: ν^{neat} 1640, 1680, 1740 cm^{-1}

uv: $\lambda_m^{C_6H_{12}}$ 223, 305 nm

log ε   4.42 2.73

Assign the ir bands and give the basis of your assignments.

4. Carbon-13 chemical shifts (in ppm relative to CS_2) of mesitylene and monoprotonated mesitylene are:

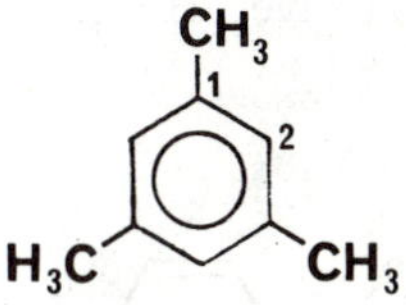

1,3,5	+56.4 ppm		1,3,5	−0.5
2,4,6	+66.1		4	+139.2
CH₃	+172.2		2,6	+58.3
			CH₃	+166.2

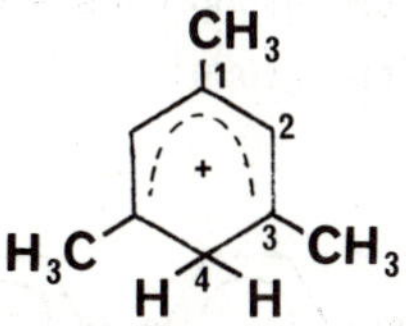

Estimate the changes in charge at the ring carbons.

ANSWERS

1. The meta fluorines experience a small inductive effect from the C^+ charge but the ortho and para fluorines are deshielded further owing to the resonance

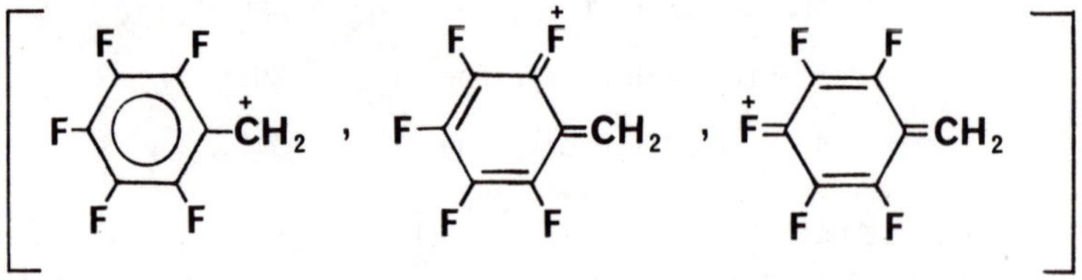

which is greater for the para fluorine.

2.

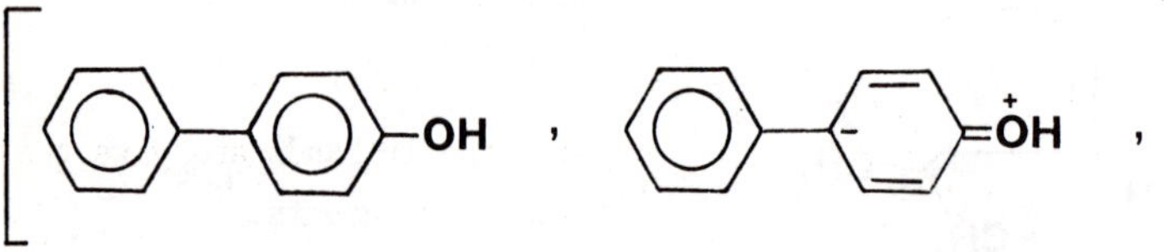

major dipolar forms

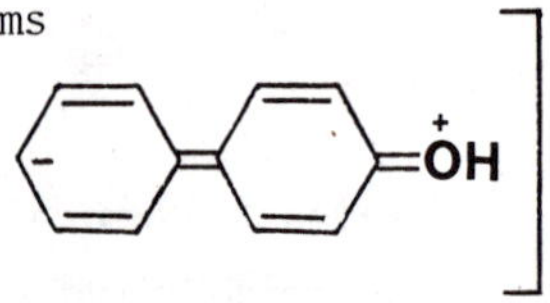

minor dipolar forms

3'-Nitro group causes only an insignificant inductive effect.
4'-Nitro group:

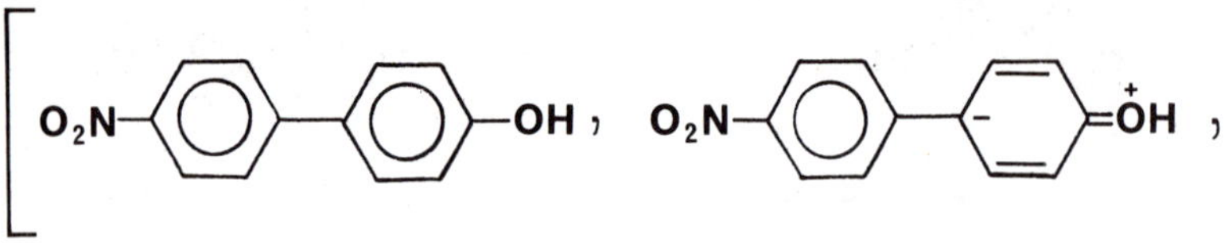

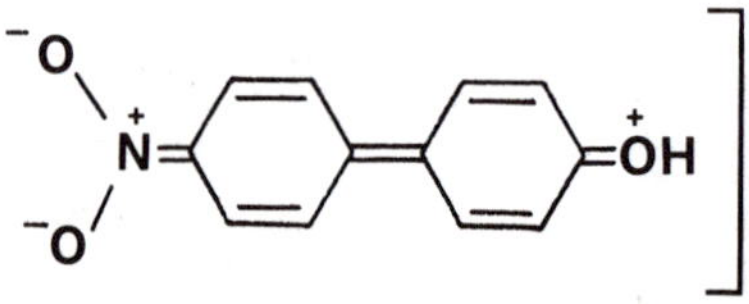

isovalent interaction resonance form

3. Possible resonance forms are

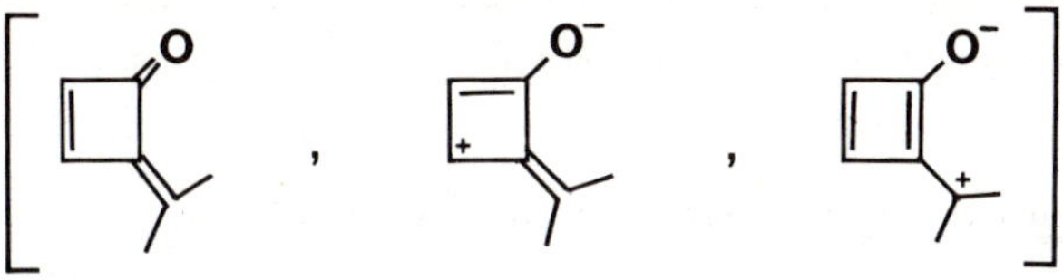

[(a) , (b) , (c)]
I

Form <u>Ic</u> makes virtually no contribution because of the instability of the cyclobutadiene ring. Consequently, the molecule acts as an α,β-enone with a nonconjugated olefin bond. The ir assignments are:

$$1640 \text{ cm}^{-1} \quad \text{endo C=C}$$
$$1680 \quad \text{exo C=C}$$
$$1740 \quad \text{C=O}$$

Form <u>Ib</u> lowers the C=O value from that of cyclobutanone (1790) and also makes the endo C=C frequency smaller than that of the exo C=C. The relative pmr shifts for ring protons reflect the charge distribution in <u>Ib</u>. The uv data bear the expected resemblance to those for 3,4-dimethylcyclobutenedione (λ_m^{EtOH} 216 and 340 nm; log ε = 4.27 and 1.41, respectively) for the cyclobutenone chromophore.

4. The change in chemical shift Δ is proportional to the total change in charge, Δq.

$$\Delta\delta_{1,3,5} = -0.5 - 56.4 = -56.9 = \underline{n\Delta q}_{1,3,5}$$
$$\Delta\delta_{2,6} = 58.3 - 66.1 = -7.8 = \underline{n\Delta q}_{2,6}$$
$$2\underline{\Delta q}_{2,6} + 3\underline{\Delta q}_{1,3,5} = 1 \text{ (unit of positive charge)}$$

Therefore we have three unkowns, $\underline{n}$, $\underline{\Delta q}_{2,6}$, and $\underline{\Delta q}_{1,3,5}$, and three equations. Solving for these three quantities gives $\underline{n}$ = 186.3; $\underline{\Delta q}_{2,6} \simeq 0.02$; $\underline{\Delta q}_{1,3,5} \simeq 0.32$. Thus, carbons 1,3, and 5 carry most of the charge, which can be accounted for by the resonance forms:

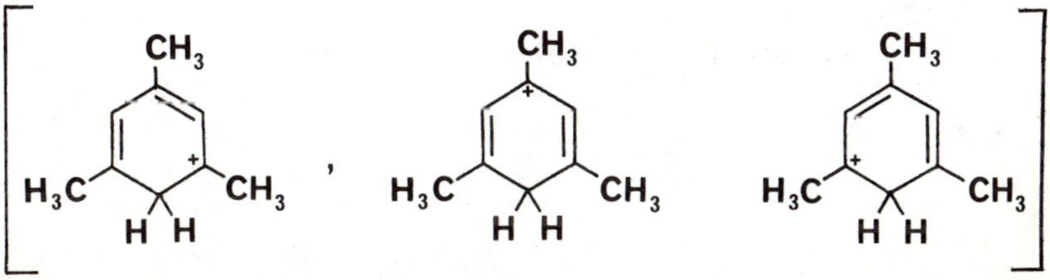

Uv spectroscopy is particularly good for distinguishing the various intramolecular relationships between chromophores, i.e., insulated, cumulated, conjugated, cross-conjugated, spiroconjugated, homoconjugated, transannular conjugated, etc.

<u>Insulated chromophores</u>. Insulated chromophores have no effect on each other and act as if they were in separate molecules. For instance, the rubber hydrocarbon has hundreds of C=C bonds per molecule but it is still colorless and has a λ_m in the region of that of ethylene. Similarly, λ_m for the <u>meta</u>-polyphenyls is essentially the same as that of biphenyl because the conjugation extends only through any two adjacent rings. Thus, for two insulated chromophores, λ_m is about the same as it is when the chromophore is single, but since there is a two-fold chance of excitation, the intensity of absorption for the insulated chromophores will be doubled.

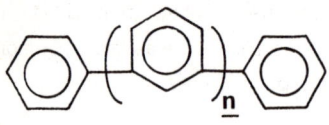

Meta-polyphenyls

n	λ_m
0	252 nm
1	252
7-12	253
13	254
14	255

	λ_m	ε
$CH_3-(CH=CH)_2-CHOH-C\equiv CH$	230 nm	28,500
$CH_3-(CH=CH)_2-CHOH-C$ $|||$ $CH_3-(CH=CH)_2-CHOH-C$	229	74,000
[2-hydroxy-3-methylphenyl-N=N-p-tolyl]	380	14,000
[2-hydroxy-3-methylphenyl-N=N-p-xylylene]$_2$	380	32,000

	λ_m	ε
Ph-CH=N-CH$_3$	247 nm	17,200
[Ph-CH=N-CH$_2$-]$_2$	247	29,800

<u>Cumulated chromophores</u>. The π orbitals in cumulated chromophores are perpendicular to each other so there is no formal overlap. However, in ketones, the lone-pair orbitals are in a position to overlap the olefin π orbital and produce red-shifts relative to the isolated chromophores.

	λ_m	ε	λ_m	ε
Et-CH=C=CH$_2$	170 nm	4,000		
Et$_2$C=C=O	227	360	375 nm	20

<u>Conjugated chromophores</u>. We have already noted the effect of conjugation on π→π*, n→π*, and benzene bands. Thus, conjugation increases λ_m, and another example can be given here.

		C=C	N=N
λ_m Isolated		185 nm	347 nm
λ_m Conjugated	π→π*	230	
	n→π*		400

One aspect of conjugated chromophores which has drawn considerable attention is the effect of lengthening the conjugated chain. Many different systems have been examined. For example, λ_m values for some linearly conjugated polynuclear hydrocarbons are given in Table 10.2. Note the difference between the <u>conjugated</u> <u>para</u>-polyphenyls and the <u>insulated</u> metapolyphenyls. In general, λ_m is a simple function of n, where <u>n</u> is the number of conjugated units, C=C, C≡C, phenyl, etc. For instance, λ_m is a linear function of <u>n</u> for the polycyanines <u>10.14</u> whereas λ_m^2 varies linearly with <u>n</u> for diphenylpolyenes and similar polyenes (19).

$$>\text{N-(CH=CH)}_{\underline{n}}\text{-CH=N}^+<$$

10.14

$CH_3-(CH=CH)_{\underline{n}}-CH_3$ $\lambda_m^2 = (2.31\underline{n} + 0.09)10^4$ nm

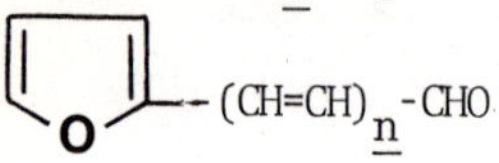

furan-$(CH=CH)_{\underline{n}}-CHO$ $\lambda_m^2 = [2.3(\underline{n} + 1.67) + 2.65]10^6$ nm

$[C_6H_5-(CH=CH)_{\underline{n}}-CH=N-CH_2]_2$ $\lambda_m^2 = [2.15(\underline{n} - 1.4) + 9.06]10^6$ nm

$C_6H_5-(CH=CH)_{\underline{n}}-CH=N$
$C_6H_5-(CH=CH)_{\underline{m}}-CH=N$ $\lambda_m^2 = [2.15(\underline{n} + 0.35\underline{m}) + 9.06]10^6$ nm

Polyenic cations (20) $\lambda_m = 75\underline{n} = 250$ nm

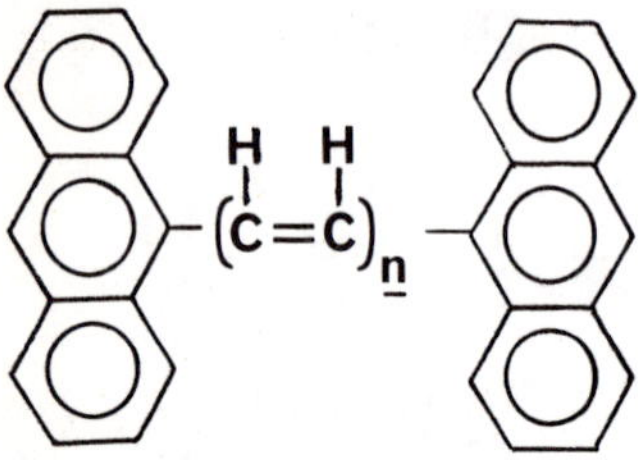

anthryl-(C=C)$_{\underline{n}}$-anthryl $\lambda_m^{cyclohexane} = 1.5\underline{n}^2 + 461$ nm (21)

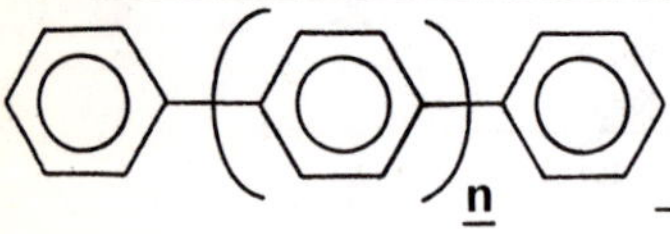

Table 10.2. λ_m of some linearly conjugated polynuclear hydrocarbons

$C_6H_5-(C_6H_4)_{\underline{n}}-C_6H_5$		Structure	Compound	Visible color	λ_m
$\underline{n}$	λ_m				
0	252 nm	benzene ring	Benzene	Colorless	203 nm
1	280				
2	300	2 fused rings	Naphthalene	White	314
3	310				
4	318	3 fused rings	Anthracene	Buff	370
		4 fused rings	Naphthacene	Yellow	460
		5 fused rings	Pentacene	Blue	580
		6 fused rings	Hexacene	Green	600

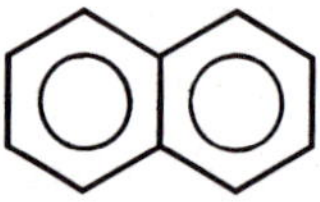

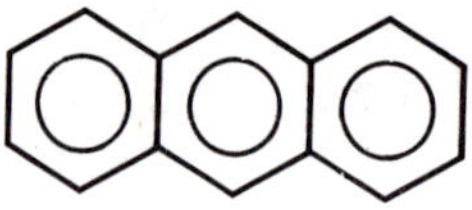

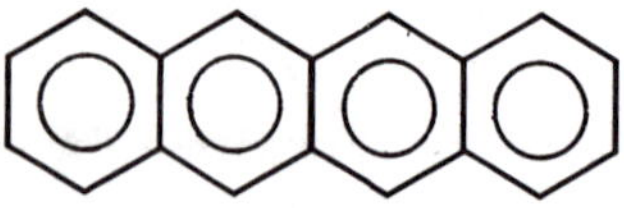

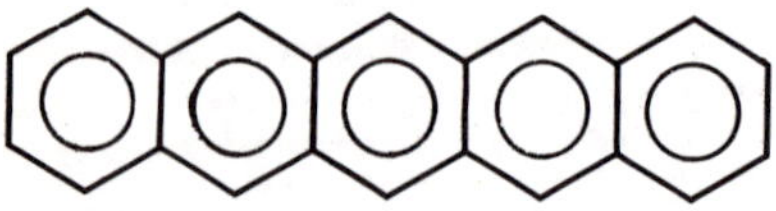

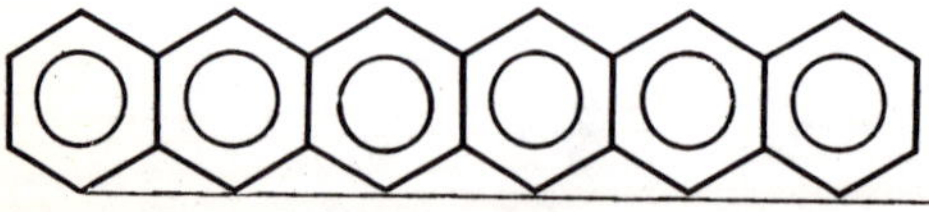

Ch. 10 SPECTRA AND ELECTRON DELOCALIZATION   341

QUESTIONS

1. Why should λ_m increase so much more rapidly with increasing $\underline{n}$ for the series of polycyanines (10.14) than for a series of polyenazines ($\underline{I}$) ?

$$C_6H_5-(CH=CH)_{\underline{n}}-CH=N-CH_3$$

$$\underline{I}$$

2. Predict the order of λ_ms for the following compounds

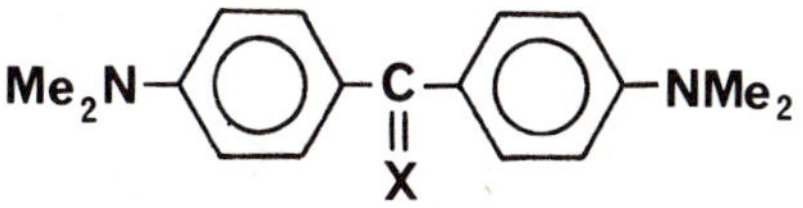

   Me$_2$N–⟨◯⟩–C(=X)–⟨◯⟩–NMe$_2$ X = S, NH, CH$_2$, O

3. To what extent will the uv spectra of the ethers below resemble one another ? Give the basis for your expectation.

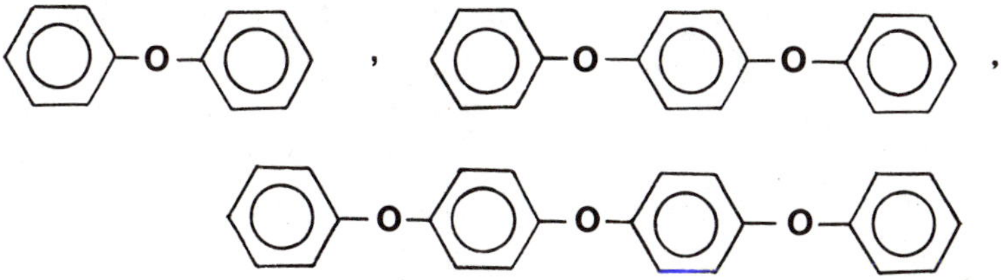

   ⟨◯⟩–O–⟨◯⟩ , ⟨◯⟩–O–⟨◯⟩–O–⟨◯⟩ ,

   ⟨◯⟩–O–⟨◯⟩–O–⟨◯⟩–O–⟨◯⟩

4. The deviation of λ_m of an unsymmetrical cyanine dye $R_2N-(CH=CH)_{\underline{n}}-CH=NR'_2{}^+$ from the arithmetic mean of the λ_ms of the two related symmetrical dyes, $R_2N-(CH=CH)_{\underline{n}}-CH=NR_2{}^+$ and $R'_2N-(CH=CH)_{\underline{n}}-CH=NR'_2{}^+$, is related to the difference in basicities of the two symmetrical dyes. Explain.

5. Since cis stilbene has a smaller λ_m than the trans isomer, how do you account for λ_m of tetraphenylethylene being close to that of trans-stilbene ?

ANSWERS

1. Resonance of the polyenazines is among dipolar structures which stabilize the excited states more than ground states. As $\underline{n}$ increases, E_1

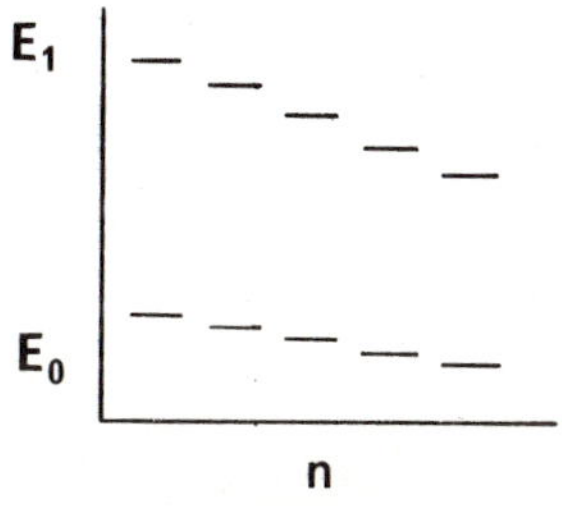

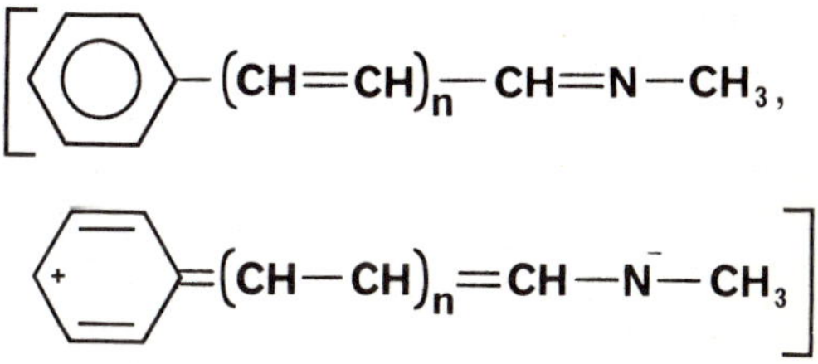

   [⟨◯⟩–(CH=CH)$_{\underline{n}}$–CH=N–CH$_3$,

   ⟨◯⁺⟩=(CH–CH)$_{\underline{n}}$=CH–N–CH$_3$]

and E_0 each get smaller, but E_1 decreases faster than E_0, to increase λ_m. In the case of the polycyanines, the major resonance is of two types, one with the positive charge on N and the other with the charge on the intervening carbons. For example,

$$R_2N\text{-CH=CH-CH=CH-CH=}\overset{+}{N}R_2, \quad R_2\overset{+}{N}\text{=CH-CH=CH-CH=CH-NR}_2,$$
$$\text{(a)} \qquad\qquad\qquad\qquad \text{(b)}$$

10.14

and $\quad R_2N\text{-CH=CH-CH=CH-}\overset{+}{\text{CH}}\text{-NR}_2, \quad R_2N\text{-CH=CH-}\overset{+}{\text{CH}}\text{-CH=CH-NR}_2, \quad R_2N\text{-}\overset{+}{\text{CH}}\text{-CH=CH-CH=CH-N}$
$$\text{(c)} \qquad\qquad\qquad\qquad \text{(d)} \qquad\qquad\qquad\qquad \text{(e)}$$

10.14

Structures <u>10.14a</u> and <u>10.14b</u> represent stable ground states of the salts, and as <u>n</u> increases there is less resonance between the two because in-- creasingly more electron delocalization must occur to go from one structure to the other. This destabilizes the ground state. However, the forms with the charge on carbon (<u>10.14c</u> to <u>10.14e</u>) make a greater contribution to the excited state and as <u>n</u> increases, more such forms are possible. Thereby, the excited states are increasingly stabil- ized with increasing <u>n</u>. Since the two energy levels approach each other with increasing <u>n</u>, λ_m for the polycyanines gets larger faster than for the poly- enazines, <u>I</u>.

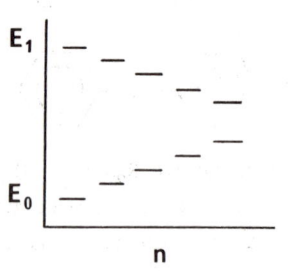

2. On a basis of the chromophoric ability of the respective chromophores, we could expect a decreasing order of λ_ms to be X = S > O > NH > CH_2. (A. Burawoy, <u>Ber</u>. 63: 3155 (1930).)

3. There is no complementary resonance between atoms attached to the benzene rings so all have the same chromophoric unit as the diphenyl ether and the same λ_m. With more such units, the chance for excitation increases thereby increasing the extinction coefficient. (E. Mayer-Pitsch, <u>Z. Elektrochem</u>. 49: 368 (1943).)

4. The degree of resonance with the positive charge on either nitrogen atom of a given cyanine dye is greater as the two NR_2 groups are more equal in basicity. As one group gets more basic, the form with the positive charge on that nitrogen makes a greater and greater contribution to the

resonance hybrid and resonance delocalization diminishes. This has a related effect on λ_m.

5. One phenyl group of each carbon rotates perpendicular to the plane of the molecule such that the chromophoric system is essentially the same as in trans-stilbene.

Molecules with extended conjugation in two dimensions have two optical axes. When electronically excited, they give rise to two absorption bands (22). It is common to associate the band of longest wavelength with the direction of greater polarizability and call it the x band. The other is called the y band. Three examples are tetracyclones (10.15) (23), benzophenones (10.16) (24), and triphenyl carbocations (25). For example,

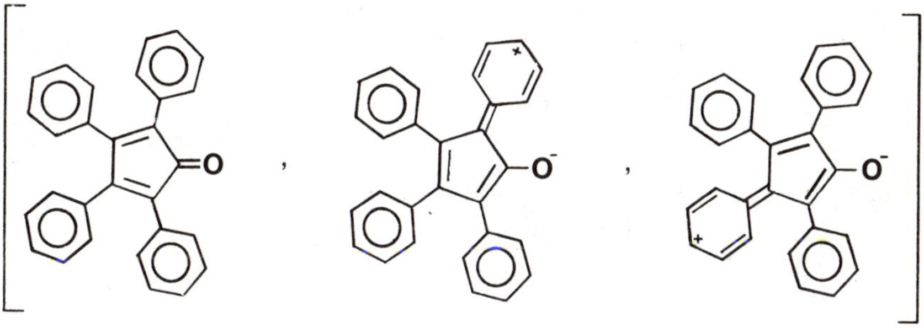

10.15

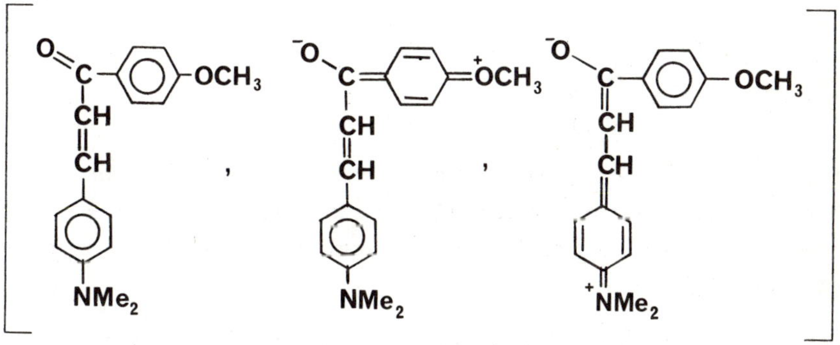

10.16

crystal violet (10.17) is symmetrical so its x and y bands coincide and it only has one band in the 600-nm range. Malachite green (10.18) should have an x band of wavelength near that of crystal violet and a y band of shorter wavelength. Methoxymalachite green (10.19) should also have an x band near that of crystal violet but a y band of longer wavelength than the y band of

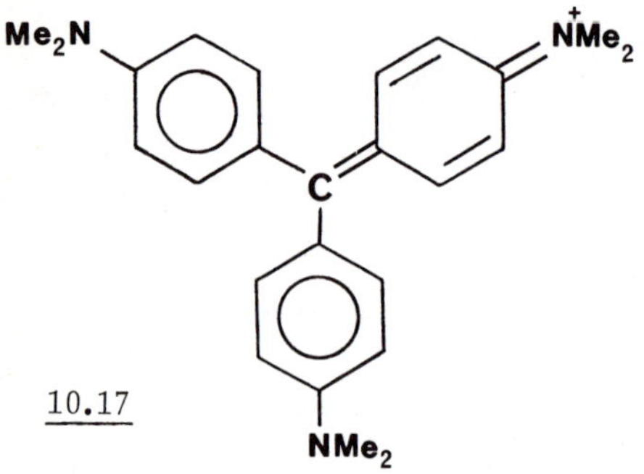

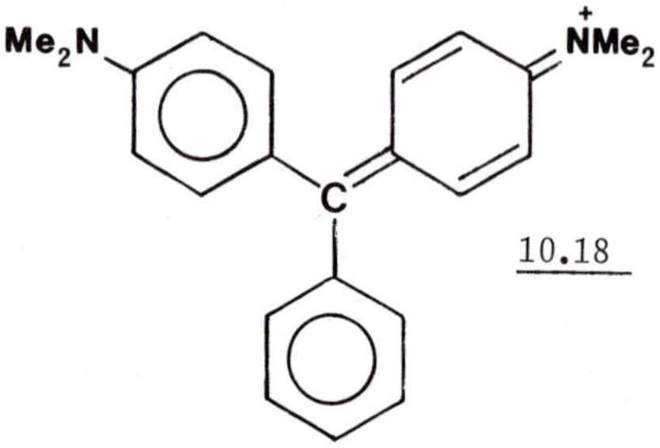

10.18 because the OCH$_3$ group increases the chromophoric power in the y direction. Finally, Michler's hydrol (10.20) should have an x band near that of crystal violet but no y band. These expectations are confirmed by the spectral data (26):

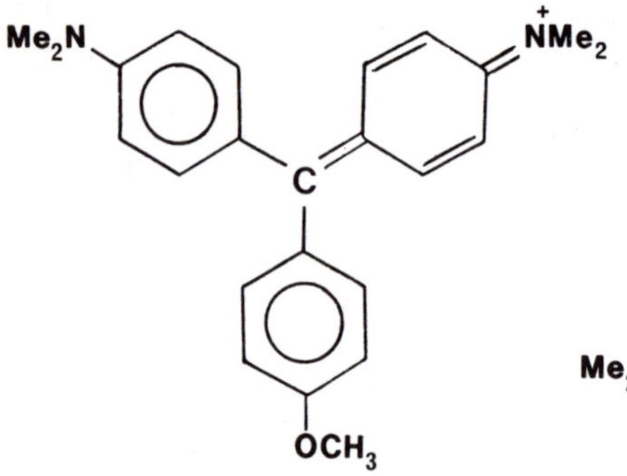

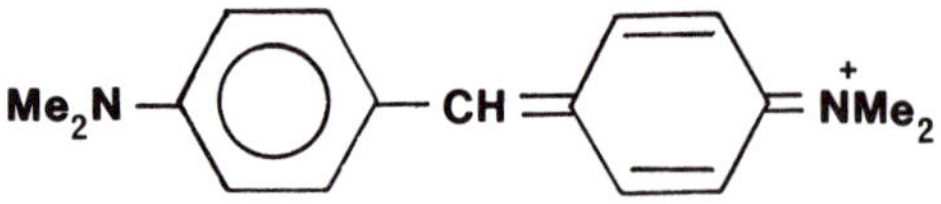

Compound	x band	y band
Crystal violet (10.17)	590 nm	
Malachite green (10.18)	621	428 nm
Methoxymalachite green (10.19)	608	465
Michler's hydrol (10.20)	595	

Cross-conjugated chromophores. Cross-conjugated chromophores are somewhat like overlapping insulated chromophores in which each gives rise to an absorption band. In some molecules one chromophore becomes excited and in other molecules it is the other chromophore which absorbs the incident light. Hence, the compound gives two bands. An example is the benzophenone system (27). A symmetrical molecule has only one benzoyl chromophore but an unsymmetrical benzophenone has two different benzoyl chromophores and exhibits a band for each. Some illustrative data are given in Table 10.3.

Another system of cross-conjugated chromophores is that of benzalanilines. Diffraction measurements show that benzalaniline has a trans, nonplanar structure in the solid state (28). If the molecule were completely planar, we might expect a spectrum similar to that of stilbene, ØCH=CHØ or azobenzene, ØN=NØ. This is not the case. What is observed are bands arising from the benzal system and other bands from the aniline fragment. Some data to illustrate this are given in Table 10.4. Bands I and III are assigned to the aniline chromophore and bands II and IV to the benzal (Ph-CH=N) chromophore. Thus, we find as Y is changed in 10.21 only bands I and III show a change whereas bands II and IV remain unaltered, and in 10.22 only bands II and IV undergo a shift as X is changed. As expected, benzalmethylamine, $C_6H_5CH=N-CH_3$, has only bands II and IV.

A simple cross-conjugated system is found in phenyl esters. The acyl group is almost perpendicular to the plane of the benzene ring and the lone-pair orbital on the phenol oxygen can overlap with either the aryl or the carbonyl orbitals but not both. Spectral, and other,

Ch. 10 SPECTRA AND ELECTRON DELOCALIZATION

Table 10.3. Spectral data for some substituted benzophenones (29)

Substituents	λ_m (nm)	λ_m' (nm)	ε	ε'
H,H		248	20,000	
4-OCH$_3$, H		247 274	10,500	17,000
4-OCH$_3$, 4'-OCH$_3$		278		27,000
4-Cl, 4'-Cl		261	27,500	
4-Cl, 4'-OCH$_3$		257 280	15,000	17,500
3-NO$_2$, H	229	246	24,000	19,500
3-NO$_2$, 4'-OCH$_3$	225	286	30,000	16,000

Table 10.4.

Uv bands for CH$_3$O—⟨O⟩—CH=N—⟨O⟩—Y, in cyclohexane (30)

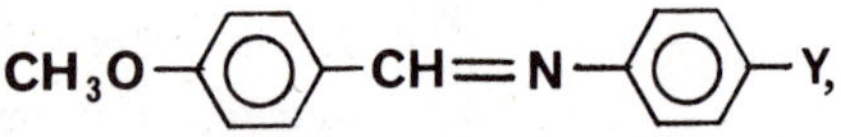

Y	I λ_m (nm)	ε	II λ_m (nm)	ε	III λ_m (nm)	ε	IV λ_m (nm)	ε
H	315	12900	280	19600	237sh		222	18600
CH$_3$	323	14500	280	20500	240sh		222	18400
OCH$_3$	330	17900	280	22300	244sh		222	20300

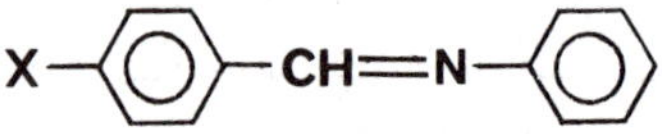

X—⟨O⟩—CH=N—⟨O⟩   (10.22 in EtOH)

X								
H	311	9000	262	16700	238sh		218	15000
OCH$_3$	311	18100	290	18600	238sh		223	17300
NMe$_2$	313sh		356	39100			238	15100

evidence indicates that it is conjugated with the carbonyl group, i.e., acyl resonance predominates over phenol resonance (31).

Ch. 10   SPECTRA AND ELECTRON DELOCALIZATION   347

<u>Spiroconjugated chromophores</u>. There still may be electronic interaction between π or lone-pair electron systems or both, when separated by a spiro carbon atom as in <u>10.23</u> or <u>10.24</u>.

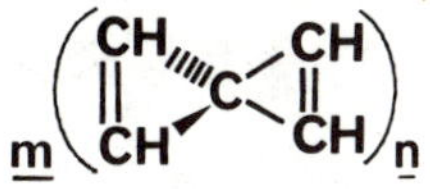

<u>10.23</u> 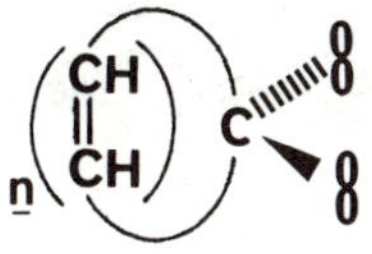   <u>10.24</u>

The effects of such spiroconjugation upon spectra, dipole moments, and other properties depend upon the values of <u>m</u> and <u>n</u> (32). For example, when the number of π electrons in <u>m</u> and <u>n</u> total 4N (N = 0, 1, 2, ...), there is observed a red-shift relative to a reference nonconjugated compound; and when <u>m</u> plus <u>n</u> = 4N + 2, a blue-shift results:

Reference system   4N + 2 system   4N system

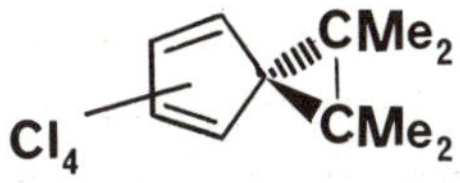

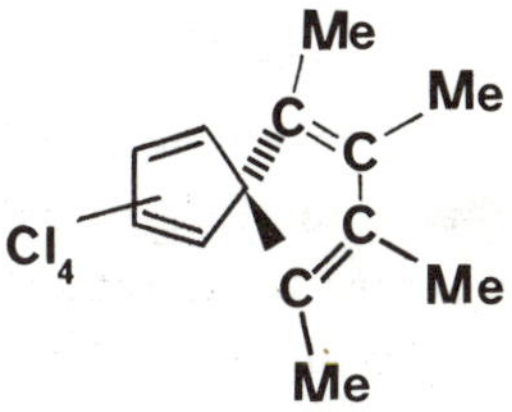

λ_m^{EtOH} (33) 296 nm   284 nm   316 nm

ε   1738   3400   51,600

                    Reference system   4N system

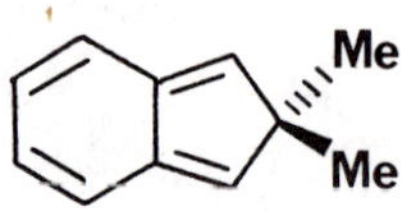

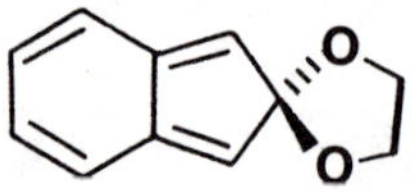

C_6H_6 (34)

λ_m 444 nm   537 nm

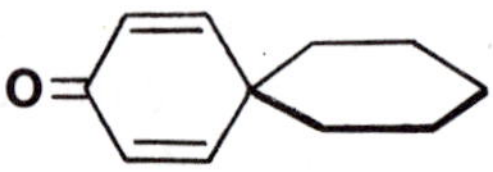

                          10.25   10.26

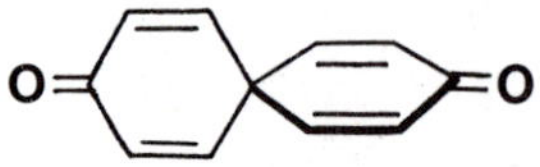

λ_m $\pi\to\pi^*$ 174 nm   200 nm
              $\pi\to\pi^*$ 233   228
              $n\to\pi^*$ 322   313

Here, the $n\to\pi^*$ and first $\pi\to\pi^*$ bands of 10.26 suffer a small blue-shift relative to 10.25, possibly attributable to the electronegative effects of the second unsaturated ring. However, the short-waved $\pi\to\pi$ band exhibits a bathochromic shift due to spiroconjugation (35).

Interestingly, no evidence was found for an unusual stability of carbanion 10.27 compared to 10.28 or 10.29 (36).

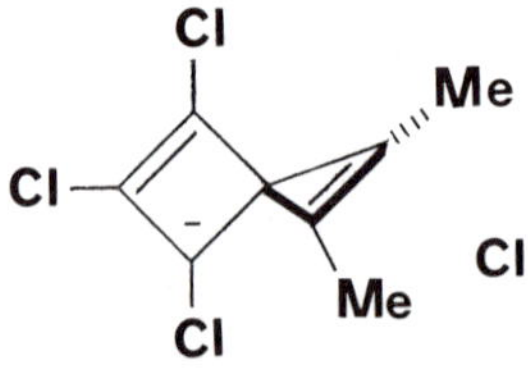

             10.27 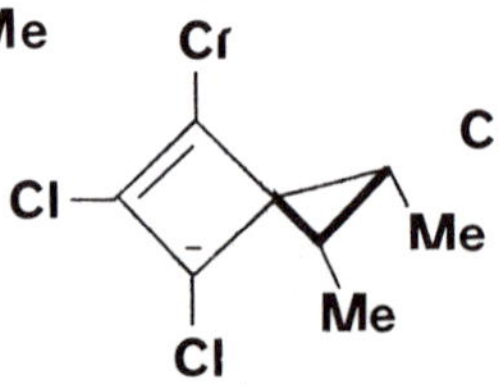   10.28 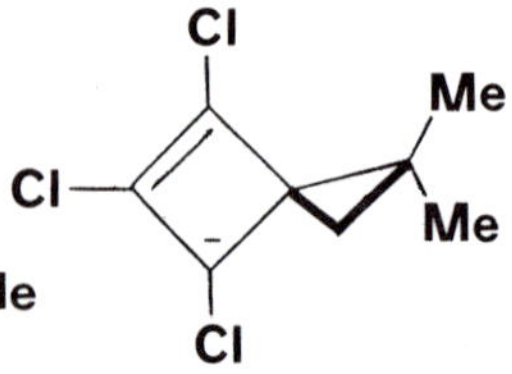   10.29

Ch. 10 SPECTRA AND ELECTRON DELOCALIZATION

That is <u>10.27</u> showed no sign of <u>spiroaromaticity</u>.

Spiroconjugation has also been detected by pe spectroscopy. The spectral characteristic sought is a splitting into doublets of the pes band or bands of a nonspiroconjugated reference molecule. This is observed for the spiranes listed below (37). For example, indene exhibits three bands with IPs of 8.13, 8.95, and 10.29 eV, respectively. For comparison, in the pe spectrum of <u>10.30</u> the three bands of indene are split into doublets at 7.80 and 8.37; 8.80, 9.10, and 10.20, 10.48 eV (37).

Spiroconjugated	Reference molecule
$(H_2C=CH-)_4-C$	$(H_2C=CH)_2-CMe_2$
	$(H_2C=CH)_2-CMe_2$
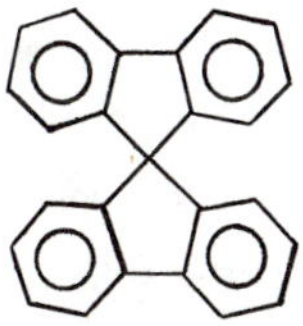	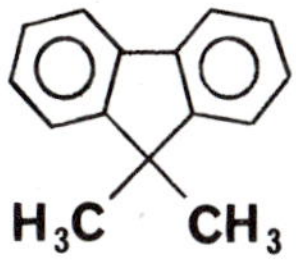 $H_3C \quad CH_3$
$(H_2C=CH)_4-Si$	$(H_2C=CH)_2-SiMe_2$
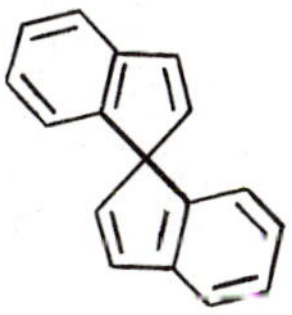 10.30	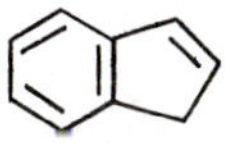 Indene

350   Ch. 10 SPECTRA AND ELECTRON DELOCALIZATION

<u>Transannular conjugated</u>. This is the π overlap (parallel orbitals) between unattached π systems as in <u>10.31</u> (38) and <u>10.32</u> (39).

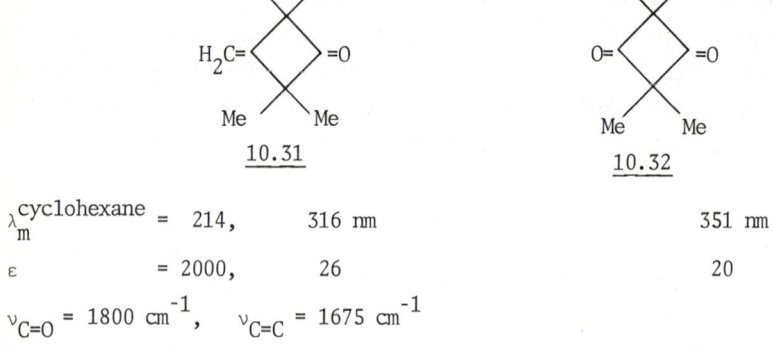

$\lambda_m^{cyclohexane}$ = 214,   316 nm   351 nm

ε   = 2000,   26   20

$\nu_{C=O}$ = 1800 cm^{-1}, $\nu_{C=C}$ = 1675 cm^{-1}

The orbital interaction in <u>10.32</u> is also confirmed by its PE spectrum but, on a firmer theoretical basis, is regarded as a "through-bond" σ-π orbital overlap (40).

<u>Homoconjugated</u>. Homoconjugation, the nonlinear σ overlap of π orbitals (Chapter 8) has been used to account for the marked stability and solvolysis products of various nonclassical carbocations (41), for enhanced solvolysis rates, and for certain photochemical and spectral properties of β, γ-enones, β,γ -ketimines, 1,4-dienes, β-arylketones, diarylmethanes, and β-ketocyclopropanes. For example, homoconjugation produces increased photochemical reactivity over the corresponding nonconjugated analogs (42) (at certain wavelengths of light) and spectral effects such as enhanced intensity of uv absorption (usually accompanied by small bathochromic shifts (43, 44)), greater rotatory power as measured by cd and ord (45), and lower photoionization potentials (46). It appears that homoconjugation occurs primarily in the excited state so that ground-state properties are unaffected. Thus, uv and PE but not significant ir spectroscopic changes are observed relative to nonconjugated systems:

Ch. 10 SPECTRA AND ELECTRON DELOCALIZATION

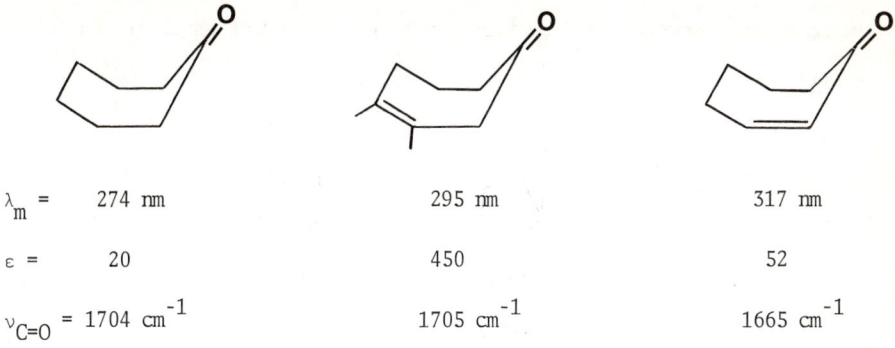

λ_m = 274 nm	295 nm	317 nm
ε = 20	450	52
$\nu_{C=O}$ = 1704 cm^{-1}	1705 cm^{-1}	1665 cm^{-1}

Similarly, cycloct-3-enone, λ_m^{EtOH} 286 nm, ε 114, exhibits homoconjugation (47). We see that the homoconjugated enone system differs from the transannular system by the intensity of the carbonyl peak, whereas both are distinguished from the classically conjugated system by relatively unchanged ir absorption. These spectral differences are summarized in Table 10.5.

Table 10.5. Spectral differences of isolated and conjugated enone systems

	Isolated	α,β-conjugated	Trans-annular conjugated	Homo-conjugated
λ_m(C=O)	~280 nm	~300 nm	~300 nm	~300 nm
ε	~20	<50	<50	>90
$\nu_{C=O}$	1700-1850	markedly lowered	little change	little change
$\nu_{C=C}$	1620-1680	markedly lowered	little change	little change

On this basis, for example, there must be transannular rather than homo-conjugation in trans-5-cyclodecenone (10.33) (see also compound 10.35). The red-shift of the C=C band indicates some type of delocalization with the C=O group and the normal intensity of the C=O band shows that it is trans-annular. Hence, the π orbitals must be parallel. Numerous other

10.33 (48)
λ_m 214.5, 270-280 nm
ε 2300, 20

Ch. 10 SPECTRA AND ELECTRON DELOCALIZATION

examples of homoconjugated systems can be given, some of which are as follows.

Enones

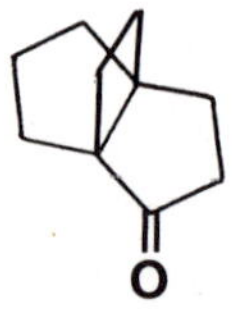

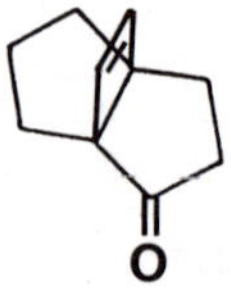

λ_m	300 nm	307 nm
ε	22	110

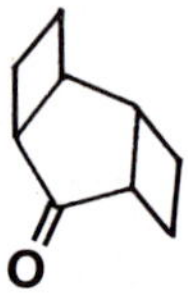

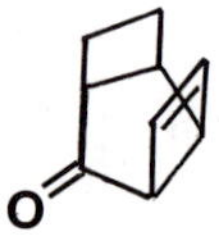

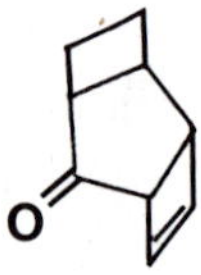

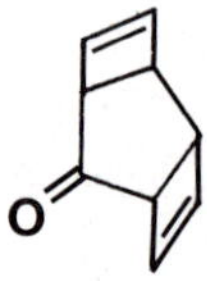

λ_m^{EtOH}	306 nm	310 nm	310 nm	306 nm
ε	22	102	146	267
$\nu_{C=O}^{CCl_4}$	1730 cm^{-1}	1725 cm^{-1}	1725 cm^{-1}	1725 cm^{-1}

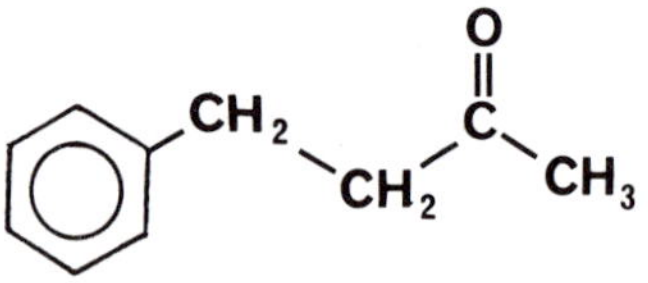

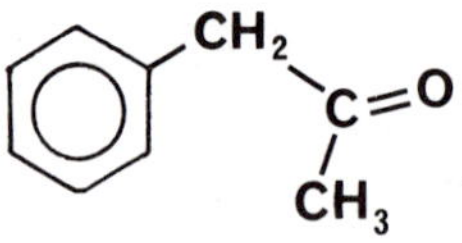

$\lambda_m^{cyclohexane}$	281 nm	289 nm
ε	32	108

$\lambda_m^{hydrocarbon}$	295 nm		215, 304 nm
ε	23		2800, 290

Ketimines

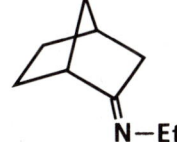

	$\pi \to \pi^*$	$n \to \pi^*$	$\pi \to \pi^*$	$n \to \pi^*$
$\lambda_m^{isooctane}$	209	249	214	249.5 nm
ε	127	200	1695	495

Homoenamines

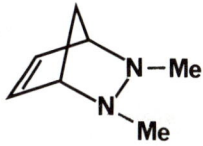

λ_m	270 nm	(This band is absent when in HCl solvent, and for the 5,6-dihydro derivative (49).)
ε	600-700	

Dienes

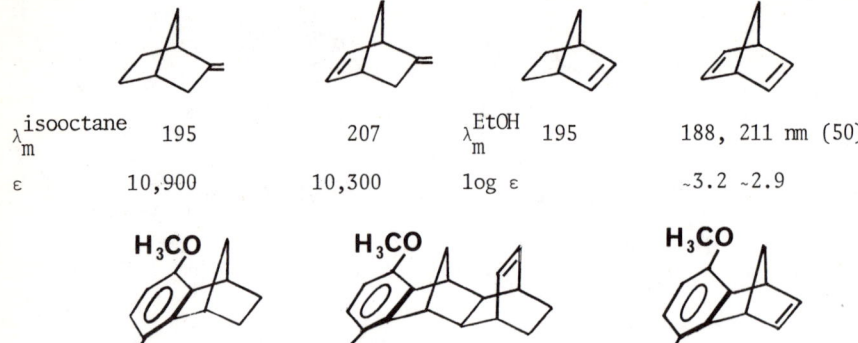

$\lambda_m^{isooctane}$	195	207	λ_m^{EtOH} 195	188, 211 nm (50)
ε	10,900	10,300	log ε	~3.2 ~2.9

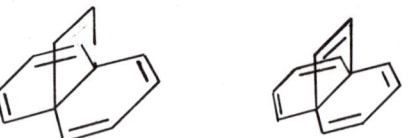

λ_m' (51)	312 nm	314 nm	328 nm
λ_m''	284	285	294

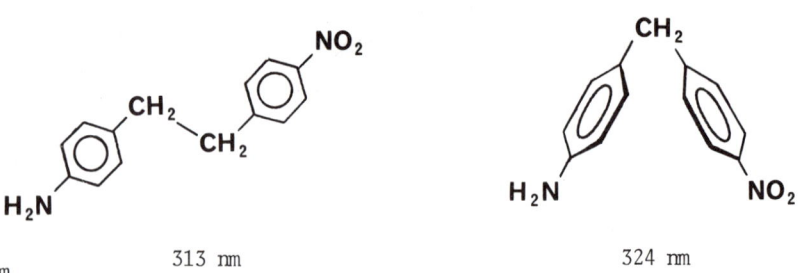

λ_m^{alkane} (52)	235, 285 nm	247, 290 nm
ε	4700, 2600	2500, 2100

λ_m 313 nm   324 nm

Ch. 10 SPECTRA AND ELECTRON DELOCALIZATION

It is also important to point out here that nonconjugated π orbitals may interact through σ bonds as well as through space (homoconjugation). A theoretical basis has been given for through-bond overlap of interacting π-electron systems (53). Experimental evidence has come from uv data. For example, some spectral properties of the diketones 10.34 to 10.36 are given below (54).

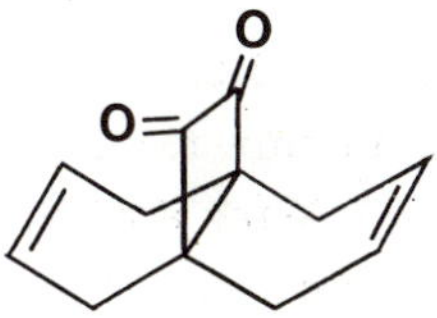

	10.34	10.35	10.36
λ_m =	538 nm	460-4, 532-5 nm	461 nm
ε =	71.7	38.8 32	73
$\nu_{C=O}$	1794, 1759 cm^{-1}	1772 cm^{-1}	1812, 1772 cm^{-1}

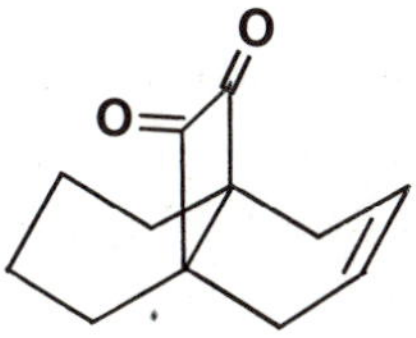

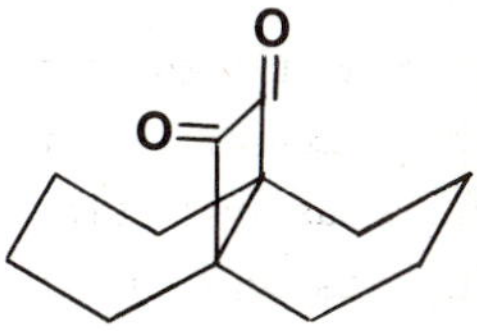

Thus, 10.34 and 10.35 exhibit a red-shift relative to 10.36 but no enhancement of the carbonyl absorption. Compound 10.35 has two peaks, one each corresponding to the chromophores in 10.34 and 10.36 and each peak in 10.35 of approximately half the intensity of those in 10.34 and 10.36. MO calculations (54) indicate that the spectral effect is through-bond rather than through-space. From this one study, one might expect that uv spectral effects of the two modes of interaction will differ in that there will be an enhancement of band intensities accompanying through-space interactions but not through-bond interactions. (See also reference 58 regarding the MO calculations on 10.51 and its pe spectrum.)

PE spectroscopy has provided the first experimental evidence for homoconjugation in cis,cis,cis-1,4,7-cyclononatriene (10.37). With a strong

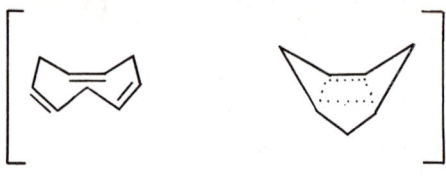

10.37

interaction between the three double bonds, the system could be regarded as a trishomoaromatic ring, a concept first advanced by Winstein (55). The triene was synthesized independently by three groups and X-ray analysis confirms its conformation. However, there is no indication of homoconjugation in 10.37 from bond lengths, nmr chemical shifts (δ_{CH_2} (exo), 2.26 ppm; δ_{CH_2} (endo), 3.95 ppm; $\delta_{=CH}$ 5.56 ppm), or heat of hydrogenation (-76.9 kcal/mol; cis-cyclononene = -23.6 kcal/mol). In fact, ΔH_{H_2} denotes that there is some strain in 10.37, probably transannular between inner C-Hs which are a little closer together (1.95Å) than accepted van der Waals distances (2.2-3.0 (57). MO calculations show that the delocalization energy (DE) in 10.37 is only 2.4 kcal/mol, which is too small to be detected by most methods in the presence of the apparent strain in the molecule. The PE spectral data (IP in eV) for 10.37 and some related hydrocarbons are as follows (56).

10.37	10.38	10.39	10.40	10.41
8.77 ⎫ π	8.97 π	8.30 π	8.36 ⎫ π	9.02 π
8.9-9.0 ⎭	10.55 σ	10.60 π	9.02 ⎭	12.12 σ
9.8 π		11.16 σ	10.55 π	
11.32 σ			11.28 σ	

The π-orbital interaction in 10.37 is a little stronger than that in norbornadiene and bicyclo(2.2.2)octadiene, compared to a much larger value in the 1,3-diene 10.40.

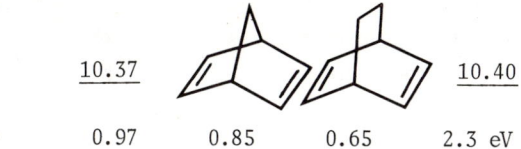

ΔIP =	10.37			10.40
	0.97	0.85	0.65	2.3 eV

From MO calculations it is deduced that the interaction in 10.37 is entirely homoconjugation with no or very little through-bond interaction and that homoconjugation also dominates the π-orbital interaction in norbornadiene and bicyclo(2.2.2)octadiene (56).

The three series below are of interest, and the first few pes IPs in eV are given for each (58). The π-orbital potentials are indicated.

10.42	10.43	10.44	10.45	10.46	10.47
10.32	9.12 π	8.80 ⎫ π	10.17	8.97 π	8.69 ⎫ π
10.98	10.66	9.80 ⎭	10.70	10.55	9.55 ⎭
11.85	11.27	11.00	11.43	11.85	11.26
12.87	12.88	11.97			12.51

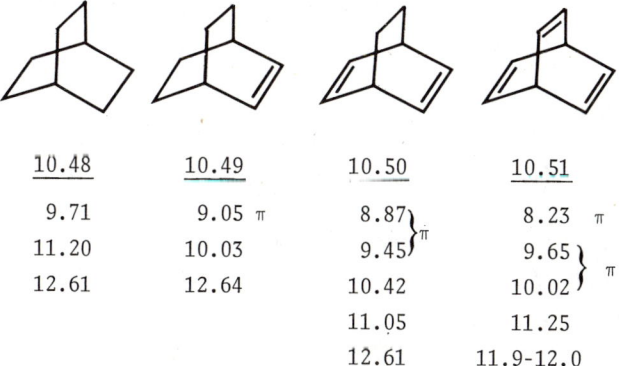

10.48	10.49	10.50	10.51
9.71	9.05 π	8.87 ⎫ π	8.23 π
11.20	10.03	9.45 ⎭	9.65 ⎫ π
12.61	12.64	10.42	10.02 ⎭
		11.05	11.25
		12.61	11.9-12.0

Observe that in each series (1) the peaks for the equivalent olefinic bonds of the dienes are split into a doublet with the average of the doublet about 0.1-0.2 eV higher than that for the corresponding monoene, and (2) the σ IPs increase with the rise in number of double bonds. These two generalizations can be illustrated by the following chart.

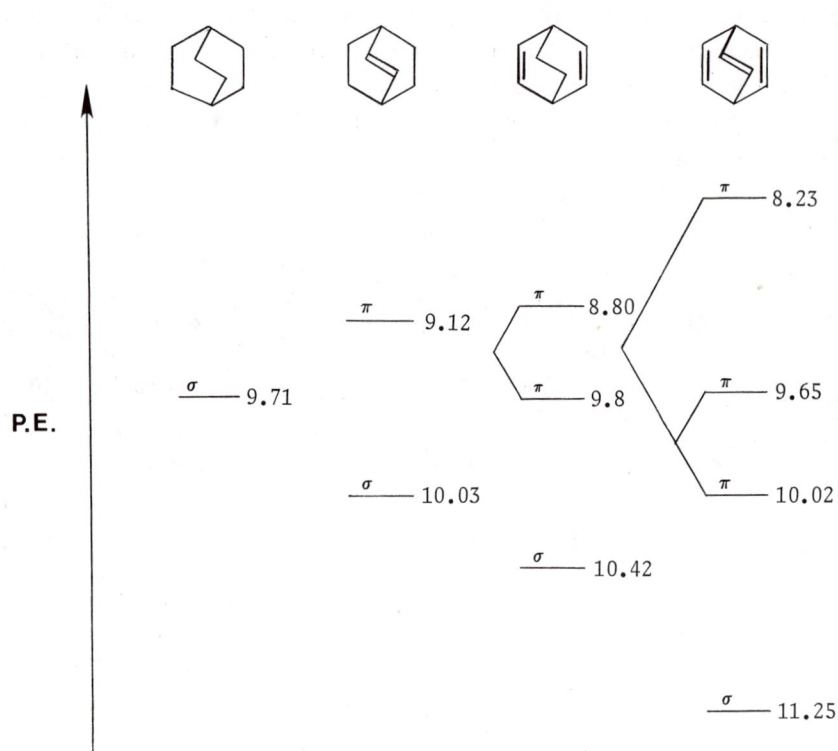

It was concluded from MO calculations on 10.50 and 10.51 that through-bond interactions of the π bonds is not negligible (58).

PE spectroscopy also provides evidence for inductive effects of sp^2-hybridized carbons on C=C or C=O ionization potentials.

Ch. 10 SPECTRA AND ELECTRON DELOCALIZATION   359

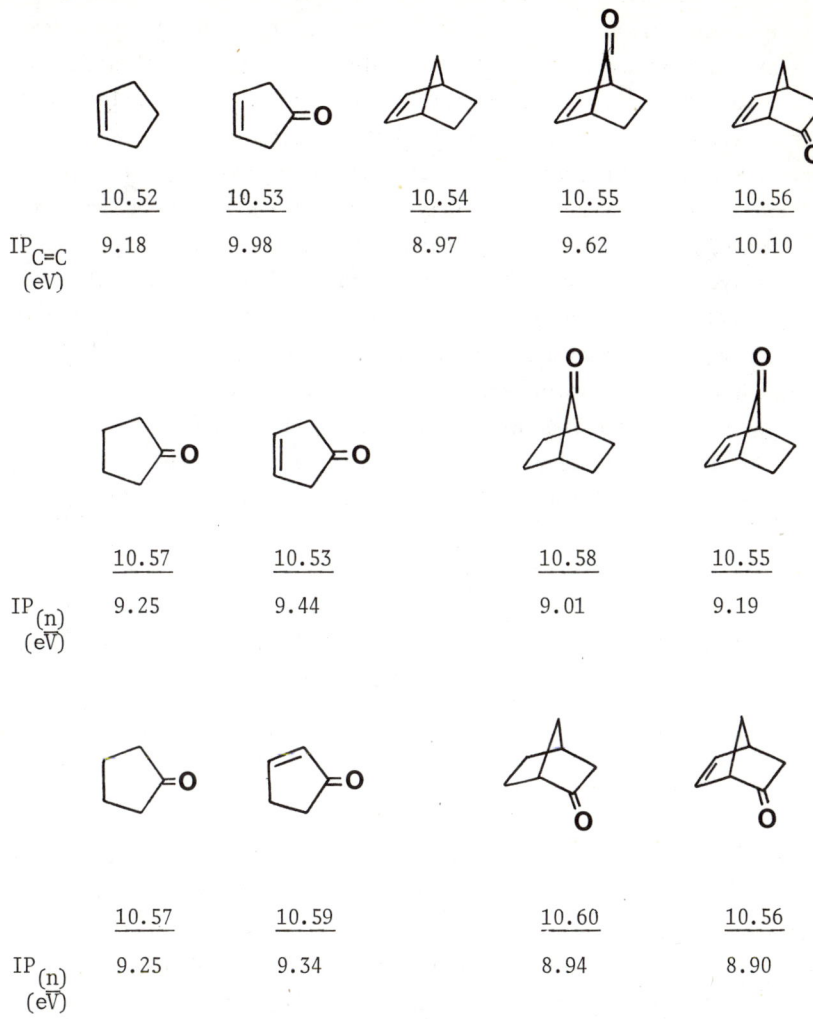

Thus, the $IP_{C=C}$ of 10.52 is increased by 0.80 eV in 10.53 and that of 10.54 is raised by 0.65 eV in 10.55 due to the β sp^2-hybridized carbonyl carbon. The combined inductive and homoconjugative delocalization effect in 10.56 increases further the $IP_{C=C}$ (by 1.13 eV). Similarly, the carbonyl n-level IPs of 10.53 and 10.55 are higher than those of the saturated ketones by virtually the same amount (0.18 eV). The carbonyl carbon has a greater IP-raising electronegative effect than two olefinic carbons because the former carries a partial positive charge ($^{\delta+}$C-O$^{\delta-}$). Delocalization lowers the n-level IP of the carbonyl group however. That is, $IP_{(n)}$ for 10.56 is lower than that of 10.60 and that of 10.59 is only 0.09 eV greater than that of 10.57. The delocalization is actually greater in 10.59 than in 10.56 but the inductive effect is also greater in 10.59 because the olefinic carbon is closer to the carbonyl group. It is tenuous to compare $IP_{C=C}$ or $IP_{(n)}$ values for different rings because variations in orbital hybridization from bond angle changes markedly affect the orbital energies.

Lastly, pe spectroscopic data have been added to nmr spectral evidence for a nonclassical structure of the 2-norbornyl cation. The 1s electron bonding energy of a charged carbenium center is much larger than that of a neutral carbon. This bonding energy difference, dE_b, almost 5 eV, shows up in the pe spectrum of carbenium ions such as the t-butyl cation. However, dE_b is significantly smaller for a nonclassical ion where the charge is delocalized on three or more carbons (59). Accordingly, the data support

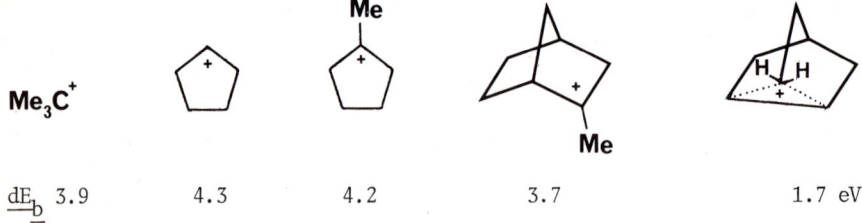

Me$_3$C$^+$				
dE_b 3.9	4.3	4.2	3.7	1.7 eV

the notion that the 2-methylnorbornyl cation is largely classical, with only a small degree of σ delocalization, whereas the 2-norbornyl cation has the nonclassical structure.* Rapidly equilibrating classical structures would give a dE_b close to that of classical structures, since pe spectra are fixed for processes of the order of 10^{-16} seconds on the time scale.

---

* The resolution in such spectra might make these decisions equivocal.

QUESTIONS

1. Predict the order of increasing $\Delta\nu_{OH}$ of phenol when equivalent quantities of phenol and **I** are mixed in CCl_4 solution, for R = H, NO_2, OCH_3.

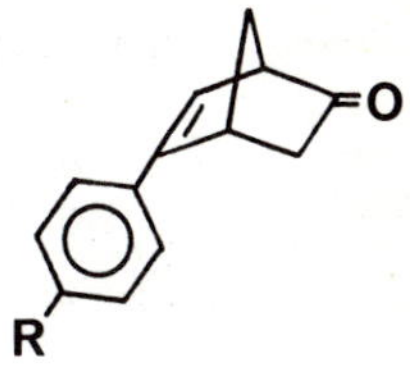

I

2. The λ_m values for some substituted naphthalenes in a common solvent are as follows.

Substituent	λ_m (nm)
H	275
2,7-Diphenyl	300
1,7-Diphenyl	302
1,3,6-Triphenyl	301
1,4,5,8-Tetraphenyl	334
1-Methyl	282
2-Methyl	276
1,4,5,8-Tetramethyl	296

Offer an explanation for the unusually large λ_m for the tetraphenyl-naphthalene.

3. Predict the relative λ_m values for **II** and **III** and give the basis for your prediction.

II   **III**

4. How might the pe spectrum of **IV** compare with that of **V** ?

IV   **V**

ANSWERS

1. Owing to homoconjugation among forms **Ia**, the O-H···O bond strength increases with greater tendency of R to accept the positive charge. $\Delta\nu_{OH}$ therefore decreases in the order R = OCH_3 > H > NO_2.

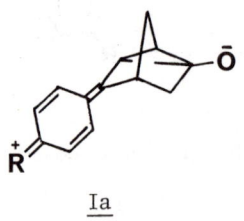

Ia

2. The phenyl groups are sterically forced to be perpendicular to the naphthalene ring and hence in a position for homoconjugation between phenyl rings, as in [3.3] paracyclophane. This increases λ_m. The increase is not due to steric strain on the naphthalene ring because the tetramethyl derivative, which would have comparable strain, does not have an unusually large λ_m.

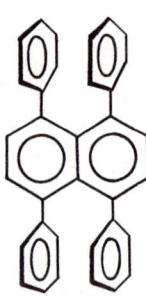

3. **II** should have the larger λ_m since it is spiroconjugated with 8 π electrons (4N) involved. The experimental values are 276 and 254 nm for **II** and **III**, respectively (M. F. Semmelhack, J. S. Foos, and S. Katz, J. Am. Chem. Soc. 95: 7325 (1973).

4. The lowest (π) IP peak of **IV** should be split into a doublet in the spectrum of **V**, attributable to homoconjugation. Probably the split would not be as large as for diene 10.47. The uv spectra (page 354) and the pes spectra confirm this: IP_{IV} = 9.02 eV; IP_V = 8.93, 9.31 eV. (S. A. Cowling, R. A. W. Johnstone, A. A. Gorman, and P. G. Smith, Chem. Commun. 627 (1973).)

Chapter 11

SPECTRA AND STERIC EFFECTS

Molecular spatial requirements affect spectra chiefly by steric hindrance of resonance or by altering the electronic ground state stability of a molecule. To illustrate the latter case, λ_m values of strained olefins ($\lambda_m \simeq$ 190 to 210 nm) are larger than those of strain-free analogs ($\lambda_m \simeq$ 175 to 190 nm). Examples are found when the spectra of methyl-ethylenes are compared with those of t-butyl- or neopentyl-ethylenes (60):

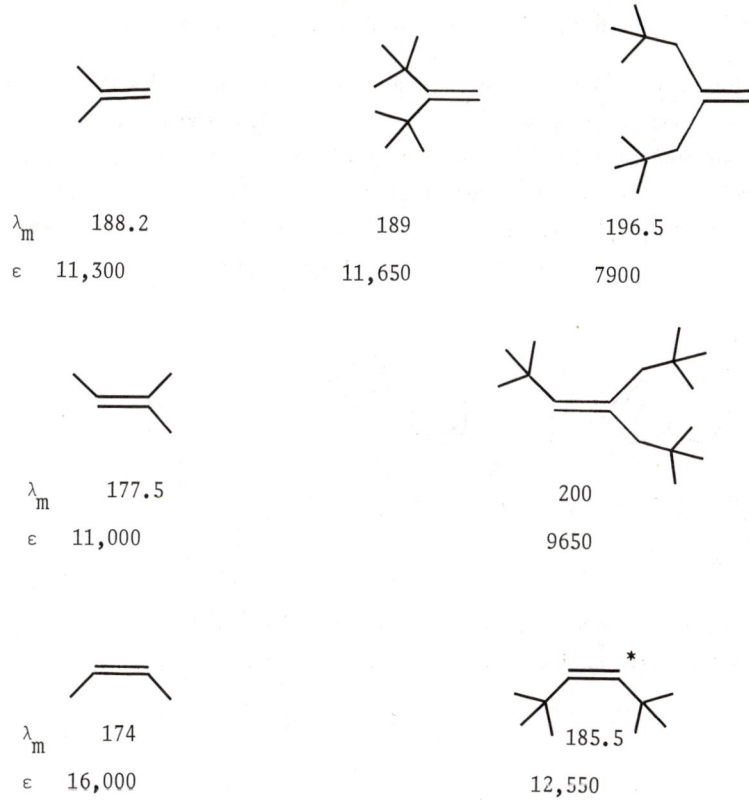

λ_m 188.2   189   196.5
ε 11,300   11,650   7900

λ_m 177.5   200
ε 11,000   9650

λ_m 174   185.5*
ε 16,000   12,550

* Models show that this compound is nonplanar.

There is a linear relationship between the frequency of the π→π* transition
and the Arrhenius activation energy for the thermal rearrangement of cyclic
olefins (61). The larger the strain, the greater is the tendency to rearrange
and the larger is λ_m.

Although 11.1 is a highly strained olfin, there is a theoretical basis

		11.1	11.2
$\lambda_m^{pentane}$	206 nm	= 190 at 300 nm	255 nm

for the long-waved absorption being due to conjugation of the double bonds
with the cyclobutane ring (62). In the case of 11.2, the 255-nm band is
probably an n→π* band in which the energy of the n level (lone pair) is
raised owing to strain. As a result of strain of the double bond at the ring
junctures, the λ_m values of such dienes are larger than those calculated
from constituent constants (234 nm for each case below) (63):

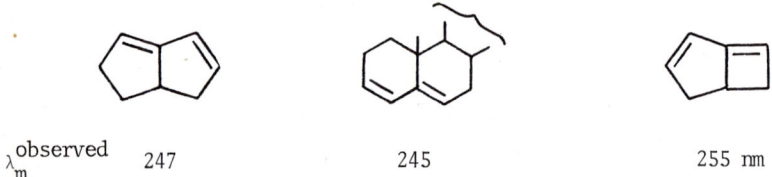

$\lambda_m^{observed}$	247	245	255 nm

Strain on the N=N bond has a surprising effect (64). The wavelength
of the n→π* transition increases with increasing NNC bond angle:

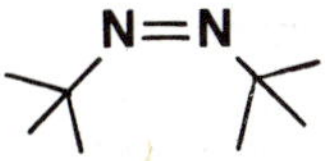

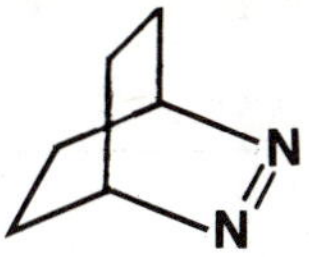

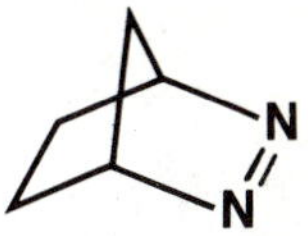

λ_m (nm)   447   385   378   341

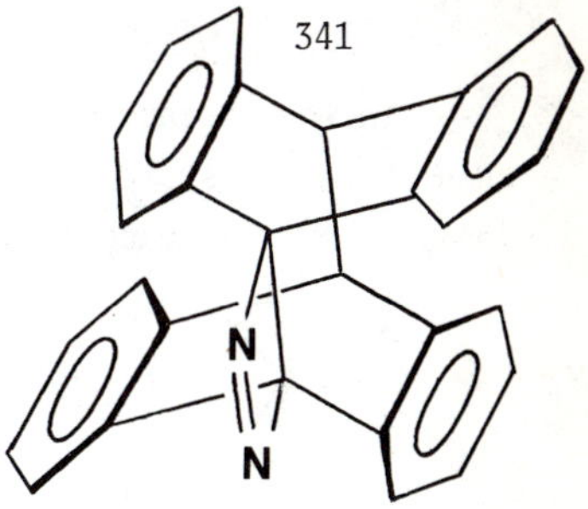

λ_m (nm)   315   <325

It has been proposed that the decrease in excitation energy with increasing NNC bond angle is due primarily to destabilization of the n orbital while the π* level remains approximately unchanged (65). Accordingly, the first IP should vary in a similar direction, and this is found to be the case (66).

There are many examples of steric hindrance shutting off conjugation, and in turn resonance, to lower λ_m. In these cases, resonance is chiefly among dipolar structures which make their greatest contribution to the excited state. Hence, their elimination destabilizes the excited state and decreases λ_m. This is the situation for most cis isomers relative to trans isomers. The cis isomer of stilbene, for instance, cannot be coplanar because of spatial requirements of the ortho hydrogen atoms. At least one phenyl group must rotate out of the molecular plane, thereby excluding itself from conjugation with the rest of the molecule.

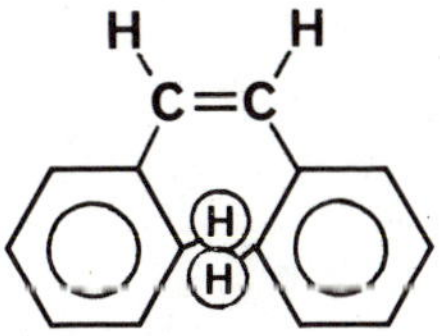

C_6H_5-CH=CH-C_6H_5 C_6H_5-N=N-C_6H_5

	π→π*		π→π*		n→π*	
	λ_m	ε	λ_m	ε	λ_m	ε
trans	295 nm	27,000	320 nm	21,300	443 nm	510
cis	280 nm	10,500	281 nm	5,260	433 nm	1518

The same situation occurs in some butadienes:

1,4-Diphenylbutadienes	λ_m	ε
trans,trans	328 nm	56,200
cis,trans	313 nm	30,600
cis,cis	299 nm	29,500

In the absence of steric effects, alkyl groups produce bathochromic changes on a π system. However, too many groups on the butadiene system leads to nonplanar conformations and a decrease in λ_m (67).

	λ_m (nm)	ε
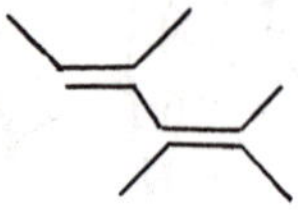	237	17,000
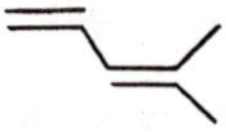	234	23,000
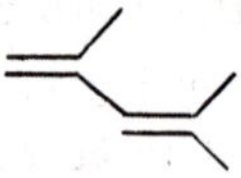	231	9,900
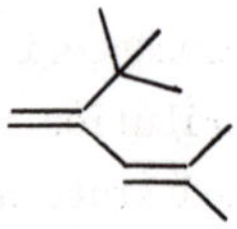	225	6,600
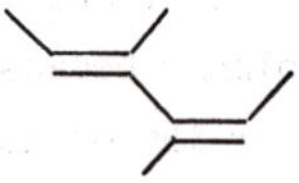	210	3,600
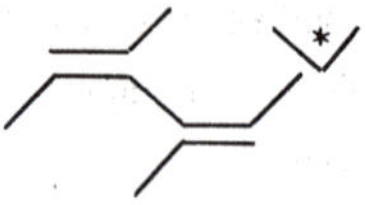 *	205	2,000
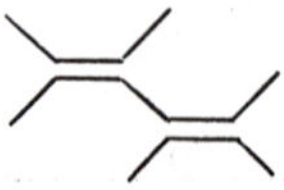	<210	

* Models show that this compound is nonplanar.

Models show that 2,3-di-t-butylbutadiene cannot be planar and indeed, its λ_m (186 nm) is close to that of ethylene.

The 1,3-cyclodienes provide another example of the blue-shift effect of strain on λ_m:

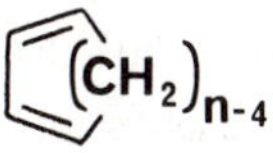

	n	λ_m (nm)	ε
	6	256	8,000
(CH$_2$)$_{n-4}$	7	248	7,400
	8	228	5,600
	10	< 215	

The fact that cis,cis-cyclodecadienes have no λ_m above 215 nm indicates a lack of conjugation of the double bonds. Models suggest a dihedral angle of 60° between planes of the double bonds (68). In contrast, the cis,trans-1,3-cyclodecadiene has a λ_m at 222 nm (log ε = 3.86) (69)

In general, we see that rotation about the single bond of a diene or α,β-enone system leads to a drop in ε_m and, for large angles of twist, a decrease in λ_m. A twisted double bond, however, owing to internal strain, produces increases in ε_m and λ_m. For example, for the series 11.3 to 11.5, λ_m is not appreciably altered but ε_m steadily decreases (70).

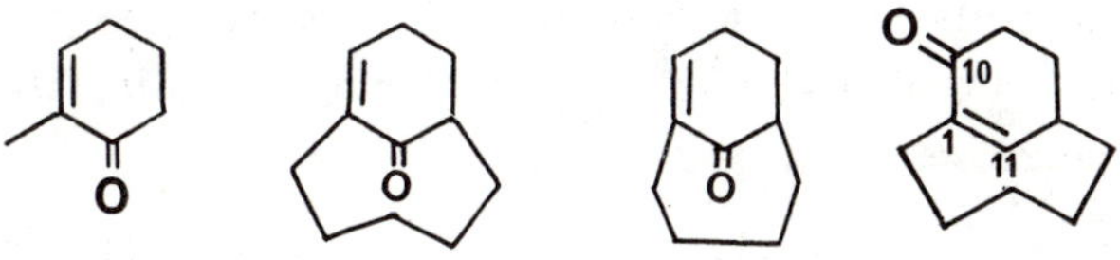

	11.3	11.4	11.5	11.6
λ_m^{EtOH}	235	238	238	254 nm
ε	10^4	5630	3000	7500
$\delta_{vinyl\ H}$	6.66	6.16	5.9	6.85 ppm

X-ray diffraction shows that the C=O bond in 11.4 is rotated 38° out of the plane of the C=C, and the rotation must be even larger in 11.5. The nmr data substantiate these deductions. We saw earlier that the enone resonance deshields the β vinyl proton. The fact that the vinyl proton signal moves upfield for the series 11.3 to 11.5 confirms that the enone resonance is increasingly inhibited. The enone resonance is fully restored, however, in 11.6 as shown by the larger λ_m, near normal ε_m, and deshielded vinyl proton. λ_m is larger than for 11.5, which could be attributed to strain in the C=C bond of 11.6.

Another α,β-enone system for which nmr and uv data have been used to infer diminished resonance is that in cycloalk-2-enones (47).

	λ_m^{EtOH} (nm)	ε	$\nu_\gamma - \nu_\beta$ (Hz)
Cyclopent-2-enone	218	12,200	101.5
Cyclohex-2-enone	225	11,400	69
Cyclohept-2-enone	228	9,600	39
Cyclooct-2-enone	229.5	8,100	24.5
Cyclonon-2-enone	233.5	7,300	18

With expanding ring size, $\nu_\gamma - \nu_\beta$ decreases, denoting diminished enone resonance (the γ proton signal is downfield from that of the β proton but as the polar $\overset{+}{C}-C=C-\overset{-}{O}$ structure makes less contribution to the resonance hybrid, the difference between the two signals decreases). The absorption band intensities in the uv corroborate this view. The significance of the small trend in λ_m is not clear.

Biphenyls have also been studied extensively with respect to steric hindrance to coplanarity of the phenyl rings. Again, increased steric hindrance decreases λ_m:

Substituent in biphenyl	λ_m	ε ε
None	249 nm	14,500
2-Me	237	10,500
2-Et	233	9,000

Similar effects have been observed for nitrobenzenes (71), dimethylanilines (71, 72), and a variety of other systems. Since the biphenyl secondary band is relatively intense whereas that of benzene is of low intensity, it is

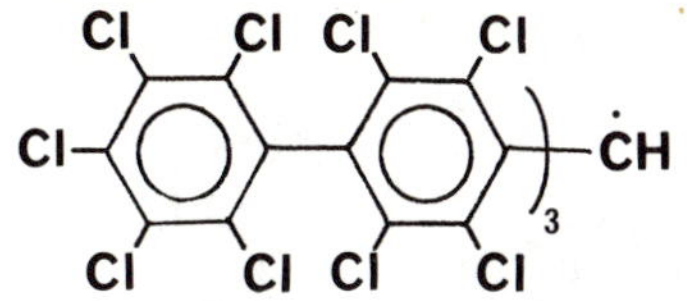

Ch. 11 SPECTRA AND STERIC EFFECTS   369

usually easy to distinguish the two. Thus, perchlorotribiphenylmethanes do not exhibit a biphenyl-type uv spectrum, confirming that the biphenyl fragments are not planar (73). This is to be expected because of steric hindrance by the ortho chlorine atoms. Nevertheless, complete coplanarity of the biphenyl system is not required for partial conjugation. For example, compound 11.7 has been resolved into d and l optical isomers (74) and optical activity can only occur if the two phenyl rings are nonplanar. Still, the compound has a λ_m at 249 nm (ε 16,980), which is of the conjugated biphenyl type.

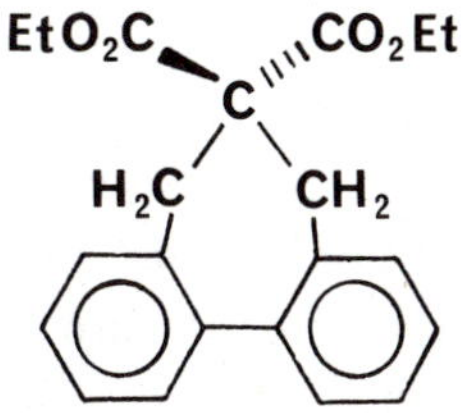

11.7

The buttressing effect (see also Chapter 12 and Section 20.1) can also affect λ_m by increasing a steric hindrance to resonance. In some acetophenones, for example, meta groups prevent the ortho substituents from bending back to allow coplanarity of the COCH$_3$ with the ring and thereby decrease λ_m:

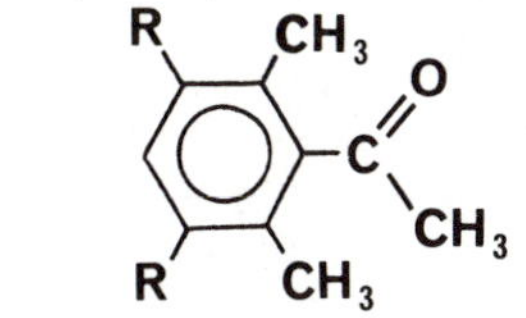

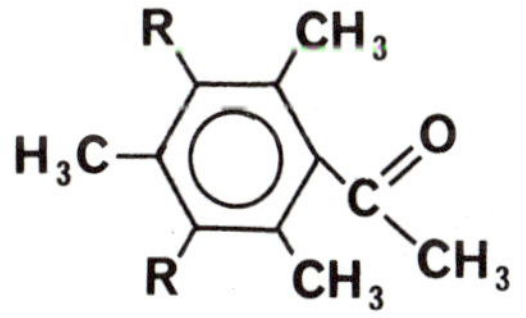

R	λ_m (nm)	ε
H	251	5,600
CH$_3$	212	11,500
H	242	3,600
CH$_3$	216	12,000

Steric crowding can manifest itself in another spectral way. As R increases in size in the sustituted 2,4-dinitrobenzenes the intensity of the $R_2\overset{+}{N}=C_1-C_2=NO_2^-$ band increases. It is proposed that as the substituent gets

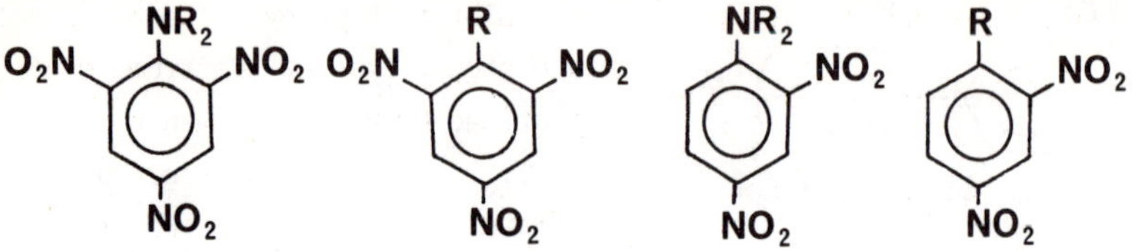

11.8

larger, it sterically forces the o-nitro group to rotate out of the molecular plane. This suppresses the o-nitro group resonance with the ring thereby decreasing its withdrawal of electrons from the amino nitrogen atom and making the lone-pair electrons more available for delocalization to the 4-nitro group. Hence, there is greater interaction resonance between the amino and 4-nitro groups, which has been described as a steric enhancement of resonance (75).

We discussed the shielding and deshielding zones of the nitro group in Section 4.2d. When a nitro group on an aromatic ring rotates out of the molecular plane, the adjacent proton falls more in the shielding zone of of the nitro group. For example, as R of **11.9** increases in size, the nitro group is forced to rotate more out of the plane of the ring. The chemical shifts of the meta and para protons (H-5 and H-6) remain fairly constant but the ortho peak (H-3) shows an upfield shift. Thus, $\Delta\delta$, the difference between the chemical shifts of H-3 in the hydrocarbon and the corresponding nitrohydrocarbon, decreases in going from R = Me to R = t-butyl (76).

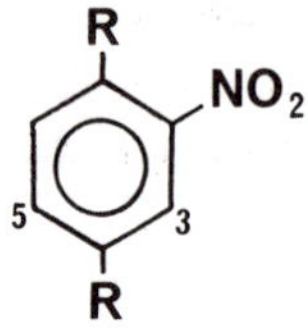

11.9

R	$\Delta\delta^{CCl_4}$	$\Delta\delta^{cyclohexane}$	ϕ*
Me	0.75	0.72	34°
Et	.63	.63	40°
i-Pr	.42	.45	47°
t-Bu	-0.02	.00	65°

* Dihedral angle of NO_2 calculated from uv spectra.

In the t-butyl compound, the H-3 signal is at even higher field than that of H-5 and H-6.

This shielding effect of nonplanar nitro groups in aromatic rings has been studied further. For instance, when two groups occupy adjacent positions in the benzene ring, each rotates some to relieve any steric crowding. Then, if a substituent is placed para to one of the ortho groups such that there is resonance interaction, as in 11.10 the double-bond character of the bond attaching it to the ring reduces its rotation and forces the other group (B) to assume a greater share of the rotation for relief of steric strain. This effect has been called electronic buttressing (77). For example, the chemical shift of the H-3 proton of 11.12 is upfield from that of 11.11 not because of the replacement of H by Me on the amino nitrogen, but because the NO_2 group of 11.12 is sterically forced to rotate more out of the molecular plane (78). This places H-3 more in the shielding zone of the nitro group and reduces o-nitro group resonance and its de-shielding effect. The difference is even larger for the 4-nitro analogs because of the electronic buttressing effect. The $R_2\overset{+}{N}=C_1-C_4=NO_2^-$ resonance interaction diminishes rotation of the amino group forcing the o-nitro group of 11.14 to rotate more for relief of steric hindrance than it does in 11.12. When the 2-NO_2 group is electronically buttressed by a 5-NMe_2 group, there is a stiffening of the 2-NO_2 group rotation and now there is little difference between the H-3 chemical shifts of the

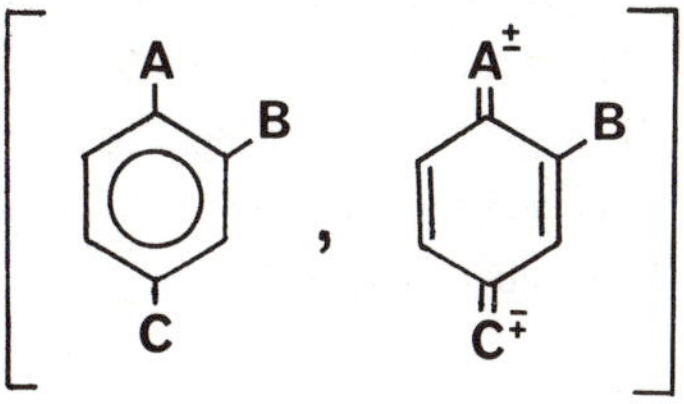

11.10

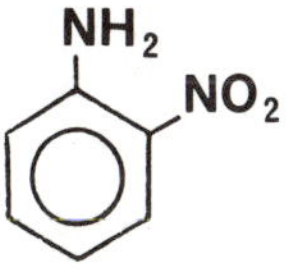

11.11
Δδ = 0.44 ppm

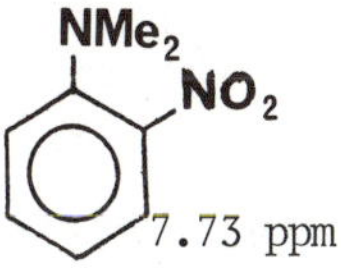

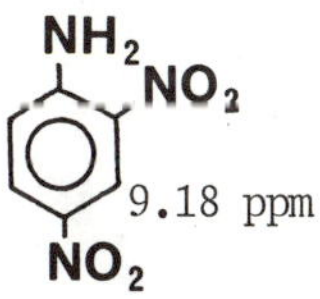

11.12
7.73 ppm

11.13
9.18 ppm
Δδ = 0.55 ppm

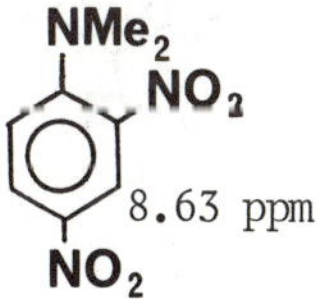

11.14
8.63 ppm

aniline and dimethylaniline compounds. Similarly, in compounds of type 11.15, as R exhibits greater resonance interaction with the 1-NO_2 group, electronic buttressing forces the 2-NO_2 to rotate more and reduce its anisotropic deshielding of H-3. This produces a larger difference between the chemical shifts of the H-3 and H-5 protons.

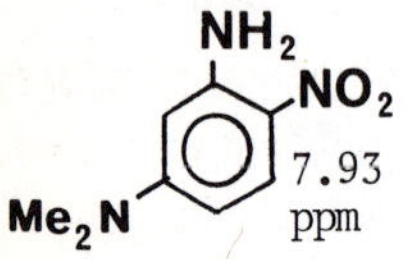

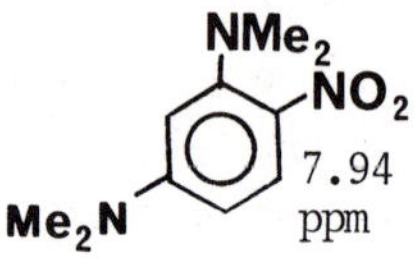

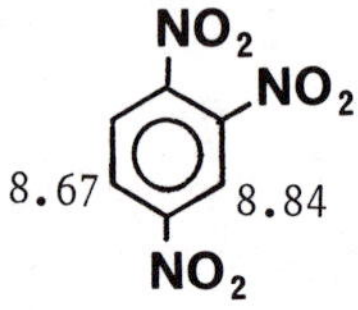

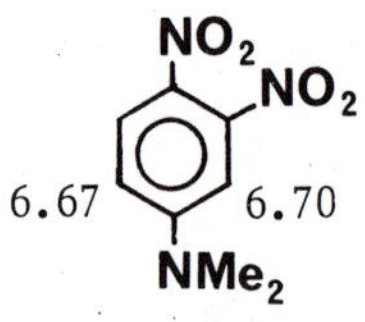

$\Delta\delta$ (ppm)   0.17   0.03

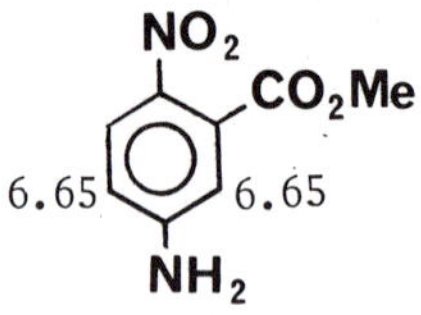

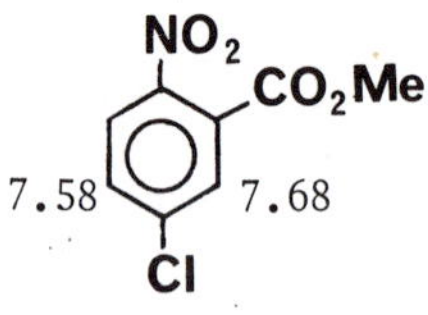

$\Delta\delta$ (ppm)   0   0.10

Indeed, Δδ is found to correlate well with σ_p^+ of R (78).

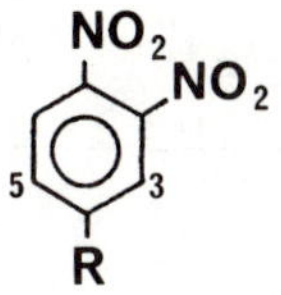

R	Δδ = $\delta_3 - \delta_5$	σ_p^+
NEt_2	0.000	-1.9
NMe_2	0.029	-1.7
OCH_3	0.080	-0.78
CH_3	0.134	-0.31
H	0.15	0
CO_2H	0.168	0.42
CHO	0.162	0.43
CO_2Et	0.146	0.48
NO_2	0.175	0.79

11.15

This shielding effect of a rotated nitro group makes the H-3 and H-5 protons of 11.16 distinct and magnetically nonequivalent at low temperatures, although they become equivalent at temperatures above -60° (79). The difference (R = Me, $\Delta\delta^{CH_2Cl_2}$ = 26.7 cps) is even greater when a larger group replaces the Me (R = $N(C_6H_5)_2$, $\Delta\delta^{CH_2Cl_2}$ = 41.7 cps). If a H-bonding solvent is used, the effect is reduced, i.e., R = $N(C_6H_5)_2$, $\Delta\delta^{acetone}$ = 15.1 cps, indicating that a H bond is involved as part of the barrier to rotation of the NHR group.

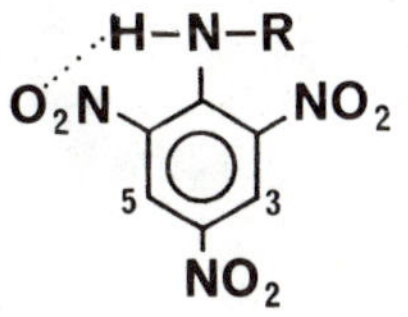

11.16

Accordingly, the chemical shift of an aromatic proton adjacent to a nitro group is sensitive to the degree of coplanarity between the nitro group and the aromatic ring. This is one reason why the nitro group sometimes shows a poor correlation when applying the Q parameter to ortho proton chemical shifts (Section 7.2).

When spatial crowding in A-B completely cuts off conjugation between A and B, then its uv spectrum will approximate the sum of the spectra of A and B. For instance, the spectrum of 2,2'-dicarboxybiphenyl is similar to that of benzoic acid (80), the spectrum of bimesityl approximates that of mesitylene (81), and the spectra of ortho-substituted benzils are close to those of the respectively substituted benzaldehydes (82). Likewise, steric hindrance in the biphenyl portion of 11.19 prevents conjugation from extending the length of the molecule as in 11.20. The result is a spectrum similar to that of the insulated chromophores of 11.18 but twice the intensity of that of the single chromophore of 11.17.

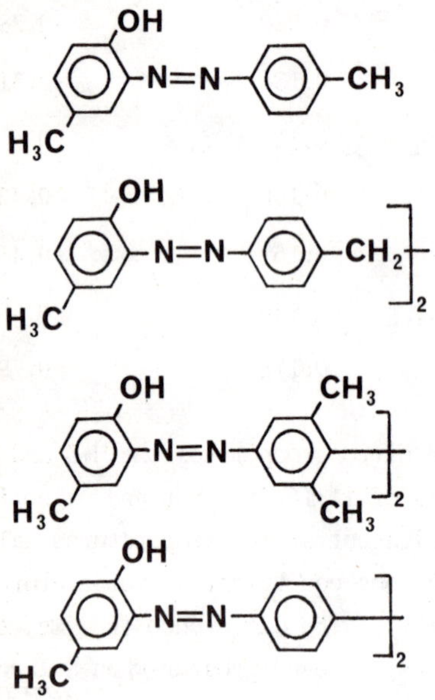

Compound	λ_m (nm)	ε
11.17	380	14,000
11.18	380	32,000
11.19	375	30,000
11.20	550	43,000

We have already seen that substantial resonance occurs even in the ground state of enone and other systems. Consequently, it is not surprising that steric suppression of resonance is reflected in ir spectra too. For instance, as methyl groups block coplanarity of an enone system, the $\nu_{C=O}$ frequency increases to approach the value of an isolated C=O group.

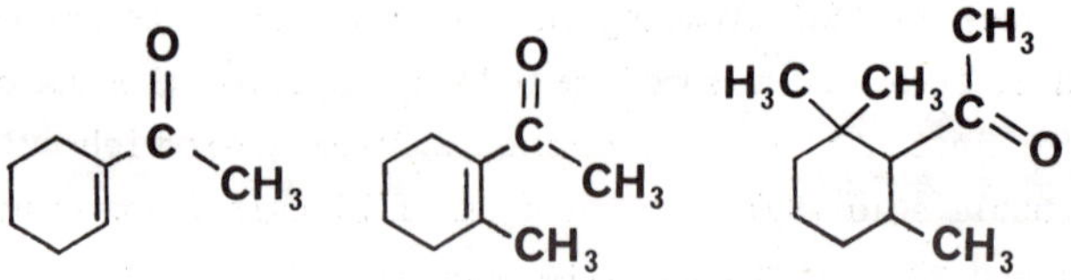

$\nu_{C=O}$ (cm^{-1})   1663   1686   1693

The carbonyl ir frequencies in the substituted acetophenones below show

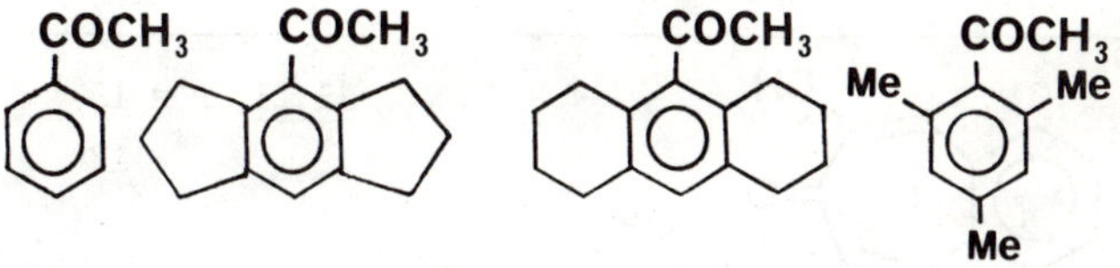

COCH₃ (phenyl)	COCH₃ (indane-fused)	COCH₃ (tetralin-fused on both sides)	COCH₃ with 2,4,6-Me₃

$\nu_{C=O}$ (cm^{-1})   1687   1687   1708   1707

that two flanking C_6 rings are as effective in blocking coplanarity of the $COCH_3$ group as are two ortho methyl groups (83). Fused C_5 rings do not appear to hinder coplanarity. A similar conclusion about the steric hindrance of fused C_5 and C_6 rings and ortho methyl groups is reached from uv spectra of N,N-dimethylanilines (84). Benzotropones of type 11.21 are resonance hybrids of the forms shown,

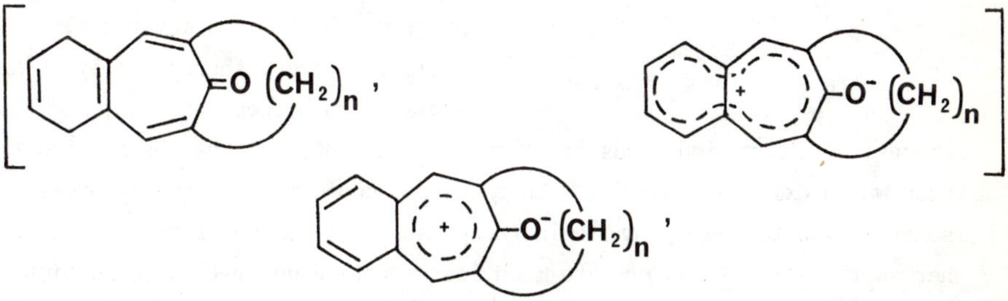

11.21

which requires the carbonyl group to be coplanar with the tropone ring. Models show that the C=O group is forced up out of the plane when $n < 7$. This is reflected in their ir spectra, where it can be seen that the contribution from the polar forms for $n < 7$ diminishes until at $n = 4$ the C=O frequency is that of an isolated carbonyl group (85).

376   Ch. 11   SPECTRA AND STERIC EFFECTS

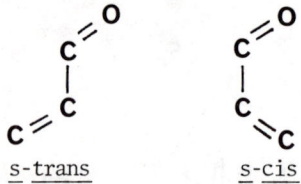

	$\nu_{C=O}$ (cm^{-1})		$\nu_{C=O}$ (cm^{-1})
11.21	1596	11.21, n = 13	1602
		12	1590
		9	1611
		8	1604
		7	1609
		6	1651
		5	1675
		4	1724

Sometimes there is a question about the s-trans or s-cis configuration of an enone system. In the solid state and most often in the liquid state, the enones exist exclusively in one or the other of the two conformations. In solution, however, there usually is an equilibrium between the two, and two carbonyl ir absorption bands are then observed (86). It has been noted that those which exist predominantly in the s-cis conformation readily undergo isomerization to the β,γ-enone in the presence of iodine or other catalysts, whereas the s-trans isomers are much less reactive under these conditions (86). As shown in Tables 14.2 and 14.4 (Chapter 14) cisoid enones absorb in the uv at longer wavelengths and have higher ir carbonyl frequencies than the transoid isomers.

$\nu_{C=O}$ (cm^{-1})	1676	1699
DM (D)	3.31	2.89

Ch. 11 SPECTRA AND STERIC EFFECTS

The s-trans or s-cis conformation can also be discerned from the r^i ratio of carbonyl to olefin band intensities. In addition, there is a larger frequency separation between C=O and C=C bands of the s-cis conformation (87).

	s-trans	s-cis
r^i	>5.2	0.6 - 3.5
$\Delta \nu$	≤ 60 cm^{-1}	≥ 70 cm^{-1}

where

$r^i = I^c \nu^c_{\frac{1}{2}} / I^o \nu^o_{\frac{1}{2}}$

I = band intensity

c = carbonyl band

$\Delta \nu = \nu_{C=O} - \nu_{C=C}$

$\nu_{\frac{1}{2}}$ = band width at half peak height

o = olefin band

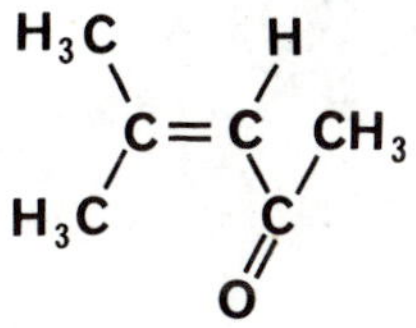

Mesityl oxide, for instance, has a $\Delta \nu$ of 62 cm^{-1} and the intensity of the C=C band is slightly greater than that of the C=O band. On this basis, it is deduced that the molecule has the s-cis conformation. In the planar s-trans conformer, there would be substantial methyl,methyl repulsion. When the above generalizations were applied to 11.22, it was concluded that 11.23 has the s-trans structure whereas 11.24 and 11.25 have nonplanar s-cis conformations (88).

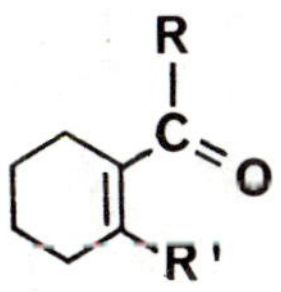

$\nu_{C=O}$ 1680 cm^{-1}

$\nu_{C=C}$ 1618 cm^{-1}

Compound	R	R'	r^i	$\Delta \nu$ (cm^{-1})
11.23	Me	H	7.7	24
11.24	Me	Me	1.3	70
11.25	Me	n-Pentyl	2.3	73

11.22

See also Chapter 17 for distinguishing s-cis and s-trans conformations of enones by nmr chemical shifts.

QUESTIONS

1. The ketosteroid $\underline{X}$ has ir bands at 1695, 1686, and 1608 cm^{-1}. How would you assign these bands to the three double bonds?

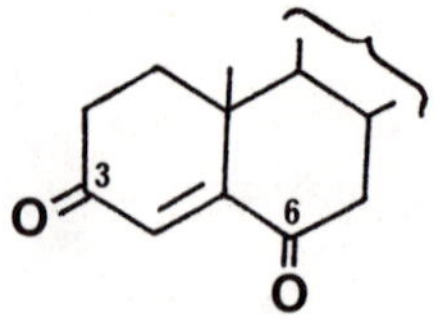

2. Which should have the larger carbonyl ir frequency, $X = NMe_2$ or $X = CH_3$?

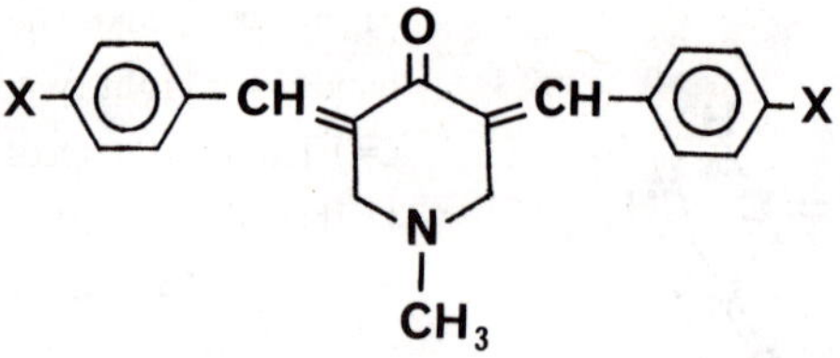

ANSWERS

1. The 1608-cm^{-1} band is readily assigned to the C=C bond. Use the generalization that $\Delta \nu = \nu_{C=O} - \nu_{C=C}$ is smaller for s-trans enones than s-cis. Accordingly, the higher frequency band must be that for the s-cis enone fragment, i.e., $\nu_{C=O}$ for the C_6 carbonyl is 1695 cm^{-1}.

2. The greater the contribution from the ionic form

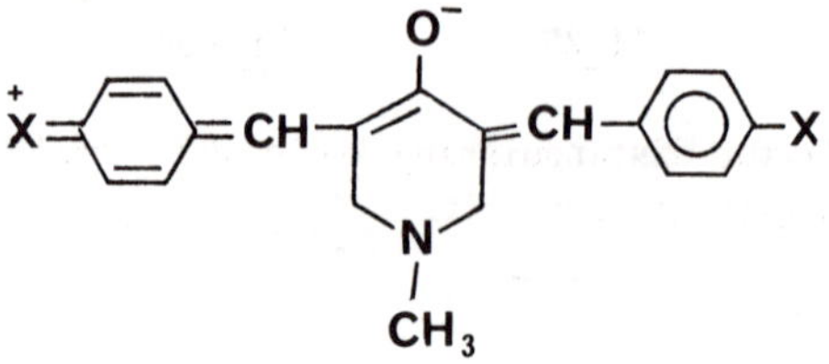

the smaller is $\nu_{C=O}$, and this would be the case when $X = NMe_2$.

There are systems where steric requirements hinder chiefly ground-state resonance, in which case red-shifts occur rather than blue-shifts. An example is 11.26. The major resonance is that between the two forms with quaternary

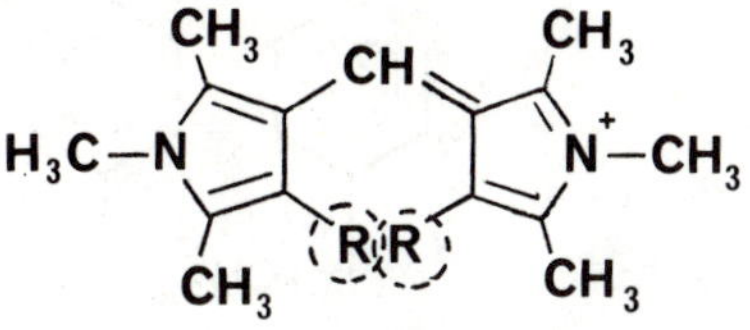

R	λ_m (nm)	ε
H	446	3.5×10^4
CH_3	479	1.25×10^4

11.26

N atoms, which requires a near-planar system. Spatial requirements prevent a planar structure when R is CH_3, which thereby produces a bathochromic shift in λ_m. This red-shift with methylation does not occur when the heterocyclic rings are farther apart.

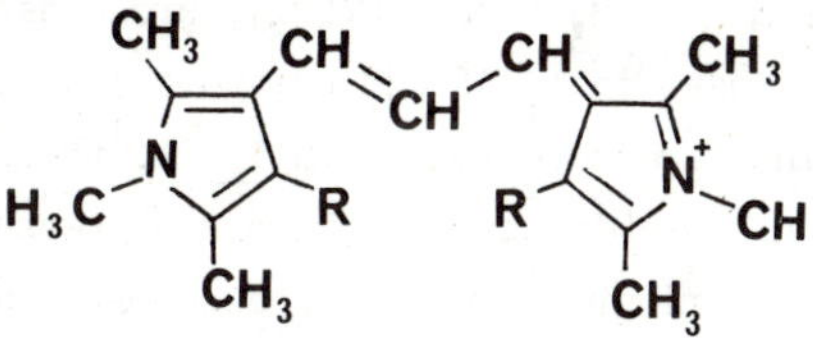

R	λ_m (nm)	ε
H	536	6.3×10^4
CH_3	534	7.6×10^4

Strain in benzene rings can be detected by uv, ir, and nmr measurements.

	Spectral effects
uv:	Strain increases λ_m and ε, with loss of fine structures.
ir (89):	Strain produces strong bands in the 1590-1630 and 710-730-cm^{-1} ranges.
nmr (90):	Strain decreases <u>meta</u> coupling constants and <u>increases</u> <u>para</u> coupling constants, until $J_{para} > J_{meta}$. Also, chemical shifts move up out of the "aromatic" range, i.e., toward the olefinic region.

For example,

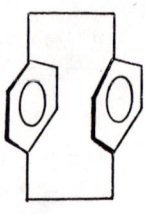

(2.2)Paracyclophane (6.6)Paracyclophane
(has 31 kcal/mol strain)

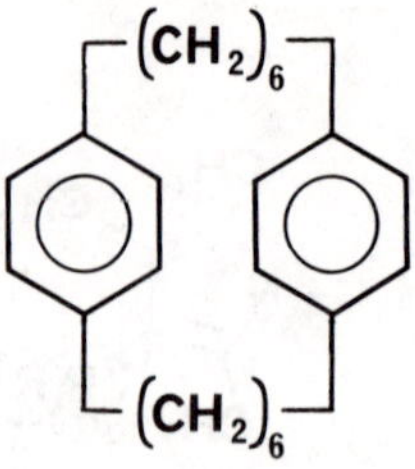

λ_m in uv: 286   266 nm

ir: 1605, 725 cm^{-1} (s)   these bands absent or weak

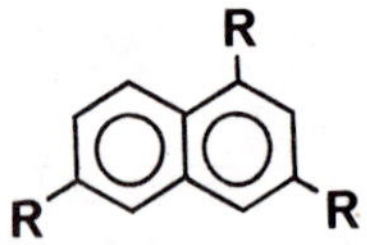

The uv spectra of 11.27 for R = CH_3 and $C(CH_3)_3$ are almost superposable. However, in the case of 11.28 the change from R = CH_3 to R = $C(CH_3)_3$ produces an increase in λ_m of ~20 nm and a loss of fine structure (91). In 11.28, there is considerable steric hindrance between the peri groups which produces strain on the rings.

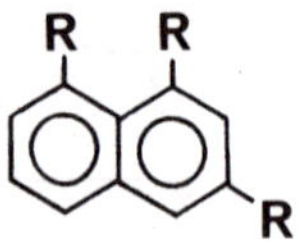

Normally J_{meta} is larger than J_{para} however, the reverse is observed for highly strained phenyl rings.

11.27

11.28

Compound	$J_{1,3}$ (Hz)	$J_{1,4}$ (Hz)
o-xylene	1.38	0.54
1,2-di-tert-butylbenzene	1.59	0.27
benzocycloalkene, n = 3	1.17	0.48
n = 2	1.00	1.03
n = 1	0.33	1.85 (92)
benzocyclobutabenzene	0.81	1.01
benzocyclobutenedione	0.77	1.24

The deformed benzene ring of benzocyclopropene, inferred by the relative $J_{1,3}$ and $J_{1,4}$ coupling constants above, is confirmed by X-ray diffraction data. Also, its strain energy, 68 kcal/mol (94a) is greater than that of cyclopropene (53 kcal/mol).

The strong interaction of face-to-face phenyl rings in (2.2)paracyclophane is well established from heats of combustion data, X-ray diffraction measurements (the rings are buckled), spectral data as cited above, and from various chemical properties (93). When the two benzene rings are held face-to-face with three 2-carbon bridges, even greater strain in the benzene rings occurs (94). Thus, the distance between the two rings at bridge points is only 2.74 Å in 11.30 (2.75 Å in (2.2)paracyclophane) and the hydrocarbon has a λ_m in the uv at longer wavelength (312 nm). However, the benzene ring carbons have not suffered rehybridization, for the ^{13}C-H coupling constants (J^{13}C-H = 162 Hz for the benzene ring and 164 Hz for the vinyl protons) are quite normal (94).

Although the mutual shielding of face-to-face phenyl rings is not a strain effect, it does result from molecular geometry. Thus, the chemical shifts for the phenyl protons of (2.2)ortho- and meta-cyclophanes are in the "aromatic range," δ = 6.5 to 9 ppm. In contrast, the protons are mutually shielded when the rings are face-to-face as in cyclophanes, 11.29 and 11.30.

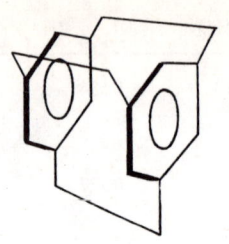 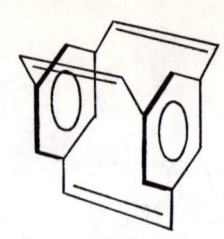

11.29m			11.30		
λ_m^{EtOH} (nm)	258,	312	λ_m^{hexane} (nm)	252,	325
ε	1200,	96	ε	1960,	90
δ		5.73 ppm	δ		6.24 ppm

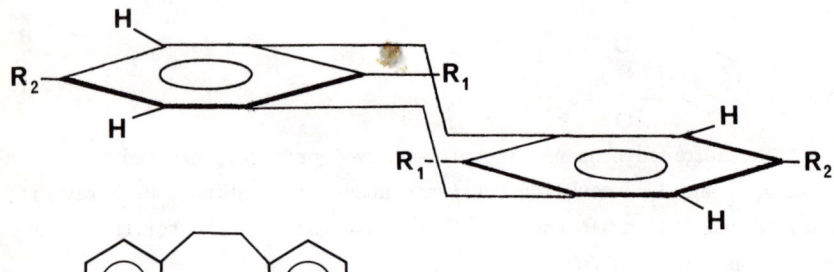

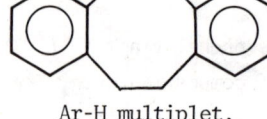

Ar-H multiplet, centered at δ = 6.9 ppm

R_1	R_2	δ_H
H	H	7.25 ppm
Me	H	6.94
H	Me	6.79
H	OMe	6.55

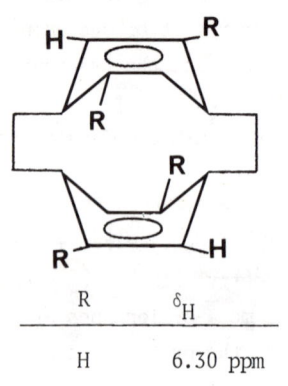

R	δ_H
H	6.30 ppm
Me	6.23

Ch. 11 SPECTRA AND STERIC EFFECTS

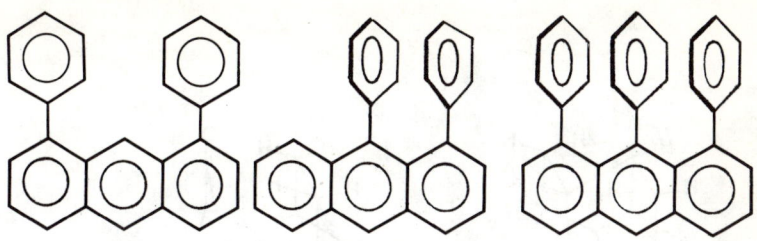

$\delta_{phenyl\ H}$ (96) 7.1-7.7 ppm   6.88-6.95 ppm   6.36 (center phenyl)

and 6.4-7.6 ppm

Furthermore, the more layers of aromatic rings enclosing a given ring, the more the protons of the sandwiched ring are shielded. This is observed for the interesting five-layered cyclophane (97)

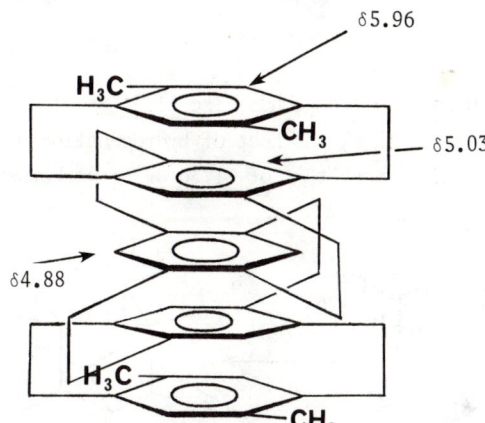

Spatial crowding may affect ir and nmr spectra in another way. For instance, crowded C-H bonds, as in 11.31 and 11.32 have unusually high ir stretching frequencies, attributed to the greater force required to vibrate against opposing atoms. In contrast, compounds 11.33 to 11.35, lacking abutting C-H bonds, have no bands above 3000 cm^{-1} (99). In addition, such abutting protons are deshielded in nmr spectra (99).

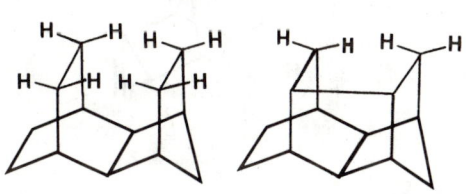

11.31   11.32

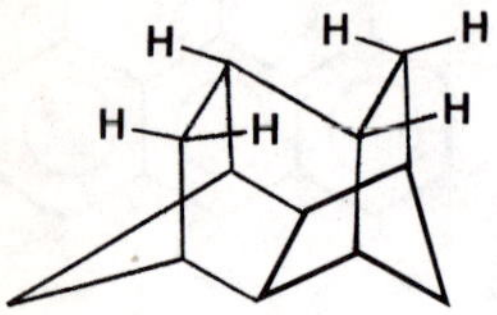

11.33 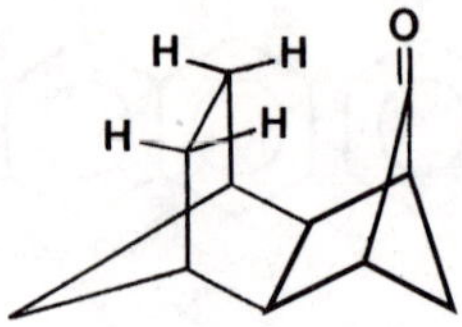   11.34 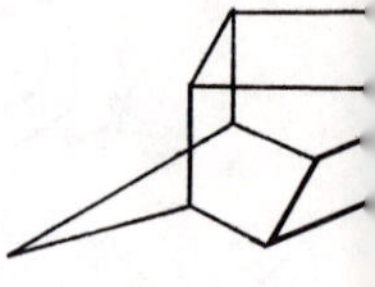   11.35

The strain from abutting C-H bonds in compounds like 11.31 is also revealed by heats of hydrogenation data. Thus, the heat of hydrogenation of the olefin 11.37 is 6-7 kcal/mol smaller than that of 11.36 or of norbornene (100):

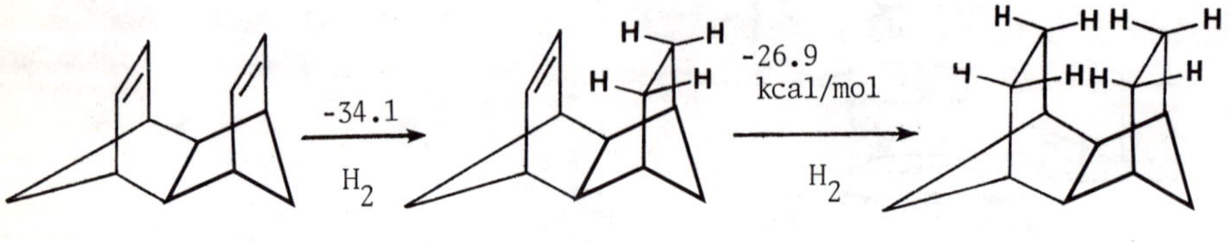

11.36   11.37   11.31

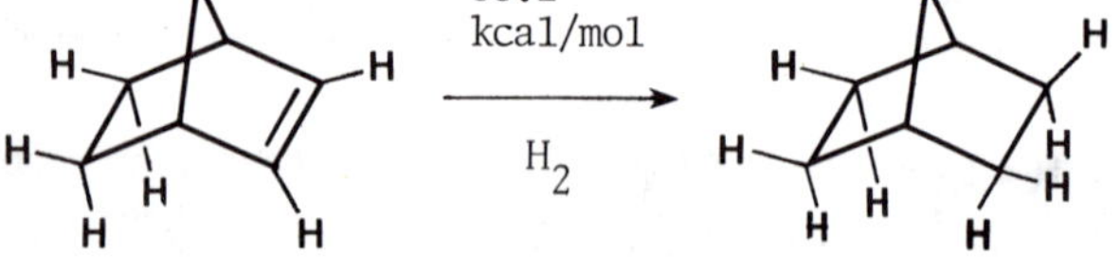

Norbornene

The repulsive interactions between the internal hydrogens of the two ethane bridges in 11.31 destabilize it and bring its heat of formation nearer that of the unsaturated precursor.

Another example of compression of a group causing higher ir frequencies is in di-o-t-butylphenols. It has been deduced that even in these hindered

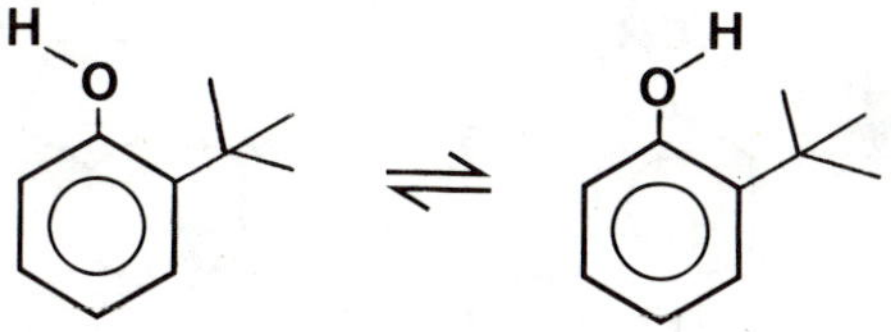

ν_{OH} = 3608 cm^{-1} 3643 cm^{-1}

~90%   ~10%

phenols, the OH is planar with the ring (101). The unusually high ν_{OH} for a phenol is attributed to the increased force necessary to vibrate against the t-butyl group. Whereas o-halphenols exist largely in the cis conformation, owing to greater stability from the intramolecular H bond, the o-alkylphenols have predominantly the trans conformation. Likewise, stretching and bending C-H frequencies for the protons on C3 and C7 of bicyclo[3.3.1]nonanes (11.38) are unusually high owing to the closeness of the hydrogens so that they push against each other as they vibrate (102).

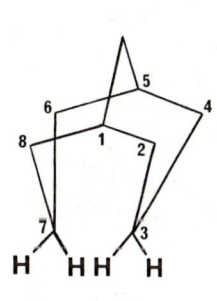

11.38

QUESTIONS

1. Which of compounds M and N should exhibit the greater para proton coupling in the nmr?

    M, n = 8
    N, n = 11

2.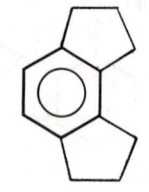

	E	F	G
λ_m^{hexane} (nm)	268	279	287
ε	470	896	470
			(no fine structure)

   What spectral support is there for the observations that E and F are unreactive toward H_2/cat or O_2 whereas G adds H_2 and adsorbs O_2? What ir spectral differences would you expect in confirmation of your answer?

3. The uv spectra of toluene and 4,4'-ditolyl are very different, whereas λ_m for mesitylene and dimesityl are almost identical. Offer an explanation.

4. How should the uv spectra of H and I compare?

5. Offer a rationalization for the following two trends of $\nu_{C=C}$.

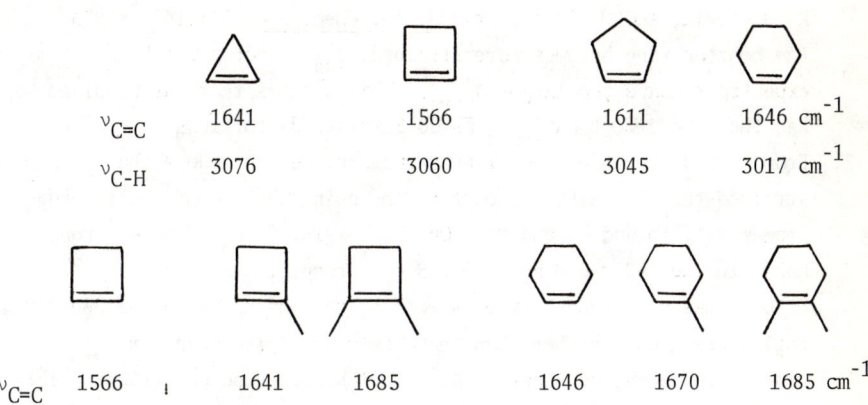

$\nu_{C=C}$	1641	1566	1611	1646 cm^{-1}
ν_{C-H}	3076	3060	3045	3017 cm^{-1}

$\nu_{C=C}$	1566	1641	1685	1646	1670	1685 cm^{-1}

6. Is it surprising that trans-2,2',4,'6,6'-pentamethyl-4-hydroxyazobenzene is red whereas the cis isomer is essentially colorless?

7. The ^{13}C nmr chemical shifts and λ_m of the n→π* band of cyclanones have an approximate parallel trend (103). Comment on these similar sequences.

Ring size	λ_m (nm)	Carbonyl $\delta^{13}C$
4	282	-79.0 ppm from benzene
5	300	-89.3
6	291	-81.0
7	292	-83.7
8	291	-86.7
9	293	-86.6
10	288	-83.8
15	286	-81.0

ANSWERS

1. Compound $\underline{M}$ is under strain, based on its uv spectrum, and has the smallest ring yet known for an isolated paracyclophane. (A. D. Wolf, V. V. Kane, R. H. Levin, and M. Jones, Jr., J. Am. Chem. Soc. 95: 1680 (1973).) As the benzene ring becomes more strained, J_{para} increases; hence, $\underline{N}$ might be expected to have the larger J_{para}. This assumes that out-of-plane bending has the same effect on J_{para} as do in-plane distortions.

2. From λ_m and the absence of fine structure, we can deduce that $\underline{G}$ is more strained than $\underline{E}$ and $\underline{F}$, at least to the point where $\underline{G}$ has lost enough aromaticity to add H_2 and O_2. On this basis, $\underline{G}$ should have strong ir bands in the 1600-1620 and 700-725 cm^{-1} ranges.

3. Ditolyl has a "biphenyl type" spectrum, the intensity in the 260-280 nm region being much higher than that of toluene. The phenyl rings are prevented from being coplanar in dimesityl however, so each ring exhibits a mesityl band.

4. There should be less fine structure and λ_m should be larger for $\underline{I}$. (See A. Iwama, T. Toyoda, T. Otsubo, and S. Misumi, Tetra. Letters (20) 1725 (1973).)

5. In the first series, increasing ring strain gives the olefinic C-H bonds more $\underline{s}$ character (ν_{C-H} increases) and the C=C bonds more $\underline{p}$ character. Accordingly, the C=C and C-H frequencies exhibit opposite trends. The $\nu_{C=C}$ value of cyclopropene is not, however, a bond frequency but a "ring" frequency, as a result of coupling of the C=C vibrations with those of other bonds in the molecule. The effect of methylation on $\nu_{C=C}$ was discussed in Section 4.2a.

6. No. We already saw a substantial difference in λ_m of cis- and trans-azobenzene. With the ortho methyl groups, steric hindrance in the cis isomer is even greater, causing a big difference in λ_m.

7. Since the nmr chemical shift is a ground state property and the two types of spectral changes are roughly parallel, they probably reflect some ground state parameter. They appear to decrease with increasing deviation (expansion or contraction) from normal C-CO-C bond angle.

Chapter 12

SPECTRA AND H BONDING

Spectroscopy has been the most powerful method for the study of H bonds and there is an abundant and still growing body of literature on the topic. You are already familiar with the general characteristics of H bonds (Section 1.2) and aware that (a) <u>inter</u>molecular H bonding is concentration dependent whereas <u>intra</u>molecular H bonding is not; (b) H bonding in a system decreases with increasing temperature; and (c) steric crowding about the proton donor or acceptor will obstruct H bonding (Section 20.2). These effects have been well demonstrated by ir spectroscopy (105, 106). H bonding also <u>lowers</u> O-H stretching frequencies and <u>raises</u> O-H bending frequencies (γ_{OH}):

Substance	Free (dilute solutions)	ν_{A-H}	$\nu_{A-H\cdots B}$	Associated (concentrated solutions)
Alcohols	O-H	3620 cm^{-1}	3355 cm^{-1}	O-H$\cdots$O
Carboxylic acids	O-H	3520	3075	O-H$\cdots$O
D_2O	O-D	2790	2500	O-D$\cdots$O
CH_3OD	O-D	2680	2500	O-D$\cdots$O
Pyrrole	N-H	3500	3425	N-H$\cdots$N

Compound	Monomeric O-H bend	Associated O-H bend
CH_3OH	1340 cm^{-1}	1420 cm^{-1}
Cyclohexanol	1370	1430

Acetic acid	Monomer	Dimer
C=O stretch	1768 cm^{-1}	1701 cm^{-1}
O-H stretch	3521	3073
C-O bend	1379	1425

On this basis, for example, it is thought that an intramolecular O-H···S bond is stronger than an O-H···O bond (see also page 12):

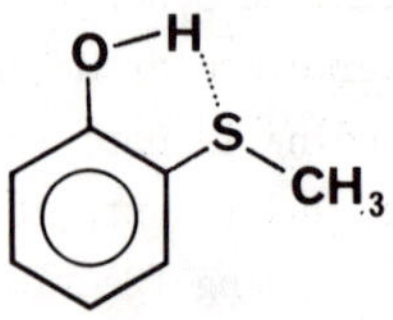

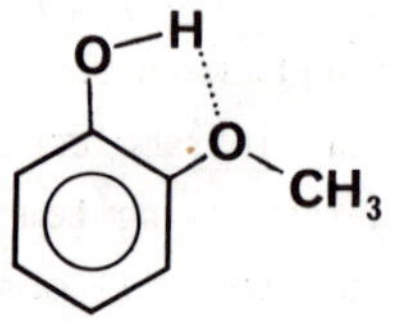

γ_{OH}	537 cm^{-1}	428 cm^{-1}
ν_{OH}	3415 cm^{-1}	3560 cm^{-1}

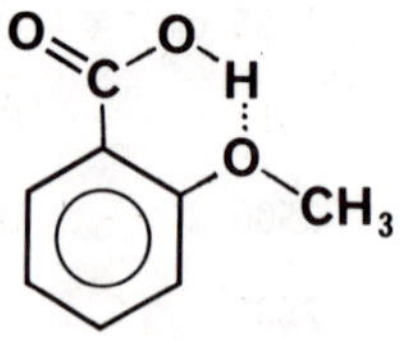

The OH stretching frequency is in the range 3531-3545 cm^{-1} for m- and p-substituted benzoic acids in CCl$_4$ at high dilution and correlates well with pK$_a$ values (107) and with Hammett substituent constants (108). With only a few exceptions, such as t-butyl groups, ν_{OH} for the ortho derivatives is in the range 3521-3538 cm^{-1}. However, when there are ortho proton acceptor groups such as OCH$_3$, intramolecular H bonds are formed and lower ν_{OH} to 3267-3386 cm^{-1}. Interestingly, electron-donating groups in the 5 position increase the basicity of the OCH$_3$ oxygen and the strength of H bond to lower ν_{OH} an additional 20 cm^{-1} (see also Section 7.1), whereas groups in the 3 position exert a buttressing effect (see also Chapter 11) which increases the H bond strength and lowers ν_{OH} an additional 73 cm^{-1}. H bonding in o-hydroxybenzoic acids is quite complex, especially in proton acceptor solvents where both inter- and intramolecular H bonding may occur. The phenolic hydroxyl appears to serve as the proton donor because the intramolecular O-H···O band in the ir is broad. The H bond is even stronger in the anion, exhibiting an ir band over the range 2800-1800 cm^{-1} (108).

	$\nu_{C=O}^{CH_3CN}$	$\nu_{O-H\cdots O}$
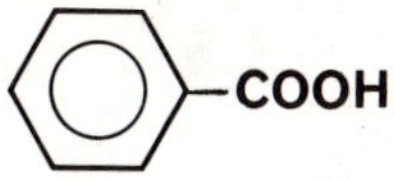 C₆H₅–COOH	1722 cm⁻¹	
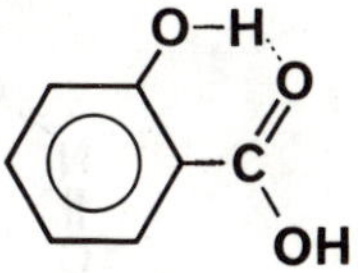 (salicylic acid, intramolecular H-bond)	1682	3200–2900 cm⁻¹

H bonding affects ir stretching frequencies more than resonance does. For instance, $\nu_{C=O}$ for enones is in the range 1630–1685 cm⁻¹ whereas it is lowered to 1605 cm⁻¹ in intramolecularly H-bonded systems (109):

	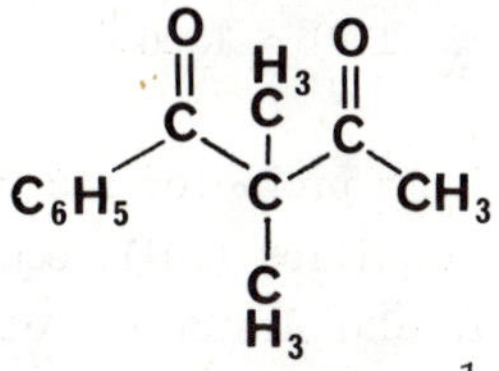 structure 1	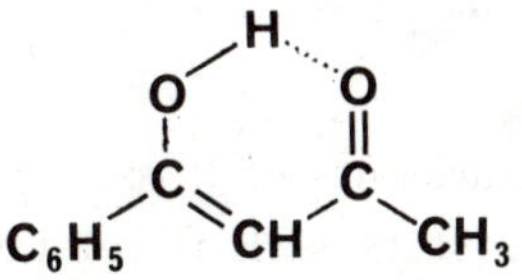 structure 2	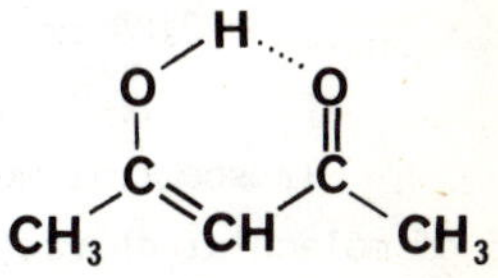 structure 3
$\nu_{C=O}$	1718, 1784 cm⁻¹	1603 cm⁻¹	1610 cm⁻¹
$\nu_{O-H\cdots O}$	none	2640	2800

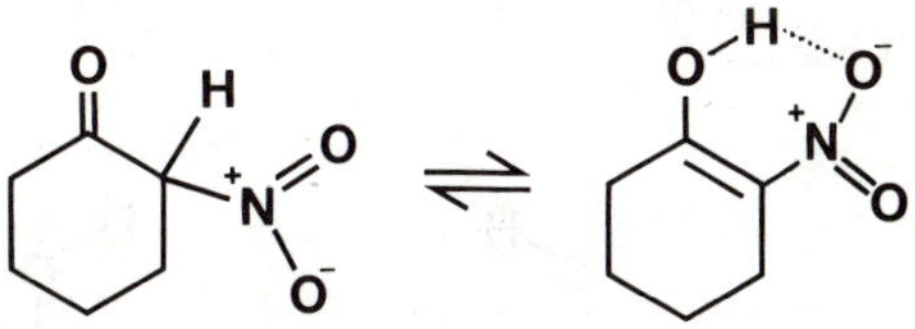

	keto form	enol form
$\nu_{C=O}$	1739 cm⁻¹	1618 cm⁻¹
ν_{NO_2}	1550	1515

Two of the lowest H-bond frequencies known for OH groups are:

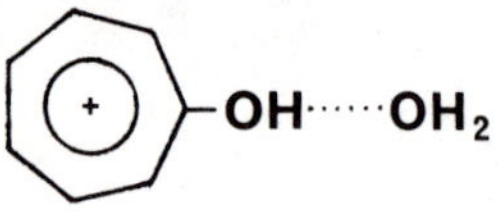

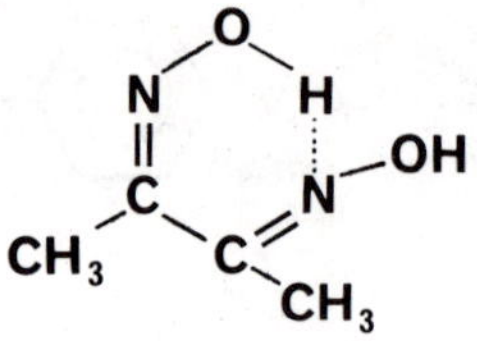

$\nu_{O-H\cdots O} \simeq 2400$ cm^{-1} $\nu_{O-H\cdots N}$ (110) $\simeq 1800$ cm^{-1}

Ir spectral measurements of H bonds provide a probe for studying a variety of molecular characteristics such as relative basicities (104), equilibria between conformations, steric effects, intramolecular distances, etc. For example, ortho substituted phenols exist in cis and trans conformations, and whereas the trans conformer predominates in o-alkylphenols (111), the cis isomer is more abundant in o-halophenols owing to the presence of an intramolecular H bond.

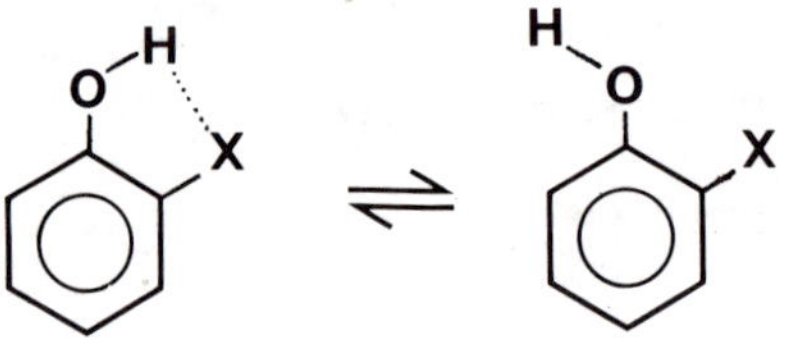

The ratio of intensities of the two OH peaks, free OH and O-H···X bands, varies with temperature, as do the dipole moments (112). Based on OH frequency shifts, ($\Delta\nu_{OH}$) in o-halophenols and in α-halohydrins, the order of decreasing H-bond strengths follows the electron donor character of the halogens (as well as their polarizabilities), I > Br > Cl > F (113). Surprisingly, there is not a similar smooth trend in the bending OH frequencies for the halophenols (114).

	$\nu_{OH}^{isooctane}$	$\gamma_{OH}^{isooctane}$
H	3610 cm	300 cm^{-1}
F	3598	366
Cl	3553	395
Br	3534	394
I	3511	379

There has been considerable controversy over the theoretical basis for a relationship between H-bond strength and OH frequency shifts (115). Empirically, a correlation is usually found between $-\Delta\underline{H}$ and $\Delta\nu_{OH}$ or with the change in the nmr chemical shift, $\Delta\delta_{OH}$ (116, 117). For example, using $(CF_3)_2CHOH$ as the proton donor toward various bases, one finds (117)

$$-\Delta\underline{H} = 0.0115\ \Delta\nu_{OH} + 3.6$$

$$-\Delta\underline{H} = 0.886\ \Delta\delta_{OH} + 3.6$$

or with phenol as the proton donor (115)

$$-\Delta\underline{H} = 0.0103\ \Delta\nu_{OH} + 3.08.$$

By the same principle, the frequency shift of a given proton donor is often used to measure the relative base strengths of a series of bases. Various proton-donor probes have been used, such as $CHCl_3$, alcohols, acetylenes (118), HCl, and phenols. From one study, for instance, it is found that the $\underline{pK_a}$ of a base can be estimated from the relationship

$$\underline{pK_a} = -16.11 + 0.0577(3687 - \nu_{MeOH})$$

where ν_{MeOH} is the O-H frequency of MeOH when mixed with the base (104). The relative base strengths for several series as determined in this fashion are given in Table 12.1. Such measurements, with phenol as the probe, indicate that acetylenes are stronger π bases than allenes and olfins, which are of the same order of magnitude (119). $\Delta\nu_{OH}$ values may be as large as several hundred wave numbers with the stronger bases; e.g., $\Delta\nu_{OH}$ for phenol toward some compounds are given below.

Table 12.1. Relative basicities of several series of Lewis bases as determined by H-bond frequency shifts

Compound	Relative basicity ($\Delta\nu_{OH}$ of phenol) (120)	Compound	Relative basicity ($\Delta\nu_{OD}$ of CH_3OD) (121)
n-Butyl ether	1.0	Acetone	1.0
n-Butyl iodide	0.27	Ethyl ketone	0.96
n-Butyl bromide	0.25	Cyclobutanone	0.91
n-Butyl chloride	0.21	Cyclopentanone	1.12
n-Butyl fluoride	0.14	Cyclohexanone	1.22
		Cycloheptanone	1.4

Compound	Relative basicity ($\Delta\nu_{OH}$ of phenol) (122)	Charge-transfer stability toward HCl (123)
Fluorobenzene	0.08	0.97
Chlorobenzene	0.60	0.87
Bromobenzene	0.79	0.85
Iodobenzene	0.96	0.84
Benzene	1.00	1.00
Toluene	1.22	1.51
p-Xylene	1.4	1.64
Mesitylene	1.55	2.61
1-Hexene	1.47	
1-Hexyne	2.0	
Cyclohexene	2.01	

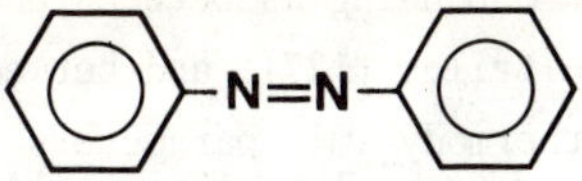

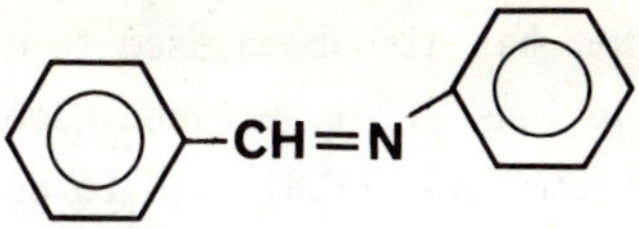

[Structure: C6H5-N=N-C6H5]   [Structure: C6H5-CH=N-C6H5]

$\nu_{OH\cdots\pi}$ 45 cm^{-1} $\nu_{OH\cdots N}$ 435 cm^{-1}
$\nu_{OH\cdots N}$ 274

[Resonance structures of 2,6-dimethyl-4H-pyran-4-one]

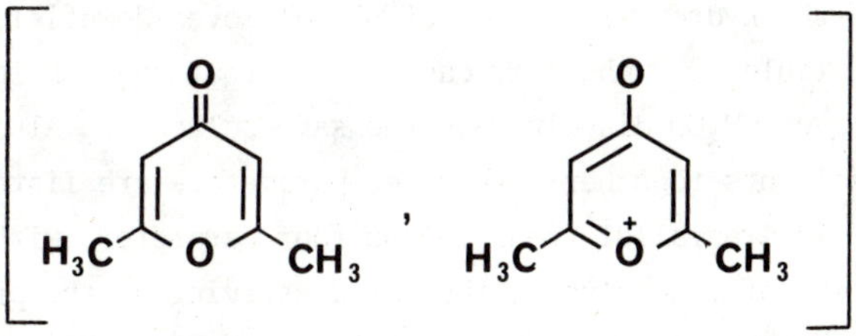

350 cm^{-1}

[Resonance structures of N-methyl-2-pyridone]

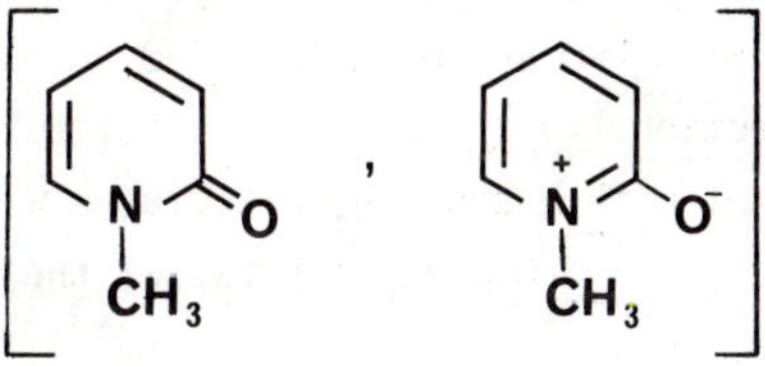

$\nu_{O-H\cdots O}$

Good correlations between uv wavelength shifts of nitrophenols or nitroanilines and the pK or σ* values for a series of solvents have been ascribed to H bonding between the solutes and the solvents (124).

As inferred above, nmr chemical shifts are very sensitive to H bonding, going as low as δ = 17 ppm for the strong H bonds in diarylacetones, Ar-C(OH)=CH-CO-Ar. In the highly enolic 3-substituted-2,4-pentanediones, there is a linear correlation between the enolic proton chemical shift and the carbonyl ir frequency (125). The more electronegative is R, the stronger is the H bond, thereby increasing $\delta_{\underline{OH}}$ and decreasing $\nu_{C=O}$.

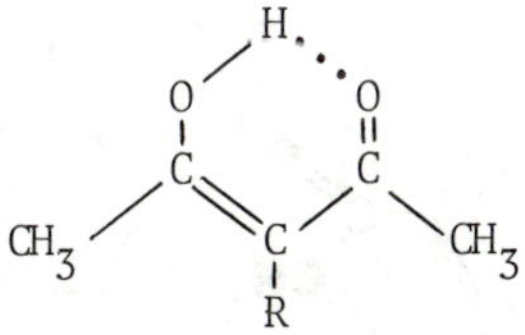

396   Ch. 12 SPECTRA AND H BONDING

Nmr has also been used to detect weak H bonding as between phenols and chloride ions (126), between alcohols and nitriles (127), and between alcohols and nitromethane (128), and to detect the thermodynamic parameters for H bonding of chloroform to Lewis bases (129). The use of ^{17}O, ^{13}C, or ^{14}N probes in the study of H bonding offers the advantage of larger shifts in δ (130).

$^{13}C=O$ and $^{13}C-OH$ shieldings have been studied as a function of their involvement in H bonding. The stronger the H bond of phenol with bases, the more $\delta^{13}C$ of the hydroxylic carbon of phenol moves downfield (131). Some data are given in Table 12.2 where it can be seen that changes in $\delta^{13}C-OH$ of phenol parallel $\Delta\nu_{OD}$ of CH_3OD dissolved in the same solvents. Also, $\delta^{13}C=O$ acetone in various solvents together with other parameters are listed in Table 12.3. We see there is a parallel trend in the four parameters with increasing H-bonding ability of the solvent, the latter serving as the proton donor. The stronger the H bond in the fragment $^+C-O^-\cdots^{\delta+}H-R^{\delta-}$, the lower is the electron density about the carbonyl carbon and the higher it is about the oxygen. Accordingly, $\delta^{13}C$ moves downfield and $\delta^{17}O$ upfield. H-bond formation produces the same change at P in $\emptyset_3P=O$ as for C in $R_2C=O$, hence $\delta^{13}C=O$ and $\delta^{31}P=O$ move in the same direction. $\nu_{C=O}$, of course, reflects the reduction in the carbonyl force constant with H bonding, associated with greater single bond character. The order of $\delta^{13}C=O$ changes parallel the H-bond energies toward acetone in the series

$$H_2O > AcOH > MeOH > EtOH > \underline{i}\text{-PrOH} > \underline{t}\text{-BuOH} > CHCl_3$$

as determined by uv spectroscopy (136), with $HCCl_3$ out of position in the $\delta^{13}C=O$ sequence. The effects of the aprotic solvents on the parameters in Table 12.3 are small and are due to minor changes in polarization of acetone from solvent dipole orientations and other Coulombic interactions.

Intramolecular H bonding shifts nmr $^{13}C=O$ signals downfield too (137):

[Structures: methyl benzoate; methyl salicylate; acetophenone; 2-hydroxyacetophenone]

$\delta^{13}CO$(ppm, rel to $CH_3C^{13}O_2Na$)

  15.8   11.4   -14.9   -22

Table 12.2.

^{13}C-OH chemical shifts of phenol and $\Delta \nu_{O-D}$ of CH_3OD in various solvents (131)

Solvent	$\delta ^{13}$C-OH (ppm from benzene)	$\Delta \nu_{O-D}$ (of CH_3OD)(cm^{-1})
Cyclohexane	-26.9	
Carbon tetrachloride	-27.0	0
Benzene	-27.9	24
Ethyl acetate	-28.4	51
Acetonitrile	-28.6	87
Ethyl ether	-29.1	96
Dioxane	-28.9	111
Tetrahydrofuran	-29.0	117
Acetone	-29.2	121
Dimethylsulfoxide	-29.8	158
t-Butylamine	-30.1	
Tri-n-butylamine	-30.3	243
Diethylamine	-30.6	268

The equilibria among the monomeric, dimeric, polymeric and other species of carboxylic acids in the gaseous, liquid, and crystalline states or in solution have been studied by a wide variety of physical methods. ^{13}C chemical shifts have been particularly revealing in these studies. From measurements of ^{13}C chemical shifts of the substituted carbon of phenol upon forming H bonds to bases and of the carbonyl carbon of ketones when serving as a proton acceptor of H bonds, we just saw that an increase in H bonding by either part of the carboxyl group should shift $\delta ^{13}$C downfield. On this basis, certain deductions may be made concerning the carboxylic acid species in various solvents (138). For example, neat acetic acid is largely dimeric, and upon dilution with cyclohexane there is only a very slight change in $\delta ^{13}$COOH to high field. This is interpreted to mean that the acid remains largely dimeric and the small upfield change is due to formation of a trace of monomer.

Table 12.3.
Comparison of solvent effects on $\delta^{13}C=O$ for acetone with other parameters (132)

Solvent	$\delta^{13}C=O$*	$\delta C=^{17}O^\dagger$	$\nu_{C=O}$ (134) (cm^{-1})	$\delta^{31}P=O$**
Alkane	2.4	-8	1723	
Ethyl ether	2.0		1721	
Carbon tetrachloride	1.3	-5	1719	-24.9
Tetrahydrofuran	0.8			
Benzene	.8	0	1717	
Acetone	0.0	0	1715	
1,4-Dioxane	0.0		1715	-24.8
N,N-Dimethylformamide	-0.7	2		
t-Butyl alcohol	-1.6			
i-Propyl alcohol	-1.9			-29.8
Acetonitrile	-2.1	4	1715	
Chloroform	-2.3	2	1712	
Ethyl alcohol	-2.9	8	1709	
Methyl alcohol	-3.7	12	1708	-32.6
Acetic acid	-6.2			-33.3
Phenol	-8.7	20		
Water	-9.1	37		
Formic acid	-9.1	40		-37.3
Dichloroacetic acid	-11.9			-41.7
Trichloroacetic acid	-13.0			
Trifluoroacetic acid	-14.1			-48.1

* ^{13}C shift in ppm with respect to shift of pure acetone (132).
† ^{17}O shift in ppm relative to shift of pure acetone (133).
** ^{31}P shift of $(C_6H_5)_3P=O$, ppm relative to H_3PO_3 (135).

Ch. 12 SPECTRA AND H BONDING   399

The addition of acetone to acetic acid produces a substantial upfield shift (7 ppm) in $\delta^{13}COOH$. This means that the COOH is less H bonded, which can be explained by the formation of a species such as 12.1

12.1
$$CH_3-C\underset{O-H\cdots O=C}{\overset{O}{\diagup}}\underset{CH_3}{\overset{CH_3}{\diagup}}$$

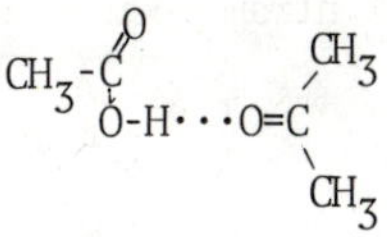

since acetone is only a proton acceptor. Chloroform too produces a fairly large upfield shift (~4 ppm) in $\delta^{13}COOH$, and in this case $CHCl_3$ is the proton donor. This suggests formation of a species such as 12.2 at the expense of the dimer.

12.2
$$CH_3-\underset{HO}{\overset{}{\diagup}}C=O\cdots H-CCl_3$$

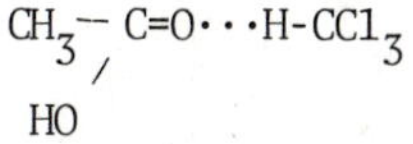

Water produces very little change in $\delta^{13}COOH$, close to that of cyclohexane. This can be attributed to the presence of a double H-bonded COOH group such as in 12.3 where the strong H bonds in dimeric acetic acid are replaced by

12.3
$$CH_3-C\underset{O-H\cdots O}{\overset{O\cdots H-O}{\diagup}}\overset{H}{\underset{H}{\diagup}}$$

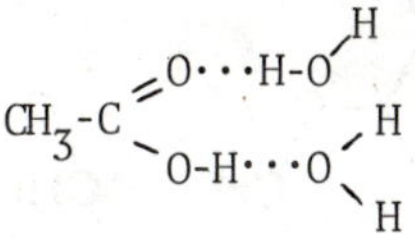

two bonds to H_2O molecules. Similar solvent effects are observed for $\delta^{13}COOH$ of acrylic acid (139), although the changes are smaller.

Esterification of a carboxylic acid also produces a shielding of the ^{13}C chemical shift of the carboxyl function (139).

	$\delta^{13}C_1$ (ppm from TMS)	
	Acid	Methyl ester
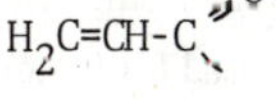 $H_2C=CH-C\diagdown^O$	168.9	166.0
(Z) 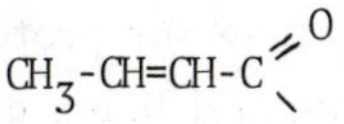 $CH_3-CH=CH-C\diagdown^O$	169.8	166.2

Replacement of the OH by OCH_3 decreases the amount of H bonding involving the carboxyl group, which has a deshielding effect.

Intramolecular H bonding has a marked visible effect on uv-visible absorption whereas intermolecular H bonding has only small effects. For instance, o-nitrophenol is yellow but the p-isomer is buff colored. Similarly, the chelated compound 12.4 is yellow whereas its methyl ether is white (140). The bathochromic effect of the intramolecular H bond can be attributed to its

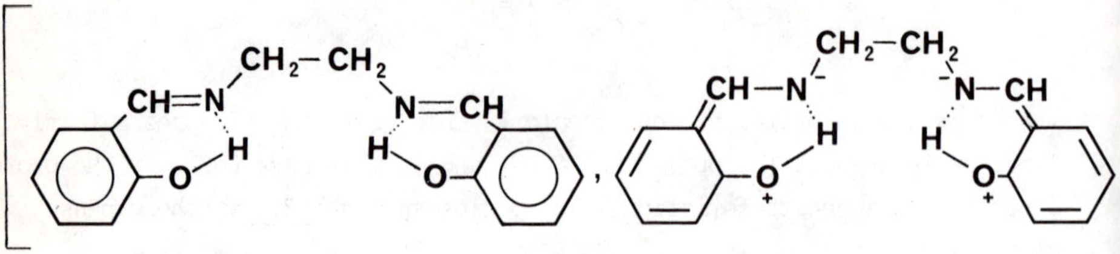

12.4
Yellow

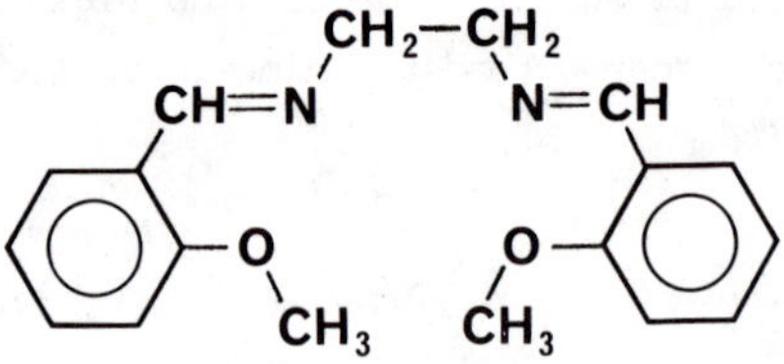

12.5
White

greater strength in the excited state as a result of a greater contribution of the dipolar structure 12.4b. This produces a net stabilization of the excited state, to diminish the transition energy, and an increase in λ_m.

Intermolecular H bonding produces only small changes in λ_m, usually positive for proton-donor compounds and negative for proton-acceptor compounds (141). For example, λ_m of the mesoionic compound below decreases with increasing H-bonding ability of the solvent (142):

Ch. 12 SPECTRA AND H BONDING   401

Solvent:	C_6H_5	Me_2SO	$CHCl_3$	t-BuOH	MeOH	H_2O
λ_m (nm):	447	421	410	370	345	325

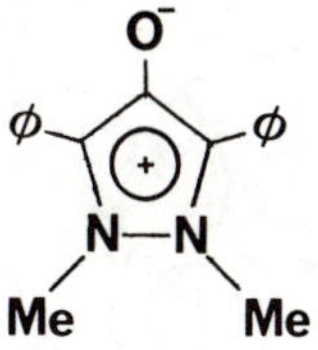

$\nu_{C=O} = 1545$ cm^{-1}

We saw in Section 4.2a that the ir band of a C=O group can be identified by plotting the changes with solvent against the changes with the same solvents for acetone. Another technique is to plot the ir changes against uv changes for a series of H-bonding solvents. H bonding stabilizes the ground states and the uv λ_m decreases (ν_m increases) with increasing proton-donor character of the solvent (143). Accordingly, one observes a linear correlation between the ir and uv frequency changes. This technique was used to identify the carbonyl ir band of the mesoionic compound above, which has an extremely low frequency (142).

QUESTIONS

1. Inasmuch as H bonding moves the chemical shift downfield, explain the upfield shifts for compounds 7 to 9 ($\delta^\circ = \delta_{OH}$ at infinite dilution) relative to 3 to 5. respectively.

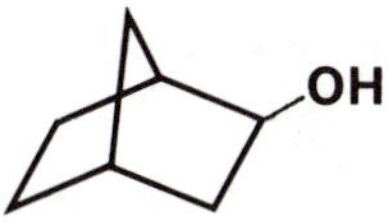

δ°

1   0.82

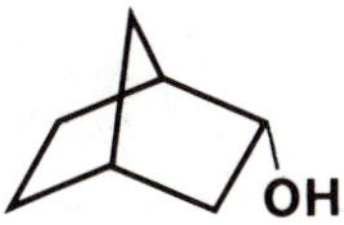

2   0.83

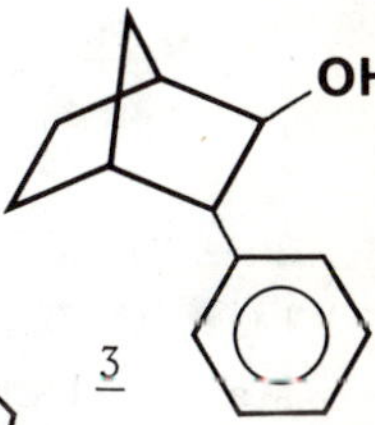

3   1.10

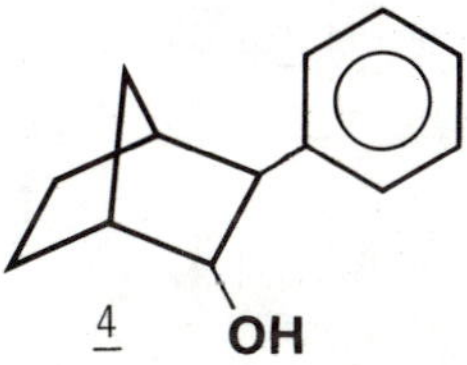

4   1.20

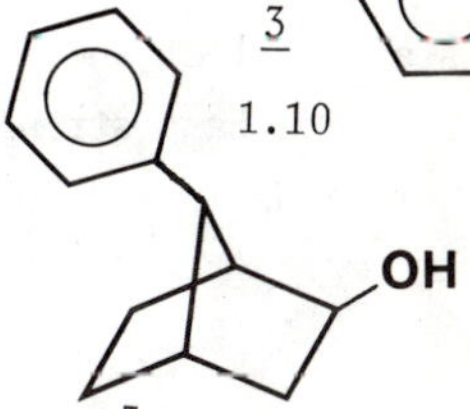

5   0.90 ppm

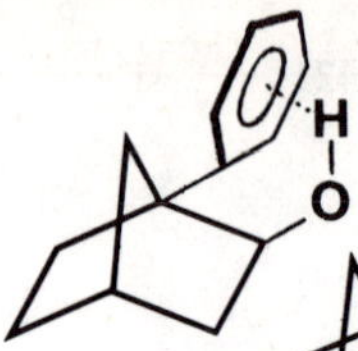

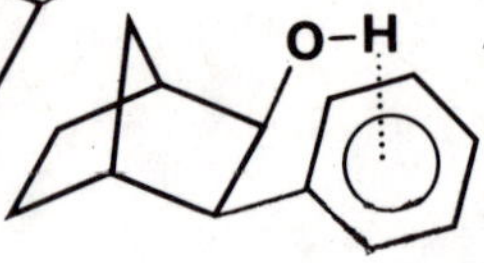

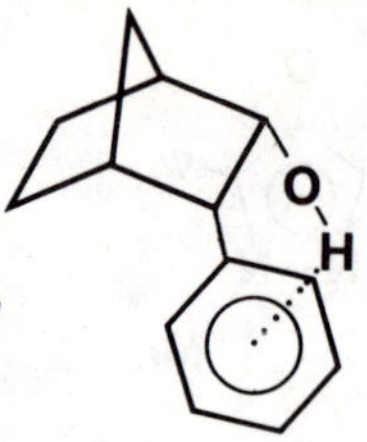

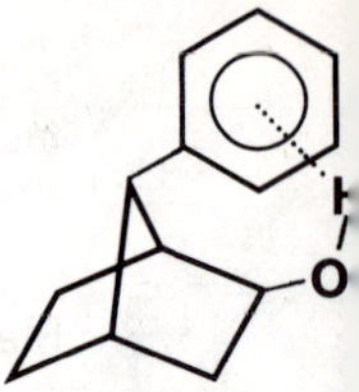

	6	**7**	**8**	**9**
δ°	0.91	0.58	1.02	0.26 ppm
$\Delta\delta^\circ$ (free less H bonded)		0.52	0.18	0.64 ppm
$\Delta\nu$ (in ir)	30	30	10	28 cm^{-1}

2. Predict relative $^+$OH proton chemical shifts for the following two couples:

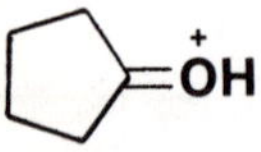

10   vs.   **11**   **12**   vs.   **13**

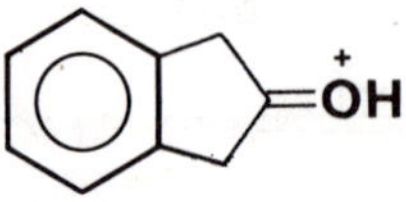

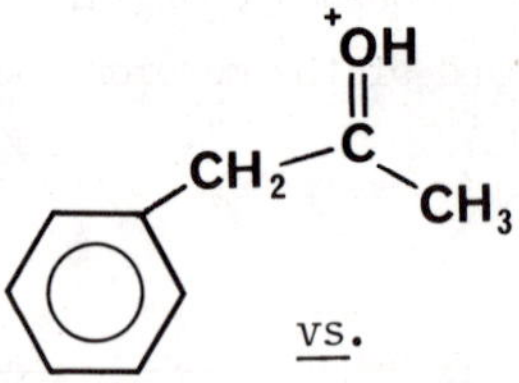

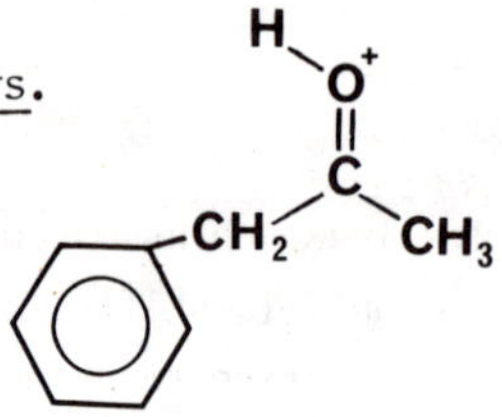

3. Some thermodynamic and ir spectral data for H bonding of phenol with some bases in CCl_4 are given below. (a) Do the data provide evidence for a steric effect in this interaction ? (b) Offer an explanation for the relative base strengths of (i) compounds **10**, **11**, **12**, and **15**; (ii) of compounds **15** and **16**; (iii) of compounds **18**, **19**, and **22**.

Compound	$\Delta\nu_{OH}$ (cm^{-1})	$-\Delta H$ (kcal/mole)	$-\Delta F$ (kcal/mole)	$-\Delta S$ cal/deg/mol
10. $(C_6H_5)_2O$		2.06	-0.14	7.4
11. Et_2O		5.41	1.29	13.8
12. $n\text{-}Bu_2O$		5.71	1.09	15.5
13. $(CH_2)_4O$	285	5.29	1.65	12.2
14. $(CH_2)_5O$		5.19	1.16	13.5
15. $t\text{-}Bu_2O$		7.31	0.75	22.0
16. $t\text{-}Bu_2S$		4.87	0.04	12.2
17. $n\text{-}Bu_2S$		4.19	0.11	13.5
18. Me_2CO	193	4.94		
19. $Me\text{-}CO_2Me$	164	4.77		
20. MeCN	150	4.65		
21. Pyridine	465	8.00		
22. $Me\text{-}CONMe_2$	345	6.84		
23. Et_3N	556	9.08		

4. When compounds 24 to 26 are O-methylated, λ_m^{MeOH} decreases by the following amounts:

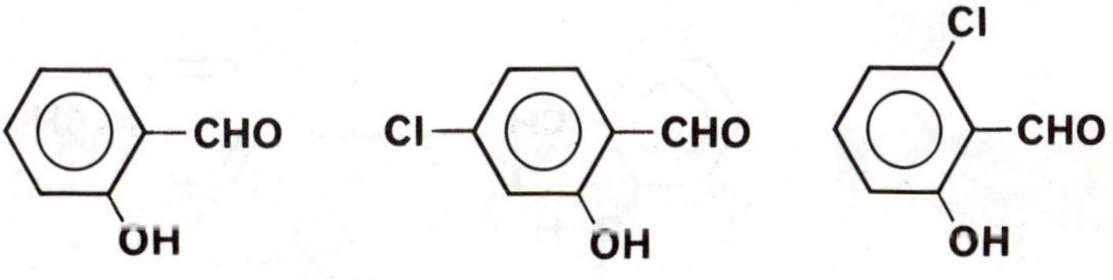

	24	25	26
$\Delta\lambda_m^{MeOH}$	-6 nm	-4 nm	-17 nm

How do you explain the blue-shifts, and for which compound should the OH proton chemical shift in the nmr be farthest downfield?

ANSWERS

1. $\delta°$ for compounds 3 to 5 compared to 1 and 2 show the deshielding inductive effect of phenyl. $\delta°$ for 7 to 9 moves upfield, even though the OH is forming an H bond, because the H is shielded by the ring current of the phenyl group. $\Delta\nu_{OH}$ and $\Delta\delta°$ are smaller for 8 because the Hs on C5 and C6 prevent the phenyl group from fully facing the OH. (See D. C. Kleinfelter J. Am. Chem. Soc. 89: 1734 (1967).)

2. δ^+_{OH} 11 > 10 and 12 > 13. For 11 there is an inductive effect of the phenyl relative to 10. In 13, the OH proton is shielded by the phenyl group and moves upfield from that of 12. The data are δ^+_{OH} = 14.17, 15.22 14.60, and 13.47 ppm for 10, 11, 12, and 13, respectively. (See M. Brookhart, G. C. Levy, and S. Winstein, J. Am. Chem. Soc. 89: 1735 (1967).

3. (a) Yes, although the H bond to t-Bu$_2$O is stronger than to the other acyclic and alicyclic ethers, there is a large negative entropy which reduces its free energy of formation. (i) The basicity, as indicated by -ΔH, is lowest for (C$_6$H$_5$)$_2$O owing to the O-phenyl resonance. The basicities of the other ethers are in the order of the relative electronegativities of the alkyl groups. (ii) This is surprising. In most cases the H bond is stronger toward S than O. The order is reversed from that expected if it were a steric effect. (iii) The order 22 > 18 > 19 reflects the electron density on the carbonyl oxygen. The inductive effect of OMe in 19 outweighs its resonance effect to decrease the basicity whereas the resonance effect of the NMe$_2$ group outweighs its inductive effect to increase the basicity. This same order shows up in $\nu_{C=O}$ frequencies of the respective compounds (see Table 14.4).

4.

Accordingly, the H bond is stronger in the excited state than the ground state (see Chapter 16). The stronger the H bond, the larger is the loss in excited-state H-bond stabilization and the larger the blue-shift upon methylation. This means 26 has the strongest H bond. Since H bonding shifts δ downfield in the nmr, the OH proton signal for 26 would be farthest downfield.

Chapter 13

SPECTRA, ELECTRONEGATIVITY, AND POLARIZABILITY

We saw in Section 3.1b that quite good correlations between electronegativity and ir stretching frequencies have been found and, indeed, this has been used as a way of determining group relative electronegativities. IR frequencies rise with increasing electronegativity of bonded atoms. Several additional sequences can be given here for illustration.

	ν_{C-H}		$\nu_{C=C}$		$\nu_{C\equiv C}$ (144)
▷-H	3009 cm^{-1}	$D_2C=CD_2$	1515 cm^{-1}	$ClC\equiv CCl$	2234 cm^{-1}
Br_3C-H	3023	$H_2C=CH_2$	1623	$BrC\equiv CBr$	2185
Cl_3C-H	3033	$F_2C=CF_2$	1872	$IC\equiv CI$	2118
F_3C-H	3062				

	$\nu_{C=O}$		$\nu_{C=O}$		$\nu_{C=O}$
$H_3C-CO-CH_3$	1715 cm^{-1}	Me-CO-H	1740 cm^{-1}	$H_3C-COOH$	1721 cm^{-1}
$H_3C-CO-CH_2Cl$	1733	Me-CO-Cl	1808	$ClH_2C-COOH$	1736
$Cl_2HC-CO-CHCl_2$	1765	Me-CO-F	1840	$Cl_2HC-COOH$	1751
$H_3C-CO-CF_3$	1769			$Cl_3C-COOH$	1764
$F_3C-CO-CF_3$	1801				
$Br_3C-CO-H$	1742	Cl-CO-Cl	1827		
$Cl_3C-CO-H$	1762	F-CO-F	1928 (highest CO frequency known)		
$F_3C-COOH$	1784				

Closely related to electronegativity is polarizability. In some cases, spectra (as well as other properties; see Section 5.3) are best interpreted in terms of polarizability. The uv spectra of <u>para</u> substituted alkyl- and halobenzenes are an example. The λ_m values correlate better with the polarizabilities of the substituents than with their delocalization ability or electronegativities. Some data are given in Tables 13.1 and 13.3 (Table 13.2 is an alternative set of data for halobenzenes which supports the same view.) In both series, λ_m increases with the polarizability of the halogen or the alkyl group (145), irrespective of whether the other substituent, G, is electron donating or electron withdrawing (X = halogen, H, or alkyl group):

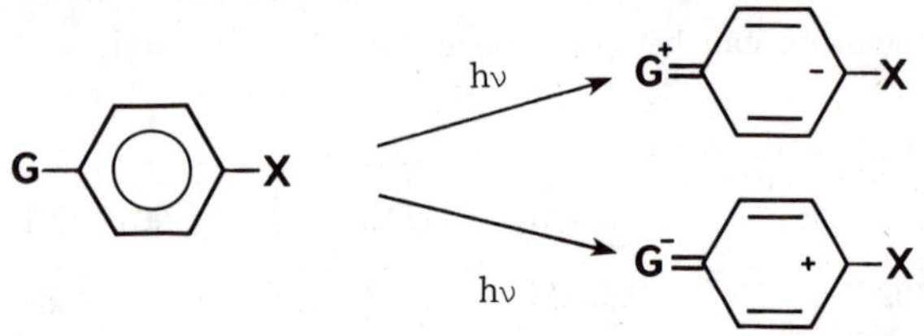

Table 13.1. λ_m values (primary band) for some halobenzenes, p-X-C_6H_4R (nm)

R	X = H	F	Cl	Br	I
H (EtOH)	203	204	210	210	207
CH_3O (gas)	215	214	223	224	230
HO (gas)	206	205	220	220	227
NH_2 (gas)	229	228	237	237	240
CH_3S (EtOH)	254	252	260	262	264
NO_2 (gas)	239	245	251	255	264
$COCH_3$ (gas)	230	233	241	245	254
Polarizability of X	0.42	0.38	2.28	3.34	5.11

Table 13.2. λ_m (first primary band) of some halobenzenes, p-X-C$_6$H$_4$R (nm)

R	X = H	X = F	X = Cl	X = Br	X = I
H (hexane) (146)	202	207	215	216	230
NH$_2$ (hexane) (146)	234	231	241	242	246
NO$_2$ (hexane) (146)	251	257	265	270	287
CHO (hexane) (147)	241	244	253	258	274
COOH (EtOH) (148)	228	227	234	239	252
Polarizability of X	0.42	0.38	2.28	2.34	5.11

The one notable exception is for fluorobenzene when G is electron-withdrawing. In this case, it is probable that the large p-π interaction resonance of F

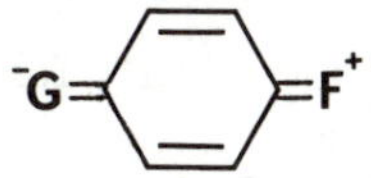

$$\text{}^-\text{G}=\!\!\left\langle\!\!\!=\!\!\!\right\rangle\!\!=\text{F}^+$$

markedly stabilizes the excited state to produce a small increase of λ_m over that of the parent (X = H). The resonance effect outweighs the polarizability factor, and as shown in Chapter 7, the resonance of fluorine is much greater than that of the other halogens and alkyl groups.

We saw in Section 5.3 that the ir spectra of saturated aliphatic ketones correlate with the polarizabilities* rather than hyperconjugative effect of the alkyl groups (see also left column below). Resonance has a much larger effect in lowering $\nu_{C=O}$ of ketones (150).

---

* To what extent the frequency decrease is due to expansion of the C-CO-C angle is not known.

Ch. 13 SPECTRA, ELECTRONEGATIVITY, AND POLARIZABILITY

	$\nu_{C=O}$		$\nu_{C=O}$
Me-CO-Me	1719 cm^{-1}	Et-CO-C$_6$H$_5$	1692 cm^{-1}
i-Pr-CO-Me	1718	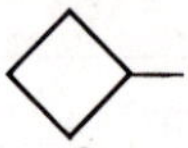 cyclobutyl-CO-C$_6$H$_5$	1686
t-Bu-CO-Me	1710	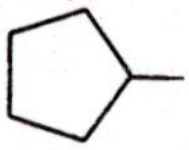 cyclopentyl-CO-C$_6$H$_5$	1687
		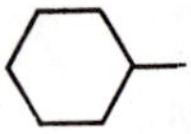 cyclohexyl-CO-C$_6$H$_5$	1686
		CH$_3$CH=CH-CO-C$_6$H$_5$	1680
		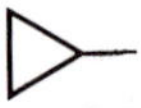 cyclopropyl-CO-C$_6$H$_5$	1677
		C$_6$H$_5$-CO-C$_6$H$_5$	1664

Table 13.3. λ_m values (primary band) for some alkylbenzenes (nm)

System	R = H	R = Me	R = Et	R = i-Pr	R = t-Bu
p-R-C$_6$H$_4$NO$_2$ (gas)	239.1	250.2	251.0	251.3	251.5
p-R-C$_6$H$_4$COCH$_3$ (gas)	231.3	238.9	239.5	239.7	239.8
p-R-C$_6$H$_4$CO$_2$H	228	237			238
p-R-C$_6$H$_4$C(CH$_3$)=$\overset{\oplus}{O}$H	295.5	312.5		315	315.5
(p-R-C$_6$H$_4$)$_3$C$^{\oplus}$	431	452		456	458
(p-R-C$_6$H$_4$)$_2$CH$^{\oplus}$	442	472			480
p-R-C$_6$H$_4$OH	206.3	216.1			216
p-R-C$_6$H$_4$OCH$_3$	215	219.8			219.8
p-R-C$_6$H$_4$NH$_2$	229.4	233.7			232.7
p-R-C$_6$H$_4$NMe$_2$	241.8	243.5			244.5
Calculated polarizability of R ($\times 10^{-25}$ cm^3)	(149)	27	46	65	84

QUESTIONS

1. Offer a rationalization for the observation that the N-H stretching frequencies of substituted anilines and N-methylanilines (P. J. Krueger and H. W. Thompson, Proc. Roy. Soc. A243: 143 (1957)) increase as the respective Hammett σ constants increase whereas the trend is reversed for the O-H frequencies of phenols. (H. W. Thompson and D. A. Jameson, Spectrochim. Acta 13: 236 (1958).)

2. How would you account for the observation that the barrier to rotation about the C-N bond in substituted benzamides increases (M. B. Shambhu, G. A. Digenus, and R. J. Moser, J. Org. Chem., 38: 1227 (1973). L. M. Jackman, T. E. Kavanagh, and R. C. Haddon, Org. Mag. Resonance 1: 109 (1969)) and the N-H ir stretching frequency decreases as the para substituents get more electron withdrawing?

ANSWERS

1. As mentioned in Section 6.1a, the nitrogen atom in anilines is largely sp^3 hybridized but resonance with the ring enhances its sp^2 character. This increases the polarity and strength of the N-H bonds. Accordingly, as the substituent becomes more electron-withdrawing, particularly by resonance, σ increases and so does ν_{N-H}. The electron density on N, of course, also decreases which accounts for the corresponding deshielding of the proton in the nmr. (B. M. Lynch, B. C. Macdonald, and J. G. K. Webb, Tetrahedron 24: 3595 (1968).) In the case of phenols, however, the substituents do not affect the geometry of the OH group and as anticipated, as the substituents gets more electron withdrawing to enhance the phenol resonance, the O-H bond gets weaker and ν_{O-H} decreases.

2. The benzoyl resonance competes with that of the amide group and the latter decreases as the electron-withdrawing resonance of the substituent increases. Any increase in amide resonance, of course, increases the double-bond character of the C-N bond and in turn the barrier to rotation. Thus, the activation energies are 14.59 kcal/mol for p-CH_3O and 16.4 for p-NO_2. Since the amide nitrogen is already sp^2 hybridized, the cross conjugation produces no change in its hybridization and as the amide resonance increases, the electron density on the N atom decreases with a lowering of the N-H ir frequency. (M. Kasai, M. Hirota, Y. Hamada, and H. Matsuoka, Tetrahedron 29: 267 (1973).)

Chapter 14

SPECTRA AND CONSTITUENT CONSTANTS

Another correlation between structure and spectra that has been developed is expressed in terms of constituent constants as was done for such properties as dipole moment (Section 3.2), parachor, diamagnetic susceptibilit (Section 4.1), etc. Generally, the λ_m for a model compound is determined and constants are given for structural changes made to the parent model structure. The most well-known set is that first proposed by Woodward for dienes and expanded by Fieser (Table 14.1). Similar tables have been constructed for enones (Table 14.2), unsaturated esters, vinylogous imides (151), and others, as well as for ir spectra of these systems (Table 14.4). Also, tables are available for uv λ_m values of aromatic series such as acetophenones (Table 14.3), anilines, nitrobenzenes, and for changes of nmr chemical shifts by benzene substituents (see Table 14.6), or simple structural fragments (Table 14.5). A few illustrations will demonstrate the use of these tables. Of course, strong steric strain or steric hindrance to planarity may void these generalizations.

Table 14.1. Uv constituent constants for dienes

Parent diene		217 nm
Heteroannular (C_6 ring)		-3
Homoannular (C_6 ring)		+36
Conjugated C=C		+36
Exocyclic C=C		+ 5
C substituent		+ 5
Auxochrome	OAc	0
	OR	+ 6
	SR	+30
	Cl, Br	+ 5
	NR_2	+60

Table 14.2.

Uv constituent constants for the $\pi \to \pi^*$ band of α,β-enones

Parent $\overset{\gamma\ \beta\ \alpha}{C=C-C=O}$	207 nm
Conjugated C=C	+39
Cisoid C=C	+30
Exocyclic C=C	+ 5
Homoannular diene	+39
C substituent or C_6 ring α	+ 8
β	+10
γ	+12
δ or higher	+18
Auxochromes: β or γ OAc	+ 6
γ OH	+30
γ SR	+85
γ NR_2	+95
β Cl	+12
β Br	+25
Solvents: EtOH	0
dioxane	+ 5
ether	+ 7
hexane	+11
water	- 8

Table 14.3. Uv constituent constants for acetophenones (152)

Parent R-C$_6$H$_4$-CO-CH$_3$			246 nm
Substitution for CH$_3$:	H		4
	OH or OR		-16
Substitution for R:			
alkyl or ring residue,		o, m	3
		p	10
OH, OR	o, m		7
	p		25
Cl	o, m		0
	p		10
Br	o, m		2
	p		15
NHMe	p		73
NMe$_2$	o, m		20
	p		85
NHAc	o, m		20
	p		45

Example 1. Methylene nmr chemical shifts.

Grouping	Calculated from Table 14.5	Observed
Br-CH$_2$-Cl	δ 5.09 ppm	δ 5.16 ppm
C$_6$H$_5$-CH$_2$-Br	4.39	4.42
-CH=CH-CH$_2$-OH	3.91	3.91
C$_6$H$_5$-CH$_2$-C$_6$H$_5$	3.89	3.93
C$_6$H$_5$-CH$_2$-CH$_3$	2.53	2.62

Table 14.4. Ir constituent constants for ketones

Parent	$\begin{matrix}\alpha\\ C\\ C\end{matrix}\!>\!C\!=\!O$		$1720\ cm^{-1}$
Substitution for α C:		H	+10
		Cl	+90
		OH	+40
		OR	+25
		NH_2	− 5
Substitution on α C:			
halogen, OR, OH, OAc			+20
alkyl			− 5
β-alkyl on conjugated C=C			− 5
Transoid conjugated C=C			−35
Cisoid conjugated C=C			−20
Second conjugated C=C			−15
Third conjugated C=C			0
Ring strain:	$C_6 \rightarrow C_5$		+35
	$C_6 \rightarrow$ medium ring		−15
	bridge		+15
H bonding:	intramolecular, weak		−10
	intramolecular, medium		−50
	intramolecular, strong		−100
	intermolecular, dimers		−45
	intermolecular, single		−15
Solvent:	hydrocarbon		+ 7
	polar ($CHCl_3$)		−15
	liquid or solid		−10

Table 14.5.

Nmr constituent constants for proton chemical shifts of methylene groups (153)

Parent a-CH$_2$-b			δ_{CH_2} = 0.23 ppm	
		Δδ		Δδ
a (or b)	Cl	2.53 ppm	R-C=O	1.70 ppm
	Br	2.33	C=C	1.32
	I	1.82	C≡C	1.44
	OR, OH	2.36	C≡N	1.70
	SR	1.64	CH$_3$ or CH$_2$	0.47
	C$_6$H$_5$	1.83		

Example 2. Ir, $\nu_{C=O}$

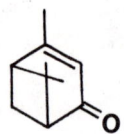

Parent	1720 cm^{-1}	Parent	1720 cm^{-1}
Bridge	+ 15	C$_6$ → C$_5$	+ 35
t conjugated C=C	− 35	OH on α C	+ 20
2 substitutions on α C	− 10	t conjugated C=C	− 35
Conjugated β-alkyl	− 5	conjugated β-alkyl	− 5
	1685 cm^{-1}	Intramol H bond	− 10
Observed	1685		1725 cm^{-1}
		Observed	1723

Therefore, the intramolecular H bond is a weak one.

Example 3. Uv of enones.

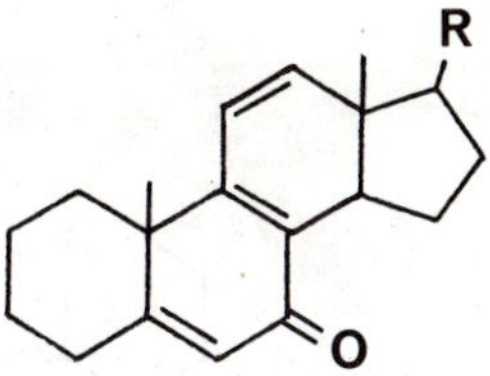

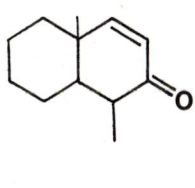

Parent	207 nm
α C (C=C)	8
β C	10
γ C	12
Higher than γ C	18
Cisoid C=C	30
Conjugated C=C	39
Exo C=C	5
	329 nm
Observed	256, and 327 nm

A compound was isolated from a particular synthesis with a structure 14.1 or 14.2. A choice could easily be made from its uv spectrum, which agrees for that calculated for 14.2 but not for 14.1. Observed λ_m values were 251 nm ($\varepsilon > 10^3$) and 321 nm ($\varepsilon \simeq 50$).

14.1

Parent	207 nm
α C	8
γ C	12
	227 nm

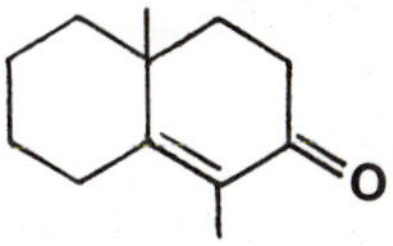

14.2

Parent	207 nm
α C	8
β C	10
2γ C	24
Exo C=C	5
	254 nm

416   Ch. 14   SPECTRA AND CONSTITUENT CONSTANTS

Example 4. Uv of dienes.

Parent	217	Parent	217
Cisoid diene	30	2 exo C=C	10
2 C substitutents	10	4 C substituents	20
	257 nm	Heteroannular	- 3
Observed	262		244 nm
		Observed	242

Example 5. Uv of acetophenones.

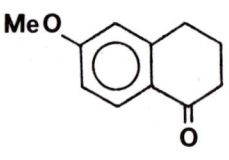

Parent	246	Parent	246
o-ring residue	3	o-ring residue	3
p-OR	25	o-OH	7
	274 nm	m-Cl	0
Observed	276		256 nm
		Observed	257

Table 14.6. NMR chemical shift substituent effects on benzene protons* (154)

Substituent	S_o (ppm)	S_m (ppm)	S_p (ppm)
H	0.00	0.00	0.00
CH_3	0.17	0.07	0.18
CH=CHR	-0.08	0.03	0.14
C_6H_5	-0.29	-0.12	0.03
CHO	-0.52	-0.20	-0.31
$C{\nwarrow}^O_R$	-0.54	-0.11	-0.23
COOH, COOR	-0.53	-0.12	-0.19
$CONH_2$, $CONR_2$	-0.60	-0.07	-0.16
CN	-0.49	-0.24	-0.32
OH	0.53	0.14	0.58
OR	0.41	0.04	0.37
OCOR	0.17	-0.07	0.11
NH_2	0.72	0.27	0.84
NR_2	0.67	0.17	0.80
$\overset{+}{N}H_3$	-0.08	-0.14	0.09
NHCOR	-0.26	0.00	0.21
NO_2	-0.78	-0.27	-0.34
Cl	-0.10	-0.07	0.03
Br	-0.24	-0.02	-0.01
SO_3H	-0.34	0.00	0.04
SO_2NH_2	-0.45	-0.21	-0.22

* In DMSO-d_6 solutions. A negative sign indicates a downfield shift.

Chapter 15

SPECTRA AND INTRAMOLECULAR GEOMETRY

Although diffraction techniques could provide the best information about molecular geometry, the interpretation of the data is still quite specialized, particularly for mobile systems and transient intermediates. The collection and interpretation of the data may take weeks, even when no special problems arise. Moreover, gas phase electron diffraction methods are only applicable to small molecules and X-ray data provide information for molecules in the crystal or solid state. Consequently, other physical methods continue to be used by most chemists. Fortunately, a large number of correlations have been developed from which information about intramolecular geometry may be deduced.

In Chapter 17 there are several spectral correlations for determining conformations of substituents attached to various ring systems, and we saw in Chapter 10 how spectral data may be used to infer planar or nonplanar chromophores and cis or trans structures. We shall discuss here several semiquantitative relationships between structure and spectra.

15.1. Dihedral angles

Nmr coupling constants are very sensitive to dihedral angles. One of the first widely used relationships was that developed by Karplus (155) which can be expressed as equation (15.1).

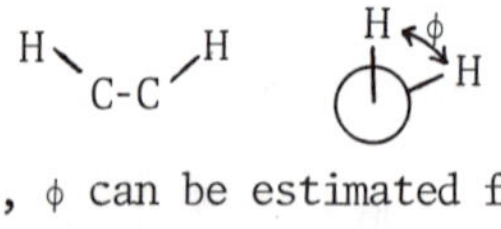

$$^3\underline{J} \text{ (Hz)} = 4.22 - 0.5 \cos \phi + 4.5 \cos 2\phi \qquad (15.1)$$

Thus, ϕ can be estimated from $^3\underline{J}$ (156). Note that $^3\underline{J}$ is a maximum when $\phi = 0°$ or $180°$. Equation (15.1) has been modified to (15.2) for application to allylic systems (157).

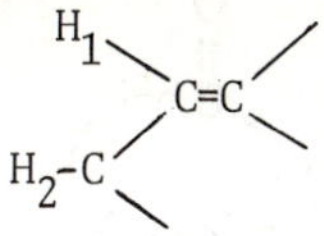

$$J_{1,2} = 10.6 \cos^2 \phi \quad (\phi = 0\text{-}90°)$$
$$= 11.4 \cos^2 \phi \quad (\phi = 90\text{-}180°) \quad (15.2)$$

(with structure: H_1, H_2–C on one carbon of C=C)

For example, it was concluded from $J_{1,2}$ for 1,4-cyclohexadiene and use of equation (15.2) that the compound is planar (158), which was confirmed by X-ray diffraction (159), Raman (160), ir (161), and other (162) measurements.

Dihedral angles between connected π systems A-B can be estimated from uv data by equation (15.3)

$$\varepsilon/\varepsilon_0 = \cos^2 \phi \quad (15.3)$$

where ε and ε_0 are the uv molar extinction coefficients of λ_m for the A-B system when the angle between the planes of A and B are ϕ and $0°$, respectively (163). For example, as the OCH_3 group is flanked by methyl groups in 15.1 it is forced out of the plane of the phenyl ring and the intensity of the "anisole" band decreases:

Structure	λ_m	ε
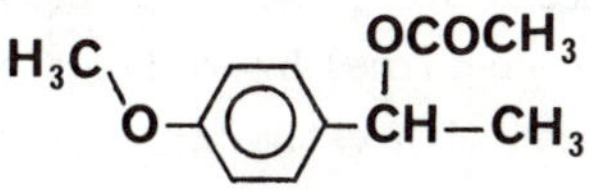 15.1 (H₃C–O–C₆H₃(CH₃)–CH(OCOCH₃)CH₃)	273 nm	4,600
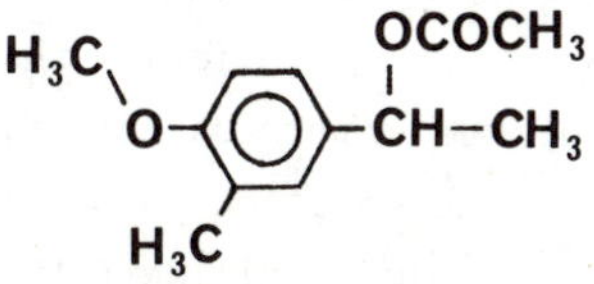 (H₃C–O–C₆H₂(CH₃)₂–CH(OCOCH₃)CH₃ with one extra H₃C)	274	4,050
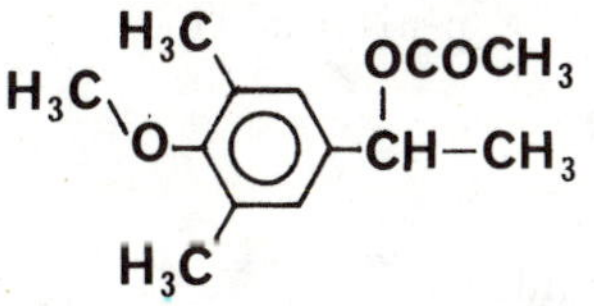 15.2 (H₃C, H₃C–O–C₆H₂(CH₃)₂–CH(OCOCH₃)CH₃ with H₃C groups)	266	690

Therefore, for 15.2, $\varepsilon/\varepsilon_0 = 690/4600 = 0.15 = \cos^2 \phi$
$\phi = 67°$.

In a similar fashion, ϕ has been estimated for a number of systems (164), including methyl vinyl ketones (165), biphenyls (166), stilbenes (167), and acetophenones (168).

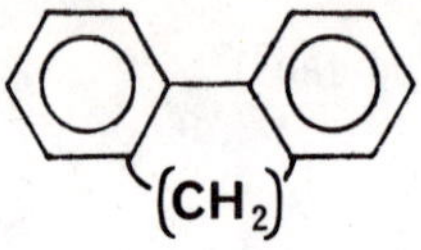

n	φ (169)
2	15°
3	49°
5	68°

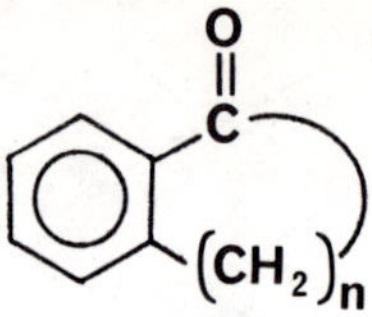

n	φ (170)
3	0°
4	29°
5	37°
6	44°

It should be added that the validity of equation (15.3) has been challenged (171). Although the relationship $RE_\phi = RE_0 \cos^2 \phi$ holds for ground-state resonance energy, where RE_ϕ and RE_0 are the resonance energies of the system A-B for dihedral angles between A and B of ϕ and 0°, respectively, there is apparently no theoretical justification for applying the $\cos^2 \phi$ function to transition energies to excited states. Nevertheless, dihedral angles based on equation (15.3) have been confirmed by diffraction measurements (172) and are in reasonable agreement with estimates from other methods (168). For instance, the angles of twist of the 2-nitro and 4-nitro groups (it was assumed and later verified by diffraction that the 6-NO_2 is not rotated and that only the 2-NO_2 group rotates as a means of relieving the steric repulsion from the NH_2 group) in 3-nitropicramide were calculated from the diminuition in intensity of the o-nitroaniline (318 nm) and p-nitroaniline (407 nm) bands (172

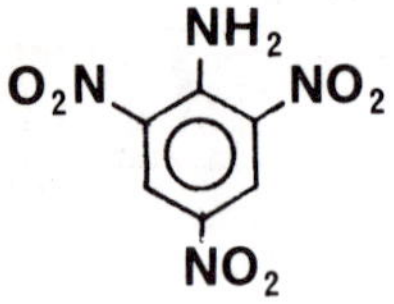

Picramide 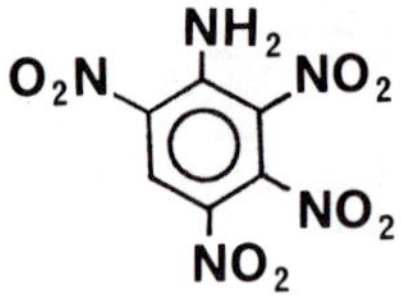   3-Nitropicramide

	ε	ε
Picramide	12,000 (318 nm)*	7,800 (408 nm)*
3-Nitropicramide	10,500 (320)	6,000 (402)
3,5-Dinitropicramide	7,600 (316)	4,800 (400)

* λ_m in parentheses.

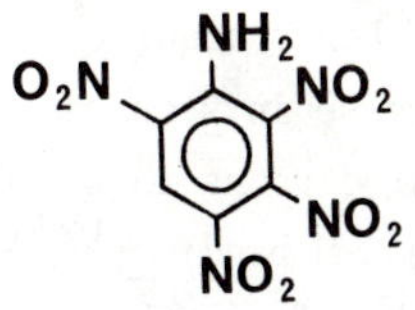

	ϕ_{calc}	ϕ_{obs}
2-NO$_2$	21°	19°
4-NO$_2$	42°	45°
6-NO$_2$	(0°)	3°

In another case, the angle of twist of the C=O bond out of the plane of the C=C bond in 15.3 is computed to be 41.5° from the $\varepsilon/\varepsilon_0 = \cos^{-2}\phi$ relationship and 55° from a $\varepsilon/\varepsilon_0 = \cos\phi$ basis (173). The former value is in reasonable agreement with that obtained by X-ray crystallographic measurements (37.7°) and the value from $\delta(^{13}C)$ nmr data (35°) (173). Also, resonance inhibition in some <u>ortho</u> substituted compounds based on absorption intensities is fairly close to values determined from Taft substituent constants (174). In the latter procedure, it is assumed that complete inhibition of resonance would reduce the Hammett substituent constant σ to the induction constant $\underline{\sigma_I}$, and the inhibition of resonance is expressed by the quantity

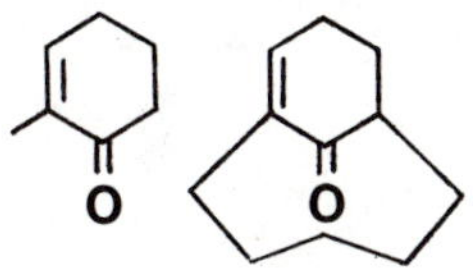

$$\frac{\sigma_{\underline{R}}^0 - \sigma_{\underline{R}}}{\sigma_{\underline{R}}^0}$$

	15.3	
λ_m^{EtOH}	235	238
ε	10^4	5630

where $\upsilon_{\underline{R}} = \sigma - \sigma_{\underline{I}}$ for the resonance-subdued group and $\sigma_{\underline{R}}^0 = \sigma^0 - \sigma_{\underline{I}}$ for the resonance uninhibited group.

15.2 Carbon bond angles

As a C-C-C bond angle decreases, the C-C bonds acquire greater strain and greater p character with a concurrent increase in s character in the attached C-H bonds. It has been shown for hydrocarbons that ^{13}C-H coupling is a linear function of the s character of the C-H bond: $\underline{J}(^{13}\text{C-H}) = 500\underline{f}_s$, where $\underline{f}_s$ is the fractional s character of the bonding carbon orbital. It has also been possible to correlate $\underline{J}(^{13}\text{C-H})$*(175) with the C-C-C bond angle (176):

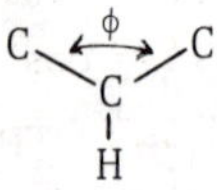

$$\phi = -1.3\underline{J}(^{13}\text{C-H}) + 269° \qquad (15.4)$$

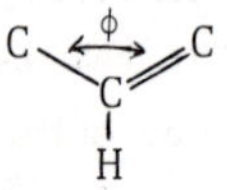

$$\phi = -0.9\underline{J}(^{13}\text{C-H}) + 264° \qquad (15.5)$$

Further, the C-C and C-H bond lengths have also been related to $\underline{f}_s$ by equations (15.6) and (15.7):

$$\underline{r}_{\text{C-H}} = 1.1597 - 0.00209\ \underline{f}_s \qquad (15.6)$$

$$\underline{r}_{\text{C-C}} = 1.692 - 0.0062\ (\text{mean } \underline{f}_s) \qquad (15.7)$$

These relationships have been used to estimate bond angles (178), bond distances (179), and other geometric conditions in molecules (180).

The C-C-C bond angle in a C-CO-C fragment can be estimated from the ir carbonyl frequency by equation (15.8):

$$\nu_{\text{C=O}}\ (\text{cm}^{-1}) = 1974.6 - 2.2\ \phi. \qquad (15.8)$$

The smaller the angle, the greater is the strain and the higher is the carbonyl frequency. For example, C-CO-C bond angles calculated from equation (15.8) are listed for some ketones in Table 15.1.

---

* The C-C=C bond angle has also been correlated with $\nu_{\text{C=C}}$: $\nu_{\text{C=C}} = 1773 - 1.1$ (175).

Table 15.1. Carbonyl ir frequencies and calculated bond angles (181, 182)

Ketone	$\nu_{C=O}^{CCl_4}$ (cm^{-1})	Calculated C-CO-C angle
Cyclopropanone	1815	72.5°
Cyclobutanone	1791	83.0°
Cyclopentanone	1749.8	102.2°
Cyclohexanone	1717.8	116.6°
Cyclooctanone	1701	123°
2-Norbornanone	1751.2	101.5°
7-Norbornanone	1780	88.5°
Adamantanone	1727 (average of doublet)	112.5° (average)

A deviation of $\nu_{C=O}$ from 1720 cm^{-1} implies a bond angle strain. The data show, for example, that there is more strain in 7-norbornanone than in the 2-isomer, and that adamantanone has a small strain in the molecule. As to be expected, the C-CO-C bond angles of di-t-butyl ketone ($\nu_{C=O}$ 1686 cm^{-1}) and t-butyl triptyl ketone ($\nu_{C=O}$ 1675 cm^{-1}) are expanded, based on their $\nu_{C=O}$ values (183). The high ir carbonyl frequency in perchlorobishomocubanone indicates, on the one hand, that there is a substantial bond angle strain in the carbonyl bridge (184).

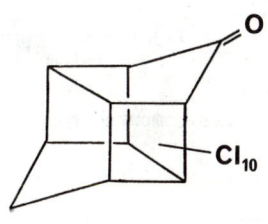

$$\nu_{C=O} = 1810 \text{ cm}^{-1}$$

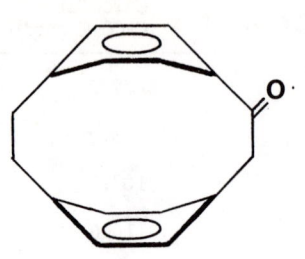

15.4

$\nu_{C=O}$ 1685 cm^{-1}

On the otherhand, 15.4 has a low carbonyl frequency because it has an expanded C-CO-C bond angle. [2,2] Paracyclophane itself has bent benzene rings, is under 31 kcal mol strain, and has uv, ir, and nmr spectral characteristics typical of strained rings.(Chapter 11) (185). The low $\nu_{C=O}$ value for 15.4 cannot be due to conjugation because the π orbitals are orthogonal.

15.3. H bond distances

It was noted in Chapter 12 that H bonding lowers O-H stretching frequencies, that this decrease is related to the strength of the H bond, which in turn is a function of the -H···O distance. Consequently, it is reasonable to expect a relationship between $\Delta\nu_{OH}$ and the H···O distance. This has been expressed in several ways, one of which is equation (15.9).

$$D_{H\cdots O} = 1.4 + \frac{42.5}{\Delta\nu(cm^{-1}) + 3.5} \text{ Å} \qquad (15.9)$$

For example, the H···O distances computed for some glycols from equation (15.9) are:

Compound	Δν (cm^{-1})	$D_{H\cdots O}$ (Å)
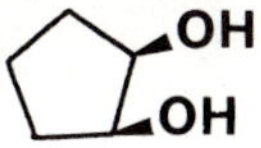 cyclopentane-1,2-diol (cis)	61	2.06
 cyclohexane-1,2-diol (diequatorial)	32	2.60
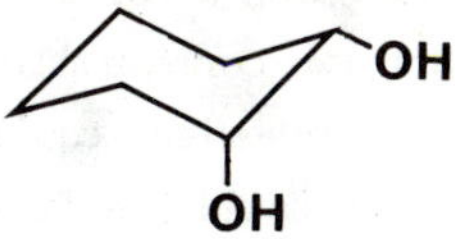 cyclohexane-1,2-diol (axial-equatorial)	39	2.40
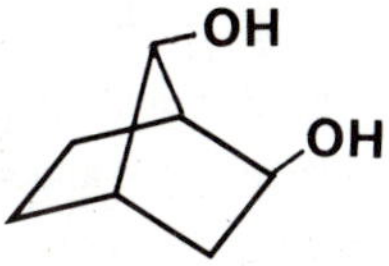 norbornane-2,3-diol (cis)	86	1.87
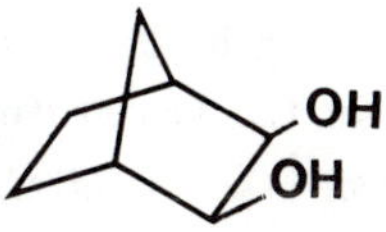 norbornane-2,3-diol	103	1.80

Observe that two axial, equatorial OH groups are closer together on a C_6 ring than when they are diequatorial. Equation (15.9) and others derived from it have been used to infer various geometric relationships in steroids and other hydroxy compounds (186, 187). Similarly, there is a correlation between the intramolecular O-H stretching frequencies of molecules possessing the β-phenylethanol structure with the dihedral angle formed by the C-C bond and the plane of the phenyl ring (188).

We see, therefore, that it is possible to derive considerable information about intramolecular geometry from simple spectral measurements without having to resort to the more elaborate diffraction techniques.

Chapter 16

SPECTRA AND EXCITED STATES

Uv spectroscopy provides convenient access to information about electronically excited states. One approach, based on a Förster cycle (189), has been outlined by Jaffe (190, 191). Consider two corresponding species of a chemical equilibrium, e.g., an acid and its conjugate acid, a redox couple, or two isomers whose energies can be designated A and A' in the ground state (GS) and A* and A*' in their electronically excited states (ES). ΔH and ΔH^* would be the heats of reaction in the two states and ΔE and ΔE^* the transition energies between the GS and ES. The latter can be obtained from the wavelengths of the uv absorption bands of fluorescence bands. According to the diagram (Figure 16.1), $\Delta E - \Delta H = \Delta E' - \Delta H^*$, which can be rearranged to $\Delta H - \Delta H^* = \Delta E - \Delta E'$. From thermodynamics, we have $\Delta E = hc\nu = RT \ln K$, where h, c, and R are Planck's constant, speed of light, and the gas constant, K is the equilibrium constant, T the absolute temperature, k is Botzmann's constant, and ν the wave number corresponding to λ_m. We can now write

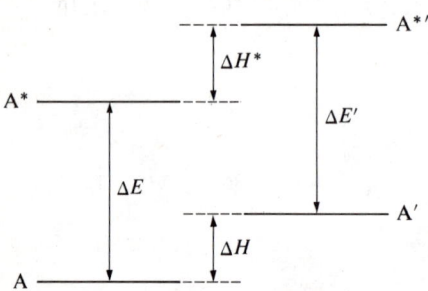

Figure 16.1. Energy diagram for a Föster cycle.

$$\ln \underline{K}^*/\underline{K} = \frac{hc}{\underline{k}T} (\nu - \nu'). \qquad (16.1)$$

Therefore, $\underline{K}^*$ can be obtained from the GS equilibrium constant $\underline{K}$ and frequencies of the absorption or fluorescence bands of the two species A and A'. In the case of dissociation constants, equation (16.1) can be rearranged to (16.2) for convenience:

$$p\underline{K}_a - p\underline{K}_a^* = \frac{0.625}{T} (\nu - \nu'). \qquad (16.2)$$

426

This method has been applied to a variety of compounds, including hydrocarbons, phenols, aromatic carboxylic acids, arylamines, azo compounds (190), ketones (192), and heterocyclic nitrogen bases, from which it is found that acidities or basicities may differ by as much as a factor of 10^6 between the GS and ES. Some illustrative data are given in Tables 16.1 and 16.2.

Table 16.1.
Ground and excited state acidities of some substituted phenols (193)

Substituent	pK_a	pK_a^*	Substituent	pK_a	pK_a^*
H	10.0	4.0			
p-F	9.91	4.4	m-F	9.21	3.8
p-Cl	9.42	3.2	m-Cl	9.13	3.0
p-Br	9.36	3.1	m-Br	9.03	2.8
p-Me	10.26	4.3	m-Me	10.09	4.0
p-Et	10.21	4.3	m-Et	10.07	4.1
p-OMe	10.21	5.6	m-OMe	9.65	4.6
p-SO$_3^-$	9.03	2.4			
p-NMe$_3^+$	8.35	1.7			

In actual practice, uv absorption occurs with transmissions to higher vibrational states and, according to the Franck-Condon principle, ΔE determined from λ_m is too large. Fluorescence transitions are also to higher vibrational states however, and ΔE determined from fluorescence is too small. Hence, the best results are obtained from an average of the two values. Obviously, wavelength values used must be taken for corresponding transitions in the two species, A and A'. A similar treatment using phosphorescence data can yield information about molecules in the triplet state. A few such values are given in Table 16.2.

Note that phenols and carboxylic acids are more acidic and amines and hydrocarbons (196) more basic in excited states than in ground states. By analogous methods, using the respective GS properties and uv or uv plus fluorescence data, other ES properties have been estimated such as dipole moments and H-bond strengths. For example, dipole moment differences between excited and ground states (ES more polar) are 0.2D for phenol and

Table 16.2. Some pK_a values in ground and excited states (193, 194, 195)

Compound		Ground state	Singlet excited state	Triplet excited state
p-R-C$_6$H$_4$OH,	R = H	10.00	4.0	8.5
	F	9.91	4.4	8.7
	Cl	9.42	3.2	8.0
	Br	9.36	3.1	7.7
	CH$_3$	10.26	4.3	8.6
	OCH$_3$	10.21	5.6	8.6
	SO$_3^-$	9.03	2.4	7.5
	$\overset{+}{N}$Me$_3$	8.35	1.7	
2-Naphthol		9.5	3.1	8.3
2-Naphthylamine		4.1	-2	2.7
Acridine		5.5	10.6	5.6
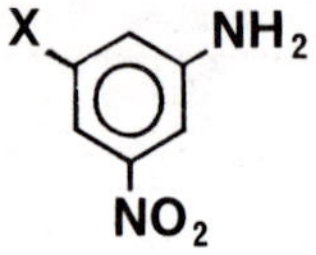 (X-C$_6$H$_3$(NH$_2$)(NO$_2$))	X = H	2.29	-2.81	
	I	1.44	-8.15	
	OCH$_3$	2.13	-7.19	
	CH$_3$	2.38	-4.88	
	NO$_2$	0.23	-8.71	
2-Phenylpyridine		4.55	10.0	
3-Phenylpyridine		4.85	12.2	
4-Phenylpyridine		5.38	14.5	
2-(2'-Thienyl)-pyridine		3.73	9.5	
3-(2'-Thienyl)-pyridine		4.87	14.0	
4-(2'-Thienyl)-pyridine		5.50	12.1	

0.85D for aniline (197), and GS and ES H-bond energies for a few compounds are (198):

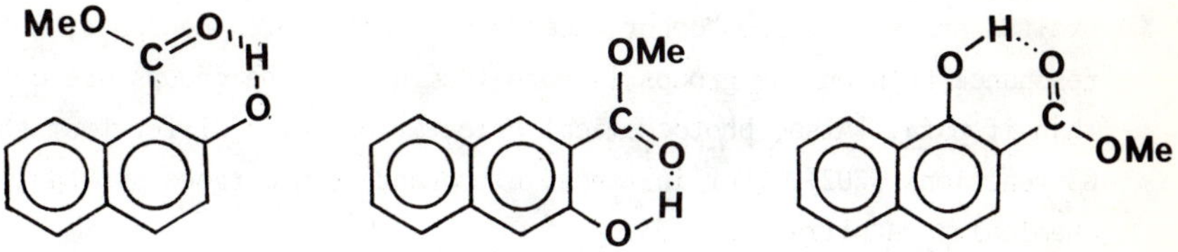

E_{GS} (kcal/mol)	9	5.7	8.4
E_{ES} (kcal/mol)	14.7	12.9	11.3

All of the data confirm the idea that resonance among dipolar structures is greater in the excited state than ground states. Even <u>meta</u>-quinoidal resonance, e.g.,

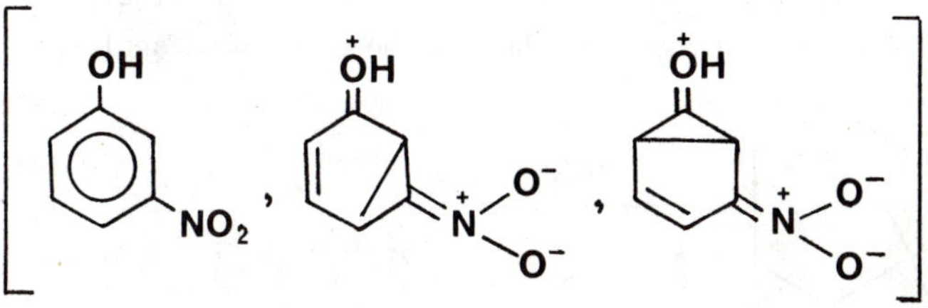

becomes significant in the excited state (193, 199). In fact, λ_m for benzene derivatives containing electron-donor and electron-acceptor groups, e.g., nitrocyclopropylbenzenes (200), nitroanisoles, and cyanophenols, are larger and more similar to each other for the ortho and meta isomers than for the para isomers (201). MO calculations (199, 201) indicate that in the first excited state of these "donor-acceptor" disubstituted benzenes, interaction resonance between the groups is more facile when the groups are ortho or meta than if para. Also, photochemical orientation rules differ from those for GS reactions (202). For instance, nitrobenzene undergoes substantial photochemical para nitration (203).

QUESTIONS

1. Predict which will exhibit the greater proton coupling, between H_1 and H_n or H_1 and H_x.

2. From ^{13}C-^{13}C and ^{13}C-H coupling, the C-C and C-H bonds in cyclopropane are regarded as sp^5- and sp^2-hybridized, respectively (204). Estimate the C-C and C-H bond lengths and the C-C-C bond angles in the compound.

3.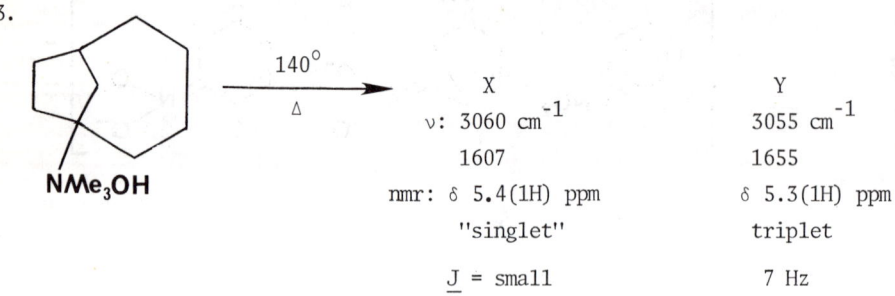

	X	Y
ν:	3060 cm^{-1}	3055 cm^{-1}
	1607	1655
nmr:	δ 5.4(1H) ppm	δ 5.3(1H) ppm
	"singlet"	triplet
	J = small	7 Hz

Suggest structures for X and Y and give the basis of your suggestion.

4. Predict the correlation between the N-H proton chemical shift and Hammett substituent constants of p,p'-substituted diphenylanilines.

5. The C=$\overset{+}{O}$H proton of protonated phenylacetone (phenylacetone in SO_2-FSO_3H-SbF_5 at low temperatures) resonates at two places, δ = 13.47 (triplet) and δ = 14.60 (quarter). How would you explain these values?

ANSWERS

1. $H_1:H_x > H_1:H_n$. From models, it can be seen that the dihedral angle between $\overline{H_1}$ and $\overline{H_n}$ is about $90°$, therefore $^3J \approx 0$. The dihedral angle between H_1 and $\overline{H_x}$ is approximately $30°$ and $^3J \approx 4-5$ Hz.

2. From equations (15.4), (15.6), and (15.7), with $\underline{f_s}$ = 16.7% for the C-C bonds and 33.3% for C-H, we get $\underline{r}_{C=C}$ = 1.589Å, $\underline{r}_{C-H}$ = 1.09Å, and ϕ = 60.7°.

3.

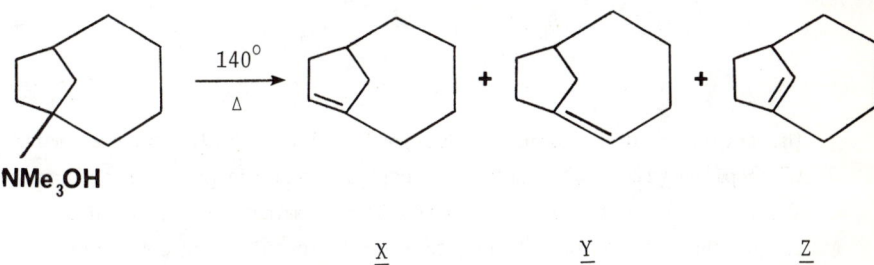

$\underline{X}$ $\underline{Y}$ $\underline{Z}$

The reaction is a common one. The question is, where is the C=C bond? The 3060 cm^{-1} band is for a =C-H stretch and the 1607 (or 1655) cm^{-1} band for a C=C stretch. $\nu_{C=C}$ is about 1611 cm^{-1} for a C_5 ring and ~1660 for a C_7 ring so X and Y are assigned on this basis. The nmr peaks are for vinyl protons. From equation (15.2), ^3J for allylic protons is smaller for a C_5 ring than a C_7 ring, so again we can assign structures to X and Y as shown. The coupling in X is so small as to appear as a singlet. The highly strained Z isomer (S = 7) is also a possible structure but in this case the vinyl proton signal would be a doublet. (See J. R. Wiseman and W. A. Pletcher, J. Am. Chem. Soc. 92: 956 (1970).)

4. The N-H proton chemical shift should increase as the substituent gets more electron withdrawing (increasing σ) on the basis of inductance as well as resonance. This is actually observed. (M. Kasai, M. Hirota, Y. Hamada, and H. Matsuoka, Tetrahedron 29: 267 (1973).)

5.

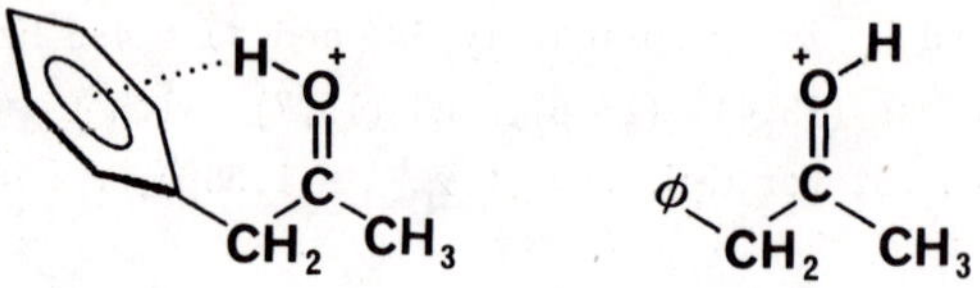

$$\text{[phenyl}\cdots\text{H-O}^+\text{=C(CH}_2\text{)(CH}_3\text{)]} \qquad \text{[}\phi\text{-CH}_2\text{-C(=O}^+\text{H)(CH}_3\text{)]}$$

The triplet peak is obviously due to coupling with CH_2 and the quartet to coupling with CH_3. H bonding deshields and the phenyl ring current shields. The latter effect apparently predominates here, even overcoming the small shielding due to the CH_3 in the other conformer.

REFERENCES
Part III

1. *Elucidation of Organic Structures by Physical and Chemical Methods*, Part II, 2nd ed., eds. K.W. Bentley and G.W. Kirby, New York: Wiley-Interscience, 1973; *Determination of Organic Structures by Physical Methods*, Vols. 3 and 4, eds. F.C. Nachod and J.J. Zuckerman, New York: Academic, 1971; I. Fleming and D.H. Williams, *Spectroscopic Methods in Organic Chemistry*, New York: McGraw-Hill, 1966.
2. S. Bank and W.D. Closson, *Tetrahedron* 24: 381 (1968).
3. P. Laszlo and P.v.R. Schleyer, *J. Am. Chem. Soc.* 86: 1171 (1964); O.L. Chapman, *J. Am. Chem. Soc.* 85: 2014 (1963).
4. D.C. Neckers, *J. Chem. Educ.* 50: 164 (1973).
5. H. Brouwer and J.B. Stothers, *Canad. J. Chem.* 50: 601 (1972).
6. G.A. Olah, Y. Halpern, and G. Liang, *J. Am. Chem. Soc.* 94: 3554 (1972).
7. D. Zook, T.J. Russo, E.F. Ferrand, and D.S. Stote, *J. Org. Chem.* 33: 2226 (1968).
8. B.G. Ramsey and R.W. Taft, *J. Am. Chem. Soc.* 88: 3063 (1966).
9. E.M. Arnett, R.P. Quirk, and J.W. Larsen, *J. Am. Chem. Soc.* 92: 3977 (1970).
10. F.A.L. Anet and G.E. Schenck, *J. Am. Chem. Soc.* 93: 556 (1971).
11. R.A. Clark and R.A. Feato, *J. Am. Chem. Soc.* 92: 4736 (1970).
12. J.F. Chiang and C.F. Wilcox, Jr., *J. Am. Chem. Soc.* 95: 2885 (1973).
13. L.D. Quin, J.J. Breen, and D.K. Myers, *J. Org. Chem.* 36: 1297 (1971).
14. D. Redmore, *J. Org. Chem.* 38: 1306 (1973).
15. H. Schindlbauer and G. Hajek, *Ber.* 96: 2601 (1963).
16. K.N. Scott, *J. Am. Chem. Soc.* 94: 8564 (1972).
17. K.B. Bates, *et al.*, *J. Am. Chem. Soc.* 95: 926 (1973).
18. G.A. Olah and G.D. Mateescu, *J. Am. Chem. Soc.* 92: 155 (1970).
19. L.N. Ferguson and G.E.K. Branch, *J. Am. Chem. Soc.* 66: 1467 (1944).
20. P. Blatz, D. Pippert, L. Sherman, and V. Balasubramaniyan, *J. Chem. Educ.* 46: 513 (1969).
21. S. Akiyama and M. Nakagawa, *Bull. Chem. Soc. Japan* 43: 3561, 3567 (1970).
22. A.I. Koprianov, *Russ. Chem. Revs.* 40: 594 (1971).
23. S.B. Coan, D.E. Trucker, and E.I. Becker, *J. Am. Chem. Soc.* 75: 900 (1953).
24. E.R. Katzenellenbogen and G.E.K. Branch, *J. Am. Chem. Soc.* 69: 1615 (1947).
25. B.M. Tolbert, G.E.K. Branch, and B.E. Berlenback, *J. Am. Chem. Soc.* 67: 887 (1945); C.C. Barker and G. Hallas, *J. Chem. Soc.* 1529 (1961).
26. G.N. Lewis and J. Bigeleisen, *J. Am. Chem. Soc.* 65: 2102 (1943).
27. R.F. Rekker and W. Th. Nauta, *Rec. Trav. Chim.* 90: 343 (1971).

28. C.H. Warren, G. Wettermark, and K. Weiss, *J. Am. Chem. Soc.* 93: 4658 (1971).
29. E.J. Moriconi, W.F. O'Connor, and W.F. Forbes, *J. Am. Chem. Soc.* 82: 5454 (1960).
30. M.A. El-Bayoumi, *J. Am. Chem. Soc.* 93: 586 (1971).
31. L.A. Cohen and S. Takahashi, *J. Am. Chem. Soc.* 95: 443 (1973).
32. H.E. Simmons and T. Fukunaga, *J. Am. Chem. Soc.* 89: 5208 (1967); R. Hoffmann, A. Imamura, and G.D. Zeiss, *J. Am. Chem. Soc.* 89: 5215 (1967).
33. E.T. McBee, *et al.*, *J. Org. Chem.* 31: 768, 4260 (1966).
34. J.M. Holland and D.W. Jones, *Chem. Commun.* 122 (1970).
35. R. Boschi, A.S. Dreiding, and E. Heilbronner, *J. Am. Chem. Soc.* 92: 123 (1970).
36. M.F. Semmelhack, R.J. DeFranco, Z. Margolin, and J. Stock, *J. Am. Chem. Soc.* 94: 2115 (1972).
37. S. Schweig, U. Weidner, R.K. Hill, and D.A. Cullison, *J. Am. Chem. Soc.* 95: 5426 (1973).
38. D.P.G. Hamon, *J. Am. Chem. Soc.* 90: 4513 (1968); P. Dowd and K. Sachdev, *J. Am. Chem. Soc.* 89: 715 (1967).
39. P.A. Leermakers and H.T. Thomas, *J. Am. Chem. Soc.* 87: 1620 (1965).
40. D.O. Cowan, *et al.*, *Angew. Chem. Int. Ed.* 10: 401 (1971).
41. See G.A. Olah, *Science* 168: 1298 (1970); C.J. Lancelot and P.v.R. Schleyer, *J. Am. Chem. Soc.* 91: 4297 (1969); I. Lillien and L. Handloser, *J. Org. Chem.* 34: 3058 (1969).
42. S. Moon and H. Bohm, *J. Org. Chem.* 36: 1434 (1971); J.K. Crandall, J.P. Arrington, and C.F. Mayer, *J. Org. Chem.* 36: 1428 (1971).
43. R.G. Warren, Y. Chow, and L.N. Ferguson, *Chem. Commun.* 1521 (1971).
44. J.C. Nnadi, A.W. Peters, and S.Y. Wang, *J. Am. Chem. Soc.* 94: 712 (1972).
45. H.T. Thomas and K. Mislow, *J. Am. Chem. Soc.* 92: 6292 (1970); D.A. Lightner and W.A. Beavers, *J. Am. Chem. Soc.* 93: 2677 (1971).
46. P. Bischof, E. Heilbronner, H. Prinzbach, and H.D. Martin, *Helv. Chim. Acta* 54: 1072 (1971).
47. N. Heap and G.H. Whitham, *J. Chem. Soc.* (B) 164 (1966).
48. E.M. Kosower, W.D. Closson, H.L. Goering, and J.C. Gross, *J. Am. Chem. Soc.* 83: 2013 (1961).
49. J. Wagner, W. Wojnarowski, J.E. Anderson, and J.M. Lehn, *Tetrahedron* 25: 657 (1969).
50. N.L. Allinger and M.A. Miller, *J. Am. Chem. Soc.* 86: 2811 (1964).
51. N. Filipescu and D.S.C. Chang, *J. Am. Chem. Soc.* 94: 5990 (1972).
52. For its pes, see R. Gleiter, *et al.*, *Tetrahedron* 29: 565 (1973).
53. R. Hoffmann, *Accts. Chem. Res.* 4: 1 (1971).
54. S.C. Neely, R. Fink, D.v.d. Helm, and J.J. Bloomfield, *J. Am. Chem. Soc.* 93: 4903 (1971).
55. S. Winstein, *J. Am. Chem. Soc.* 81: 6524 (1959).

56. P. Bischof, R. Gleiter, and E. Heilbronner, *Helv. Chim. Acta* 53: 1425 (1970).
57. W.R. Roth, *et al.*, *J. Am. Chem. Soc.* 86: 3178 (1964).
58. P. Bischof, J.A. Hashmall, E. Heilbronner, and V. Hornung, *Helv. Chim. Acta* 52: 1745 (1969).
59. G.A. Olah, G.D. Mateescu, and J.L. Riemenschneider, *J. Am. Chem. Soc.* 94: 2529 (1972).
60. G.J. Abruscato, R.G. Binder, and T.T. Tidwell, *J. Org. Chem.* 37: 1787 (1972).
61. J.E. Baldwin and A.H. Andrist, *J. Am. Chem. Soc.* 93: 3289 (1971).
62. R. Hoffmann and R.B. Davidson, *J. Am. Chem. Soc.* 93: 5699 (1971).
63. G.N. Fickes and C.B. Rose, *J. Org. Chem.* 37: 2898 (1972).
64. H. Rau, *Angew. Chem. Int. Ed.* 12: 224 (1973).
65. N.C. Baird, P. de Mayo, J.R. Swenson, and M.C. Usselman, *Chem. Commun.* 314 (1973).
66. R.J. Boyd, J.C. Bünzli, J.P. Snyder, and M.L. Heyman, *J. Am. Chem. Soc.* 95: 6478 (1973).
67. W. Reeve and D.M. Reichel, *J. Org. Chem.* 37: 68 (1972); W.F. Forbes, R. Shilton, and A. Balasubramanian, *J. Org. Chem.* 29: 3527 (1964).
68. J. Sicher, *et al.*, *Tetrahedron* 22: 659 (1966).
69. A.T. Blomquist and A. Goldstein, *J. Am. Chem. Soc.* 77: 998 (1955).
70. G.L. Buchanan and G. Jamieson, *Tetrahedron* 28: 1129 (1972).
71. B.M. Wepster, *Progr. in Stereochem.* 2: 99 (1958), eds. W. Klyne and P.B.D. de la Mare; W.G. Brown and H. Reagen, *J. Am. Chem. Soc.* 69: 1032 (1947).
72. W.R. Remington, *J. Am. Chem. Soc.* 67: 1838 (1945).
73. M. Ballester, *et al.*, *J. Am. Chem. Soc.* 93: 2215 (1971).
74. D.C. Iffland and H. Siegel, *J. Am. Chem. Soc.* 80: 1947 (1958).
75. M.J. Kamlet, J.C. Hoffsommer, R.R. Minesinger, and H.G. Adolph, *J. Org. Chem.* 33: 3070 (1968).
76. R.W. Franck and M.A. Williamson, *J. Org. Chem.* 31: 2420 (1966).
77. M.J. Kamlet, H.G. Adolph, and J.C. Hoffsommer, *J. Am. Chem. Soc.* 86: 4018 (1964).
78. R.I. Herrmann and I.D. Rae, *Australian J. Chem.* 25: 811 (1972).
79. J. Heidberg, J.A. Weil, G.A. Janusonis, and J.K. Anderson, *J. Chem. Phys.* 41: 1033 (1964).
80. B. Williamson and W.H. Rodebush, *J. Am. Chem. Soc.* 63: 3018 (1941).
81. L.W. Pickett, G.F. Walter, and H. France, *J. Am. Chem. Soc.* 58: 2296 (1936).
82. N.J. Leonard, R.T. Rapala, H.L. Herzog, and E.R. Blout, *J. Am. Chem. Soc.* 71: 2997 (1949).
83. R.T. Arnold and E. Rondestvedt, *J. Am. Chem. Soc.* 68: 2176 (1946).

84. R.T. Arnold, V.J. Webers, and R.M. Dodson, *J. Am. Chem. Soc.* 74: 368 (1952).
85. A. Eschenmoser, *et al.*, *Helv. Chim. Acta* 39: 786 (1956).
86. See K. Noack and R.N. Jones, *Canad. J. Chem.* 39: 2225 (1961) and references cited there.
87. A. Bienvenue, *J. Am. Chem. Soc.* 95: 7345 (1973); F.H. Cottee, *et al.*, *J. Chem. Soc.* (B) 1146 (1967); D.D. Faulk and A. Fry, *J. Org. Chem.* 35: 364 (1970).
88. J.K. Groves and N. Jones, *Tetrahedron* 25: 223 (1969).
89. D.T. Longone and C.L. Warren, *J. Am. Chem. Soc.* 84: 1507 (1962); S.W. Chow, L.A. Pilato, and W.L. Wheelwright, *J. Org. Chem.* 35: 20 (1970).
90. M.A. Cooper and S.L. Manatt, *J. Am. Chem. Soc.* 92: 1605 (1970).
91. R.W. Franck and E.G. Leser, *J. Org. Chem.* 35: 3932 (1970).
92. W.E. Billups, A.J. Blakeney, and W.Y. Chow, *Chem. Commun.* 1461 (1971).
93. M. Sheehan and D.J. Cram, *J. Am. Chem. Soc.* 91: 3544 (1969); for a summary, see D.J. Cram and J.M. Cram, *Accts. Chem. Res.* 4: 204 (1971).
94. V. Boekelheide and R.A. Hollins, *J. Am. Chem. Soc.* 95: 3201 (1973).
94a. W.E. Billups, *et al.*, *J. Am. Chem. Soc.* 95: 7878 (1973).
95. V. Boekelheide and R.A. Hollins, *J. Am. Chem. Soc.* 92: 3513 (1970).
96. H.O. House, D.G. Keepsell, and W.J. Campbell, *J. Org. Chem.* 37: 1003 (1972).
97. T. Otsubo, S. Mizogami, Y. Sakata, and S. Misume, *Tetra. Letters* (27) 2457 (1973).
98. D. Kivelson, S. Winstein, P. Bruck, and R.L. Hanson, *J. Am. Chem. Soc.* 83: 2938 (1961); L. DeVries and P.R. Ryason, *J. Org. Chem.* 26: 621 (1961).
99. S. Winstein, P. Carter, F.A.L. Anet, and A.J.R. Bourn, *J. Am. Chem. Soc.* 87: 5247 (1965); see also C.D. Poulter, R.S. Boikess, J.I. Brauman and S. Winstein, *J. Am. Chem. Soc.* 94: 2291 (1972).
100. See reference 98 above.
101. K.U. Ingold and D.R. Taylor, *Canad. J. Chem.* 39: 481 (1961).
102. G. Eglinton, J. Martin, and W. Parker, *J. Chem. Soc.* 1243 (1965).
103. G.B. Savitsky, K. Namikawa, and G. Sweitel, *J. Phys. Chem.* 69: 3105 (1965).
104. See D.N. Glew and N.S. Rath, *Canad. J. Chem.* 49: 837 (1971); I.M. Kolthoff and M.K. Chantooni, Jr., *J. Am. Chem. Soc.* 93: 3843 (1971).
105. See M. Tichy, *Adv. Org. Chem.* 5: 115 (1965); E. Osawa, T. Kato, and Z. Yoshida, *J. Org. Chem.* 32: 2803 (1967).
106. See L.N. Ferguson, *The Modern Structural Theory of Organic Chemistry*, Englewood Cliffs, N.J.: Prentice-Hall, 1963, pp. 530 ff.
107. H.A. Lloyd, K.S. Warren, and H.M. Fales, *J. Am. Chem. Soc.* 88: 5544 (1966).
108. I.M. Kolthoff and M.K. Chantooni, Jr., *J. Am. Chem. Soc.* 93: 3843 (1971).

109. J.U. Lowe, Jr., and L.N. Ferguson, *J. Org. Chem.* 30: 3000 (1965); S. Bratoz, D. Hadzi, and G. Rossmy, *Trans. Farad. Soc.* 52: 464 (1956); I.M. Hunsberger, *J. Am. Chem. Soc.* 72: 5626 (1950).
110. L.E. Godycke and R.E. Rundle, *Acta Cryst.* 6: 487 (1953).
111. N.L. Allinger, J.J. Maul, and M.J. Hickey, *J. Org. Chem.* 36: 2747 (1971).
112. R. Linke, *Z. Physik. Chem.* B46: 251, 261 (1940) and 47: 194 (1940); L.R. Zumwalt and R.M. Badger, *J. Chem. Phys.* 7: 87 (1939).
113. A.W. Baker and W.W. Kaeding, *J. Am. Chem. Soc.* 81: 5904 (1959); P.v.R. Schleyer and R. West, *J. Am. Chem. Soc.* 81: 3164 (1959); A. Nickon, *J. Am. Chem. Soc.* 79: 243 (1957).
114. R.A. Nyquist, *Spectrochim. Acta* 19: 1655 (1963).
115. G.C. Vogel and R.S. Drago, *J. Am. Chem. Soc.* 92: 5347 (1970); E.M. Arnett, et al., *J. Am. Chem. Soc.* 92: 2377 (1970); A.D. Sherry and K.F. Purcell, *J. Am. Chem. Soc.* 94: 1853 (1972); E.R. Lippincott and R. Schroeder, *J. Chem. Phys.* 23: 1099 (1955).
116. L.W. Reeves, E.A. Allan, and K.O. Stromme, *Canad. J. Chem.* 38: 1249 (1960).
117. F.L. Slejko and R.S. Drago, *J. Am. Chem. Soc.* 95: 6935 (1973); L.J. Bellamy and R.J. Pace, *Spectrochim. Acta* 25A: 319 (1969).
118. M. Goldstein, C.B. Mullins, and H.A. Willis, *J. Chem. Soc.* (B) 321 (1970).
119. Z. Yoshida, N. Ishibe, and H. Ozoe, *J. Am. Chem. Soc.* 94: 4948 (1972).
120. P.v.R. Schleyer and R. West, *J. Am. Chem. Soc.* 81: 3164 (1959).
121. M. Tamres and S. Searles, Jr., *J. Am. Chem. Soc.* 81: 2100 (1959).
122. G.A. Olah, S.J. Kuhn, and S.H. Flood, *J. Am. Chem. Soc.* 83: 4581 (1961); R. West, *J. Am. Chem. Soc.* 81: 1615 (1959); A.W. Baker and A.T. Shulgin, *J. Am. Chem. Soc.* 81: 1523 (1959).
123. H.C. Brown and J.D. Brady, *J. Am. Chem. Soc.* 74: 3570 (1952).
124. M.J. Kamlet, R.R. Minesinger, and W.H. Gilligan, *J. Am. Chem. Soc.* 94: 4744 (1972).
125. Z. Yoshida, H. Ogoshi, and T. Tokumitsu, *Tetrahedron* 26: 5691 (1970).
126. H.M. Bailes, *J. Org. Chem.* 35: 4280 (1970).
127. P.A. Clarke and N.F. Hepfinger, *J. Org. Chem.* 35: 3249 (1970).
128. N.F. Hepfinger and P.A. Clarke, *J. Org. Chem.* 34: 2572 (1969).
129. G.R. Wiley and S.I. Miller, *J. Am. Chem. Soc.* 94: 3287 (1972).
130. H. Saito, *J. Am. Chem. Soc.* 93: 1072, 1077 (1971).
131. G.E. Maciel and R.V. James, *J. Am. Chem. Soc.* 86: 3893 (1964).
132. G.E. Maciel and J.J. Natterstad, *J. Chem. Phys.* 42: 2752 (1965).
133. H.A. Christ and P. Diehl, *Helv. Phys. Acta* 36: 170 (1963).
134. L.J. Bellamy and R.L. Williams, *Trans. Farad. Soc.* 55: 14 (1959).
135. G.E. Maciel and R.V. James, *Inorg. Chem.* 3: 1650 (1964).
136. A. Balasubramanian and C.N.R. Rao, *Spectrochim. Acta* 18: 1337 (1962).
137. G.E. Maciel and G.B. Savitsky, *J. Phys. Chem.* 68: 437 (1964).

138. G.E. Maciel and D.D. Traficante, *J. Am. Chem. Soc.* 88: 270 (1966).
139. H. Brouwer and J.B. Stothers, *Canad. J. Chem.* 50: 601 (1972).
140. L.N. Ferguson and I. Kelly, *J. Am. Chem. Soc.* 73: 370 (1951).
141. M.J. Kamlet, E.G. Kayser, J.W. Eastes, and W.H. Gilligan, *J. Am. Chem. Soc.* 95: 5210 (1973).
142. M.J. Nye, M.J. O'Hare, and W.-P. Tang, *Chem. Commun.* 402 (1973).
143. M. Ito, K. Inuzuka, and S. Imanishi, *J. Am. Chem. Soc.* 82: 1317 (1960).
144. E. Kloster-Jensen, *J. Am. Chem. Soc.* 91: 5674 (1969).
145. W.M. Schubert, R.B. Murphy, and J. Robins, *Tetrahedron* 17: 199 (1962).
146. A. Burawoy, *Tetrahedron* 5: 340 (1959).
147. J.C. Dearden and W.F. Forbes, *Canad. J. Chem.* 36: 1362 (1958).
148. W.F. Forbes and M.B. Sheratte, *Canad. J. Chem.* 33: 1829 (1955).
149. T.L. Brown, *J. Am. Chem. Soc.* 81: 3229 (1959).
150. N. Fuson, M.-L. Josien, and E.M. Shelton, *J. Am. Chem. Soc.* 76: 2526 (1954).
151. D.L. Ostercamp, *J. Org. Chem.* 35: 1632 (1970).
152. A.I. Scott, *Interpretation of the Ultraviolet Spectra of Natural Products*, Oxford: Pergamon, 1964.
153. L.M. Jackman and S. Sternhell, *Applications of Nuclear Magnetic Resonance Spectroscopy in Organic Chemistry*, 2nd ed., New York: Pergamon, 1969.
154. J.L. Gove, *J. Org. Chem.* 38: 3517 (1973).
155. M. Karplus, *J. Am. Chem. Soc.* 85: 2870 (1963).
156. For ^5J across allenic bonds, see M. Barfield and S. Sternhell, *J. Am. Chem. Soc.* 94: 1905 (1972); M. Santelli, *Chem. Commun.* 939 (1971). For a review of long-range coupling, see M. Barfield and B. Chakrabarti, *Chem. Revs.* 69: 757 (1969). For ^3J alpha to trigonal carbons in rigid C_6 rings, see K.L. Williamson and W.S. Johnson, *J. Am. Chem. Soc.* 83: 4623 (1961). The extent of long-range coupling which occurs "through space" is an unsettled question: E. Abushanab, *J. Am. Chem. Soc.* 93: 6532 (1971).
157. M. Barfield, *J. Chem. Phys.* 48: 4463 (1968); K.A.W. Parry, *et al.*, *J. Chem. Soc.* (B) 700 (1970).
158. J.L. Marshall, K.C. Erickson, and T.K. Folsom, *J. Org. Chem.* 35: 2038 (1970).
159. B.A. Shoulders, *et al.*, *J. Am. Chem. Soc.* 90: 2992 (1968).
160. Private communication from Prof. Joseph Bragin, California State University, Los Angeles.
161. J. Laane and R.C. Lord, *J. Mol. Spectroscopy* 39: 340 (1971).
162. P. Dowd, T. Dyke, R.M. Newman, and W. Klamperer, *J. Am. Chem. Soc.* 92: 6327 (1970); H.D. Stidham, *Spectrochim. Acta* 21: 23 (1965).
163. For references on the use of pes for estimating dihedral angles, see C. Batich, O. Ermer, E. Heilbronner, and J.R. Wiseman, *Angew. Chem. Int. Ed.* 12: 312 (1973).

164. H. van Bekkum, P.E. Verkade, and B.M. Wepster, *Rec. Trav. Chim.* 78: 815 (1959); J.E. Bloor and A. Burawoy, *Tetrahedron* 20: 861 (1964); H.H. Jaffe and M. Orchin, *Theory and Applications of Ultraviolet Spectroscopy*, New York: Wiley, 1962, p. 384.
165. S. Searles, Jr., R.A. Sanchez, R.L. Soulen, and D.G. Kundiger, *J. Org. Chem.* 32: 2655 (1967); J.K. Groves and N. Jones, *Tetrahedron* 25: 223 (1969).
166. H. Suzuki, *Bull. Chem. Soc. Japan* 32: 1340, 1350 (1959).
167. H. Suzuki, *Bull. Chem. Soc. Japan* 33: 613, 619 (1960).
168. E.A. Braude and F.J. Sondheimer, *J. Chem. Soc.* 3754 (1955).
169. K. Mislow, S. Hyden, and H. Schaefer, *J. Am. Chem. Soc.* 84: 1449 (1962).
170. R. Huisgen, *et al.*, *Chem. Ber.* 90: 1946 (1957).
171. N.L. Allinger and E.S. Jones, *J. Org. Chem.* 30: 2165 (1965).
172. C. Dickinson, J.R. Holden, and M.J. Kamlet, *Proc. Chem. Soc.* 232 (1964).
173. G.L. Buchanan and G. Jamieson, *Tetrahedron* 28: 1123 (1972).
174. R.W. Taft, Jr., and H.D. Evans, *J. Chem. Phys.* 27: 1427 (1957).
175. S. Bank and W.D. Closson, *Tetrahedron* 24: 381 (1968).
176. C.S. Foote, *Tetra. Letters* (9) 579 (1963); P.L. Laszlo and P.v.R. Schleyer, *J. Am. Chem. Soc.* 86: 1171 (1964).
177. M.J.S. Dewar and H.N. Schmeising, *Tetrahedron* 5: 166 (1959).
178. B. Dischler, *Z. Naturforsch.* 19a: 887 (1964); K. Mislow, *Tetra. Letters* 1415 (1964).
179. R.A. Alden, J. Kraut, and T.G. Traylor, *J. Am. Chem. Soc.* 90: 74 (1968).
180. I.C. Paul, *et al.*, *J. Am. Chem. Soc.* 91: 7542 (1969).
181. P.v.R. Schleyer and R.D. Nicholas, *J. Am. Chem. Soc.* 83: 182 (1961).
182. C.S. Foote, *J. Am. Chem. Soc.* 86: 1853 (1964); P.v.R. Schleyer, *J. Am. Chem. Soc.* 86: 1854 (1964).
183. P.D. Bartlett and M. Stiles, *J. Am. Chem. Soc.* 77: 2806 (1955).
184. R.G. Pews and C.W. Roberts, *J. Org. Chem.* 34: 2029 (1969).
185. D.J. Cram and R.C. Helgeson, *J. Am. Chem. Soc.* 88: 3515 (1966); A.G. Pinkus and H.C. Custard, Jr., *J. Phys. Chem.* 74: 1042 (1970).
186. F.V. Brutcher, Jr., and W. Bauer, Jr., *J. Am. Chem. Soc.* 84: 2236 (1962).
187. L. Joris and P.v.R. Schleyer, *J. Am. Chem. Soc.* 90: 4599 (1968); P.v.R. Schleyer, *J. Am. Chem. Soc.* 91: 3965 (1969); H. Kwart and G.C. Gatos, *J. Am. Chem. Soc.* 80: 881 (1958).
188. H. Iwamura, *Tetra. Letters* 2227 (1970).
189. T. Förster, *Z. Elektrochem.* 54: 42 (1950).
190. H.H. Jaffe and H.L. Jones, *J. Org. Chem.* 30: 964 (1965).
191. R.H. Ellerhorst, H.H. Jaffe, and A.L. Miller, *J. Am. Chem. Soc.* 88: 5342 (1966).

192. A.C. Hopkinson and P.A.H. Wyatt, *J. Chem. Soc.* (B) 1333 (1967); C.C. Greig and C.D. Johnson, *J. Am. Chem. Soc.* 90: 6453 (1968).
193. E.L. Wehry and L.B. Rogers, *J. Am. Chem. Soc.* 87: 4234 (1965).
194. G. Jackson and C. Porter, *Proc. Royal Soc.* (London) 260A: 13 (1961).
195. J.P. Idoux and C.K. Hancock, *J. Org. Chem.* 32: 1935 (1967); E. Bouwhuis and M.J. Janssen, *Tetra. Letters* (3) 233 (1972).
196. J.P. Colpa, C. MacLean, and E.L. Mackor, *Tetrahedron* 19: Suppl. 2, 65 (1963); R.L. Flurry, Jr., and P.G. Lykos, *J. Am. Chem. Soc.* 85: 1033 (1963).
197. J.R. Lombardi, *J. Am. Chem. Soc.* 92: 1833 (1970).
198. B.A. Zadorozhnyi, *Chem. Abstracts* 63: 16191f (1965).
199. R.L. Letsinger, O.B. Ramsay, and J.H. McCain, *J. Am. Chem. Soc.* 87: 2945 (1965); H.E. Zimmerman, *et al.*, *J. Am. Chem. Soc.* 85: 915, 922 (1963); S. DeVries and E. Havinga, *Rec. Trav. Chim.* 84: 601 (1965); E. Havinga, R.O. deJongh, and M.E. Kronenberg, *Helv. Chim. Acta* 50: 2550 (1967).
200. R.C. Hahn, P.H. Howard, and G.A. Lorenzo, *J. Am. Chem. Soc.* 93: 5816 (1971).
201. R. Grinter and E. Heilbronner, *Helv. Chim. Acta* 45: 2496 (1962).
202. D.A. deBie and E. Havinga, *Tetrahedron* 21: 2359 (1965).
203. C.S. Foote, P. Engel, and T.W. DelPesco, *Tetra. Letters* (31) 2669 (1965).
204. See description in L.N. Ferguson, *Highlights of Alicyclic Chemistry*, Palisades, N.J.: Franklin, 1973, p. 210.

Part IV: INTRAMOLECULAR FORCES AND MOLECULAR PROPERTIES

That steric hindrance may have profound effects on chemical and physical properties of molecules is a well-established fact. For example, hindered esters are extremely slow to hydrolyze; hindered acids cannot be esterified in the usual fashion with alcoholic hydrogen chloride; hindered ketones do not add bulky Grignard reagents and are slow to form Schiff bases if at all; and optical isomers of biphenyls with three or four substituents in the ortho positions can be isolated. Now we shall see how intramolecular forces greatly influence molecular geometries. A simple example is the effect on bond distances. The C-C bond lengths between crowded carbons are longer than normal (1.54 Å in H_3C-CH_3):

	$(CH_3)_3C-C(CH_3)_3$	1,1'-Biadamantyl	$Me_3C-\underset{H}{\overset{CMe_3}{\underset{|}{C}}}-CMe_3$ 1.611 Å, 1.016 Å
	Bi-t-butyl	1,1'-Biadamantyl	Tri-t-butyl-methane (1)
C-C length	1.57 Å	1.578 Å	

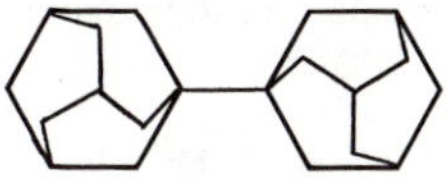

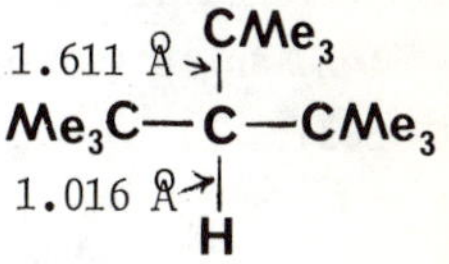

We saw in Table 1.4 that the C-C bond in bi-t-butyl is unusually weak.

One aspect of conformational analysis is the study of how molecules accommodate strain. It is observed, for instance, that acyclic molecules generally assume conformations with bulky groups trans, and that a substituted cyclohexane prefers that conformation which has the most bulky groups equatorial.

Chapter 17

STATIC CONFORMATIONAL ANALYSIS (2)

As was the case when H-bond forces and inductive effects were discussed, it is useful to partition intramolecular forces into several categories: <u>bond angle</u> strain, <u>torsional</u> strain, <u>dipole-dipole</u> forces, and <u>van der Waals</u> forces. You already know that bond angles depend upon the type of bonding orbitals used and that any distortion of bond angles destabilizes a molecule. For instance, the C-C-C bond angles in cyclopropane are much smaller than the tetrahedral angle of 109.5° and, as von Baeyer first proposed, the resulting b bond angle strain causes cyclopropane to have olefinic properties (Section 5.3).

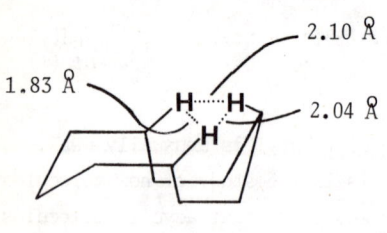

Cyclodecane

Some of the hydrogen atoms across the ring of medium size rings are unusua ly close together, closer than the sum of their van der Waals radii (2.8 Å) (3) Consequently, there is a substantial repulsion between the protons and the molecules are under large internal strain (Table 18.4). They also undergo a varie of transannular reactions (4). A simple example is the addition of bromine to cyclodecene, which gives the 1,6-dibromo product instead of the usual 1,2-product

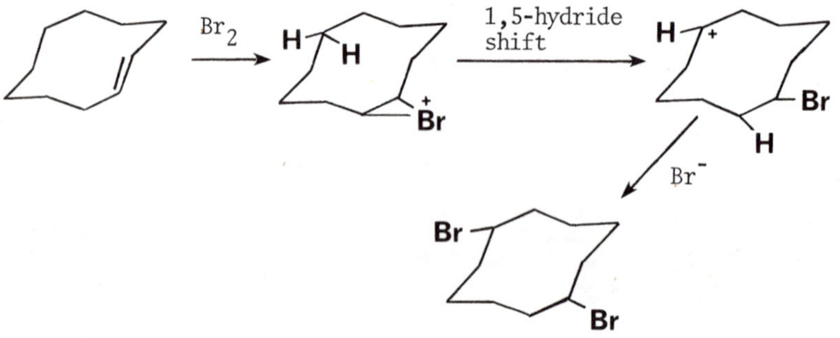

Transannular repulsion is also responsible for the greater stability of <u>exo</u>- over <u>endo</u>-substituted norbornanes.

The repulsion between coparallel, adjacent C-H bonds, as in the eclipsed conformation of ethane, is about 1 kcal/mol per $\overset{H}{C}$-$\overset{H}{C}$ interaction. For example, the barrier to rotation of CH_3 groups about the C-C bond in ethane is 3.04 kcal/mol. Called <u>torsional strain</u>, it is believed to be a combination of repulsive van der Waals forces and repulsion between bonding electrons (see also Chapter 18). A clear understanding of the barrier to internal rotation in ethane has been a long-standing problem and articles on this topic continue to appear unabated (5). Part of the elusive nature of the barrier is the fact that the observed value is the difference between relatively large energy quantities.

In planar cyclopentane there are ten eclipsed C-H bonds plus five $C\diagdown_C\diagup^C$ eclipsed fragments. Skewed or <u>gauche</u>-oriented C-C bonds exert about 0.7 kcal/mol torsional strain. Hence, in planar cyclopentane there would be a torsional strain of about 14 kcal/mol. It can be relieved by a puckering of the ring, but this creates some bond angle strain. The result is that the ring is puckered to the extent (<u>ca.</u> 40°) that the total strain is a minimum at ~7 kcal/mol (6)*. Owing to torsional strain and a little transannular strain in the boat structure of cyclohexane (make scale models to convince yourself), the chair structure is more stable by about 5.5 kcal/mol (7). Similarly, 1,3-diaxial torsional forces normally make equatorial conformations more stable than axial. For instance, compounds <u>17.1</u>, <u>17.2</u>, and <u>17.3</u> are more stable with the substituents equatorial rather than axial.

<u>17.1</u>, X = CH_3 <u>17.3</u>, X = CH_3

<u>17.2</u>, X = halogen   <u>17.4</u>, X = Br

---

* The puckering is not rigid but is like a ripple around the ring as one atom after another moves out of the plane of the other four, a movement called "pseudo rotation". See Adams, Geise, and Bartell (6).

The difference in stability between 17.5 and 17.6 is 2.6 kcal/mol, determined from heats of combustion measurements (8). Part of this differential strain in 17.5 is due to the normal repulsion between ortho methyl groups, estimated to be 0.6 kcal/mol. The 1.4 kcal balance is due to the buttressing effect of the outer methyl groups which prevent the two inner methyl groups from bending back to relieve the strain, as they can in 17.6.

Two ortho t-butyl groups have about 11 kcal/mol strain. Two such groups are apparently bent back from each other in the molecular plane, with too little distortion of the ring to be detectable by ir, uv, or nmr spectroscopy (Chapter 11). There is a reduction of magnetic susceptibility of the ring, however. When there are three adjacent t-butyl groups, as in 1,2,3-tri-t-butyl-benzene, the van der Waals repulsion cannot be sufficiently relieved by bending back of t-butyl groups because of the buttressing effect, and there is a distortion of the benzene ring. The attendant loss in aromaticity is revealed by uv spectroscopy (Chapter 11).

When there are polar groups attached to adjacent carbons, bond dipole-dipole repulsion may become dominant (9). For instance, diequatorial dimethyl cyclohexane 17.3 is much more stable than the diaxial isomer, but in 17.4 the dipolar repulsion of diequatorial bromines outweighs the van der Waals repulsion of diaxial bromines to make the latter conformer the more stable. The diaxial ⇌ diequatorial equilibria for the dihalocyclohexanes are solvent sensitive owing to electrostatic solvation, whereas the equilibrium of a dialkylhexane, which involves only nonpolar forces such as van der Waals repulsion, is virtually independent of solvent. In the case of 1,2-dichloroethane, the combination of dipole-dipole and van der Waals repulsion makes the cis isomer with eclipsed chlorines less stable than the trans isomer by 4.5 kcal/mol. Esters, on the other hand, are "locked" in an s-trans conformatio with opposed bond dipoles (10) as shown by microwave, ir, electron diffraction and dipole moment data. Just as torsional strain forces are not completely clear, the attractive forces between lone-pair electrons on nonbonded atoms are somewhat of a mystery. For example, 1,1-dihaloethylenes have rather small geminal angles,

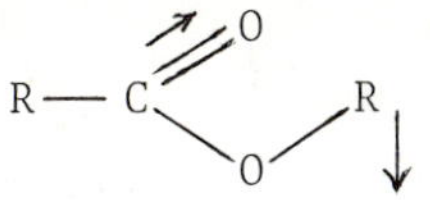

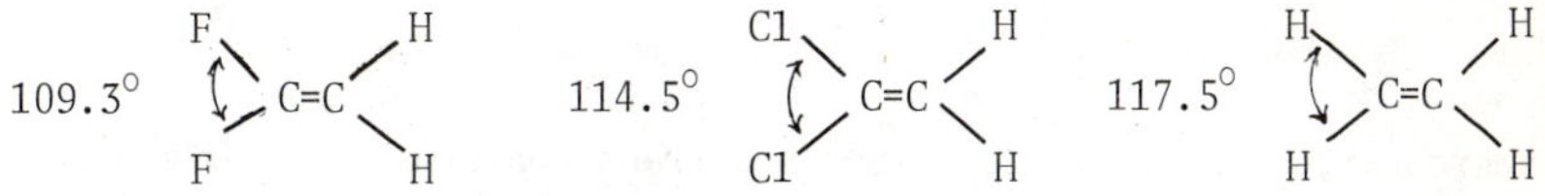

109.3° (F,F / H,H)   114.5° (Cl,Cl / H,H)   117.5° (H,H / H,H)

and various compounds with vicinal dipolar groups favor the cis or gauche over the trans conformation (11, 12):

X,Y	% cis at equilibrium for XHC=CHX (13)
F,F	63
F,Cl	70
F,Br	70
F,I	67
Cl,Cl	61
Br,Br	50
CH_3,Cl	76
CH_3,Br	68
CH_3,OC_6H_5	65
CH_3,OC_2H_5	81

Thus, cis-dihaloethylenes are about one kcal/mol more stable than the trans, despite a greater dipole-dipole repulsion in the cis isomer of about one kcal/mol (14). The situation is just the reverse when there are hydrocarbon groups in place of halogens. For instance, the cis isomers are less stable than the trans by 1.0, 1.9, and 9.3 kcal/mol for dimethyl-, diisopropyl-, and di-t- butylethylenes, respectively (15).

In FN-NF, the cis isomer is 3.0 kcal/mol more stable than the trans (16). Similarly, the 2,2'-dihalobiphenyls prefer a cisoid conformation in the gas phase.

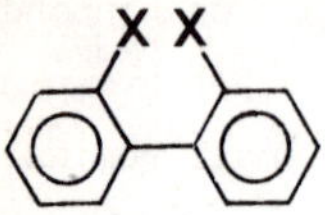

X,X	Dihedral angle between phenyl rings (17)
H,H	42°
I,I	79°
Br,Br	75°
Cl,Cl	74°
F,F	60°

Whereas 1,2-dihaloethanes assume the <u>trans</u> conformation (although nmr measurements of the liquid state and in solution suggest that the <u>gauche</u> conformation of 1,2-difluoroethane is preferred) and <u>trans</u>-1,2-dihalocyclohexanes prefer the diaxial conformation, the corresponding 1,2-halohydrins have the <u>gauche</u> and diequatorial conformations, respectively. Although it is tempting to attribute this inversion of relative stabilities to intramolecular H bonding (10a), this probably is not so, because 2-chloro- and 2-methoxyethyl acetate as well as 1,2-dimethoxyethane and gaseous 1,2-dicyanoethane also have the <u>gauche</u> conformation. Each of these compounds illustrates the so-called "gauche" effect": the tendency to adopt that structure which has the maximum number of <u>gauche</u> interactions between the adjacent lone pairs or polar bonds or both (18). For instance, <u>gauche</u> conformations of the simple molecules below predominate.

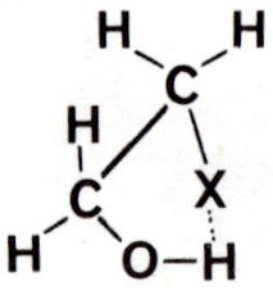

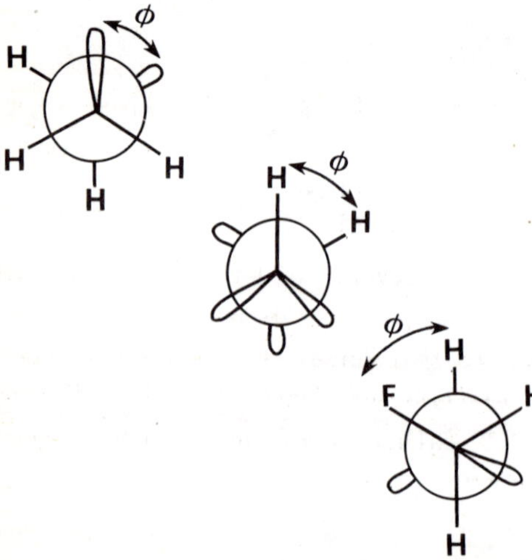

	ϕ
N_2H_4	90-95°
P_2H_4	90-110°
H_2O_2	111°
FCH_2OH	60°

Theoretical discussions of these interactions between lone pairs have appeared (14).

Factors responsible for the favored conformations of 2-alkylcyclohexanones are not clear. It is noticeable that the percentage equatorial isomer does not change monotonically with increasing size of alkyl group (19).

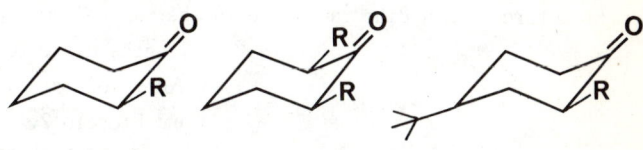

R	Equatorial R (%)	Equatorial R (%)	Equatorial R (%)
Me	89.7	91.6	93.1
Et	85.5	79.4	86.4
i-Pr	76.6	56.7	73.0
t-Bu	93.0	86.6	94.1

An additional type of intramolecular strain is that produced by π-electron clouds or lone-pair electrons. In Section 1.1 mention was made of the electron-pair repulsion theory to account for bond angles. In a compound such as 17.7, R = Me or Et, the C-R carbon-carbon bond is longer than normal:

17.7

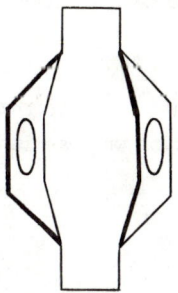

C-Me, 1.564 Å

C-Et, 1.573 Å

A reasonable explanation is that the π cloud of the aromatic ring pushes the R groups out from the ring to lengthen the C-R bond (20). Similarly, repulsion between the face-to-face rings of [2.2]-paracyclophane causes the benzene rings to buckle (strain energy, 33 kcal/mol), and the thermal rearrangement of 1,8-divinylnaphthalene has been attributed to the "π strain" between the two peri-vinyl groups (21).

448   Ch. 17 STATIC CONFORMATIONAL ANALYSIS

As a result of the various types of ring strains -- bond angle, torsional, transannular, and sometimes dipole-dipole repulsion -- cyclic structures have preferred conformations, depending upon the ring size. The preferred conformations for some cyclic structures are as follows.

<u>Cyclobutanoids</u>. Most cyclobutanoids, even bicycloderivatives, have a folded structure with a dihedral angle in the range from 20 to 35° and C-C bond lengths close to 1.55 Å (22). Among the exceptions are cyclobutanone, cyclobutene, and bicyclo[2.1.0]pentane. The planar structure has eclipsed bonds and puckering reduces the torsional strain but is counterbalanced by increased bond angle strain brought on by folding of the ring. In the case of disubstituted cyclobutanes, the <u>trans</u> isomer is the more stable for 1,2-substitution and the <u>cis</u> for 1,3-substitution. 1,3-Dihalocyclobutanes have a puckered ring in which the dihedral angle increases for the <u>cis</u> isomers in going from Cl to I (32°, 38°, and 48°, respectively) but decreases for the <u>trans</u> isomers from Cl to I (37°, 32°, and 24°, respectively) (23).

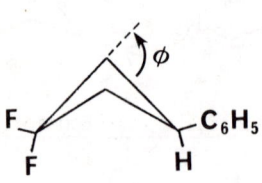

φ = dihedral angle
  = 27°
ΔG = 0.95 kcal/mol

<u>Cyclopentanoids</u>. Most derivatives of cyclopentane have the envelope conformation, which has a minimum torsional strain. Notice that the ir and nmr conformational generalizations (1) and (4) stated on page 453 hold for the cyclopentane ring. Infrared and dipole moment data indicate that the D ring of 17-ketosteroids assumes the envelope conformation whereas it has a half-chair structure in 16-ketosteroids.

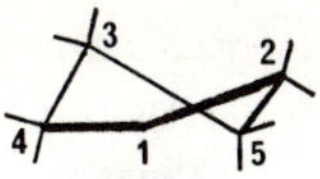

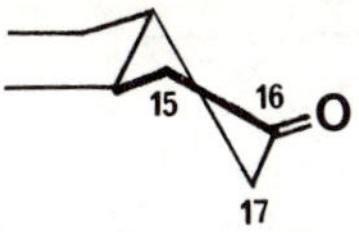

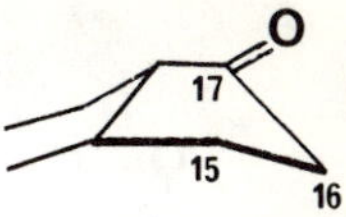

Half-chair conformation
Carbons 4, 1, and 2 in a plane, C3 above and C5 below this plane.

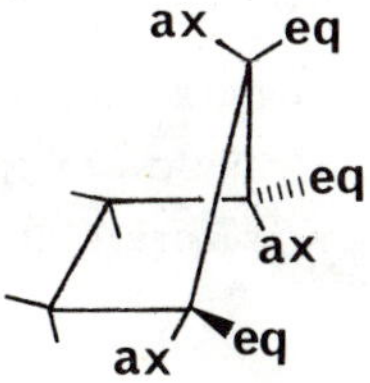

Envelope conformation

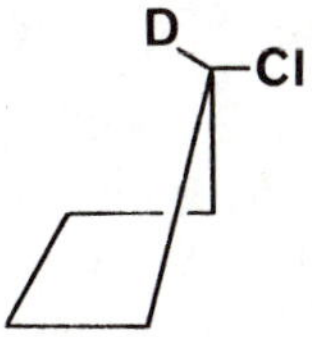

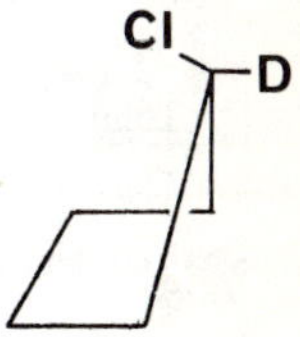

ν_{C-Cl} 614 cm^{-1} 588 cm^{-1}
ν_{C-D} 2168   2190

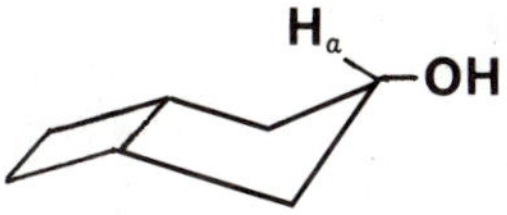

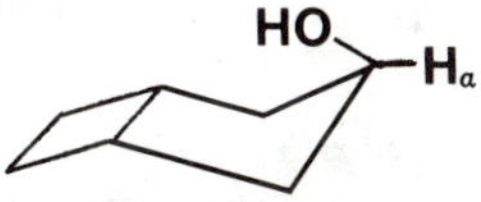

δ_{H_α} 8.00   8.08 ppm

<u>Cyclohexanoids</u>. Every first-year organic chemistry student knows that the boat conformation of cyclohexane is so much less stable than the chair that the amount of the boat conformer present at room temperature is negligible. Bridging, as in norbornanes, or H bonding in some heterocycles (24), produces boat conformations of six-membered rings. An assortment of intramolecular strains force most substituted C_6 rings to have twisted or other nonchair

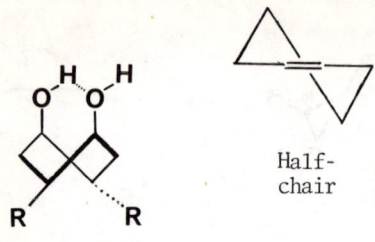

Half-chair

structures. Thus, most 4,4-di-substituted cyclohexanones have flattened chair structures, and the combined effects of H bonding and torsional strain makes cis,cis,cis-2,5-dialkyl-1,4-cyclohexanediols assume nonchair conformations (25). Cyclohexenes have half-chair structures.

Cycloheptanoids. A combination of experimental data (26) and calculations (27) indicates that the twist-chair conformation of many cycloheptanes is more stable than the full chair, and the full chair is more stable than the boat.

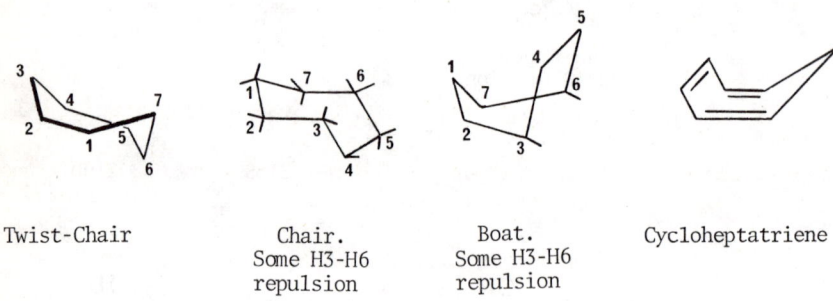

Twist-Chair

Chair.
Some H3-H6 repulsion

Boat.
Some H3-H6 repulsion

Cycloheptatriene

Cycloheptatrienes, however, have a boat structure.

Cyclooctanoids. Most cyclooctane derivatives assume preferably one of three conformations, a crown, boat-chair, or saddle confromation.

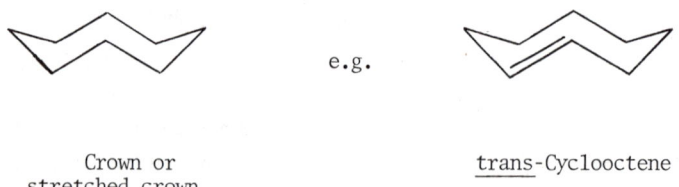

Crown or stretched crown

e.g.

trans-Cyclooctene

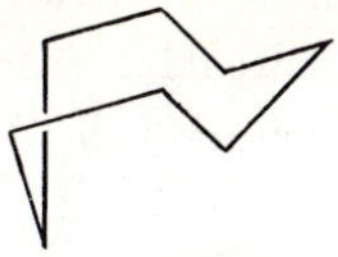

Boat-chair

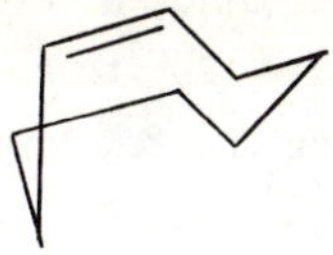

cis-Cyclooctene

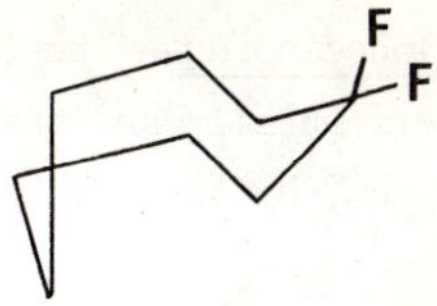

1,1-Difluoro-
cyclooctane

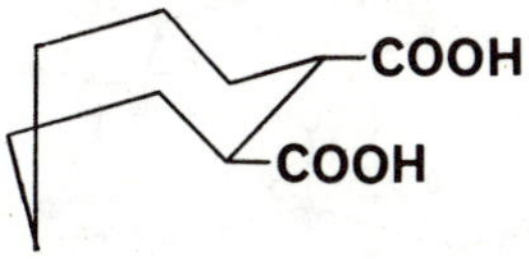

trans-1,2-Dicarb-
oxylic acid

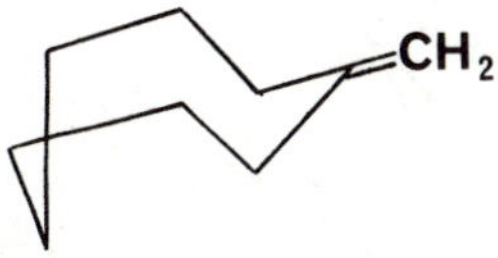

Methylenecyclo-
octane

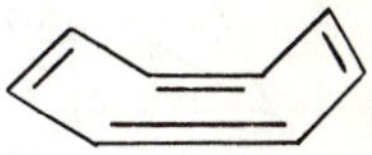

Cyclooctatetraene
is an exception in
having a boat
conformation

Cyclononanoids. The lowest energy conformation, experimentally and theoretically, is the twist-boat-chair (TBC) (28).

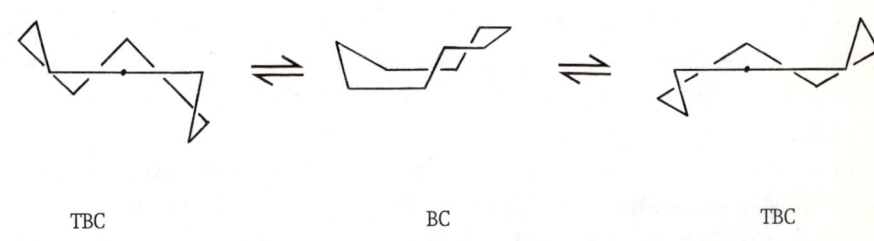

TBC   BC   TBC

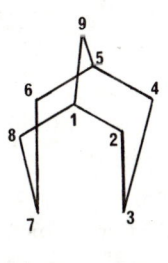

Saddle
(dichair)

The bicyclononane system is generated by placing a methylene bridge across the cyclooctane ring. The two cyclohexane rings may invert to give rise to a chair-boat or diboat conformation. The dichair conformation is the most stable but when there are endo substituents at C3 or C7, the boat-chair conformation becomes the major conformer.

Cyclodecanoids. The BCB conformation of cyclodecane rings is found experimentally and theoretically (29) to be a distinct potential energy minimum, by several kcal, among the few lowest energy conformations.

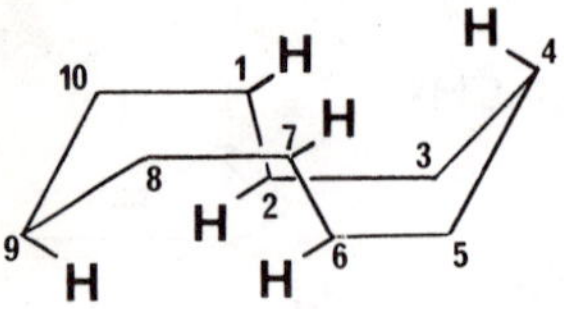

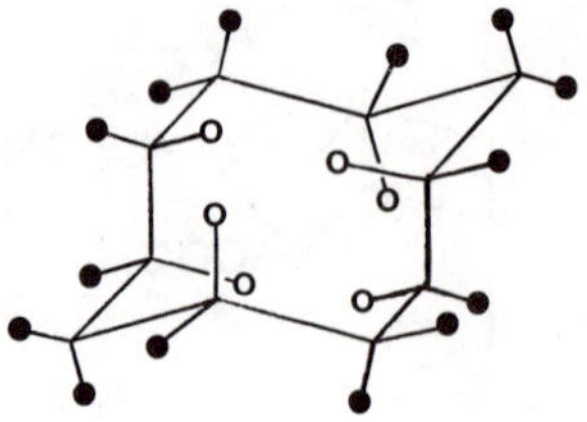

BCB ≡ 17.8

However, there is still considerable repulsion between Hs on C1, C4, and C7 on one side of the ring and between Hs on C2, C6, and C9 on the other side. Any reaction which reduces this transannular strain is energetically favored, such as formation of trigonal carbons at any of these positions. For example, cyclodecanone, which has such a trigonal carbon, does not add HCN or sodium bisulfite, in which the C=O carbon would revert to a tetrahedral carbon and build up transannular strain. In contrast, cyclohexanone has more torsional strain than cyclohexane and therefore addition reactions to the ketone are favored (see Section 19.3). Moreover, substituents preferably occupy the extraannular positions (black balls in 17.8) because much greater transannular strain would form if they held innerannular positions (open balls in 17.8). Accordingly, only carbons 3, 5, 8, or 10 can be geminally disubstituted under normal conditions (30).

Chemists have learned in the past twenty years how valuable it is to be able to determine conformations of molecules. The significance of such information was demonstrated by Barton, who was inspired ty the papers of Hassel, and together they shared the 1969 Nobel Prize in chemistry for their work in conformational analysis.

Diffraction techniques could provide the best information about molecular structure, but unfortunately analysis of the data is so time consuming that other methods are usually sought (31). Other experimental methods, particularly spectral, can provide the information sought very quickly. In most cases, an empirical approach is used in which measurements are made on a number of compounds of known structure to establish a generalization. Then the relationship is used for compounds of unknown structure.

We saw in Chapter 15 how empirical correlations can be developed to provide semiquantitative information about intramolecular geometry, e.g., dihedral angles, etc. The following qualitative correlations regarding equatorial and axial groups were developed from using C_6 rings but apply in a limited extent to other size rings.

Ir: (1) $\nu_{C-X}^{eq} > \nu_{C-X}^{ax}$ (2) $\nu_{C=O}^{eq\ \alpha-X} > \nu_{C=O}^{ax\ \alpha-X}$

Uv: (3) $\lambda_{C=O}^{ax\ \alpha-X} > \lambda_{C=O}^{eq\ \alpha-X}$

Nmr: (4) $\delta_H^{eq} > \delta_H^{ax}$ (5) $W_{\frac{1}{2}}^{ax} > W_{\frac{1}{2}}^{eq}$ (6) $\delta_{OH}^{eq} > \delta_{OH}^{ax}$
(except α to C=O)

(7) $^3J_{\underline{OH}:H}^{eq} > {^3J}_{\underline{OH}:H}^{ax}$

Relationship (1) means that the ir frequency of a C-X bond is larger for equatorial than for axial groups. (3) Means that the wavelength of a carbonyl band in the uv is larger when an alpha group is axial than when it is equatorial. $W_{\frac{1}{2}}$ is the band width at half intensity in relationship (5), and in (7), $^3J_{OH:H}$ is the 3-bond coupling between the hydroxyl proton and the vicinal proton for equatorial or axial OH groups. Relationship (1) may be related to the energy it takes to stretch a ring compared to energy required to bend a ring and (2) may be a dipole-dipole field effect (32). To reduce the dipole-dipole repulsion, the dipolar character of the carbonyl group is decreased and the resulting greater double-bond character raises the $\nu_{C=O}$. For instance, the C=O frequency is greater when the bond is coparallel to polar groups:

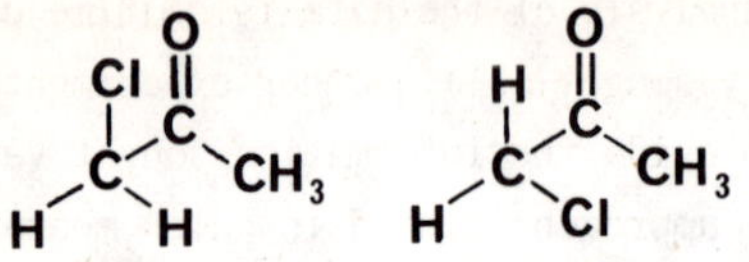

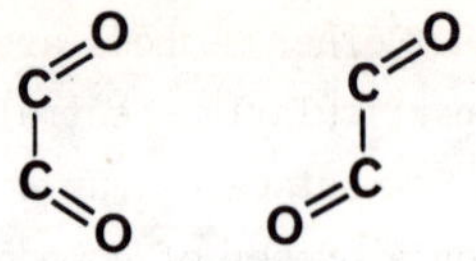

Ch. 17 STATIC CONFORMATIONAL ANALYSIS

$\nu_{C=O}$ 1743   1725 cm^{-1} $\nu_{C=O}^{\text{s-cis}} > \nu_{C=O}^{\text{s-trans}}$

(see Section 3.2)

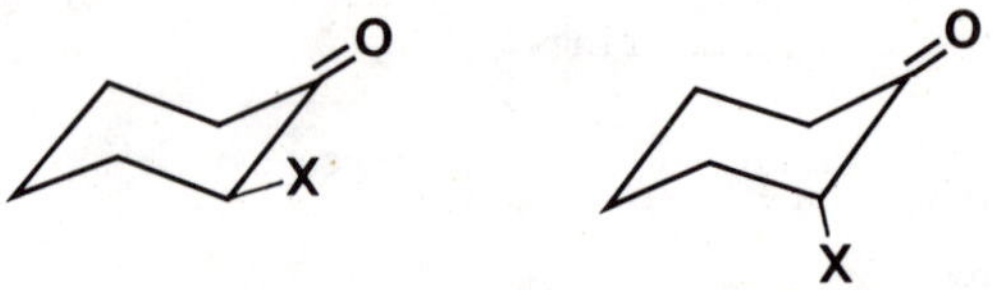

$\nu_{C=O}^{\text{eq }\alpha\text{-x}} > \nu_{C=O}^{\text{ax }\alpha\text{-x}}$ X = halogen, OCH$_3$, etc.

A few examples of the use of these generalizations can be given.

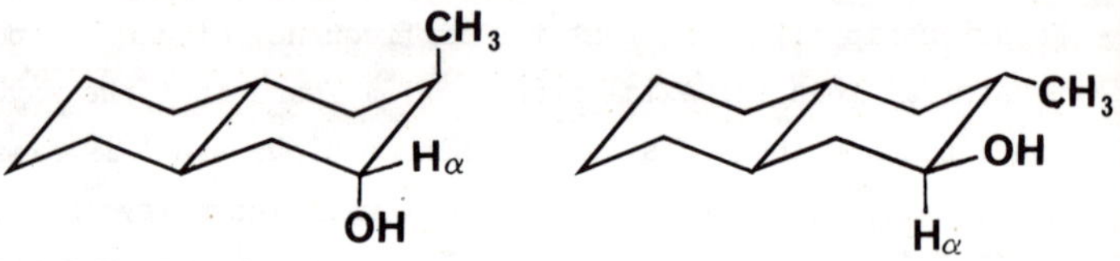

ν_{C-OH}	983	1010 cm^{-1}
δ_{H_α}	5.44	3.18 ppm
$W_{\frac{1}{2}}$	6	20 Hz

Thus, the conformations could be assigned on the basis of relationship (1), (4), or (5), and all three agree on the assignment (33).

$\nu_{C=O}^{KBr}$	1715	1728 cm^{-1}
λ_{m}^{MeOH}	307	283 nm
δ_{H_α}	4.34	4.72 ppm
$W_{\frac{1}{2}}$	5	19 Hz

Again, there is complete agreement on conformational assignment from any of the spectral parameters. Notice that relationship (4) is reversed for protons α to a C=O group.

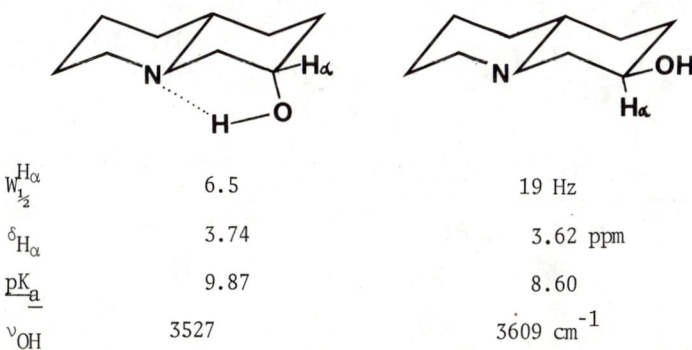

$W_{\frac{1}{2}}^{H_\alpha}$	6.5	19 Hz
δ_{H_α}	3.74	3.62 ppm
pK_a	9.87	8.60
ν_{OH}	3527	3609 cm^{-1}

As expected, the intramolecular H bond reduces the acid strength of the axial OH (34).

Incidentally, as close to cyclohexane as piperidine might seem, the conformation of the proton on nitrogen is very much in doubt (35). Convincing arguments have been offered for an axial as well as for an equatorial conformation.

Relationship (5) is based on the fact that ^3J coupling, i.e., 3-bond coupling between geminal protons, is a maximum for dihedral angles of 0° or 180° (see Chapter 15). It is very useful in assigning conformations. For

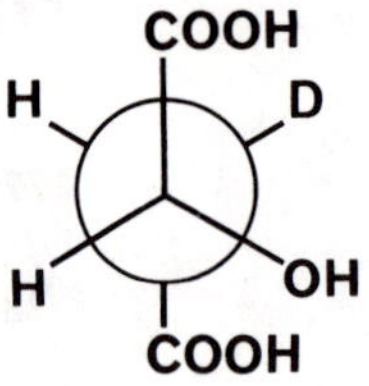

threo
17.9

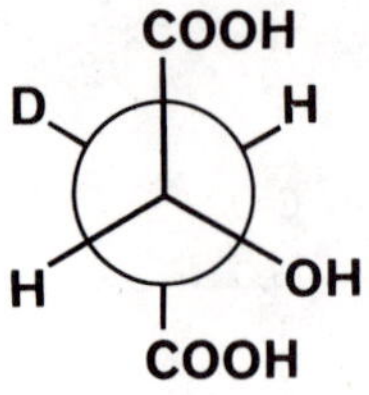

erythro
17.10

instance, L-malic acid-3-d was prepared enzymatically and assigned the threo structure 17.9. It had a ^3J coupling of 9 Hz. Later, a sample of 17.9 was synthesized by an unequivocal method and a ^3J coupling of 3 to 4 Hz was observ Since $J_{180°} > J_{60°}$, the structure of the enzymatic product was changed to 17.1 Moreover, ^3J across the double bond is sensitive to the inplane angles ϕ and ϕ'. As a result, ^3J is affected by the ring size of cyclenes and the following ranges have been determined:

Ring size	Range of ^3J
3	0.5-1.5 Hz
4	2.5-4
5	5.1-7
6	8.8-11
7	9-12.5
8 to 10	10.3-12.8

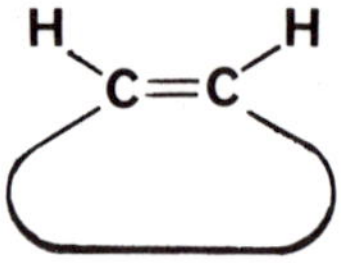

On this basis it is a simple matter to ascertain the ring size of an alicyclic ring containing a double bond (36).

Relationships similar to (1) to (5) have also been drawn for bicyclo ring systems. For instance, for norbornanes and norbornenes:

$\delta\frac{exo}{H_\alpha} > \delta\frac{endo}{H_\alpha}$ (note that the exo and endo terms refer to H_α rather than as usual to the substituent).

X	exo H_α	endo H_α	exo H_α	endo H_α
OH	4.17 ppm	3.66 ppm		
OAc	4.87	4.54		
Cl			4.29	3.77 ppm

Many other parameters may be used for assigning conformations, such as anisotropic and solvent effects on nmr chemical shifts, ORD and CD rotary dispersion, dipole moments, nuclear Overhauser effects (NOE) (37) and a host of others (4). The NOE effect is the enhancement in the intensity of an nmr proton signal when the peak of a nearby proton is irradiated. This increase in intensity is dependent on the distance between the two protons, which in turn provides a basis for structural assignment in certain situations. Electron spin resonance (esr) spectroscopy is particularly useful for studying radicals (38). The hyperfine splitting constant α (analogous to J in nmr spectroscopy) of a proton in a radical is dependent upon the dihedral angle ϕ between the axis of the p_z orbital and the C-H bond,

$$\alpha = \delta(\underline{B_0} + \underline{B} \cos^2 \phi)$$

where δ is the spin density in the carbon p_z orbital and $\underline{B_0}$ and $\underline{B}$ are constants near 0 and 50 G, respectively (39). Thus, α_β for cyclopropyl hydrogen atoms in several conjugated systems is substantially smaller than for the corresponding atom of the isopropyl group because the hydrogen atom lies close to the nodal plane of the attached π system.

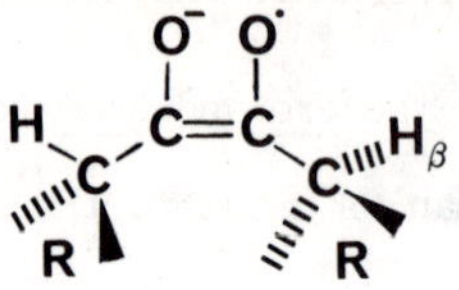

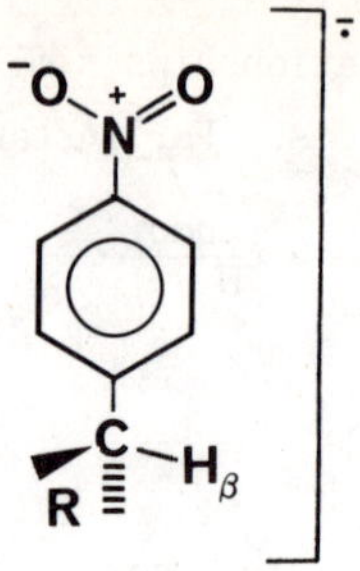

Ch. 17 STATIC CONFORMATIONAL ANALYSIS

R	a_β (40)	a_β (41)
Isopropyl	1.92 G	1.74 G
Cyclopropyl	0.57 G	1.39 G

When steric effects block the bisected conformation of the cyclopropyl group, a_β is much larger.

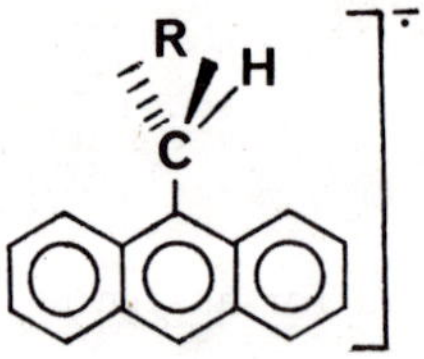

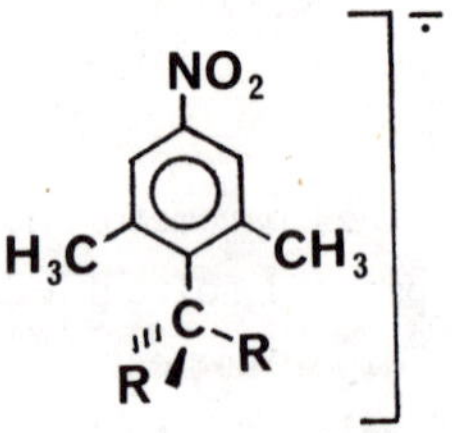

R	a_β	a_β
Isopropyl	0.62 G (42)	
Cyclopropyl	6.64 G	5.68 G (41)

These data, of course, support the bisected conformation of the cyclopropyl group over the perpendicular conformation (Section 5.3).

Dipole moment data have also been used for conformational analysis (43). For example, the dipole moment of 17.11 should be larger than that of 17.12 since the C=O and C-Cl dipoles are in the same general direction in 17.11 but opposed in 17.12. An assignment based on the observed moments is confirmed by the ir data (relationship (3)).

	17.11	17.12
μ_{obs}	4.29 D	3.17 D
$\nu_{C=O}$	1748 cm^{-1}	1735 cm^{-1}

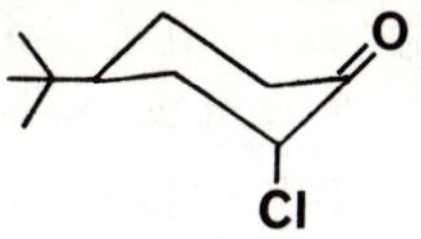

Although phenyl and carbonyl groups may shield or deshield protons, depending upon the molecular geometry, <u>cis</u>-located groups are often shielded by phenyl and alkyl groups and deshielded by ester and carbonyl groups. Some illustrations are:

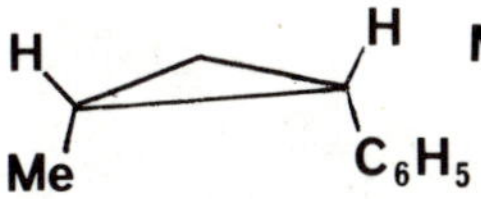

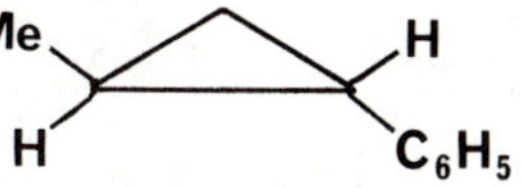

δ_{Me} 6.1   6.6 ppm

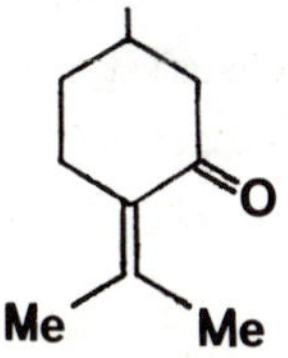

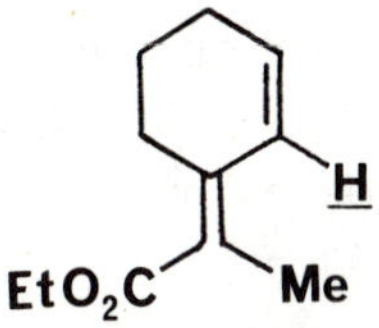

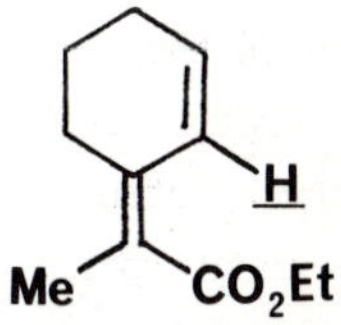

δ_{Me} 1.77   1.95 ppm $\delta_{\underline{H}}$ 6.45   7.0 ppm

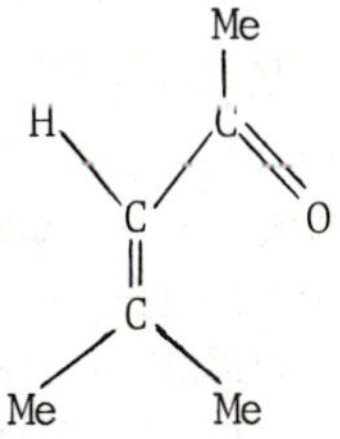

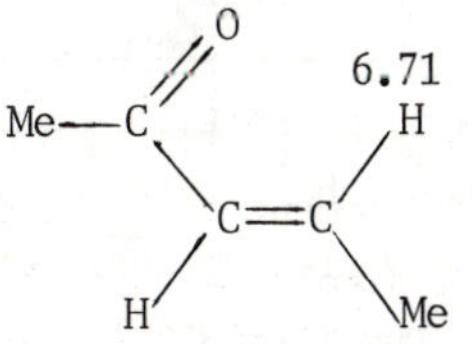

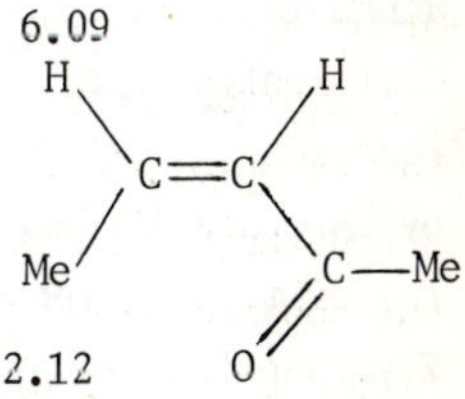

$\delta^1_{\underline{H}}$ (ppm) (44)   1.86   2.11   1.90   2.12

^{13}C chemical shifts are also sensitive to molecular geometry, similar to those of the related protons. Particularly useful has been the so-called "γ effect", wherein carbons <u>deshield</u> other carbons three bonds away (45). For example, β-located methyl carbons <u>cis</u> to an alkyl or even a carboxyl group resonate upfield of those in the <u>trans</u> orientation (46).

	H\C=C/CO$_2$Me H$_3$C/ \CH$_3$		H$_3$C\C=C/CO$_2$Me H/ \CH$_3$	
δ^{13}CH$_3$ (ppm from TMS)	13.9	11.8	20.5	20.5

	H\C=C/H H$_3$C/ \CO$_2$Me	H\C=C/CO$_2$Me H$_3$C/ \H
δ^{13}CH$_3$ (ppm from TMS)	18.9	27.0

Moreover, <u>gauche</u>-oriented γ carbons are shielded relative to <u>anti</u>-located γ carbons. Hence, an axial carbon bonded to a cyclohexane ring absorbs at higher field than its equatorial counterpart and even the C3 and C5 cyclohexyl carbons are shielded in the axial conformer (47). The proton chemical shifts in the above α,β-unsaturated esters follow the usual trends in that the β protons resonate at lower fields than the α protons, and β protons or β-methyl protons <u>cis</u> to the ester group are deshielded relative to their <u>trans</u> counterparts. (See Chapter 11 for methods of determining <u>s-cis</u> and <u>s-trans</u> conformations of enones.)

An interesting question is what conformation do diarylacetylenes assume? Resonance theory would predict a planar structure in which the <u>p</u> orbitals of both aromatic rings overlap the same set of <u>p</u> orbitals of the alkyne carbons to give conjugation across the entire molecule. Or, they could have a nonplanar structure in which the <u>p</u> orbitals of the phenyl rings overlap different sets of alkyne <u>p</u> orbitals. Should these two conformations be of equal energy, there would be freely rotating aryl groups. MO calculations by CNDO and INDO methods predict unhindered rotations or the nonplanar conformer to be slightly the more stable (48). Experimentally, it is found by X-ray diffraction that diphenylacetylene is planar in the solid state and by uv spectroscopy that it is also planar in solution (48).

Another conformational aspect often sought is the absolute configuration of a dissymmetric molecule. Because the X-ray diffraction method is lengthy and difficult, absolute configurations are usually determined by optical rotatory dispersion (ord) or circular dichroism (cd) (49) measurements, by comparisons with compounds of known stereochemistry (50), or by chemical methods. There are several empirical rules relating molecular rotation to absolute configuration. Details will not be given here but reference can be made to the benzoate sector rule (51), Brewster's method (52), the octant rule (53), which is the best known), Lowe's rule for allenes (54), and others (55).

In the comparison method for determination of absolute configurations, optical properties of the sample are compared with those of a reference compound whose absolute structure is known or synthesized from the reference compound by reactions known to preserve optical configuration (56). For example, the structures of biphenyls have been assigned from comparisons with 17.13, whose absolute configuration was determined by anomalous X-ray diffraction (57). Mislow and coworkers have studied extensively the ord (58) and cd (59) of several substituted 1,1-binaphthyls, recognizing the twisted systems as inherently dissymmetric chromophores. They offered the generalization that a positive Cotton effect centered at 285 nm corresponds to the R configuration in this series. Since the parent compound shows a negative Cotton parent compound shows a negative Cotton effect in the cd spectrum (60) in this region, it has the S configuration. This agrees with the assignment based on a chemical correlation with 17.13. The sign of the Cotton effect and the chirality of cyclic α-diketones is, however, a controversial issue (61).

R-(+)

17.13

S-(+)

The absolute configurations of some norbornyl derivatives are based on that of (-)-1-methyl-2-norbornanone (17.14) (62).

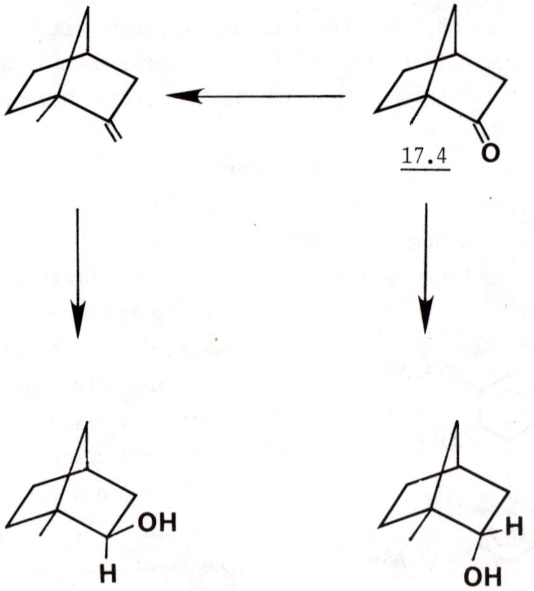

There are three chemical methods of determining the absolute configuration at C* of an asymmetric alcohol, R_1R_2*CHOH. In Prelog's method (63), a series of reactions is used to convert the alcohol into an atrolactic acid ϕ*C(Me)OH-COOH and the rotation of the latter is measured. In Horeau's procedure (64), the alcohol is converted into phenylethylacetic acid. In Brewster's method (52), the alcohol is converted to its benzoate ester and the rotations of the ester and alcohol are compared. For an alcohol with the absolute configuration 17.15, R_1 larger than R_2, Prelog's method will give

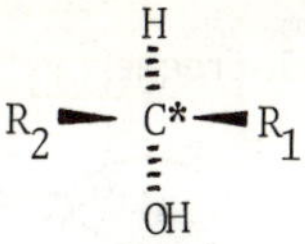

Ch. 17 STATIC CONFORMATIONAL ANALYSIS   463

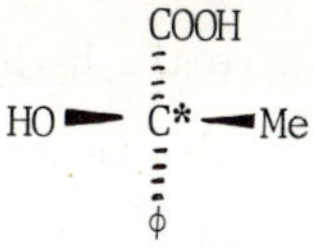

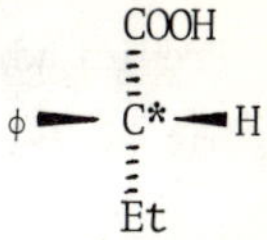

$$\underset{17.15}{R_2-\overset{H}{\underset{OH}{C^*}}-R_1} \qquad \underset{17.16}{HO-\overset{COOH}{\underset{\phi}{C^*}}-Me} \qquad \underset{17.17}{\phi-\overset{COOH}{\underset{Et}{C^*}}-H}$$

excess (+)-atrolactic acid 17.16, Horeau's method will give excess (-)-phenylethylacetic acid 17.17, and according to Brewster's method, the molar rotation of the benzoate will be larger than that of the alcohol (or acetate (65)).

QUESTIONS

1. Predict the relative rates of oxidation with CrO_3, the relative ease of esterification, and $J_{OH:H}$ for the following two pairs.

2. The $J_{1,2}$ and $J_{2,3}$ values for the three isomers $\underline{X}$, $\underline{Y}$, and $\underline{Z}$ are (a) 2.5 and 9.2 Hz, (b) 8.4 and 4.8 Hz, and (c) 9.1 and 9.1 Hz. Assign the coupling constants of (a), and (b), and (c) to their respective isomers $\underline{X}$, $\underline{Y}$, and $\underline{Z}$.

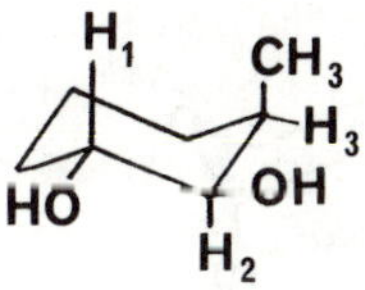

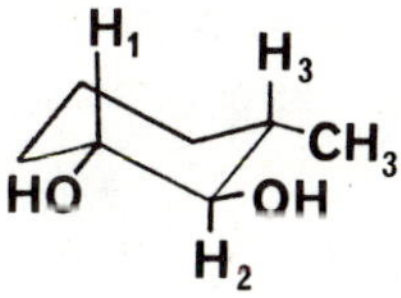

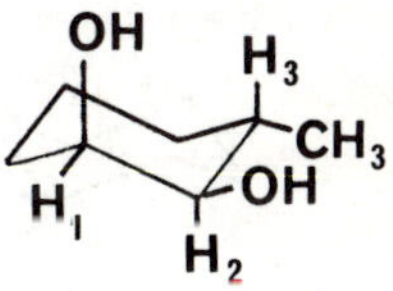

3. Predict which should have the higher carbonyl frequency, R or W.

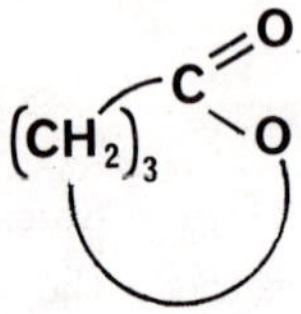

R
DM 4.09 D

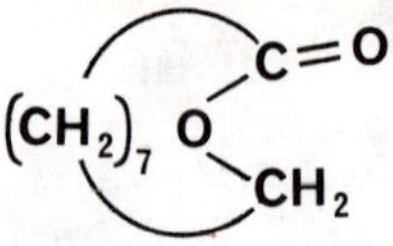

W
2.01 D

4. Predict which designated proton would resonate at the higher field in M and N.

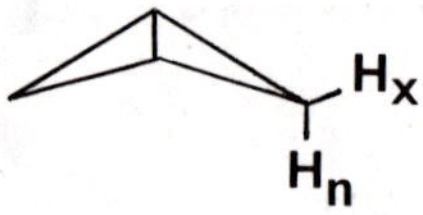

M

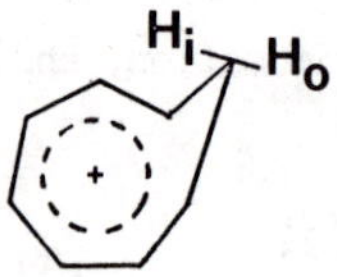

N

5. How would you expect the chemical shifts for the NOH proton and the CH=N proton of aldoximes to compare for the syn and anti isomers?

6. Predict the more stable conformer, P or Q.

P

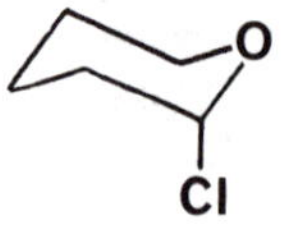

Q

7. Which isomer, S or T, whould have the larger $W_{\frac{1}{2}}$ for the epoxide methylene protons ?

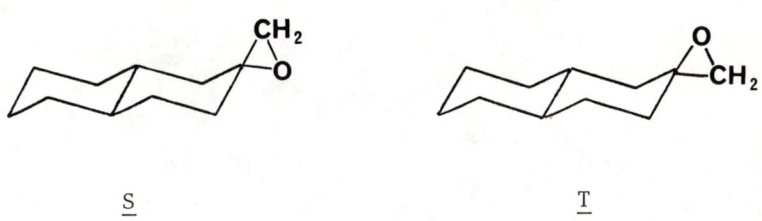

      S   T

8. How do you rationalize the relative positions of the following two equilibria ?

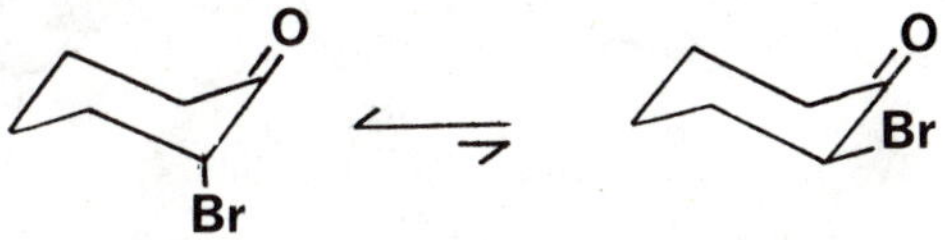

      U   V

I.e., axial more stable

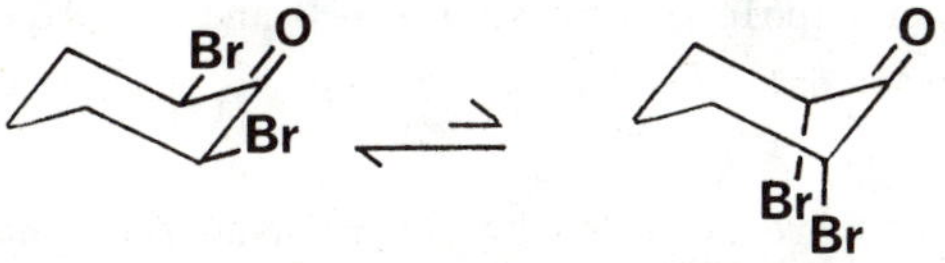

      F   G

I.e., diequatorial more stable

9. How could you determine spectroscopically whether a compound has structure J or K ?

      J   K

10. The $W_{\frac{1}{2}}$ values for the α methine proton (C$\underline{H}$OH) of cis- and trans-2-t-butylcyclohexanols are 7.7 and 22.3 Hz. Assign these values to the respective isomers.

ANSWERS

1. The observed data are:

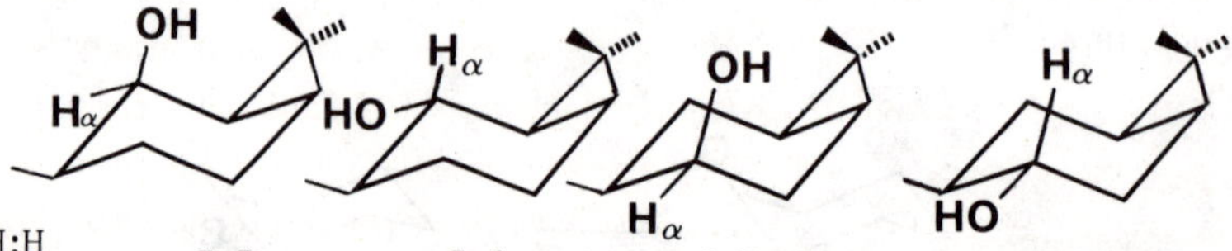

$J^{OH:H}$	3.3	5.6	3.3	5.5 Hz
k_{rel} (CrO$_3$ oxid)	40.7	2.6	10.0	1.0
% yield of ester	21	79	18	86

2. (a) for $\underline{Z}$; (b) for $\underline{X}$; (c) for $\underline{Y}$.
3. $\underline{R}$ has the larger dipole moment so the C=O and Ö dipoles must be co-parallel. Since coaligned poles increases $\nu_{C=O}$, $\underline{R}$ should have the higher $\nu_{C=O}$.
4. The inner protons are shielded by the ring currents and resonate upfield of the outer protons. The chemicals shifts, in ppm, are:

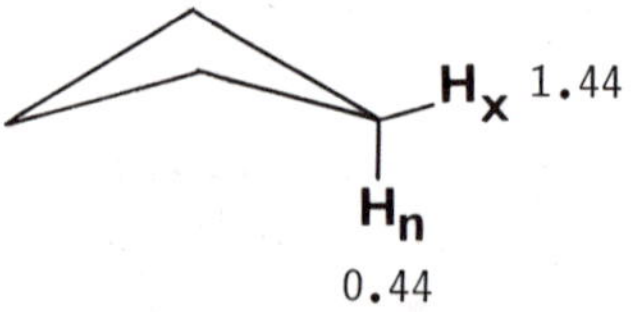

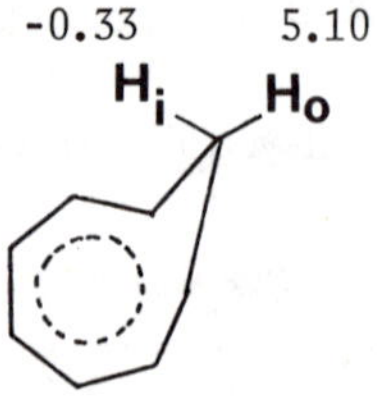

5.

	syn	anti
Δδ	3.21	4.26 ppm

R = (furan-2-yl)

The N-OH proton should be downfield of the HC=N proton. In the syn isomer, the C-H proton is deshielded by the OH and the two signals get closer together. In the anti isomer, the methine proton may be shielded by the lone pair on N, or at least is not deshielded by the OH group, so its chemical shift is farther from the N-OH signal than for the syn isomer. This difference as well as solvent effects, has been used to assign conformations of oximes. (See D. R. Boyd, et al., Tetra. Letters (18) 1747 (1972); E. M. Acton, M. A. Leaffer, S. M. Oliver, and H. Stone, J. Agr. Food Chem. 18: 1061 (1970); G. G. Kleinspehn, J. A. Jung, and S. A. Studniarz, J. Org. Chem. 32: 460 (1967); J. V. Burakevich, A. M. Lore, and G. P. Volpp, J. Org. Chem. 36: 1 (1971); R. Wasylishen and T. Schaeffer, Canad. J. Chem. 50: 274 (1972).)

6. P has no diaxial van der Waals repulsion which Q has, whereas Q has less dipole-dipole repulsion than P. Based on the relative stabilities of α-halocyclohexanone, where axial halogen is the more stable, one would predict Q to be more stable. Experimentally, this is the case (G. E. Booth and R. J. Ouellette, J. Org. Chem. 31: 544 (1966)).

7. $W_{1/2}$ is larger for S(1.41 Hz) because the CH_2 is closer to axial than in T(0.15 Hz).

8. The dipole-dipole forces in V outweigh the 1,3-diaxial van der Waals repulsion in U to make U the major conformer present. In G, however, the 1,3-diaxial van der Waals repulsion plus dipole-dipole forces between the C-Br bonds outweigh the dipole-dipole forces in F to shift the equilibrium toward the diequatorial conformer.

9. If a sample of each were available, the one with the methyl proton signal in the nmr farther downfield would be K, where the methyl is deshielded by the carbonyl group. If only one sample is in hand, one could use the solvent effect technique. In this method, CH_3 or protons "in front" of the C=O group give a negative $\Delta = \delta_{CCl_4} - \delta_{C_6H_6}$, where

these are the chemical shifts in CCl_4 and benzene, and CH_3 or protons behind the C=O group give a positive Δ. Thus, Δ would be positive for J and negative for K. (See J. D. Connally and R. McCrindle, Chem. and Ind. (London), 379 (1965).)

10. Since the t-butyl group is equatorial in each isomer, the trans isomer must have an axial CHOH methine proton and hence the larger $W_{\frac{1}{2}}$ value.

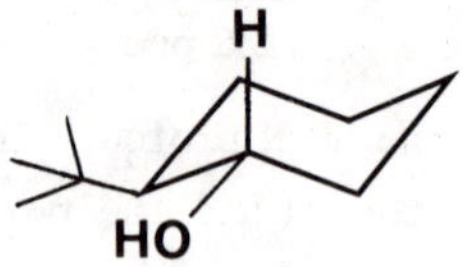

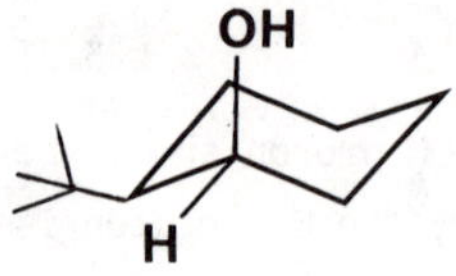

Chapter 18

DYNAMIC CONFORMATIONAL ANALYSIS

After having developed methods for the determination of conformations in molecules it is only natural to ask, what are the energy differences between isomers and what are the energy barriers to isomerization of one to another ? Again, a variety of methods is available, particularly spectral. By and large they are based on the observation that conformers usually exhibit different absorption bands whose relative intensities are directly related to the relative amounts of the two conformers. For a mixture of two conformers, for instance <u>gauche</u> and <u>trans</u> or equatorial (eq) and axial (ax), the equilibrium constant is

$$\underline{K} = C_{eq}/C_{ax} = \frac{A_{eq}\varepsilon_{ax}}{A_{ax}\varepsilon_{eq}}$$

where C is concentration, ε is molar extinction coefficient, and A is absorbance. Extinction coefficients change a little with temperature but the change in their ratio is negligble. Hence, using the van't Hoff equation

$$\frac{d(\ln \underline{K})}{d(1/T)} = \frac{-\Delta H}{R} = \frac{d(\ln A_{eq}/A_{ax})}{d(1/T)}$$

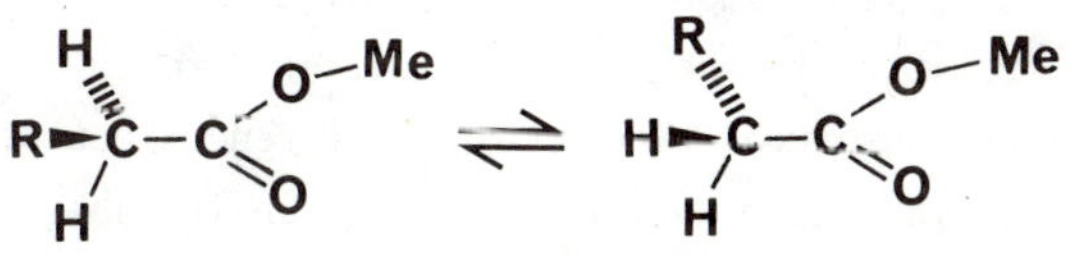

when the ratio of absorbances is plotted against 1/T, the slope of the line will give a value for the enthalpy difference and $\underline{K}$ can be obtained from the intorcept at 1/T = 0. For example, in the ir spectra of propionate and butyrate esters, there is a pair of bands near 1195 and 1170 cm^{-1} whose relative intensities vary with temperature. This can be attributed to an equilibrium between conformations which affects the C-C-O antisymmetrical stretch. From the temperature coefficient of the band intensities the enthalpy differences were found to be 78 ± 50 cal/mol for R = Me and 182 ± 6 for R = Et (66).

Some properties, such as nmr chemical shifts and dipole moments, of a rapidly equilibrating mixture of two conformers are related to the mole fractions of the components by the equation

$$\underline{P} = p_1 N_1 + p_2 N_2 \tag{18.1}$$

where $\underline{P}$ is the value of some property observed for the mixture, N_1 and N_2 are the mole fractions, and p_1 and p_2 are the characteristic values for the pure components. In the case of nmr chemical shifts of equatorial and axial conformers,

$$\underline{E} \qquad \underline{A}$$

$$\delta = N_{\underline{E}} \delta_{ax} + N_{\underline{A}} \delta_{eq} \tag{18.2}$$

where δ is the observed chemical shift of the mixture, N_E and N_A the mole fractions of $\underline{E}$ and $\underline{A}$, and δ_{ax} and δ_{eq} are the chemical shifts for the axial and equatorial protons. Equation (18.2) can be rearranged to equation (18.3),

$$N_{\underline{E}}/N_{\underline{A}} = \underline{K} = \frac{\delta_{eq} - \delta}{\delta - \delta_{aq}} \tag{18.3}$$

from which $-\Delta G^\circ$, the free energy difference between $\underline{E}$ and $\underline{A}$, may be obtained, $-\Delta G^\circ = RT \ln \underline{K}$. It is necessary to know δ_{ax} and δ_{eq}, which are best obtained from a measurement at a temperature low enough that the interconversion of $\underline{E}$ and $\underline{A}$ is slow relative to the nmr time scale (50 to 100 sec^{-1}) and two separate signals are observed. The energy differences between some equatorial and axial groups in cyclohexane rings are given in Table 18.1 and the energy differences between some *gauche* and *trans* disubstituted ethanes are given in Table 18.2.

Table 18.1.
Conformational energies of some groups in cyclohexane rings (4)

Group	ΔG° (kcal/mol)	Group	ΔG° (kcal/mol)
Me	1.7	F	0.15
Et	1.8	Cl	0.43
i-Pr	2.0	Br	0.38
t-Bu	>5	I	0.43
$H_2C=CH$	1.35	OH	0.97
HC≡C	0.18	OCH_3	0.60
CF_3	2.4-2.5	COOH	1.35
NH_3^+	1.7-2.0	NO_2	1.10
NH_2	1.15-1.7	CO_2Et	1.20
ND_2	1.2 (68)		

Table 18.2.
Energy differences between gauche and trans disubstituted ethanes (69)

1,2-substituents	ΔG° (kcal/mol)	1,2-substituents	ΔG° (kcal/mol)
F, F	0.2	Br, Br	1.6-1.9
F, Cl	0.2-0.5	I, I	2.6
F, Br	0.3-1.0	CH_3, F	-0.5
Cl, Cl	1.2	CH_3, Cl	-0.1
Cl, Br	1.43	CH_3, CH_3	0.8

In addition to energy differences between conformers, the energy barriers to interconversion of conformers have been of special interest. The lack of powerful physical equipment available today prevented early chemists from detecting more than one isomer in a substance like 1,2-dichloroethane and led them to postulate <u>free rotation</u> about single bonds. The thermodynamic properties of molecules can be calculated through statistical mechanics in which the contributions from translation, rotation, and vibrations are considered (67). Such calculations are in good agreement with experimental measurements when applied to rigid molecules. However, based on the concept of free rotation about single bonds, the statistical method gave poor results for some molecules, and it was first shown by Kemp and Pitzer (70) that the discrepancies can be attributed to the existence of barriers to rotation about single bonds. In general, barriers less than 10 kcal/mol are detected and measured only by sophisticated techniques such as microwave, far infrared, and Raman spectroscopy using temperature variation methods (71). For barriers in the 10 to 20 kcal/mol range, the conformers are not normally isolable, and nmr spectroscopy is especially useful (72, 73). When there are barriers between 20 and 30 kcal/mol, the isomers may be isolable but they easily interconvert, especially upon heating. Isomers separated by barriers of over 30 kcal/mol are stable in the ordinary sense.

Reversible processes, such as ring inversions, rotations about bonds with partial double-bond character, and rotations against sterically hindering groups, produce changes in nmr spectra with variations of temperature. Consider the inversion of chlorocyclohexane. At room temperature there is a single peak for the α C-H bond because the ring is inverting too rapidly to give separate peaks for the axial and equatorial α C-H protons. As the temperature is lowered, the rate of inversion decreases and the nmr peak broadens. Finally, at a sufficiently low temperature when the inversion rate is slow compared to the nmr oscillation frequency, two peaks appear, the broader one for the axial and the other for the equatorial α protons. At the coalescence temperature, the rate constant $\underline{k}$ for interconversion of two conformers can be expressed by equation (18.4),

$$\underline{k} = \sqrt{2}\pi(\nu_a - \nu_e) \qquad (18.4)$$

where the νs are the frequencies of the chemical shifts of the two conformers, say for instance, the axial and equatorial proton signals. From reaction rate theory, $\underline{k}$ can also be expressed by equation (18.5) in which c is the trans-

$$\underline{k} = \frac{c\underline{k}_B T}{\underline{h}} e^{-\Delta G^{\ddagger}/RT} \qquad (18.5)$$

mission coefficient (usually taken as unity), $\underline{k}_B$ and $\underline{h}$ are Boltzmann's and Planck's constants, T is the absolute temperature, and $\Delta G^{\ddagger}$ is the activation energy. Thus, the frequency difference for the two conformers at the coalescence temperature T_c gives $\underline{k}$ from equation (18.4), and $\Delta G^{\ddagger}$ may then be calculated from $\underline{k}$ by equation (18.5).

If Δt represents the lifetime of the molecule in one conformation, then

$$\Delta t = \frac{1}{2\pi(\nu_a - \nu_e)} \qquad (18.6)$$

and when one plots log $(1/2\Delta t)$ against $1/T$ (Arrhenius plot), the slope of the line will give a value for $\Delta H^{\ddagger}$ (74). Lastly, $\Delta S^{\ddagger}$ may be computed from $\Delta G^{\ddagger}$ and $\Delta H^{\ddagger}$ (75). These thermodynamic quantities for inversion of several common rings are listed in Table 18.3 and similar data for the barrier to rotation about the C-C bond of substituted ethanes are given in Table 18.4. The barriers to internal rotations in some other molecules are given in Table 18.5.

Table 18.3.
Thermodynamic quantities for some ring inversions

Compound	$\nu_a - \nu_e$ (Hz)	T_c	$\Delta G^{\ddagger}$ (kcal/mol)	$\Delta H^{\ddagger}$ (kcal/mol)	$\Delta S^{\ddagger}$ (e.u.)
Cyclohexane (C_6HD_{11})	18	$-63°$	10.2	10.8	2.8
Cyclohexene ($C_6H_4D_6$)	24	$-164°$	5.2	5.3	1.3
Cyclohexanone		$-170°$	~4.9		
1,1-Difluorocyclohexane	884	$-46°$	9.7	10.4	3.0
Methylenecyclohexane		$-105°$	8.4	8.6	1.4
cis-Inositol hexaacetate	19	$19°$	15.4	6.6	-30.1
2-Carbomethoxy-(3.2)-meta-cyclophane	23	$59.3°$	16.8	16.1	-2.4

Table 18.4.

Barriers to rotation about the C-C bond of some substituted ethanes (76)

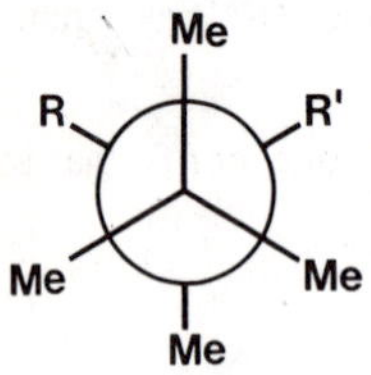

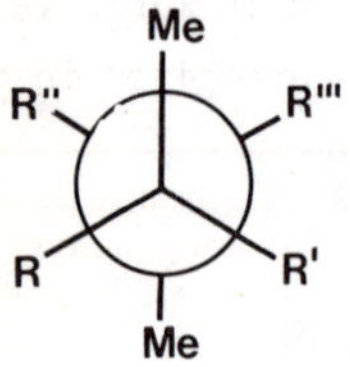

		R	R'	$\Delta G^{\ddagger}$ (kcal/mol)	$\Delta H^{\ddagger}$ (kcal/mol)	$\Delta S^{\ddagger}$ (e.u.)	T_c
	(1)	Cl	CH$_3$	9.82	11.4	8.1	−81°
	(2)	Br	CH$_3$	10.80	13.2	11.9	−68°
	(3)	Cl	Cl	10.81	10.5	−1.6	−61°
	(4)	Br	Br	12.27	11.4	−3.8	−38°
	(5)	Cl	Br	11.90	11.8	−0.4	−45°
	(6)	Cl	C$_2$D$_5$	10.87	9.9	−5.0	−71°
	(7)	Cl	t-Bu	11.42	13.7	10.4	−58°

	R	R'	R''	R'''				
(8)	Me	Br	Br	Br	12.61	11.8	−3.2	−30°
(9)	Cl	Cl	Cl	Cl	13.47	14.1	2.6	−31°
(10)	Br	Br	Br	Br	15.96	13.5	−8.4	+18°

Table 18.5.
Energy barriers restricting internal rotations in some molecules (69, 77)*

Compound	Barrier (kcal/mol)	Compound	Barrier (kcal/mol)
H_3C-CH_3	2.93	H_3C-NO_2	~0
$H_3C-CH_2CH_3$	3.33	$H_3C-CO_2^-$	~0
$H_3C-CH(CH_3)_2$	3.9	$H_3C-Zn-CH_3$	0
$H_3C-C(CH_3)_3$	4.3	$H_3C-C\equiv C-CH_3$	0
$H_3C-CF(CH_3)_2$	4.3	$H_3C-O-CH_3$	3.1
$H_3C-CCl(CH_3)_2$	4.5	$H_3C-S-CH_3$	2.0
$H_3C-CBr(CH_3)_2$	3.9	$H_3C-CO-CH_3$	1.0
H_3C-CH_2F	3.33	H_3C-CF_3	3.45
H_3C-CH_2Cl	3.71	F_3C-CF_3	4.35
H_3C-CH_2Br	3.68	H_3C-CCl_3	5.49
$Cl_3Si-SiCl_3$	0	Cl_3C-CCl_3	10.8

* All of the values are not directly comparable because they are not all for the gaseous state.

The low temperature-high temperature spectral change usually occurs over a small temperature range from which one can estimate a coalescence temperature T_c, the temperature when the broad band just begins to break into two peaks. In the case of 18.1, the O-Me groups give a singlet peak at δ 1.92 ppm at room temperature owing to rapid rotation of the mesityl ring, but at -81° they give two singlets at δ 2.4 and 1.1 ppm with a T_c at -44° (78).

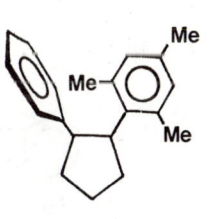

18.1

Rotations about single bonds, especially C-C bonds in substituted ethanes, have been the subject of a vast number of experimental and theoretical investigations concerning conformational energies and barriers to their interconversion. In spite of this tremendous effort, the only quantitative aspect known about the barriers is their magnitude (79). There is little agreement about the relative importance of the several factors contributing to the barriers (5). Nevertheless, several qualitative observations can be noted. From Tables 18.2 and 18.4, for example, we make the following observations. (1) gauche halogen-halogen and methyl-methyl interactions are more destabilizing than a gauche halogen-methyl repulsion (except for fluorine) (2) Substitution of Cl or Br for Me increases the barrier, e.g., (1) vs. (3), (1) vs. (5), (2) vs. (4), (2) vs. (5), (4) vs. (8), (3) vs. (9), (4) vs. (10), and (8) vs. (10) in Table 18.4. (3) Barriers increase generally with increasing size or close approach of groups. Thus, $\Delta G^{\ddagger}$ increases upon substituting Br for Cl, e.g., (1) vs. (2), (3) vs. (4), (5) vs. (4), and (9) vs. (10) in Table 18.4, but not uniformly so for the halogens (see Table 18.5). Rotation of $SiCl_3$ groups about the Si-Si bond is virtually unhindered (Table 18.5) because the bond is long enough to avoid close approach of the Cl atoms as rotation occurs. The barrier is quite large in Cl_3C-CCl_3 because now the nonbonded Cl atoms are very close together. (4) Barriers to rotation of t-butyl groups are in the range 6 to 10 kcal/mol (73, 80). Some examples are:

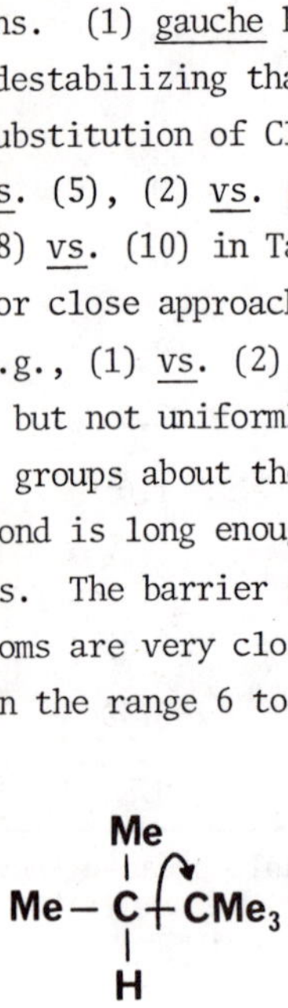

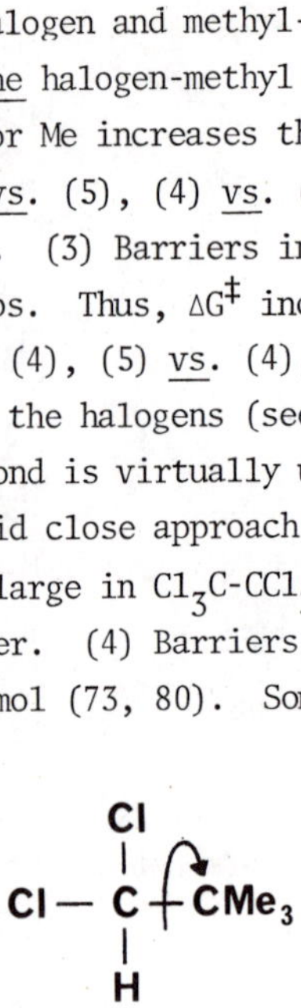

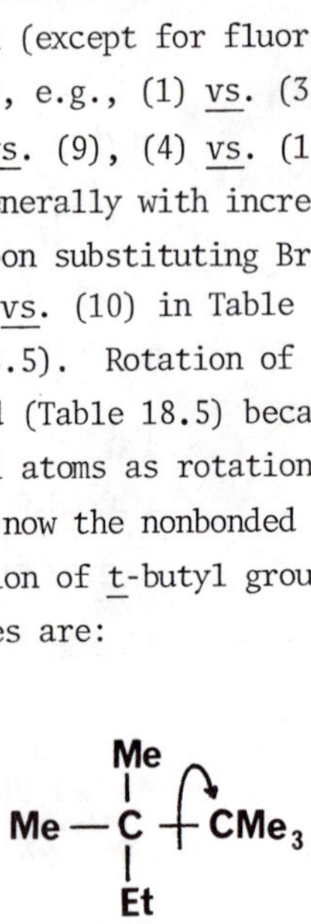

Me–C(Me)(H)–CMe₃	Cl–C(Cl)(H)–CMe₃	Me–C(Me)(Et)–CMe₃
$\Delta G^{\ddagger}$ 6.9	9.0	8.3 kcal/mol
$(CH_2)_n$–C(CMe₃)(H) n = 4-7	cyclohexane with two CMe₃	fluorenyl–C(OH)(CMe₃)
$\Delta G^{\ddagger}$ 6-8	10.1	9.4 kcal/mol

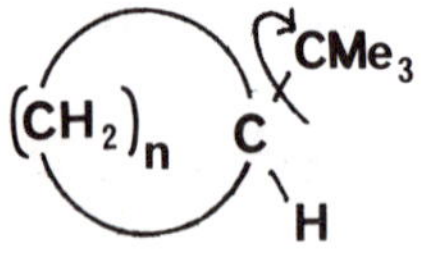

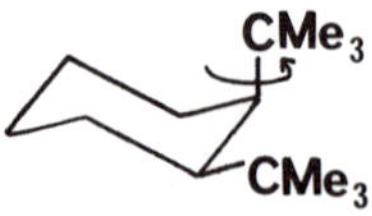

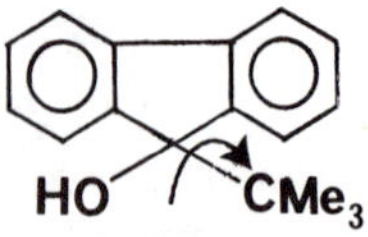

(5) Phenyl groups are often oriented so as to have about the same, or smaller, steric requirements as methyl or ethyl (80).

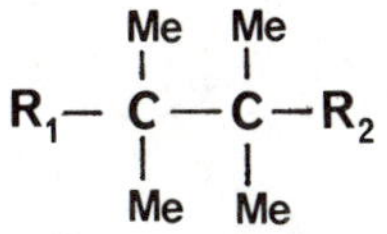

	R_1	R_2	$\Delta G^{\ddagger}$ (kcal/mol)
	Me	Et	8.3
	Et	Et	10.7
	Phenyl	Phenyl	8.0
	Me	H	6.9
	Phenyl	H	6.7

There is a restricted rotation about the C-phenyl bond in phenylcarbinols and phenylhalomethanes. The methane substituents extend out on each side of the benzene ring in order to avoid a steric interaction with the ortho atoms. The barriers to rotation from one conformation to the other in two cases are:

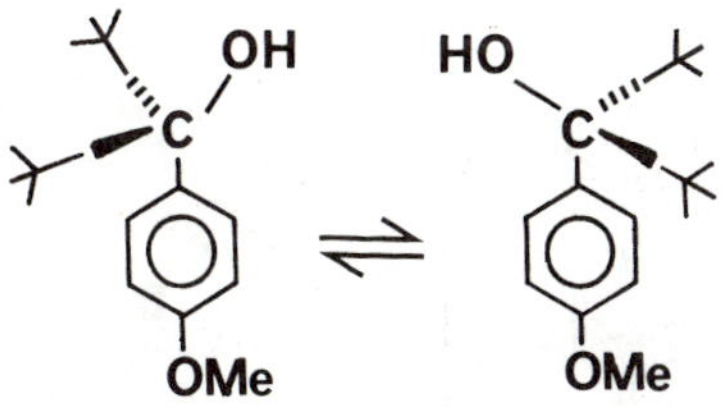

20 kcal/mol (81) 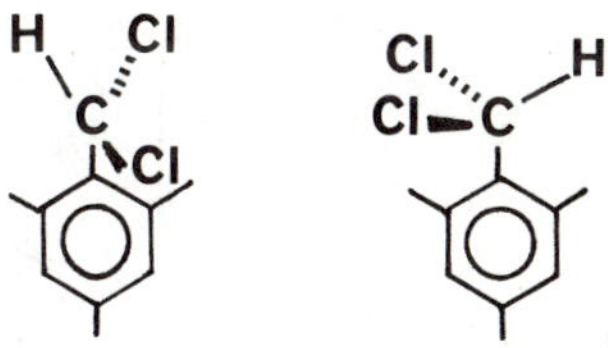   14 kcal/mol ((82)

Another large barrier to rotation about a single C-C bond is that in chloromethyltriptycene (83). The value for a methyl group in this position is only about 7 kcal/mol (83a).

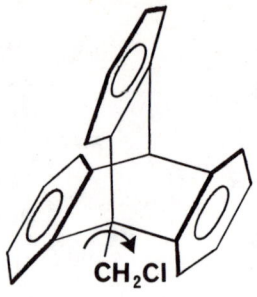

$\Delta G^{\ddagger}$ = 16 kcal/mol

Barriers to rotations about single bonds having double-bond character have been measured to provide some insight to the resonance involved. For example, $\Delta G^{\ddagger}$ increases as the dipolar structure assumes greater importance to the resonance hybrid of p-substituted benzaldehydes (84).

R =	H	OCH$_3$	NMe$_2$
$\Delta G^{\ddagger}$ =	7.9	9.2	10.8 kcal/mol

Rotation about the C-N bond of amides is already frozen at room temperature so nmr measurements of the barrier are made at elevated temperatures (85). The thermodynamic data for rotation in dimethylformamide are $\Delta G^{\ddagger}$ = 20.9 kcal/mol, $\Delta H^{\ddagger}$ = 24.3 kcal/mol, and $\Delta S^{\ddagger}$ is 8.8 e.u. (86). Some illustrative $\Delta G^{\ddagger}$ values are as follows (73):

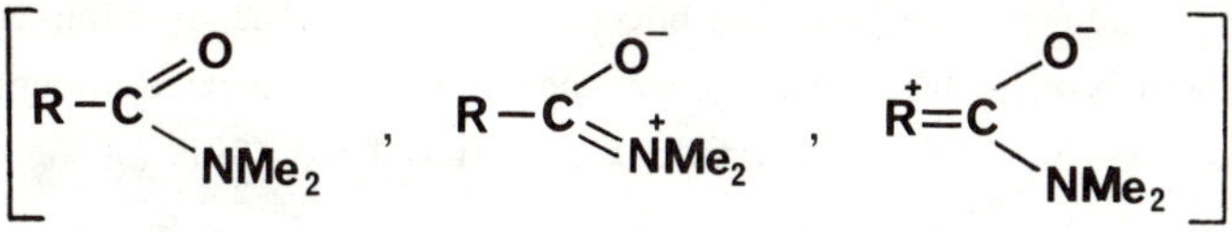

$$\left[R-C\begin{matrix}\nearrow O\\ \searrow NMe_2\end{matrix} \quad , \quad R-C\begin{matrix}\nearrow O^-\\ \searrow \overset{+}{N}Me_2\end{matrix} \quad , \quad \overset{+}{R}=C\begin{matrix}\nearrow O^-\\ \searrow NMe_2\end{matrix} \right]$$

<u>a</u>   <u>b</u>   <u>c</u>

<u>18.2</u>

R	$\Delta G^{\ddagger}$ (kcal/mol)	R	$\Delta G^{\ddagger}$ (kcal/mol)	R	$\Delta G^{\ddagger}$ (kcal/mol)
H	21.7	CCl_3	14.9	 phenyl	14.4
Me	18.9	CF_3	17.6	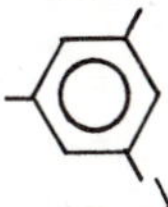 mesityl	22.5
Et	16.7	Cl	16.8	2,4,6-tri-t-Bu-phenyl	~30
		OCH_3	14.8	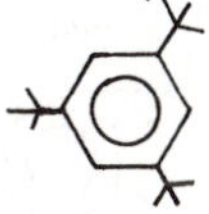	
		$H_2C=CH$	16.1		

Resonance of the R group with the carbonyl (<u>18.2c</u>) decreases the amide group resonance (<u>18.2b</u>). This decreases the double bond character of the C-N bond and lowers $\Delta G^{\ddagger}$ (H < Me < $H_2C=CH$ < OCH_3 < phenyl). There are, however, counteracting inductive and steric effects so that a simple analysis of the sequence of $\Delta G^{\ddagger}$ values for these groups cannot be given. Barriers to rotations about single bonds having double-bond character have been measured for a variety of other systems. Examples are:

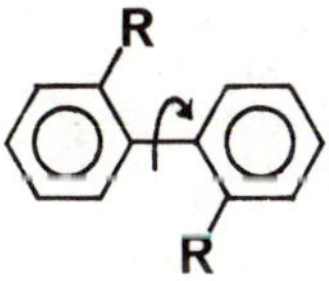

R	$\Delta G^{\ddagger}$ (73)
Me	17.4 kcal/mol
i-Pr	26
OCH_3	17.7
OAc	18.5
CH_2OAc	20.2

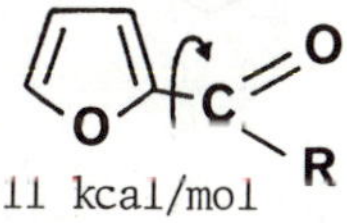

furan-2-C(=O)R: 10-11 kcal/mol

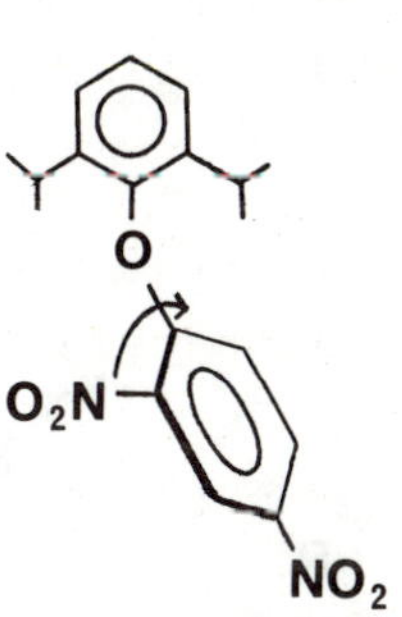

2,6-diisopropylphenyl 4-nitrophenyl ether with NO₂: 17.8 kcal/mol (87)

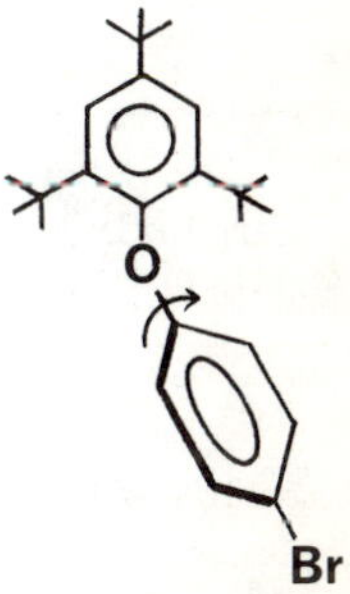

2,4,6-tri-t-butylphenyl 4-bromophenyl ether: 16.5 kcal/mol (87)

The positions of equilibrium between conformations arising from restricted rotations about single bonds have been estimated from dipole moment data for years (88). Some scattered results can be cited here (89).

			μ_{obs}	% isomer on right
μ_{calc}	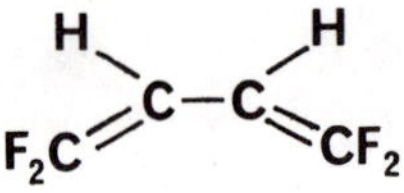 2 D (F₂C=CH–CH=CF₂, s-trans-like)	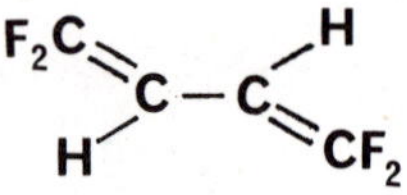 0 D (F₂C=CH–CH=CF₂, other)	0.4 D (90)	96
μ_{calc}	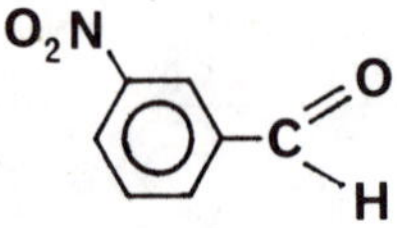 5.21 (o-nitrobenzaldehyde, one rotamer)	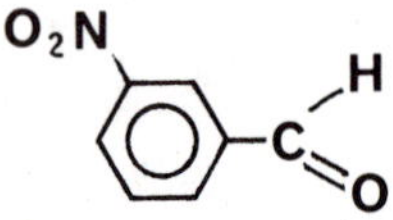 1.81 (other rotamer)	3.22 (89)	70
μ_{calc}	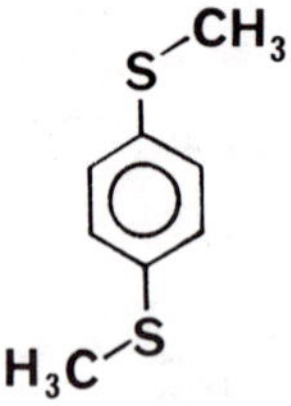 0 (1,4-bis(methylthio)benzene, anti)	 2.48 (syn)	1.81 (91)	53
μ_{calc}	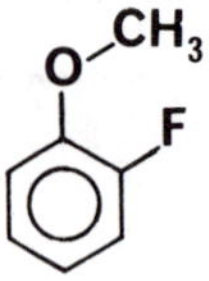 1.14 (o-fluoroanisole, one rotamer)	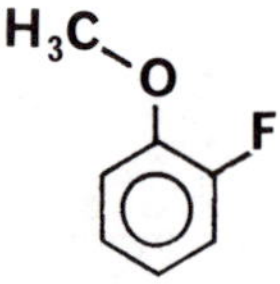 2.50 (other)	2.31 (92)	82

Similarly, esters have the s-trans conformation as shown by their dipole moments

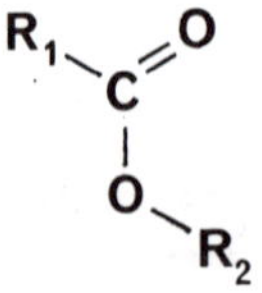

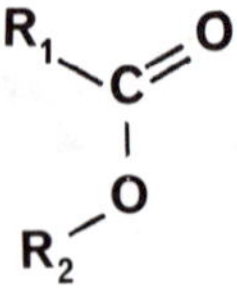

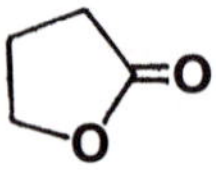

s-trans   s-cis

$\mu_{calc} \approx$ 1.5 D   3.5 D μ_{obs} = 3.82 D

usually between 1.45 and 2.0 D, which are temperature independent. For comparison, the γ-butyrolactone, which is confined to an s-cis conformation, has a dipole moment of 3.82 D (93).

When the dipole moment is temperature independent, it is assumed that the molecule has a fixed, noncentrosymmetric structure rather than being an equilibrium mixture of two conformations. For example, the moment of biacetyl (1.08 D) can correspond to an equilibrium mixture of the s-cis and s-trans conformations

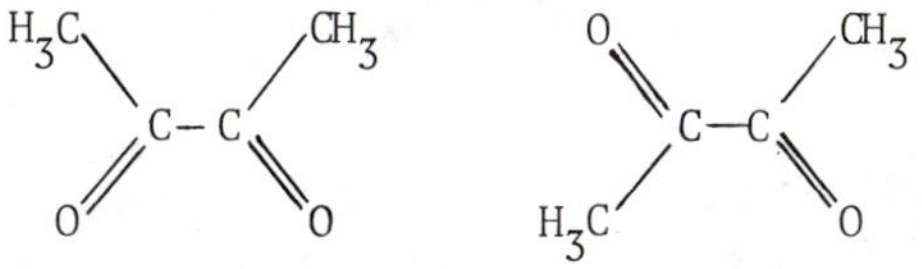

μ_{calc} =   4.73 D   0 D

or to a rigid nonpolar structure with an angle ρ between the planes of the acetyl groups. If the latter, ρ can be calculated from the equation

$$\mu_{obs}^2 = 2\mu_{COMe}^2 \sin^2 \phi (1 + \cos \rho)$$

where μ_{COMe} is the group moment and ϕ is the angle of inclination of the group vector to the C-C axis. Angle ρ was calculated to be ~160°, i.e., a skew s-trans conformation (94). A rigid nonplanar structure was favored over an equilibrium mixture on the basis of Kerr constants (94). Cases in which the barriers to restricted rotation are large enough to permit isolation of optical isomers are discussed in Section 20.1.

As mentioned earlier in this section, barriers to ring inversions are readily measured by variable temperature nmr spectroscopy. The inversion barriers for some rings are:

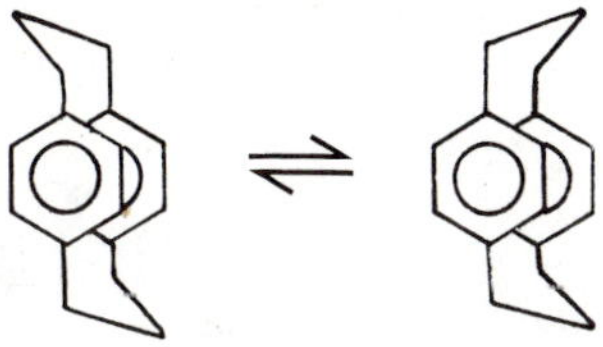

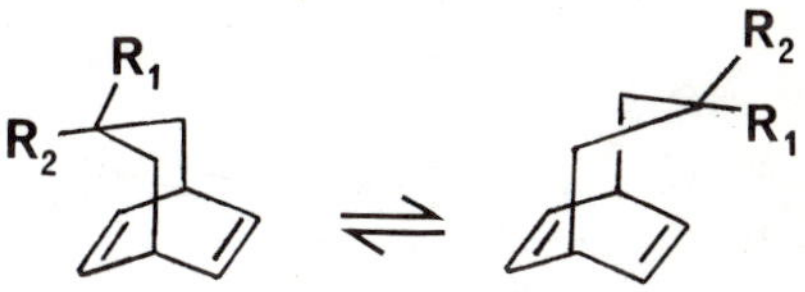

$\Delta G^{\ddagger}$ = 11.7 kcal/mol   Too fast even at -140° to measure by nmr.

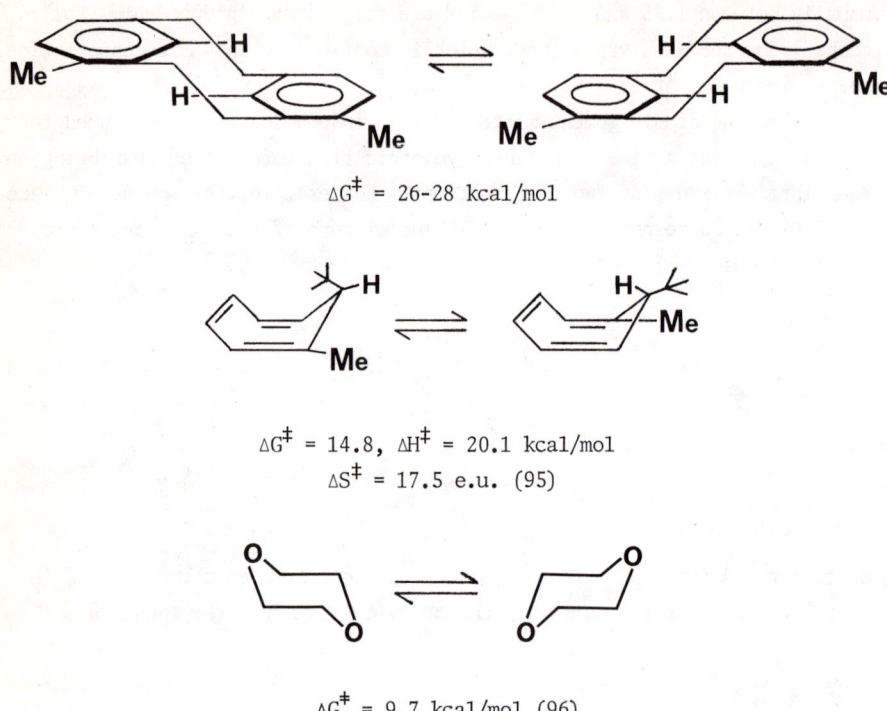

$\Delta G^\ddagger$ = 26-28 kcal/mol

$\Delta G^\ddagger$ = 14.8, $\Delta H^\ddagger$ = 20.1 kcal/mol
$\Delta S^\ddagger$ = 17.5 e.u. (95)

$\Delta G^\ddagger$ = 9.7 kcal/mol (96)

Incidentally, ring inversion of nitrogen heterocycles is even more complex because nitrogen atom inversion occurs as well as ring inversion (97). For instance, two of the many inversions possible for 18.3 are:

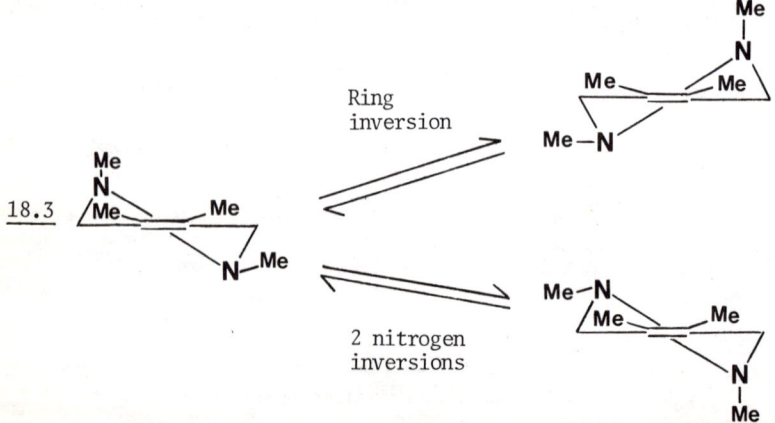

$\Delta G^{\ddagger} = 7$ for ring inversion (73)	8.8	20.5	19.1 kcal/mol

At -99° and above, the methyl groups in 18.4 are equivalent and appear as a doublet with coupling with the methine proton. At -110°, inversion of the nitrogen atom is frozen and the two methyls are no longer equivalent so they give two peaks, each a doublet (98).

18.4

$\Delta G^{\ddagger}$ (N inversion) = 9.1 kcal/mol

Although the concept of the "effective size" of lone-pair electrons is a nebulous one, theoretical (99) and experimental work continues along this line. Most experimental data support the view that a lone pair on a hetero atom is smaller than a bonded hydrogen atom. Thus, diaxial repulsion in m-dioxane is smaller than in cyclohexane (100). In another case, as X changes from S, O, N, to lone pair, in the compound 18.5 for which one may write the resonance -N=C=X, $-\overset{+}{N}\equiv C-X^{-}$ for the substituent, the contribution from the polar form to the resonance hybrid decreases. There is a simultaneous increase of the electron density of nitrogen to increase diaxial repulsion. This is reflected in an increase in the $\Delta G^{\ddagger}$ value (101).

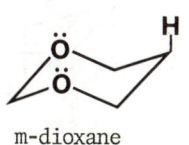

m-dioxane

18.5

X	$\Delta G^{\ddagger}$
NC_6H_{11}	1.00 kcal/mol
O	0.51
S	0.28
: (lone pair)	0.21

Chapter 19

RING STRAIN AND MOLECULAR PROPERTIES (102)

The synthesis of highly strained compounds has been a challenge to chemists (103) and their efforts have been rewarded by the fascinating properties exhibited by many of the products. Even in their nomenclature, there has been an unusual flair for catchy trivial names:

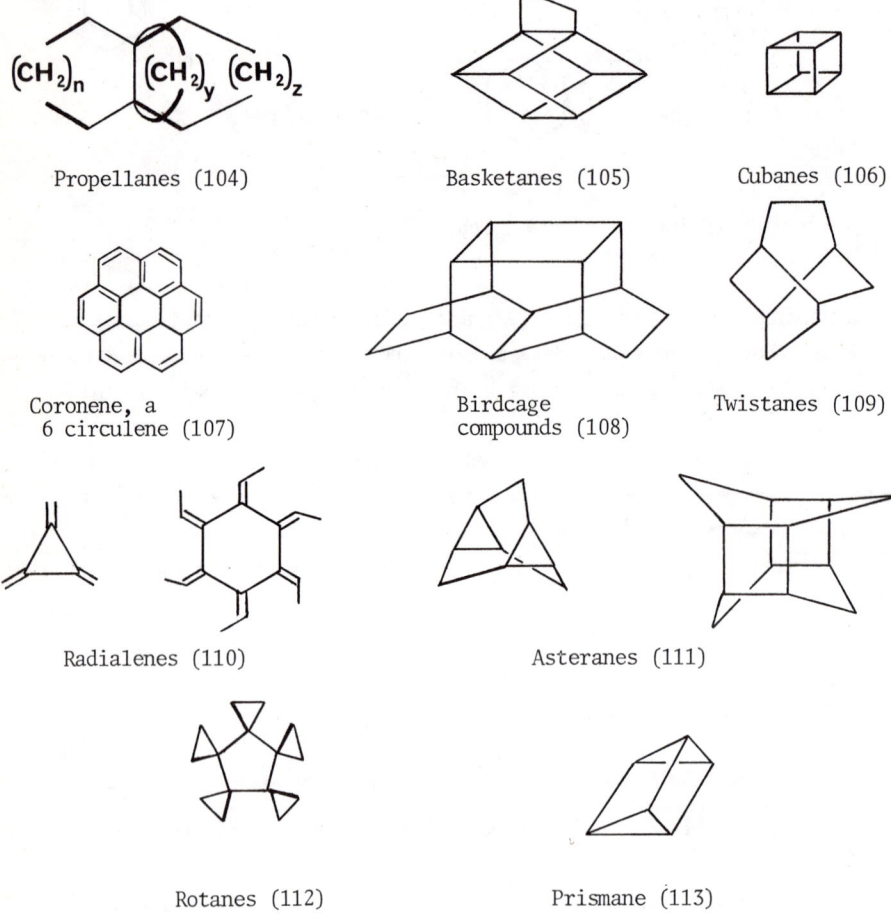

Propellanes (104) Basketanes (105) Cubanes (106)

Coronene, a 6 circulene (107) Birdcage compounds (108) Twistanes (109)

Radialenes (110) Asteranes (111)

Rotanes (112) Prismane (113)

Most of their chemistry has been learned in the past decade. One aspect of their chemistry which has been of special interest is the influence that ring strain has upon their properties.

19.1. Ring strain energies

Ring strain energies have been determined mostly from thermochemical data such as heats of combustion or hydrogenation. In one approach, cyclanes have been compared with the respective n-alkanes, whose intramolecular strains are a minimum. The strain energies so determined for cyclanes from heats of combustion are given in Table 19.1. Ring-size classification is based on their ring strain:

Small rings	3, 4	Very large strain, mostly bond angle.
Regular rings	5 to 7	Moderate or no strain, mostly torsional.
Medium-size rings	8 to 11	Large strain, mostly transannular, van der Waals, and bond angle.
Large rings	≥12	Small or no strain, mostly torsional.

Heats of hydrogenation have been useful in determining ring strain in cyclenes and cyclynes. Thus, the double-bond strain in trans-cyclooctene is about 9 kcal/mol and the triple-bond strain in cyclooctyne is at least 10 kcal/mol, whereas the triple-bond strain in cyclodecyne and cyclododecyne is negligible.

A second procedure for estimating strain energies is from semitheoretical calculations of heats of formation (114). A more widely used method makes use of group increments (116). These are empirically-chosen values assigned

Table 19.1. Strain energies of cyclanes (115)

Ring size	Total ring strain (kcal/mol)	Ring size	Total ring strain (kcal/mol)
3	27.6	9	12.6
4	26.2	10	12.0
5	6.5	11	11.0
6	0	12	3.6
7	6.3	14	0
8	9.6	16	1.6

to hydrocarbon fragments, plus corrective constants for skew interactions, determined from thermochemical data on minimum-strain model compounds (values in kcal/mol):

$$CH_3 = -10.05 \qquad {>}CH = -2.16 \qquad \text{skew } +0.70$$
$$\diagup CH_2 = -5.13 \qquad -\underset{|}{\overset{|}{C}}- = -0.30$$

For example, the heat of combustion of cubane is 1156 kcal/mol. From this quantity and its heat of sublimation computed from its vapor pressure, the heat of formation of gaseous cubane at 0° is 148.7 kcal/mol (117). The value calculated from group increments for eight ${>}CH$ units is -17.28 kcal/mol. Hence, the strain energy in cubane is 148.7 - (-17.28) = 166 kcal/mol. Strain energies for a variety of alicycles have been determined by the use of group increments and are listed in Table 19.2. Considering the steps involved in their determination, the values are in good agreement with those obtained directly from heats of combustion. Two recent comparisons, for example, are

	Direct values (118)	Schleyer group increment values
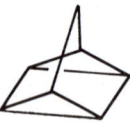	9.3	11.0 kcal/mol
	9.7$_{liq}$	12.2 kcal/mol

Note that the strain energy of quadricyclene (101 kcal/mol) is greater than the energy of a C-C bond. Fortunately the strain is distributed over several bonds, but the large strain does give it olefinic properties. Bicyclo(2.2.2)-octane has six pairs of eclipsed C-C bonds and the distance between bridgehead carbons (2.54 Å) is less than the sum of the carbon van der Waals radii. As anticipated, an analysis of the strain in this hydrocarbon indicates that the strain arises chiefly from torsional and van der Waals repulsions (119).

Quadricyclene

Bicyclo(2.2.2)octane

Sec. 19.1 RING STRAIN ENERGIES

Table 19.2. Strain energies based on group increments (116)

Compound	Strain energy (kcal/mol)*	Compound	Strain energy (kcal/mol)*
Cyclopropane	28.1 (27.6)	Bicyclo[1.1.0]butane	66
Cyclopropene	54.5	Bicyclo[2.1.0]pentane	56.5
Cyclobutane	26.9 (26.2)	Bicyclo[2.2.1]heptane (norbornane)	17.6 (17.6)
Cyclobutene	30.6		
Cyclopentane	7.2 (6.5)	Bicyclo[2.2.2]octane	11
Cyclohexane	1.4 (0)	Norbornene	27.2 (17.6)
Cycloheptane	7.8 (6.3)	Norbornadiene	34.7 (25.6)
Cyclooctane	10-11.7 (9.6)	cis-Cyclooctene	7.4
Cyclononane	14.4 (12.6)	trans-Cyclooctene	16.7
Cyclodecane	15 (12.0)	Bicyclo[3.3.0]octane, cis	12.0
Spiropentane	65	Bicyclo[3.3.0]octane, trans	18.4
cis-Decalin	4.9	Nortricyclene	47.0 (38.8)
		Quadricyclene	101 (78.7)

* Thermochemical values in parentheses (H. K. Hall, Jr., C. D. Smith, and J. H. Baldt, J. Am. Chem. Soc. 95: 3197 (1973); ref. 115).

QUESTIONS

1. The calculated dipole moments of s-cis and s-trans-o-bromo-anisole are 1.19 and 2.58 D, respectively. Calculate the percentage of s-trans in the equilibrium mixture if the observed dipole moment is 2.47 D.
2. Which isomer should have the larger heat of combustion, cis or trans-decalin?

3. Which isomer should have the greater strain, M or N ?

M N

4. trans-Cyclooctene has an unusually large dipole moment (0.8 D) for an olefin, and has been isolated in an optically active form. How would you account for these properties ?

ANSWERS

1. Using equation (18.1) in Chapter 18 with dipole moment data, we get 90% s-trans.
2. The cis isomer, because it is under more strain.
3. Strain energies are often approximately additive. On this basis, one might expect a strain of 7.2 + 7.2 + 28.1 ≃ 43 kcal/mol for M (with its 5 + 5 + 3 membered rings) and 1.4 + 26.9 + 28.1 ≃ 56 kcal/mol for N (with its 6 + 4 + 3 membered rings).
4. The molecule is chiral and rigid. Racemization would involve passage of the olefinic or methylenic protons or both through the ring. There

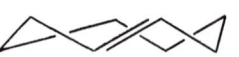

is a large energy barrier against this, making the isomer optically stable. The ring is under considerable strain and this affects the hybridization of the trigonal carbons to make them more electronegative than usual, thereby increasing the -C-C= bond polarity and giving the molecule a dipole moment.

19.2. Ring strain and bonding

With decreasing C-C-C bond angles, ring bonds acquire greater p character and exo bonds greater s character. This bond character change is reflected in many ways. The C-C bonds of the ring become olefinic, i.e., undergo addition reactions, exhibit hyperconjugation, and delocalize electrons to or from conjugated π orbitals. At the same time, exo C-C or C-H bonds have increased s character, which makes the carbon more electronegative. For instance, as ring strain increases, it becomes increasingly hard to extract a hydrogen or iodine atom from carbon (120),

	Relative rates of reaction per CH_2 toward chlorine	t-butylhypochlorite	Relative rates of abstraction of I by phenyl radicals (121)	
Cyclohexane	1.0	1.0	t-Butyl	1.08
Cyclopentane	0.95	0.89	Adamantyl	0.75
Cyclobutane	0.84	0.51	Bicyclo(2.2.2)octyl	0.62
Cyclopropane	0.05	0.01	Bicyclo(2.2.1)heptyl	0.19

C-H stretching frequencies of cyclanes increase,

Cyclane ring size	ν_{C-H}	Cyclane ring size	ν_{C-H}
6	2925 cm^{-1}	4	2970 cm^{-1}
5	2945	3	3080

dipole moments of the cycloalkyl bromides decrease (122), and ir carbonyl frequencies increase (both properties reflect less electron migration from the ring carbon).

Ring size	C-Br bond moment in cycloalkyl bromide (122)		$\nu_{C=O}$ for cyclanones (123)
3	1.69 D		1815 cm^{-1}
4	2.09		1791
5	2.16		1750
6	2.31		1718
Isopropyl bromide	2.05	Acetone	1719

Similarly, the increased strain from bridging across the cyclohexene ring is shown by a much larger heat of hydrogenation of norbornene and a larger $\nu_{=CH}$.

$\nu_{=CH}$ 3024 cm^{-1}
ΔH_{H_2} 28.2 kcal/mol

$\nu_{=CH}$ 3070 cm^{-1}
ΔH_{H_2} 33.2 kcal/mol

Moreover, with increasing strain, the C-H bonds become more acidic, and since J_{CH} (124) (as well as J_{NH} (125)) increases with greater electronegativity of the adjacent atom, C-H bonds of strained rings exhibit significant ^{13}C-H coupling. We saw earlier (Section 15.2) that the s character of C-H bonds may be calculated from ^{13}C-H coupling: $J(^{13}C\text{-}H) = 500\, \underline{f}_s$, where $\underline{f}_s$ is the fractional s character of the carbon bonding orbital. Similarly, the s character of carbon orbitals of C-C bonds may be determined from $^{13}C-^{13}C$ coupling (126): $J(^{13}C-^{13}C) = 550\, \underline{f}_{s1}\underline{f}_{s2}$, where $\underline{f}_{s1}$ and $\underline{f}_{s2}$ are the fractional s character of the two bonded carbon atoms (actually the proportionality constant ranges from 500 to 575, but most often the value 550 has been used). Since the latter equation has two variables, $\underline{f}_{s1}$ and $\underline{f}_{s2}$, an assignment is normally given to one in order to determine the other. For example, some calculated $\underline{f}_s$ values are given below based on the indicated assigned $\underline{f}_s$ values and the experimental $J(^{13}C\text{--}^{13}C)$ data (127).

Compound	$J(^{13}C-^{13}C)$		$\underline{f}_s$ (assigned)		$\underline{f}_s$ (calculated)	
$\overset{1}{CH_2}=\overset{2}{C}=\overset{}{CH_2}$	$\underline{J}_{1,2}$	98.7 Hz	$\underline{f}_{s2}$	0.5	$\underline{f}_{s1}$	0.36
(bicyclobutane)	$\underline{J}_{1,3}$	20.2	$\underline{f}_{s3}$	0.25	$\underline{f}_{s1}$	0.15
(spiropentane)	$\underline{J}_{1,2}$	29.8	$\underline{f}_{s1} = \underline{f}_{s2}$		$\underline{f}_{s2} = \underline{f}_{s1} = 0.23$	
	$\underline{J}_{3,5}$	36.0	$\underline{f}_{s5}$	0.25	$\underline{f}_{s3}$	0.26
(cyclopropane)	$\underline{J}_{1,2}$	21.0	$\underline{f}_{s2}$	0.15	$\underline{f}_{s1}$	0.25
(bicyclo)	$\underline{J}_{1,2}$	36.7	$\underline{f}_{s2}$	0.23	$\underline{f}_{s1}$	0.29
	$\underline{J}_{4,5}$	16.0	$\underline{f}_{s5}$	0.15	$\underline{f}_{s4}$	0.19
(bicyclo)	$\underline{J}_{2,3}$	18.2	$\underline{f}_{s3}$	0.15	$\underline{f}_{s2}$	0.22

Thus, one gets a value of $\underline{f}_s$ = 0.15 for methylene (CH_2) carbons in C_3 rings and 0.23 for C_4 rings. Inasmuch as the $\underline{s}$ character of the C-H carbon in bicyclobutane is 0.40, and 0.25 for each $C-CH_2$, this leaves 0.1 for the $\underline{s}$ character in the zero bridge bond. That is, it has almost pure $\underline{p}$ character (calculated to be $\underline{sp}^{24}$) (128), which is consistent with its properties (see later in this section). Likewise, the carbons of the zero bridge bond of bicyclopentane have only 0.16 $\underline{s}$ character.

The $\underline{s}$ character of C-H bonds, when calculated as above, parallel kinetic acidities of the bonds as determined from rates of reaction with methyllithium (129) or rates of tritium exchange in the presence of N-tritiated cesium cyclohexylamide (CsCHA) (130) (see Table 19.3).

Table 19.3. Polar character of some C-H bonds

Cyclane	$J(^{13}C-H)$	$\underline{s}$ character (%)	$\underline{k}_{rel}$ toward CsCHA (130)
C_7 to C_{14}	118 to 126 Hz	~25	0.4 to 1.0
C_6	123	25	(1.0)
C_5	128	26	5.7
C_4	134	27	28
C_3	161	32.2	7×10^{14}

Hydrocarbon	$J(^{13}C-H)$	$\underline{s}$ character (%)	$\underline{k}_{rel}$ toward MeLi (129)
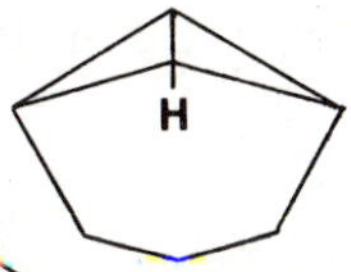 (cyclopentane with H)	200 Hz	40	1.0
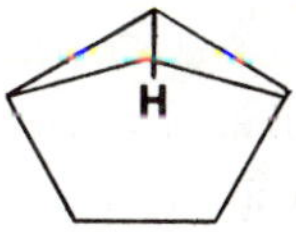 (bicyclic with H)	206	41.2	12
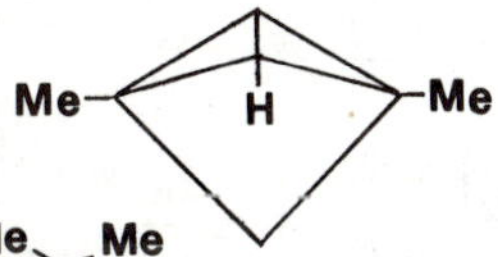 (Me–/H/–Me bicyclic)	212	42.4	65
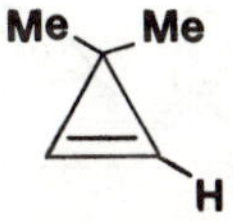 (Me, Me cyclopropene with H)	221	44.2	2,500

Increased ring strain also increases the acidity of attached O-H bonds (131).

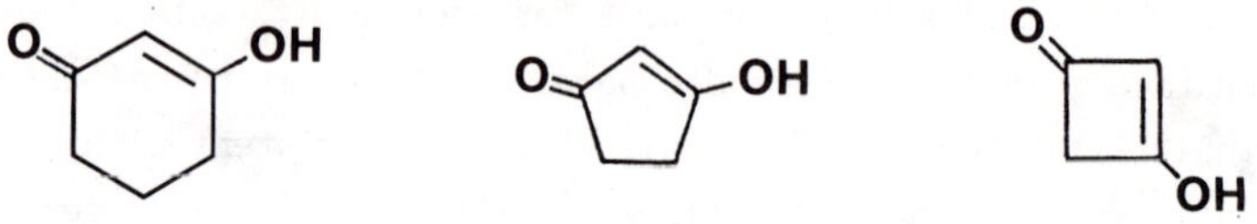

pK_a 5.25   4.5   3

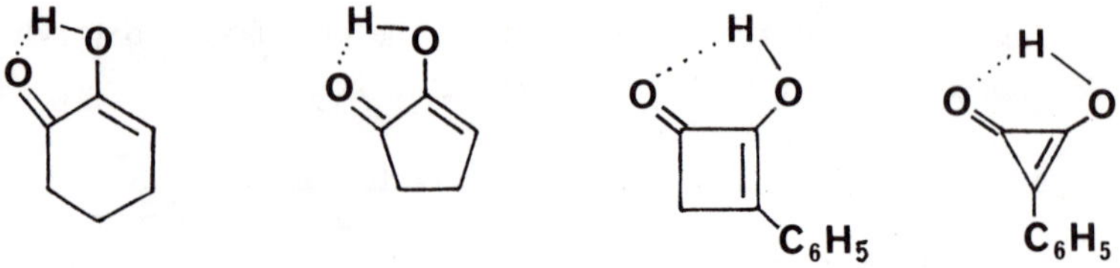

pK_a 10.3   9.1   6.3   2

Part of the increase in the lower series is due to successively weaker H bonds.

A less convenient index of s character in the exo bond of strained rings, but nevertheless a reliable correlation, is the sum of the three internal bond angles. In a tetrahedral arrangement, this is 328.5°. The total about the bridgehead C-H of 19.1 is 308° (132). This implies greater p character in the C-C bonds and greater s character in the bridgehead C-H bond.

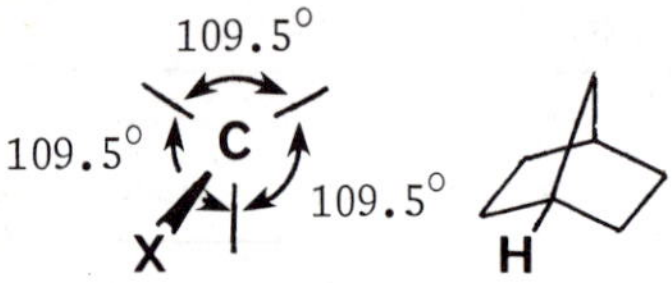

19.1

The increased s character and hence electronegativity of exo carbon bonds with increasing ring strain is also reflected in the increased acidity of attached COOH groups and diminished basicity of amino groups (Table 19.4). For example, three comparisons are (133):

pK_a =   6.62   6.37   6.35

Sec. 19.2 RING STRAIN AND BONDING   493

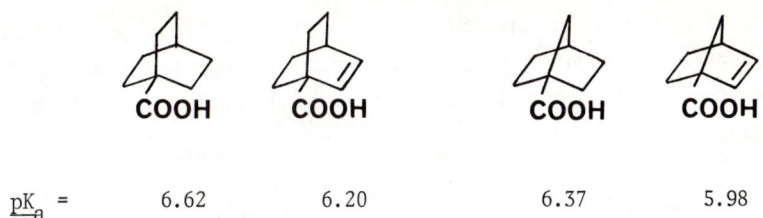

pK$_a$ =   6.62   6.20   6.37   5.98

Table 19.4.
Dissociation constants of some strained-ring acids and bases (134)

Compound	Ring strain (kcal/mol)	$10^5 K_a$ (X = COOH)	$10^5 K_b$ (X = NH$_2$)
◇—X	26.9	1.64	11
(norbornyl)—X	17.6	3.48	2.0
X—(bicyclic)	~60	8.05	0.38
(cube)—X	166	11.1	

However, as was pointed out in Section 3.2, dissociation constants are not a good index for relative electronegativities because solvation has a big influence. For instance, the cyclobutyl derivatives in Table 19.4 do not follow the order of increasing ring strain.

The effects of a strained, fused ring on the properties of aromatic rings has been the subject of many studies. We have noted that increasing

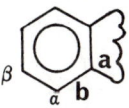

ring strain accompanying decreasing bond angles gives the ring bond "a" greater p character and the exo bond "b" greater s character. This is reflected in preferred electrophilic substitution at the β atom and increased acidity of the α proton. Various physical properties are consistent with this model, such as (135) nmr and ir data, polarographic reduction potentials, and esr data. For one example, the basicities of substituted quinolines decrease with greater ring strain in fused rings (136).

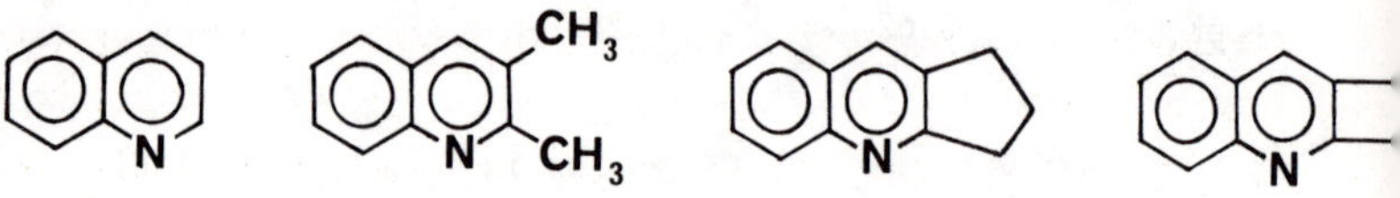

pK$_a$ 5.06   5.99   5.45   4.55

Alkyl groups or strain-free rings increase the basicity but the strained, fused C$_4$ ring markedly decreases the basicity.

The olefinic character of C-C bonds of strained rings is shown by their tendency to undergo addition reactions (137):

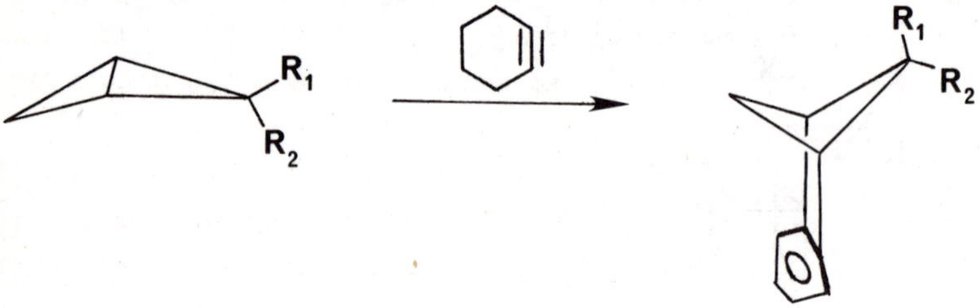

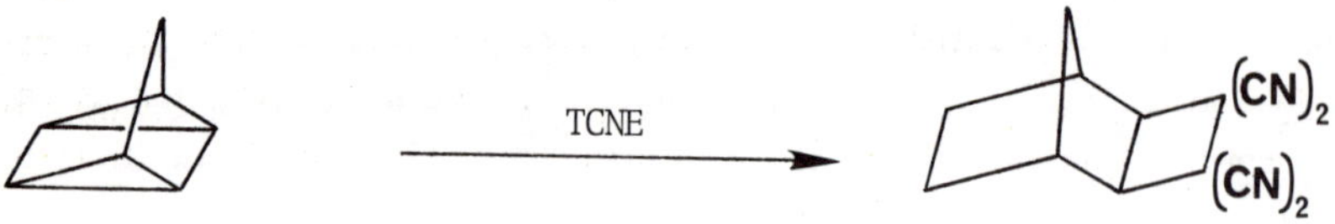

Indeed, bicyclobutanes may be polymerized like olefins (138):

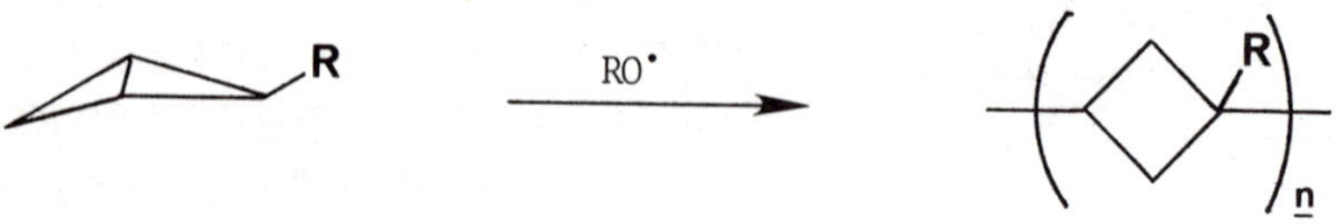

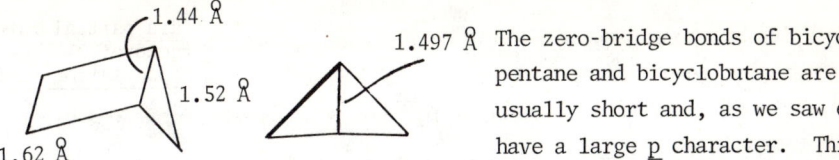

1.497 Å The zero-bridge bonds of bicyclopentane and bicyclobutane are unusually short and, as we saw earlier, have a large p character. This view is supported by the observation that the uv spectra of bridgehead derivatives of bicyclobutane indicate that there is a conjugation across the zero-bridge bond (139). Also, 19.2 has a λ_m in the uv at about 190 nm whereas 19.3 exhibits only end absorption in this region (140). The question of

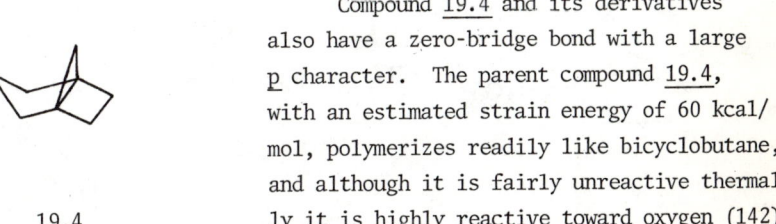

19.2   19.3

conjugation across a cyclopropane ring, that is, can the cyclopropyl group extend conjugation, has been studied by a variety of experimental techniques including uv, ir, nmr, ord, cd, and reaction kinetics. It appears that the electronic interaction is small but does occur in some systems (141).

Compound 19.4 and its derivatives also have a zero-bridge bond with a large p character. The parent compound 19.4, with an estimated strain energy of 60 kcal/mol, polymerizes readily like bicyclobutane, and although it is fairly unreactive thermally it is highly reactive toward oxygen (142).

19.4

Twist bent bonds, as in trans-fused cyclopropanes, are more reactive, having even less orbital overlap than the symmetrically bent bonds of cis-fused cyclopropane rings (143).

The greater p character of C-C bonds in strained rings is accompanied by greater hyperconjugation. The C-C bonds of cyclopropane are sp^5 hybridized (144) and the ability of cyclopropyl groups to delocalize electrons via hyperconjugation was discussed in Section 5.3. Similarly, the hyperconjugative ability of the norbornyl group is revealed by its activating effect on electrophilic substitution:

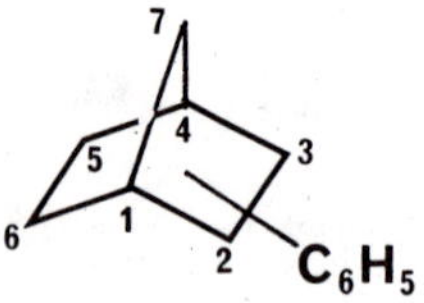

	Para partial rate factor for benzoylation (145)
t-Bu-C$_6$H$_5$	398
i-Pr-C$_6$H$_5$	519
Et-C$_6$H$_5$	563
Me-C$_6$H$_5$	633
7-Phenylnorbornane	822
endo-2-Phenylnorbornane	1040
exo-2-Phenylnorbornane	1630
1-Phenylnorbornane	1790

The tertiary C-H bond of norbornane has 28% s character, hence the C-C bonds have greater than 75% p character.

This delocalizing ability of strained rings via hyperconjugation has been observed for other groups too (146). For instance, the relative rates of solvolysis of tosylates RCH$_2$OTs is related to the ability of R to stabilize the transition state, and the rates correlate well with ring strain.

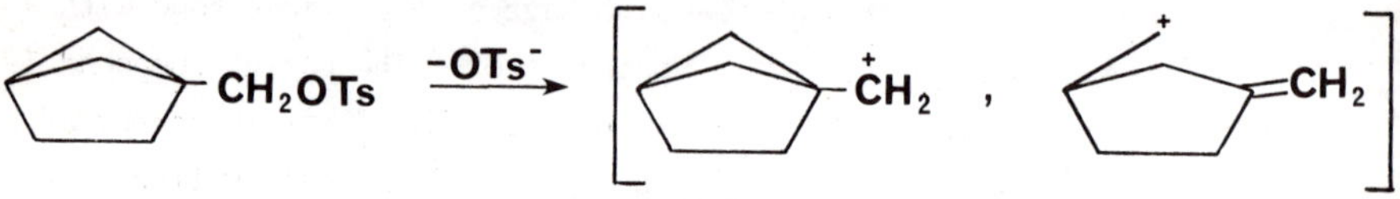

Also, the basicity of ketones RCOCH$_3$ is related to the ability of R to stabilize the conjugate acid (147):

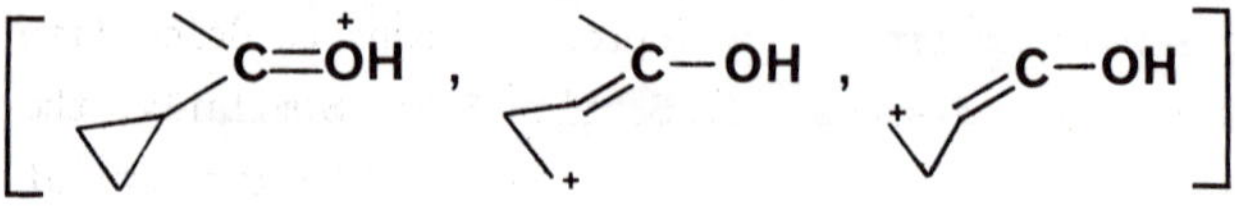

R	pK_{BH^+}
Isopropyl	-7.42
Cyclohexyl	-7.03
Cyclobutyl	-6.86
Cyclopropyl	-6.52
▷—CH₃	-5.47
(bicyclic structure)	-4.06

Semiquantitative correlations of this type were discussed in Section 7.1 through Hammett equations

19.3. Ring strain and reactivity

Ring strain affects the reactivity of cyclic compounds in a variety of ways such as the ease of thermal rearrangements, rates of solvolysis, and addition reactions of double bonds.

An early correlation of ring strain with reactivity was made by H. C. Brown. He expressed the total ring strain of alicycles as <u>internal strain</u>, I, and related the reactivity to the difference between I of the reactant and the reaction transition state. He correlated the relative reactivities of cyclane derivatives with ring size as shown in Figures 1.4 and 1.5 depending upon the change in hybridization of the reacting carbon (148). For example, the relative rates of ethanolysis of cycloalkyl chlorides (149) and of acetolysis of cyclanol tosylates (150) ($C_{sp^3} \rightarrow C_{sp^2}$ to reach transition state) follow the curve in Figure 19.1, whereas the rates of reaction of cyclanones with diazomethane (151) and borohydride ($C_{sp^2} \rightarrow C_{sp^3}$ to reach transition state) and the log K/K_0 (K_0 for acetone) ratios for acetal and cyanohydrin formation of cyclanones (152) follow the curve in Figure 19.2. These relative reactivities can be rationalized as was done for the failure of cyclodecanone to add HCN (Chapter 17).

Other studies have also identified activation energies with differences in ring strain between reactants and transition states. Thus, the log of the rates of solvolysis of some neopentyl p-nitrobenzoates (<u>19.5</u>) is linearly related to the strain released upon reaction (153).

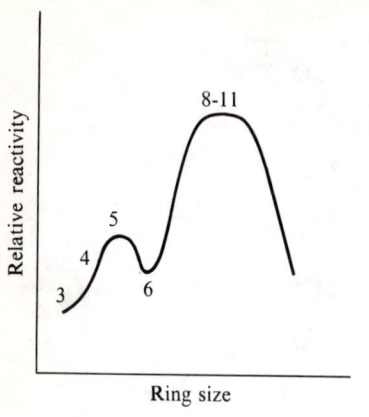

Figure 19.1. $C_{sp^3} \rightarrow C_{sp^2}$

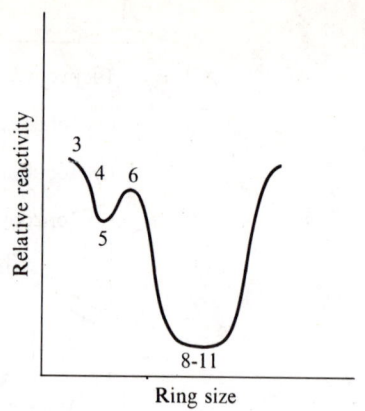

Figure 19.2. $C_{sp^2} \rightarrow C_{sp^3}$

19.5

The log of the rates of solvolysis of some bridgehead bromides is linearly related to the strain energy differences between respective hydrocarbons and carbocation intermediates. The more strain in the carbocation because of deviation from planarity, the slower is the reaction. Bridgehead reactivities via radical intermediates somewhat parallel carbocation reactivities but the radical reactions are much faster:

	k_{rel} (via C^+) (152)	k_{rel} (via $C^{\cdot}$) (1
Me_3CBr	1.0	1.0
	0.1	

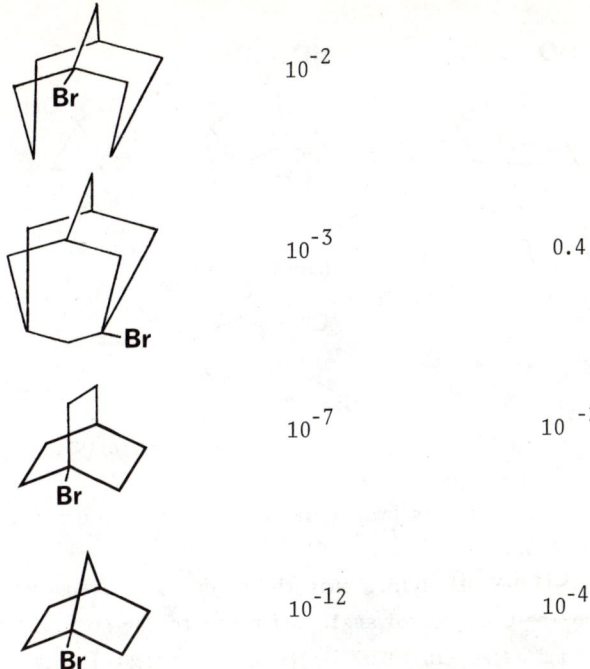

Thus, the adamantyl radical is essentially as reactive as an aliphatic tertiary radical. Esr alpha hyperfine coupling constants for alkyl and cycloalkyl radicals support the contention that these radicals are planar (155). However, beta hfc constants from esr measurements on bridgehead adamantyl and bicyclo(2.2.2)octyl radicals indicate that these two radicals are pyramidal (156). Hence, although carbon radicals strive to achieve a trigonal structure, they may still be fairly stable if only slightly distorted from planarity.

Carbonyl groups are destabilized by ring strain and are easily quaternized when highly strained. Thus, the following hydrates and acetals are readily isolated:

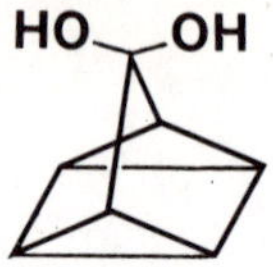

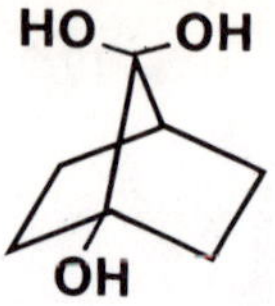

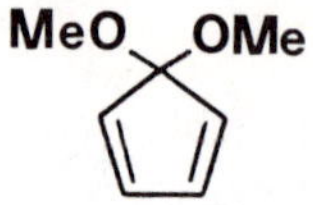

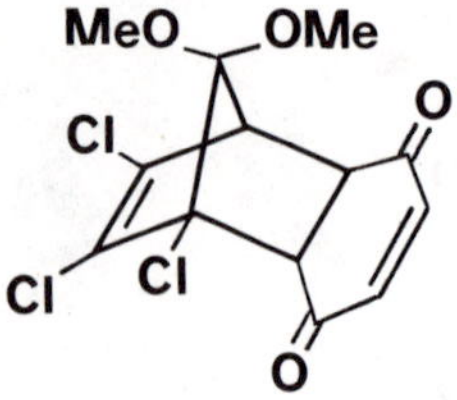

and the last one resists hydrolysis to the ketone under a variety of conditions (157).

We saw in Section 15.2 that the ir carbonyl frequency is directly related to the C-CO-C bond angle and hence to the bond angle strain. The carbonyl bond angle, in turn, is related to rates of solvolysis of secondary tosylates because both, the transition state for S_N1 solvolysis and $v_{C=O}$, depend upon the trigonal character of the carbon (158),

$$\log k_{rel} \text{ to cyclohexyl} = 0.132(v_{C=O} - 1720).$$

Hence, ketone carbonyl frequencies not only provide a convenient measurement for estimating carbonyl bond angles but also serve as a simple index for predicting rates of solvolysis (159).

Another way in which the effect of ring strain on reactivities is shown is the nucleophilicity of double bonds. Several studies have been made of rates of addition to cyclenes, e.g., hydrogenation, diimide reduction, etc., and generally, the greater the ring strain, the more reactive is the cyclene (160). There is a parallel trend, for instance, in the rates of addition of diisoamylborine and the stabilities of silver ion complexes of cyclenes (161). Similarly, norbornenes add HCl quantitatively at -78° in CH_2Cl_2 whereas unstrained chloroalkenes are inert (162).

An empirical rule known as Bredt's rule states, in effect, that if the number of atoms in the bridges of a bicyclic system is designated S, then compounds with double bonds to bridgehead carbons will be isolable only when S ≥ 9. This generalization challenged chemists to synthesize compounds which defy Bredt's rule but for years the smallest bicyclic compounds with a bridgehead double bond had an S value of 9. Recently, however, several exceptions to Bredt's rule have been synthesized (163) such as 19.6 and 19.7 and the rule was modified by Wiseman to the effect that the strain of a bridgehead

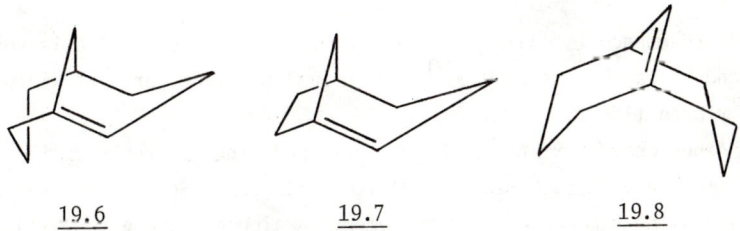

19.6   19.7   19.8

double bond should be related to the stability of the corresponding trans cyclene. Whereas Bredt's rule would not distinguish the strain of 19.6 from that of its isomer 19.8, Wiseman's modification would predict 19.8 to be under considerably more strain because it incorporates a trans cyclohexene structure (164). Even compound 19.6 has properties of an extremely strained double bond. Thus, it readily adds water, acetic acid, atmospheric oxygen, and phenyllithium, it is unusually easy to hydrogenate, and it reacts as a dienophile toward 1,4-cyclohexadiene.

As another illustration, 19.9 is a very unstable amide because it lacks resonance stabilization from a polar structure. Structure 19.9b is prohibited by Bredt's rule. Accordingly, the combination of ring strain and no single-bond character gives the amide an anomalously high $\nu_{C=O}$ frequency (1799 cm^{-1}) (165). However, by placing a nitrogen atom at the other α position as in 19.10, the amide resonance is again possible. As to be expected, 19.10 is very stable and has a $\nu_{C=O}$ at 1650 cm^{-1}, the normal position for a strainless tetraalkylurea (166).

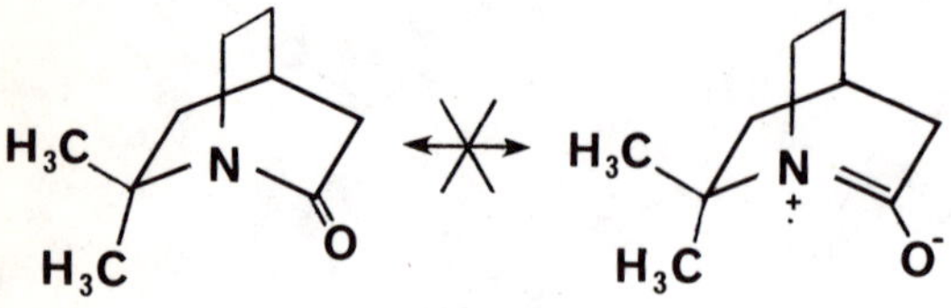

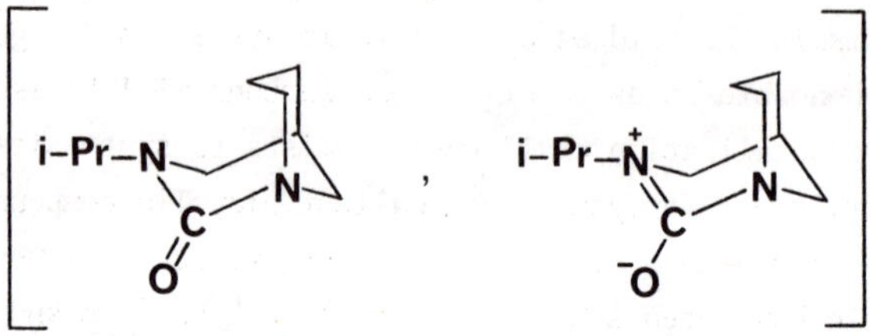

Another characteristic of highly strained rings is their facile thermal (167) and photochemical (168) isomerization. Several rearrangements occur, including sigmatropic rearrangements, valence tautomerism, and 1,3- and 1,5-hydrogen transfers. One technique especially suited to the study of highly reactive molecules is flash thermolysis (169) -- brief heating to high temperatures followed by rapid thermal quenching of the products. The method offers the advantage that the spectra of the transitory products can be measured at the time of their formation. Among the chemically interesting compounds studied in this way are cyclobutadiene, 1-methylpentalene, and sulphene (170).

$H_2C=SO_2$

Sulphene

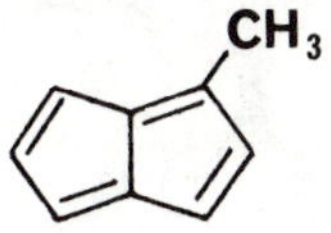

1-Methylpentalene

Sigmatropic rearrangements. A sigmatropic rearrangement is an uncatalyzed, intramolecular shifting of σ and π bonds, and one of the most well-known examples is the Cope rearrangement.

Factors which stabilize the product, e.g., resonance between R and the double bond, or which destabilize the reactant, e.g., ring strain, will shift the equilibrium toward the right. Divinylcyclopropane, for example, rearranges at room temperature (171).

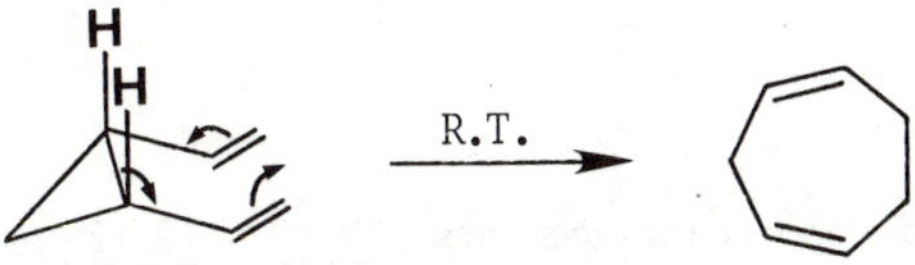

This concerted rearrangement of bonds is called <u>valence tautomerism</u>. The reaction is degenerate in the case of homotropylidene (19.11), and the degenerate isomers are said to have <u>fluxional</u> structures.

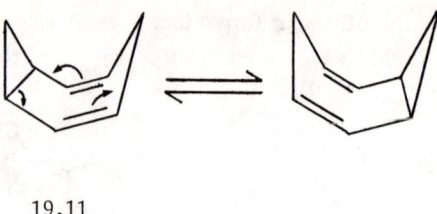

19.11

The activation energy is even lower for bullvalene (19.12).

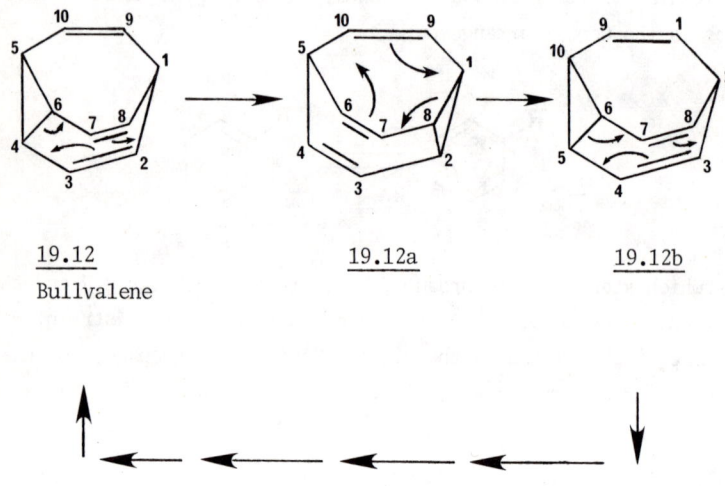

19.12 19.12a 19.12b
Bullvalene

By a succession of Cope rearrangements, 19.12 → 19.12a → 19.12b → → 19.12, each carbon becomes attached at some time to every other carbon. Thus, when the reaction occurs rapidly, all carbons are virtually equivalent. At -25°, where the rearrangement is frozen, bullvalene has two peaks in the nmr with an area ratio of 6:4. This is in accord with any one of the possible structures having 6 vinyl protons and 4 allylic protons. However, at 100°, the rearrangement occurs rapidly and all protons become equivalent to give a single nmr peak (172).

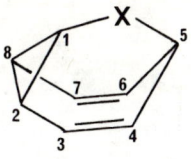

19.13

The tricyclononadienes (19.13, X = 1 carbon) and tricyclooctadienes (19.13, X = null) have successively smaller barriers to rearrangements, which is attributed to a greater release of strain in the transition state to rearrangement in the smaller rings (173).

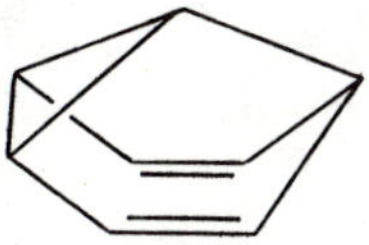

	Activation energy for rearrangement (173)
1,5-Hexadiene	35 kcal/mol
1,3,5,7-Tetramethyl-homotropylidene	13.6
Bullvalene (19.13, X = -CH=CH-)	12.8
Barbaralone (19.13, X = C=O)	9.6
Barbaralane (19.13, X = CH$_2$)	7.8
Octamethylsemibullvalene (19.14 system)	6.4

19.14
Semibullvalene

Hypostrophene (from Greek, a <u>turnabout</u>, a <u>recurrence</u>) is another system which apparently undergoes a degenerate Cope rearrangement slightly above room temperature (174).

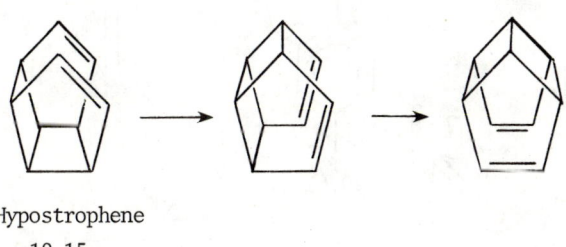

Hypostrophene
19.15

1,3-H shifts (175)

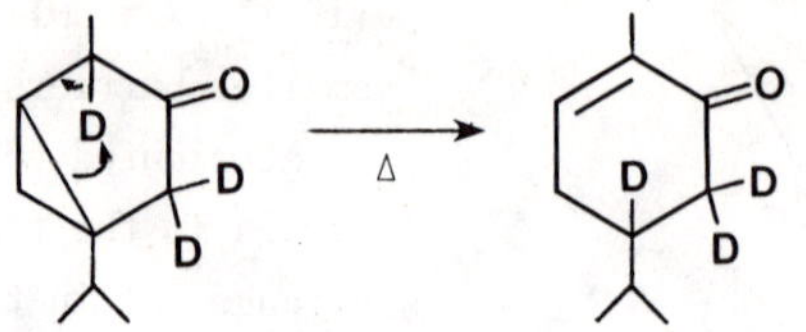

1,5-H Transfer (176)

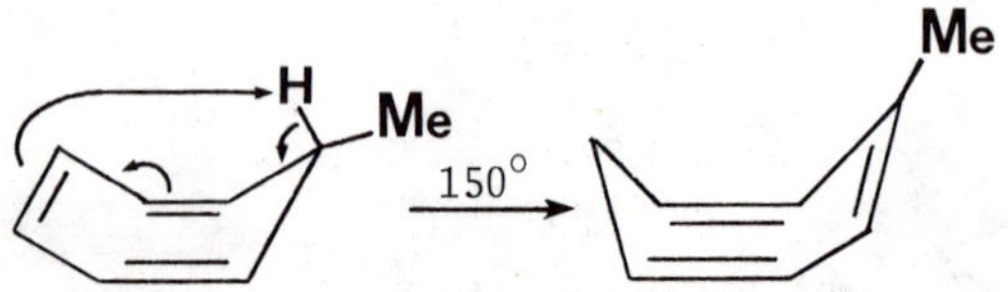

Photolytic isomerization (177)

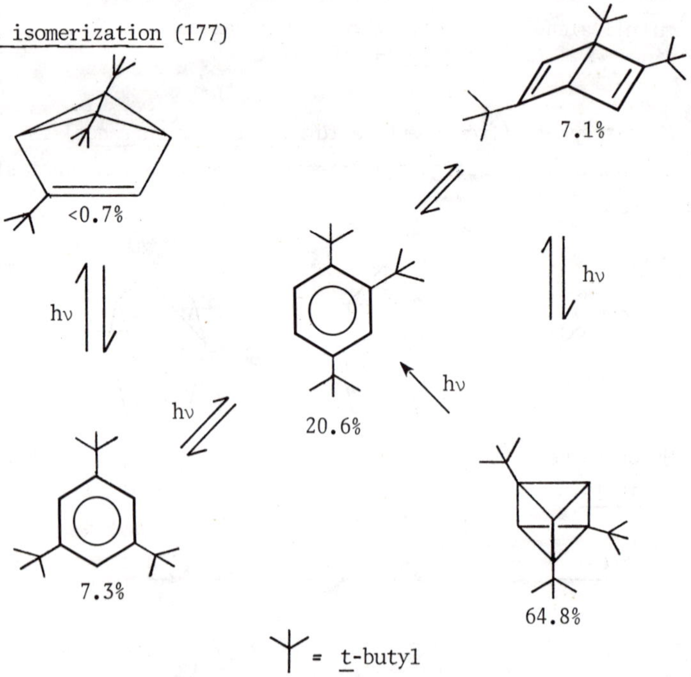

<0.7%

7.1%

20.6%

7.3%

64.8%

Y = t-butyl

Note that the energy profile for benzene valence isomer interconversions has been found to be as shown in Figure 19.3 (178).

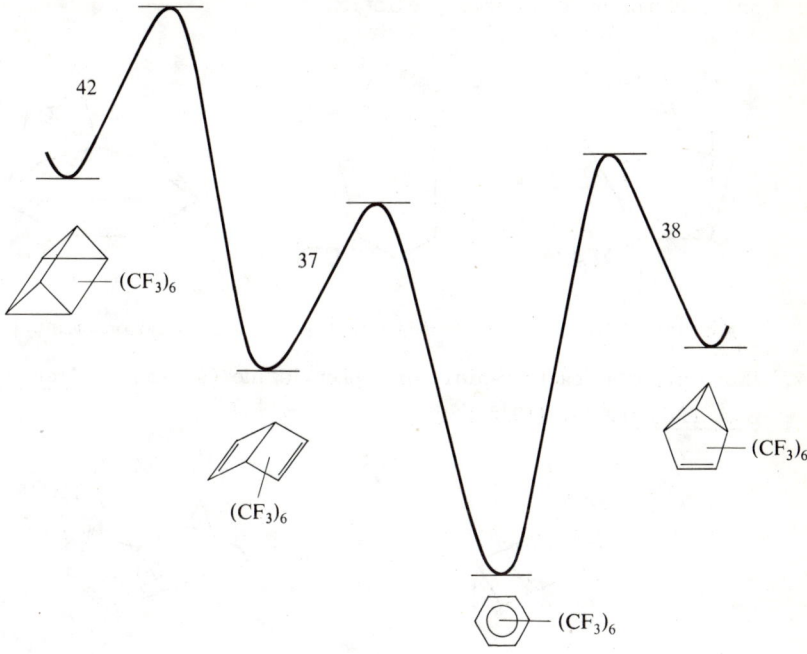

Figure 19.3. Energy profile for benzene valence isomer interconversions ($\Delta H^{\ddagger}$ in kcal/mol).

The material in this section can be summarized by the following list of properties of highly strained rings, which become more pronounced with increasing ring strain:
1. Large heats of combustion and hydrogenation.
2. Olefinic character of C-C bonds, although formally saturated, as shown by addition reactions and electron delocalization (hyperconjugation).
3. Olefinic C-H bonds, exhibiting large $\underline{J}(^{13}\text{C-H})$ coupling constants, acidity, and resistance to H atom abstraction.
4. Relatively facile thermolytic and photolytic rearrangements.

508   Ch. 19 RING STRAIN AND MOLECULAR PROPERTIES

QUESTIONS

1. Predict the relative rates of acetolysis of the three tosylates here and give the basis of your prediction.

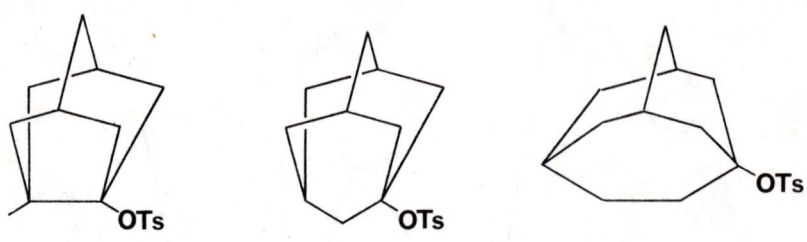

Noradamantyl   Adamantyl   Homoadamantyl

2. Hydrocarbon A reacts rapidly with phenylazide (D. H. Aue, Amer. Chem. Soc. PRF Rpt. 187 (1972)).

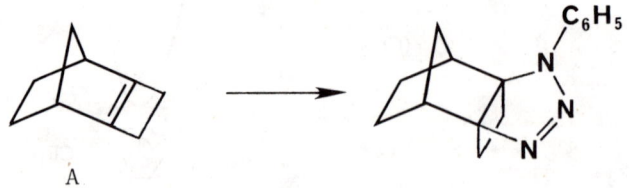

   Of what significance is this observation ?

3. Strong peaks at m/e 52 and m/e 26 appear in the mass spectrum of

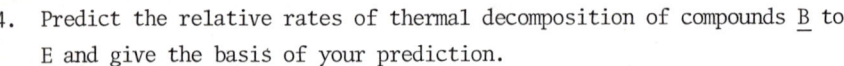

pyridazine. What do these peaks suggest ?

4. Predict the relative rates of thermal decomposition of compounds B to E and give the basis of your prediction.

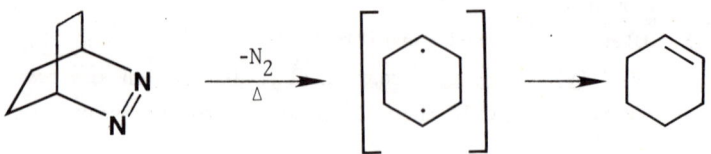

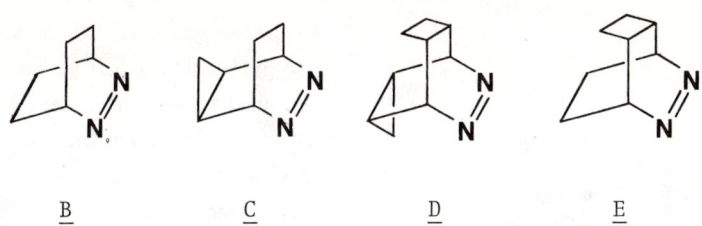

      B   C   D   E

5. The fact that phenylacetylene reacts with sodium metal or sodium amide to form a salt supports the idea that sp-hybridized carbon atoms are much more electronegative than sp^2 or sp^3-hybridized carbons. What other types of data add to this view?

ANSWERS

1. Solvolyses via carbocations increase with increasing stability of the intermediate ions. Deviation from planarity is destabilizing, and on this basis the relative rates of solvolysis should increase in the order drawn. The experimental ratios are $10^{-11}:10^{-3}:1$ (see P. v. R. Schleyer and E. Wiskott, Tetra. Letters (30) 2845 (1967)).
2. It confirms the expectation that A is a highly strained olefin, since this is a typical reaction of strained rings.
3. These peaks suggest the formation of singly and doubly charged cyclobutadiene following the loss of molecular nitrogen. (See M. M. Bursey, et al., J. Am. Chem. Soc. 90: 1557 (1968).)
4. On the basis of strain and/or neighboring group participation, a likely increasing order is B < E ≈ D << C:

    Exptl k_{rel} = 1 10^5 10^3 10^{17}.

In D and E, the cyclobutyl orbitals are aligned for disrotatory opening and orbital overlap in the transition state. The cyclopropyl orbitals are orthogonally oriented in D and actually produce a small retarding effect relative to

E. In C however, openings of the cyclopropyl orbitals are perfectly aligned and one observes one of the largest rate factor enhancements (10^{14}) reported between the two (exo vs. endo) cyclopropyl configurations. (E. L. Allred and K. J. Voorhees, J. Am. Chem. Soc. 95: 620 (1973); L. A. Paquette and M. J. Epstein, J. Am. Chem. Soc. 95: 6717 (1973); K.-W. Shen, J. Am. Chem. Soc. 93: 3064 (1971).)

5. The observation that phenylacetylene serves as a proton donor in H-bonds, nmr ^{13}C-H coupling constants, the ≡C-H bond length, and ir stretching frequencies, among others.

Chapter 20

MISCELLANEOUS STERIC EFFECTS

Steric hindrance affects a multitude of intramolecular and intermolecualr properties, chemical as well as physical.

20.1. Restricted rotations

Energy barriers to internal rotations were discussed in Chapter 18. There is a variety of molecular environments in which steric hindrance prevents planarity, distorts normal bond angles, and even permits the isolation of optical isomers. A scattering of examples can be given for illustration. The two substituents are bent back a few degrees from the normal 120° in the plane and the COOH group is rotated 14° out of the plane.

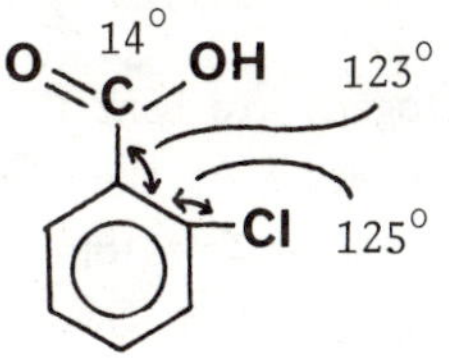

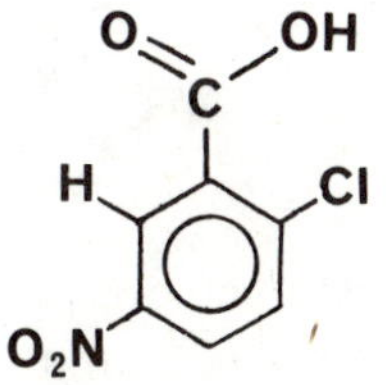

The buttressing effect (Chapter 11) of the NO_2 group is shown, because the COOH group is rotated 23° out of the plane. The NO_2 group hinders bending back of the H to reduce repulsion toward the COOH group.

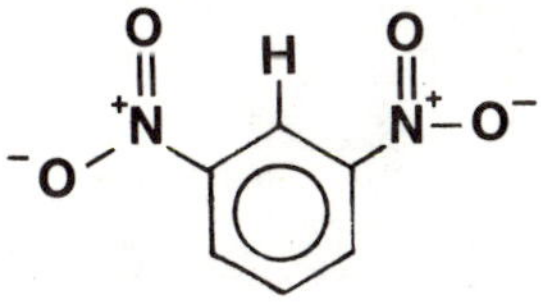

Each NO_2 group is rotated 11° out of the plane of the ring.

Several properties reflect the steric strain between <u>peri</u> substituents in a naphthalene ring.(179). As a means of relieving this strain, such groups preferably rotate, or may bend away from each other in or out of the plane of the naphthalene ring:

Each NO₂ group is rotated 49° out of the molecular plane and bent away from the peri H.

The COOH group is rotated 12° out of the plane and the O=OH bond angle is reduced from the normal 120° to 110°. In contrast, the COOH group in β-naphthalene carboxylic acid is planar with the ring.

The α halogens are bent 0.3 to 0.4 Å out of the plane of the ring, whereas, for comparison, 2,6-dichloronaphthalene is planar.

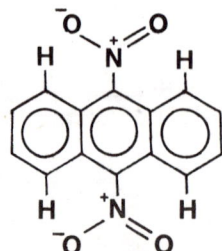

The dihedral angle between planes is 73°. Optical isomers may be isolated when there are substituents in the 8 and/or 8' positions.

The buttressing effect (Chapter 11) of two peri Hs increases the angle of rotation of the NO₂ groups to 65°. The same situation exists in nitromesitylene.

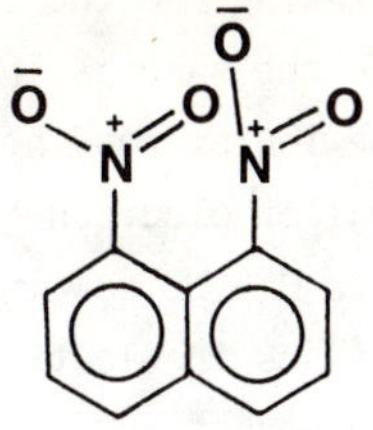

Sec. 20.1 RESTRICTED ROTATIONS

The NO_2 groups are rotated 45° in one direction.

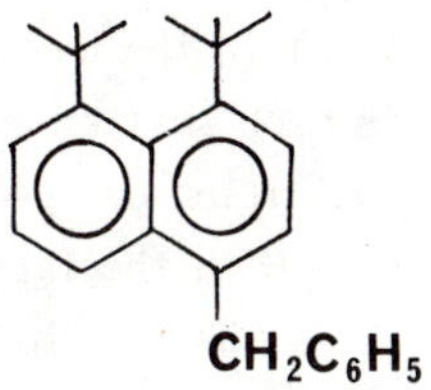

The t-butyl groups are bent in opposite directions out of the molecular plane and the reduced steric strain between groups lowers the t-butyl rotation barrier to 6.5 kcal/mol (180).

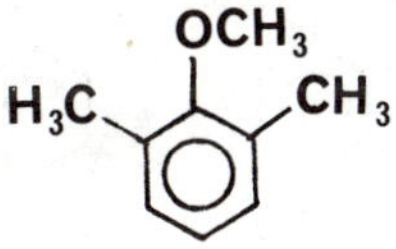

Whereas the OH group of di-o-t-butyl-phenol is coplanar with the ring, the OCH_3 group of di-o-methylanisole is forced almost perpendicular to the ring (181). Tropolones have planar rings; however, the rings are distorted and nonplanar in the heavily substituted 3,7-dibromo- and 3,5,7-tribromo-4-isopropyl tropolones (182).

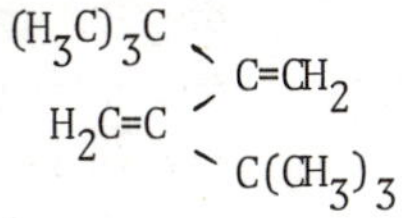

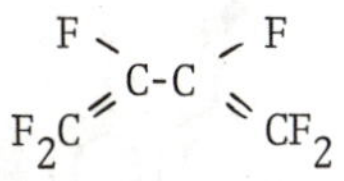

λ_m ($\pi \to \pi^*$)   186   197 nm

Both compounds are nonplanar, 2,3-di-t-butylbutadiene (183) because of van der Waals repulsion and hexafluorobutadiene (184) as well as other halogen-substituted butadienes (185), because of bond dipole repulsion. The wavelengths of their $\pi \to \pi^*$ band is unusually short (λ_m butadiene = 210 nm; see also Chapter 10). Although the C_2-C_3 bond of 2,3-dimethylbutadiene is longer than that (1.48 Å) in butadiene, the bond lengthening is not attributable to repulsion between the CH_3 and CH_2 groups because the CH_3 groups are

514   Ch. 20 MISCELLANEOUS STERIC EFFECTS

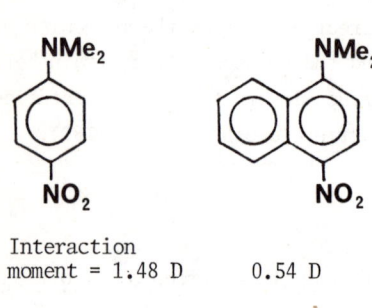

freely rotating and the C=C-C bond angles are the same as in butadiene (186) Hence, the longer bond here is ascribed to the smaller electronegativity of CH_3 compared to H. We saw in Section 1.1 that increases in electronegativity are bond contracting.

The interaction moment (Section 6.1b) in the naphthalene compound is much smaller owing to steric inhibition or resonance by the peri H in the latter compound.

The red shift in the uv accompanying increased resonance of unrestricted NMe_2 groups over NH_2 is not observed in naphthalene compounds because of the steric hindrance of the peri H.

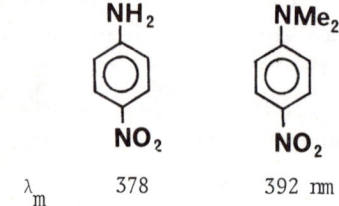

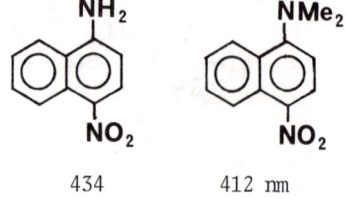

The uv spectra of 20.1 for R = Me or t-Bu are almost superposable, indicating that the difference in strain from the peri H is not detectable by uv spectroscopy. However, the change from Me to t-Bu in 20.2 produces a bathochromic shift of ca. 20 nm and a loss of fine structure (187). This is indicative of a substantial strain in 20.2, R = t-Bu (see Chapter 11).

The dihedral angles in biphenyls have been studied by a variety of methods, one of which is dipole moment measurements. For example, from the equation

$$\mu_{obs}^2 = 2\mu_{\underline{x}}^2 \cos^2 30°(1 + \cos\rho)$$

where μ_{obs} and $\mu_{\underline{x}}$ are the observed and group moment of $\phi\text{-}\underline{x}$ and ρ is the dihedral angle between aryl groups. The dihedral angles for a few biaryls are estimated to be as follows (188).

Biphenyl	μ_{obs} (D)	ρ_{calc}	$\rho_{x\text{-}ray}$
2,2'-Difluoro	1.88	79°	86°
2,2'-Dinitro	5.10	80°	86°
2,2'-Dimethyl-5,5'-dichloro	2.37	90°	91°
9,9'-Bianthryl			
2,2'-Difluoro	2.18	85°	
2,2'-Dichloro	2.51	62°	

Optical isomers may be isolated when barriers to rotation in asymmetric molecules are large enough. One such system which has received considerable attention is again, that of biphenyl (189). With substituents in the ortho positions, rotation about the $C_1\text{-}C_1'$ bond is hindered and when neither phenyl ring is symmetrically substituted, the conformers are mirror images and nonsuperposable, i.e., optical isomers. If rotation does occur, either isomer is converted into the other, that is, racemization takes place. The racemization has been found to be a reversible,

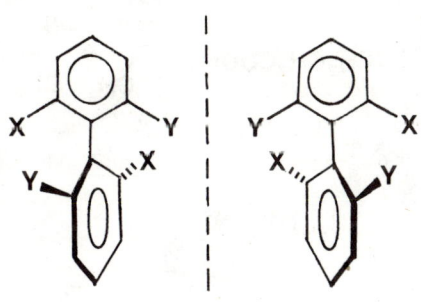

Nonsuperposable mirror images of a biphenyl

first-order reaction with activation energies in the range 18-25 kcal/mol. In a classical semitheoretical calculation of this activation energy for 20.3, assuming that the planar conformation is the activated complex, Westheimer and Mayer (190) got a calculated value of 18 kcal/mol, which compares well with the experimental value of 19 (191). The interesting result of this calculation is that most of this energy barrier is shown to be due to van der Waals repulsion (6 kcal) and bending back of the bromine atoms (7.2 kcal). Thus, groups in the 3, 5, 3', and 5' positions exert a buttressing effect* by hindering this bending back of <u>ortho</u> groups to allow rotation. For instance, the buttressing effect (Chapter 11) of the 3,3' iodine atoms in 20.4 is about 7 kcal/mol (192).

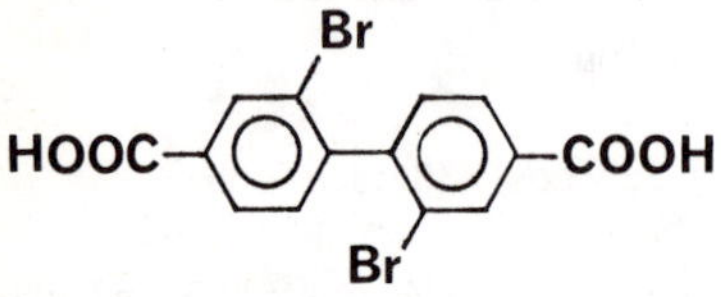

20.3

Since the early work on the biphenyls, a large variety of compounds has been studied for their optical isomerism resulting from sterically hindered rotations. Among these are 20.5 to 20.10.

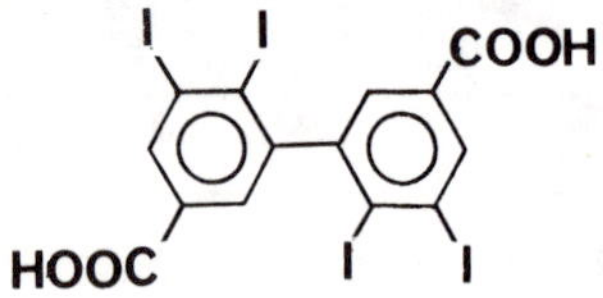

20.4

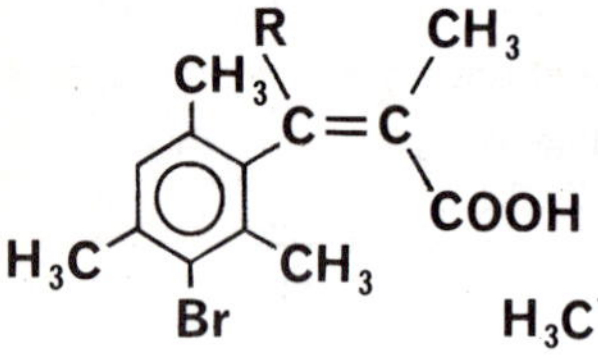

20.5

R = Cl, CH$_3$, OCH$_3$

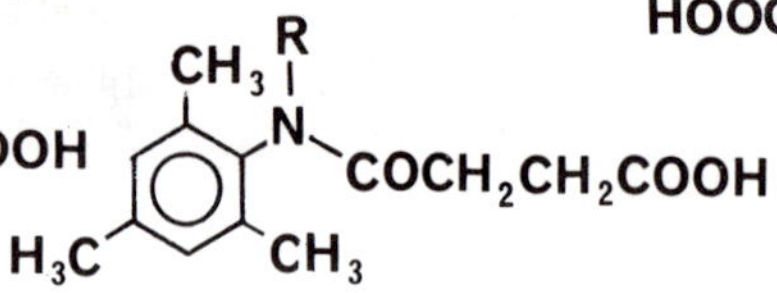

20.6

R = I, Br, CH$_3$, OCH$_3$, NO$_2$

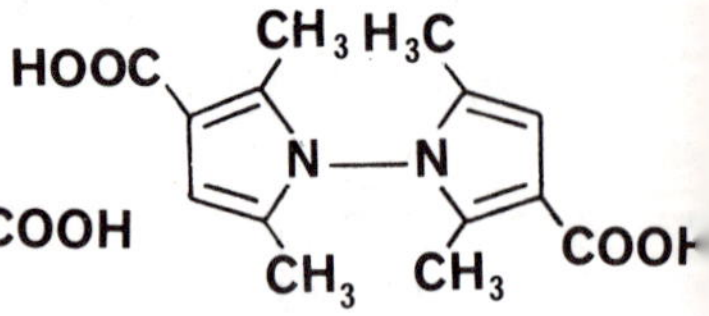

20.7

*See the Index for references to other examples of the buttressing effect.

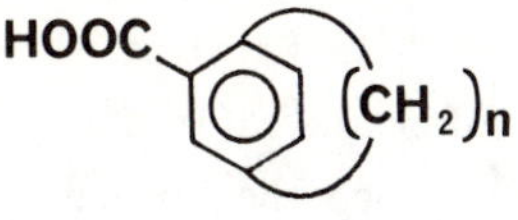

20.8   20.9 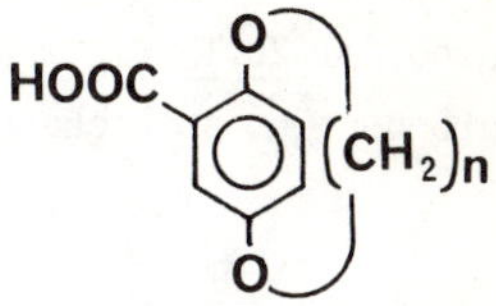   20.10

n = 8, 9   n = 8, 9

Hindered rotation in biphenyl fragments is also responsible for the isolation of some biradicals (see also Section 20.5). The hydrocarbon 20.11 would be expected to dissociate in solution as does hexaarylethane (Section 6.2a) to give the biradical 20.12.

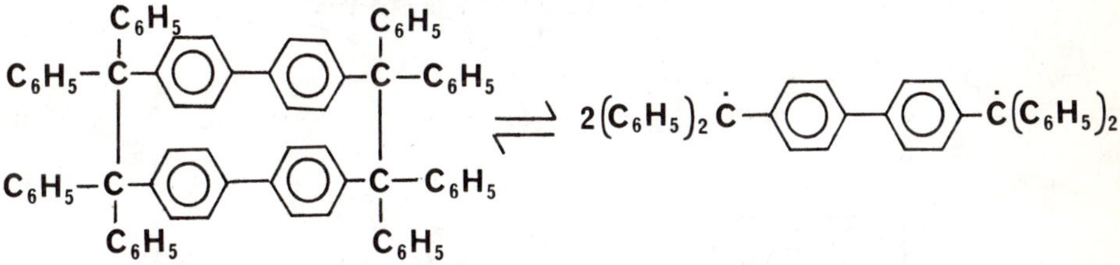

20.11   20.12

However, there is only 4 per cent biradical in benzene solution (193). This relatively small biradical character is attributed to the conversion of the paramagnetic 20.12 to the diamegnetic 20.13, resulting in an apparent small degree of dissociation.

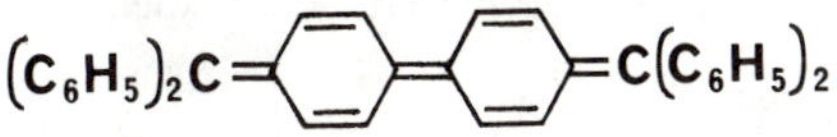

20.13

If there were any structural feature to block 20.12 from going into the quinoidal 20.13, it would increase the biradical character of the hydrocarbon. One such change is to make the meta isomer, which is found to be 6 to 8 per cent dissociated into the biradical 20.14. Since the biphenyl fragment of 20.12 must be coplanar to give 20.14, any groups which block this coplanarity should also raise the amount of biradical produced from 20.11-type compounds. This is found to be the case when four chlorine atoms are placed in the ortho positions. Thus, the respective compounds yield the biradicals 20.15 and 20.16=in 28 and 67 to 83 per cent, respectively (194).

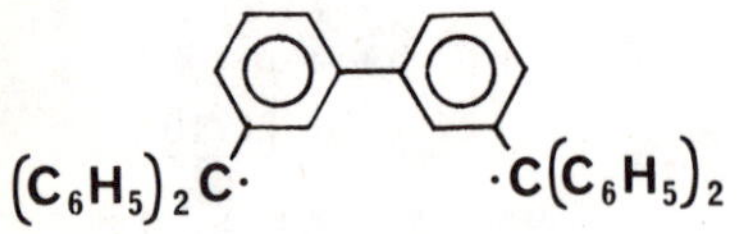

$(C_6H_5)_2\dot{C}$—[meta-biphenyl]—$\dot{C}(C_6H_5)_2$

20.14

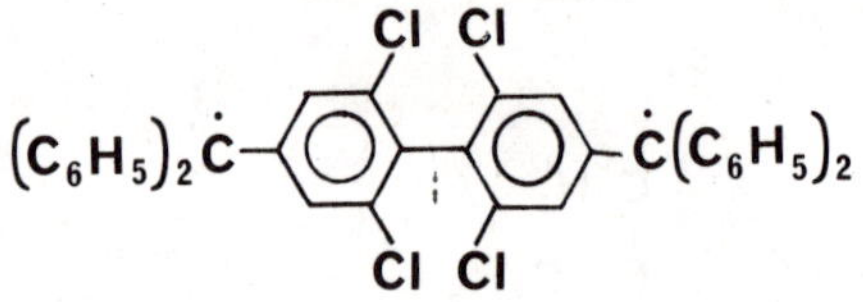

$(C_6H_5)_2\dot{C}$—[tetrachlorobiphenyl]—$\dot{C}(C_6H_5)_2$

20.15

$\left(\text{[biphenyl]} \right)_2 \dot{C}$—[tetrachlorobiphenyl]—$\dot{C} \left(\text{[biphenyl]} \right)_2$

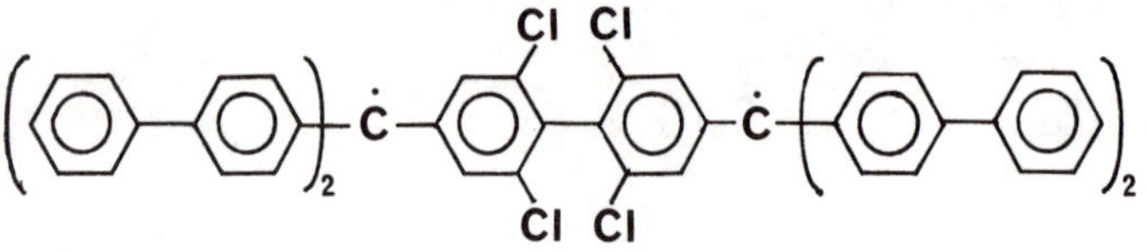

20.16

QUESTIONS

1. Describe the electron shifts that occur when biradical 20.12 is changed into 20.13.

2. Predict the relative proton chemical shifts for OCH_3 in the nmr for 20.17 and 20.18 and give the basis of your reply.

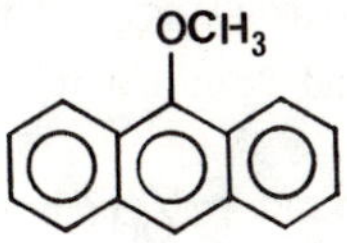

20.17

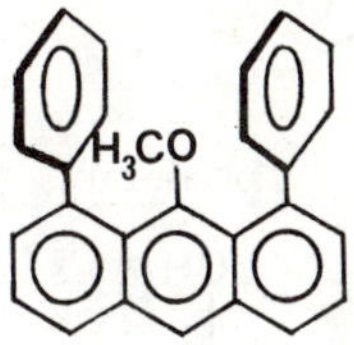

20.18

ANSWERS

1.

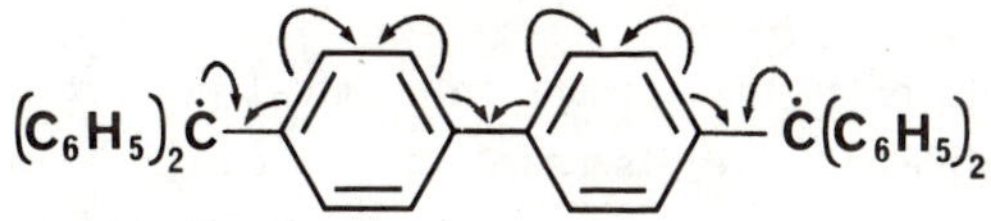

Each arrow represents a shift of one electron.

2. The methoxy protons of 20.18 should resonate upfield of those of 20.17 because the former are shielded by the 1,8 phenyl groups. The actual values are δ 2.37 ppm for 20.18 and δ 4.10 ppm for 20.17. (H. O. House, D. Koepsell, and W. Jaeger, J. Org. Chem. 38: 1167 (1973).)

20.2. Steric hindrance to H bonding

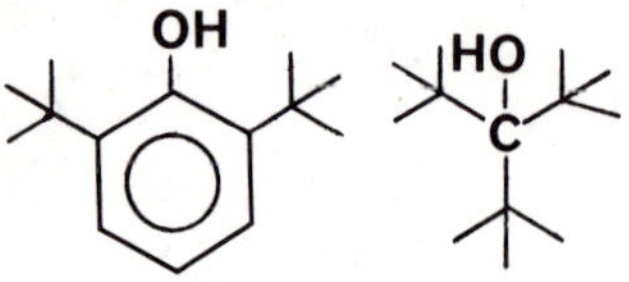

Steric requirements often prohibit H bonding, paricularly intermolecular H bonding. Large flanking groups prevent the close approach of the OH groups for H bonding. Thus, tri-t-alkyl-carbinols and di-o-t-butylphenols are monomeric.

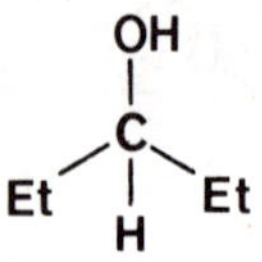

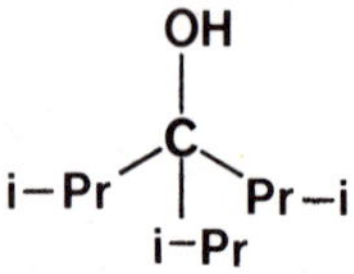

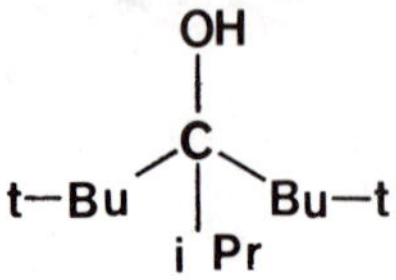

	Monomeric		
	Et—C(OH)(H)—Et	i-Pr—C(OH)(Pr-i)—i-Pr	t-Bu—C(OH)(Bu-t)—i-Pr
ν_{OH} =	3350 cm^{-1} (st)	3350 cm^{-1}	3640 cm^{-1}
	3650 (wk)	3640	
	Mostly associated	About equally associated and monomeric	Monomeric

The steric effects of t-butyl groups on chemical properties and acidities of 2,6-di-t-butylphenols are discussed in Sections 20.4 and 20.5.

Intramolecular H bonding too can be blocked by steric crowding. The COOH group forms intramolecular H bonds with electron-pair donating ortho groups

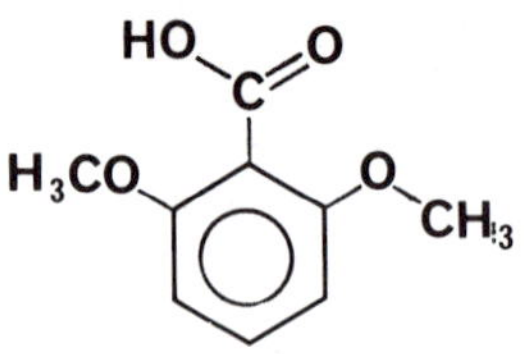

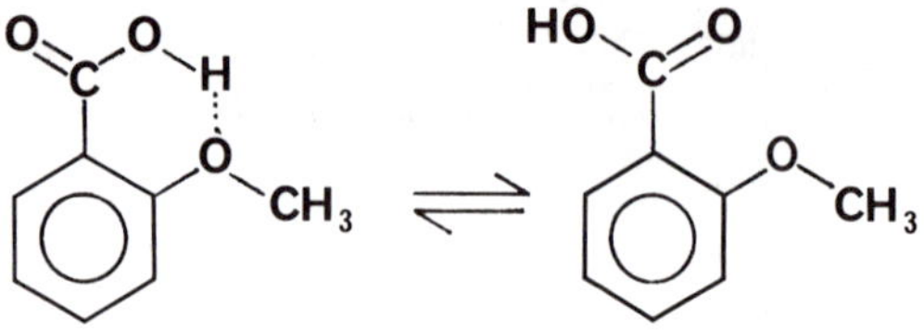

ν_{OH} =	3525 cm^{-1}	3366 cm^{-1}	3533 cm^{-1}

but in 2,6-dimethoxybenzoic acid, the COOH group cannot lie coplanar with the ring. This prevents formation of an H bond and the compound exhibits only a free OH ir band (195).

20.3. Base strengths of aliphatic amines*

Ammonia and simple aliphatic amines are weak bases, $K_b \approx 10^{-5}$, but strong enough to give a basic reaction to litmus paper. Since H atoms are more electron-withdrawing than alkyl groups in aqueous media, thereby pulling in bonding electrons from nitrogen in ammonia and making N less willing to share its lone pair with an acid, the amines should be more basic than ammonia. This is the case toward most acids. In the other direction, the strongly electronegative fluorine atoms in perfluoroamines make them such weak bases that they do not form salts even with very strong acids.

Brown showed that the reference acid used is of tremendous importance in determining relative base strengths. For example, in accordance with the rule of thumb that "stronger base liberates weaker base from its salt," trimethylamine liberates pyridine from its salts with HCl, BH_3, and BF_3, indicating that trimethylamine is the stronger base.

$$Me_3N: + Py:HCl \rightleftharpoons Me_3N:HCl + Py$$
Pyridine hydrochloride
$$Me_3N: + Py:BH_3 \rightleftharpoons Me_3N:BH_3 + Py$$
$$Me_3N: + Py:BF_3 \rightleftharpoons Me_3N:BF_3 + Py$$

Pyridine, however, is the stronger base toward borontrimethyl.

$$Py: + Me_3N:BMe_3 \rightleftharpoons Py:BMe_3 + Me_3N:$$

This reversal in relative base strengths is due to steric hindrance in the $Me_3N:BMe_3$ salt. The crowding at the interface in the salt makes it unstable

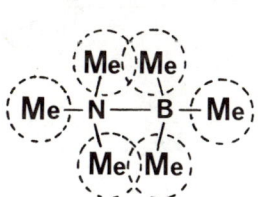

and reduces its tendency to form, thereby resulting in a low apparent basicity of Me_3N. Brown called this F-strain (for frontal strain) (196). Similarly, lutidine is a stronger base than pyridine toward HCl but toward larger acids such as BF_3, F-strain renders lutidine the weaker base. Lutidine does not even form a salt with borontrimethyl. By competitive measurements toward BF_3 using

---

*Steric effects in aromatic amines are discussed in Section 6.2a.

522   Ch. 20 MISCELLANEOUS STERIC EFFECTS

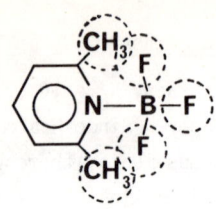

Lutidine-boron
trifluoride

^{1}H, ^{19}F, and ^{13}C nmr measurements, the order of decreasing basicities in non-aqueous solvents is found to be (197):

4-MePy > 3-MePy > Py > 2-MePy > 2,6-Me$_2$Py

whereas toward HCl$_{aq}$, the order is

2,6-Me$_2$Py > 4-MePy > 2-MePy > 3-MePy > Py

where Py = pyridine.

This view finds quantitative support in the data of Table 20.1. It can be seen there that an alkyl group increases the basicity of the nitrogen toward protonic acids and that 2-t-butylpyridine is only a slightly weaker base than 2-picoline. Toward BF$_3$, F-strain in 2-picoline is apparent, becomes substantial in 2,6-lutidine, and amounts to 10 kcal/mol in the salt of 2-t-butylpyridine. Finally, toward BMe$_3$, the F-strain is over 17 kcal in 2,6-lutidine and 2-t-butylpyridine.

Thus, the relative base strengths of amines can be altered by changing the sizes of the alkyl groups on nitrogen or on boron, to give rise to several sequences, some of which are:

1. $R_3N > R_2NH > RNH_2 > NH_3$
2. $R_2NH > RNH_2 > R_3N > NH_3$
3. $RNH_2 > R_2NH > NH_3 > R_3N$
4. $RNH_2 > NH_3 > R_2NH > R_3N$
5. $NH_3 > RNH_2 > R_2NH > R_3N$

Table 20.1. Relative base strengths of some pyridine derivatives

Base	pK$_a$	Heat of reaction (kcal/mol)		
		CH$_3$SO$_3$H	BF$_3$	B(CH$_3$)$_3$
Pyridine	5.17	17.1	25	15.3
4-Picoline	6.02	18.8	25.5	17.6
3-Picoline	5.68		25.3	17.6
2-Picoline	5.97	18.3	23.3	10
2,6-Lutidine	6.58	19.5	17.5	<0
2-t-Butylpyridine	5.76	17.3	14.8	<0

Sequence 1 is that expected on the basis of polarizability and 2 is that observed for simple amines in aqueous media. Indeed, Brown observed sequences 2 to 5 by varying the reference acids and the sizes of the alkyl groups in the amines (Table 20.2).

Table 20.2.

Relative base strengths of aliphatic amines toward various acids (198) (designated with respect to sequences 1 to 5)

R in amine	$H_3O^{\oplus}_{aq}$	BMe_3	BEt_3	$B(\underline{i}\text{-}Pr)_3$	$B(\underline{t}\text{-}Bu)_3$	$Ag^{\oplus}_{aq}$
Me	2	2	3	3	4	
Et	2	3	4	4	5	4
$\underline{i}$-Pr	2	4	4	5	5	
$\underline{t}$-Bu		5	5	5	5	

In nonaqueous solvents (199) or the gas phase (200), the order of base strengths of $\underline{n}$-alkylamines toward protonic acids is that expected from polarizability or inductive effects, or both:

$$R_3N \gg R_2NH > RNH_2 > NH_3.$$

Solvation in water upsets this order to give sequence 2, in which the tertiary amine is out of order. In aqueous media, several types of H bonding may occur: (1) N-H$\cdots$OH$_2$ bonding may occur between the free base and water, which increases water solubilities of primary and secondary amines over that of the respective tertiary amines. Statistically this occurs most for NH$_3$ because there are more N-Hs available. (2) HO-H$\cdots$N bonding may occur between water and the free base which, owing to steric hindrance between the alkyl groups and the aggregated solvent molecules, takes place with NH$_3$ to the largest extent. (3) $\overset{+}{N}$-H$\cdots$OH$_2$ bonding may occur in the salts, which occurs the most for the $\overset{+}{NH_4}$ ion and the least for $R_3\overset{+}{N}$-H. Consequently, H bonding of types (1) and (2) stabilize the free base in the order NH$_3$ > RNH$_2$ > R$_2$NH > R$_3$N, which in the absence of any other factor would give just the opposite order of relative base strengths. H bonding in the salts stabilizes the conjugate acids in the order $\overset{+}{NH_4}\cdots R_3\overset{+}{NH}$, which in the absence of any other effect would give a relative order of base strengths NH$_3$ > RNH$_2$ > R$_2$NH > R$_3$N. H bonding in the free bases and the conjugate acids gives opposite orders of relative base strengths and the net order observed in aqueous media, coupled with inductive and steric effects, is sequence 2, $R_2NH > RNH_2 > R_3N > NH_3$.

Solvation effects, whether through H bonding or dipole-dipole forces, produces many differences from gas phase properties. For example, we saw in Section 3.3c that there is a reversal in the acidity order of alcohols. In the gas phase, toluene and propene are stronger acids than water, and ΔH° for HCl dissociation is +333 kcal/mol in the gas phase compared to -16 kcal/mol in water (201).

The solvation effect even shows up with pyridine and aniline (202). The lower base strength of pyridine compared with that of ammonia and aliphatic amines in aqueous solution is usually attributed to a larger $\underline{s}$ character of the nitrogen molecular orbital. The nitrogen atom of pyridine is therefore less willing to share its lone pair. However, in the gas phase, pyridine is indeed more basic than ammonia, with proton affinities of 225 and 207 kcal/mol, respectively (203). The proton affinities of several amines relative to ammonia are (202):

	Gas phase	Aqueous
Ammonia	0 kcal/mol	0 kcal/mol
Aniline	-8.9	6.3
Methylamine	-10.8	-1.9
N-Methylaniline	-15.1	6.1
Dimethylamine	-18.3	-1.9
Pyridine	-18.6	5.4
Trimethylamine	-23.3	-0.8

Thus, although aniline is inherently a weaker base than methylamine, the difference is much, much smaller than in water solution. Hence, resonance in aniline is not as base-weakening as implied from aqueous basicities. Comparison of aniline with an amine of comparable size, however, shows that aniline is a much weaker base than cyclohexylamine in the gas phase.

The inductive order is observed for alkylphosphines, which can be explained in terms of hyperconjugation as well as inductive effects (204).

In summary then, the order of the relative base strengths of amines
1. depends upon the size of the reference acid used as well as the sizes of the alkyl groups;
2. toward small acids is $3° > 2° > 1° > NH_3$ in nonaqueous solvents and the gas phase;
3. in aqueous media and toward small acids is $2° > 1° > 3° > NH_3$, in which the order in (2) is upset by solvation (205);
4. may give any mixed order, depending upon the size of the acid and the alkyl groups in the amines.

20.4. Acidity of hindered acids

The dissociation constants of hindered carboxylic acids and phenols are distinctly reduced, e.g.,

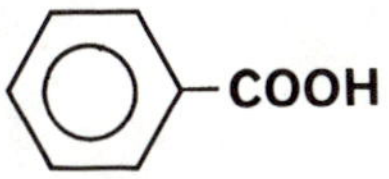

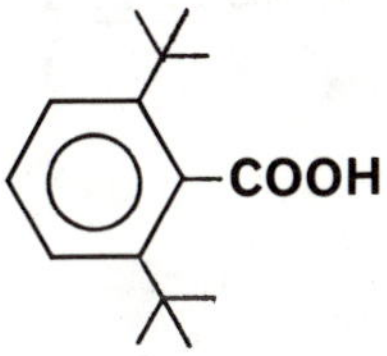

Ph-COOH   2,6-di-t-butyl-C₆H₃-COOH
pK_a 5.05   6.25

Substituent	pK_a of phenol	pK_a of 2,6-di-t-butylphenol (206)
p-CHO	8.81	11.3
p-CN	9.10	11.4
p-CO$_2$Et	9.61	11.8

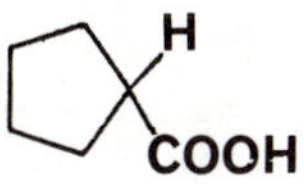

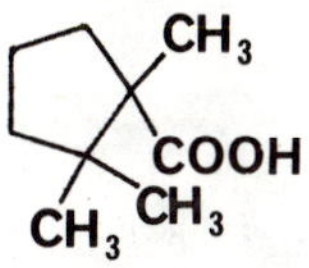

cyclopentyl-H-COOH   cyclopentyl(CH₃)₂-COOH with CH₃ groups
$10^{-6} K_a$ (207)   1.26   0.348

The K_a of 20.19 is 1/25th of that of acetic acid (108) and 2,6-di-t-butyl-phenols are about 1/1000th as strong acids as the corresponding phenols.

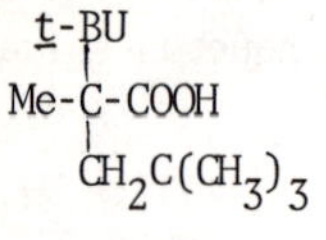

```
    t-BU
     |
   Me-C-COOH
     |
   CH₂C(CH₃)₃

     20.19
```

This reduction in ionization of these acids and phenols is due partly to lowering of the dielectric field about the ion and partly to shielding the ion from hydration by the solvent*. This is shown by the fact that o-t-butylphenol is 1.8 kcal/mol less acidic than phenol in aqueous solution but in the gas phase it is 3.4 kcal/mol *more* acidic than phenol (208a). We saw in Section 3.3c that the relative acidities of alcohols as well as relative rates of solvolysis of alkyl substrates are reversed for gaseous conditions compared to aqueous solutions. 2,6-Di-t-butylphenols of the type 20.20 do not dissolve in alkali or give other phenolic tests whereas analogs of type 20.21 do. The solubility of phenols 20.21 can be attributed

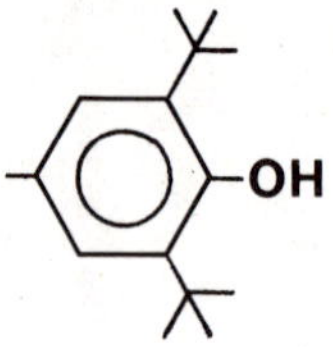

20.20 R = alkyl, alkoxyl
20.21 R = CN, CHO, CO_2Et

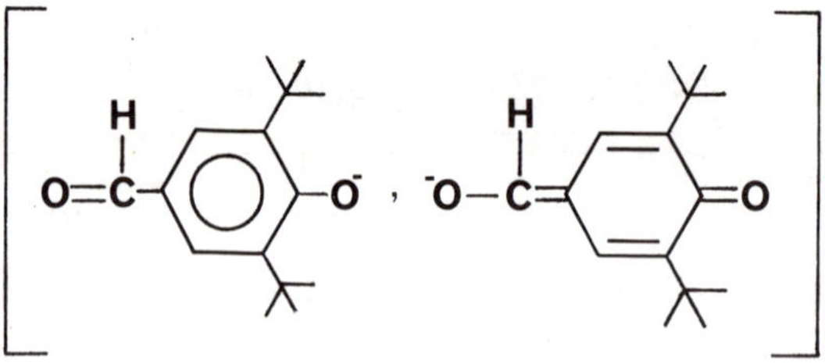

20.22

---

* The steric effect on dissociation constants in aqueous media is complex and is not fully understood, particularly in terms of sigma induction, field effects, inhibition of resonance, and solvation. For further discussions see E. A. McCoy and L. L. McCoy, *J. Org. Chem.* 33: 2354 (1968); J. Steigman and D. Sussman, *J. Am. Chem. Soc.* 89: 6406 (1967).

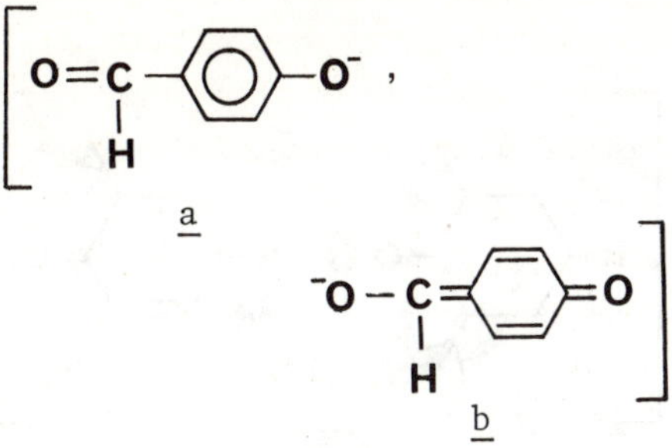

to the delocalizing effect of R, as in 20.22, which distributes negative charge to an unshielded side of the ion for hydration. This notion is supported by ir and uv spectra. The ir spectra of 20.22 and 20.23 show that the carbonyl group of 20.23 has the larger double bond character, i.e., 20.23a makes a greater contribution to the resonance hybrid of 20.23 than 20.22a does in 20.22. The phenols 20.21, nevertheless, are slow to react, if they react at all, with sodium metal, methyl Grignard, or diazomethane, do not give a color with $FeCl_3$, and do not form H bonds, so the OH group is highly shielded.

20.5. Hindrance to conventional reactions

Many reactions are severly inhibited merely because substituents sterically prevent the molecules from getting close enough at the right sites for reaction. This is the origin of the idea of steric hindrance: the inertness of sterically hindered benzonitriles and esters to hydrolysis by conventional methods (209). In some cases this steric hindrance is desirable as a means of stabilizing a compound for its isolation. For example, t-butyl groups have been used to block the dimerization or polymerization of compounds. For a period, only t-butyl derivatives could be isolated from the highly reactive ring systems, cyclopropeneone, cyclopentadieneone, and others. 2,6-Di-t-butylphenols form radicals which are unusually inert and isolable as the monomers, simply because of steric hindrance to dimerization (210).

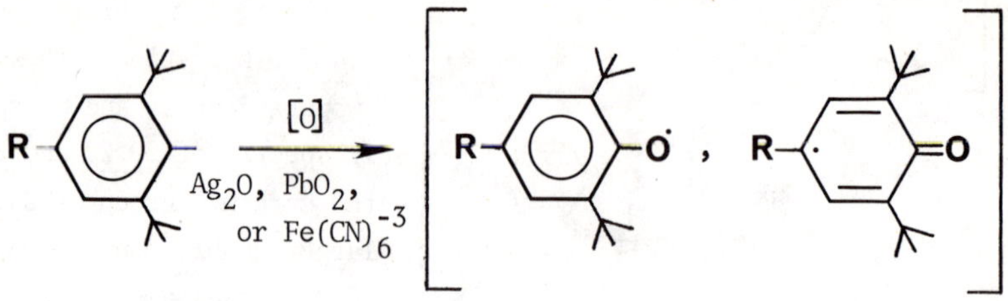

R = alkyl, alkoxyl

20.24

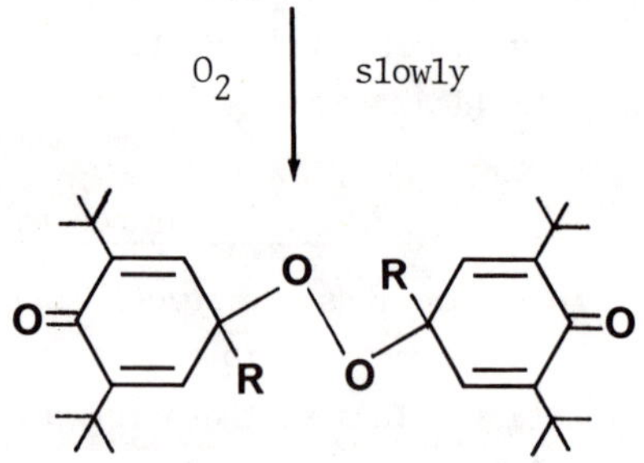

The free radicals 20.24 are deep blue and paramagnetic. A related stable biradical is 20.25. It may be recrystallized from benzene in deep purple, paramagnetic prisms and melts at 280° (211).

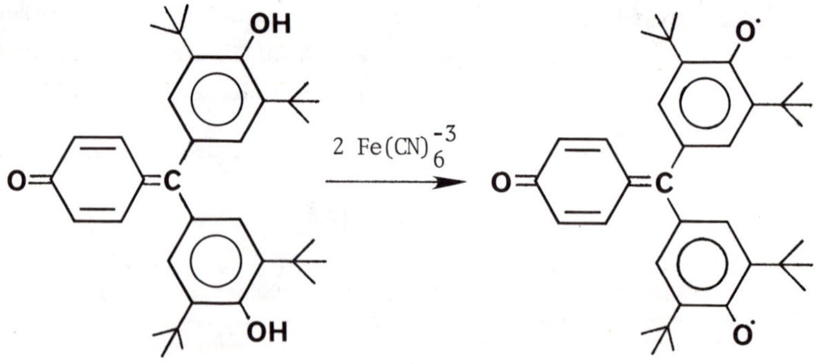

20.25

Sec. 20.5 HINDRANCE TO CONVENTIONAL REACTIONS   529

These stable radicals are used as antioxidants in tires.

Among the factors responsible for the formation and existance of free radicals are resonance and steric hindrance. We just saw the effect of steric hindrance, preventing the dimerization of radicals, and we discussed the stabilizing effect of resonance in the triarylmethyl radicals in Section 6.2a. An example where both effects combine to make some radicals chemically inert are the perchlorotriarylmethyl radicals (212).

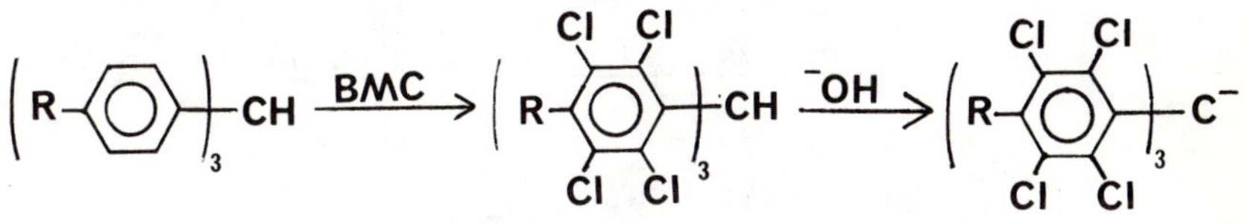

$(R-C_6H_4-)_3CH \xrightarrow{BMC} (R-C_6Cl_4-)_3CH \xrightarrow{^-OH} (R-C_6Cl_4-)_3C^-$

BMC is an unusually strong chlorinating agent (S_2Cl_2-SO_2Cl_2-$AlCl_3$)

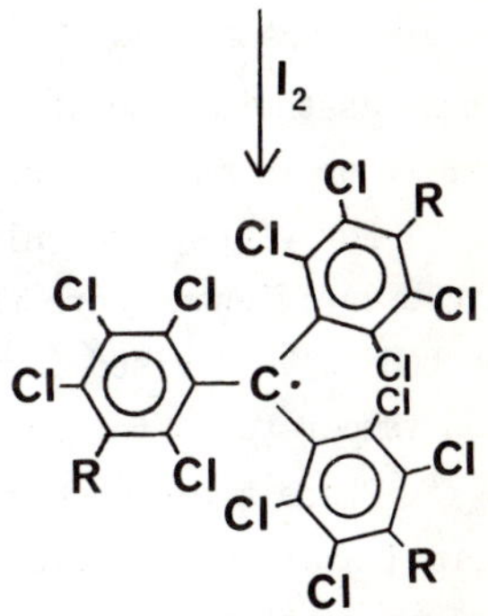

$\downarrow I_2$

$(R-C_6Cl_4-)_3C\cdot$

20.26 R = Cl

20.27 R = C_6H_5

The perchlorophenyl groups radiate out from the central carbon like propeller blades and are just planar enough to permit a resonance stabilization of the radical (Section 6.2a) but bulky enough to block dimerization of the radical. As a result they are completely radical even in the solid state. They are also chemically inert toward radical scavengers such as NO, hydroquinone, and boiling toluene, and are unaffected by concentrated nitric and sufuric acids at room temperature in the dark. The contribution that the inductive effect of the chlorine atoms makes to the stability of these radicals has not been determined. It is well known, for instance, that simple perchlorocarbanions, e.g., $^-CCl_3$, and perchloro radicals, e.g., $Cl_3\overset{\cdot}{C}$, are much more stable than the corresponding hydrocarbon species.

2,6-Di-t-butylphenols also exhibit steric hindrance to ether formation. Alkyl halides react with hindered phenols in alcoholic KOH to give C-alkyl and O-alkyl products (213).

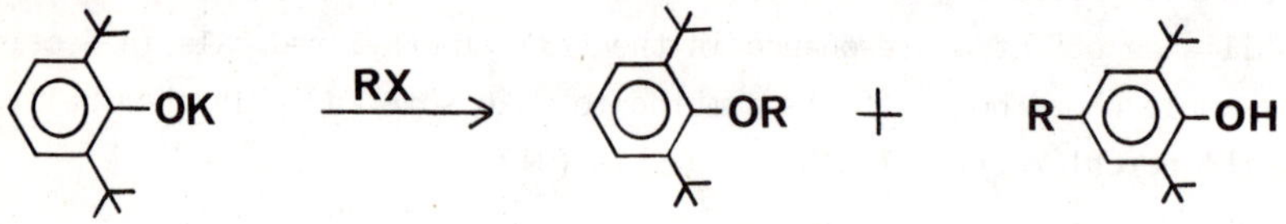

$$\text{(2,6-di-t-butylphenol-OK)} \xrightarrow{RX} \text{(2,6-di-t-butylphenol-OR)} \;\; \mathbf{A} \;+\; \text{(R-2,6-di-t-butylphenol-OH)} \;\; \mathbf{B}$$

R	%A	%B (or other C-alkylation products)
Me	88	6
Et	11	66
i-Pr	0	100

Thus, as the size of R increases, it is less able to reach the OH oxygen for ether formation, and isopropyl halides give no ether.

For years, the only isolable derivatives of cyclopropanone (214) and cyclopentadienone were their di-t-butyl derivatives or their ketals. Finally, methods were devised for their synthesis and isolation, albeit only at very low temperatures for cyclopentadienone (215).

There are numerous isolated examples of steric hindrance blocking normal addition reactions. For example, 2,6-disubstituted styrenes resist polymerization, hindered ketones such as isopropyl ketones do not add isopropyl Grignards (216), only aldehydes and a few methyl ketones form isolable sodium bisulfite addition products, and hindered ketones form Schiff bases only under "forcing conditions" (217). Similarly, di-t-butylketene is inert toward the usual methods of methylene transfer:

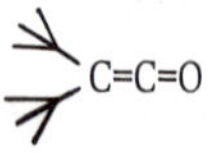

$$\text{(di-t-butyl)C=C=O} \xrightarrow{\text{no reaction}}$$

CH_2I_2, Zn/Cu

Cl_3C-CO_2Et, NaOMe

$C_6H_5HgCCl_3$, NaI

$C_6H_5HgCCl_3$, refluxing benzene

$HCBr(CN)_2 + Et_3N$

CH_2N_2

Sec. 20.5 HINDRANCE TO CONVENTIONAL REACTIONS

Aromatic substitution is well known to be subject to steric hindrance (218). For illustration, from the data in Table 20.3 it can be seen that

Table 20.3. The steric effect in aromatic substitution (219)

A. Effect of size of entering group

Entering group	ortho, %	meta, %	para, %
Me	54	17	29
Et	45	30	25
i-Pr	38	30	33
t-Bu	0	7	93

B. Effect of size of substituent on mononitration

Compound	ortho, %	meta, %	para, %
Me-C_6H_5	59	4	37
Et-C_6H_5	45	7	49
i-Pr-C_6H_5	30	8	62
t-Bu-C_6H_5	16	12	73

the size of a substituent in a monosubstituted benzene as well as the size of the entering group markedly alters the amount of ortho substitution. Thus, di-o-t-butylbenzenes cannot be made by a substitution reaction (220). Similarly, mesitylene will react with a methyl halide but not with a t-butyl halide in a Friedel-Crafts reaction. Chlorine will react with m-t-butylphenol at the 2, 4, and 6 positions, bromine will substitute at the 4 and 6 positions but not the 2 position, and iodine will enter position 6 but not positions 2 and 4 (221).

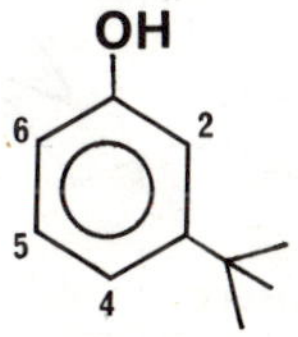

Another type of reaction whose rate is correlated fairly well from steric considerations is the rate of oxidation of secondary cyclanols with chromic acid. The conversion of a tetrahedral carbon to a trigonal carbon reduces 1,3-diaxial strain, and the greater this relief of strain, the faster is the oxidation of the alcohol to the ketone. For example, cholestane-3α-ol has two 1,3-diaxial OH:H interactions whereas there are none in the 3β isomer. The 3α isomer, therefore, undergoes oxidation faster than the 3β isomer.

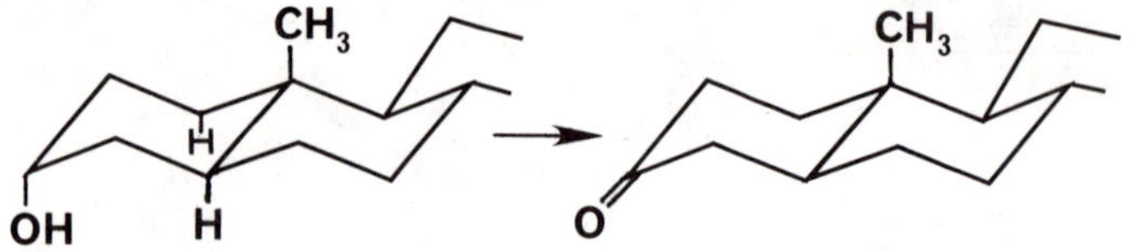

Cholestane-3α-ol → Cholestanone-3
k_{rel} = 3

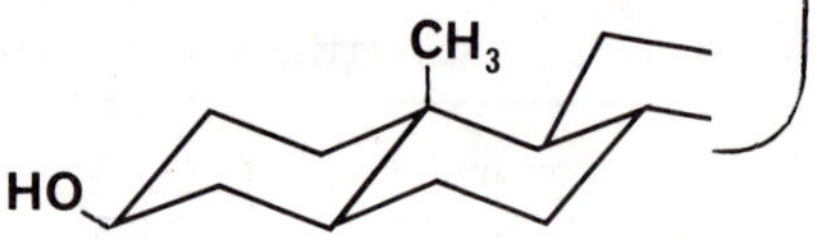

Cholestane-3β-ol
k_{rel} = 1.0

The relative rates of oxidation of other cholestanols are in accord with this view (222):

1,3-diaxial interactions:	1α-ol 3 OH:H
k_{rel}	13

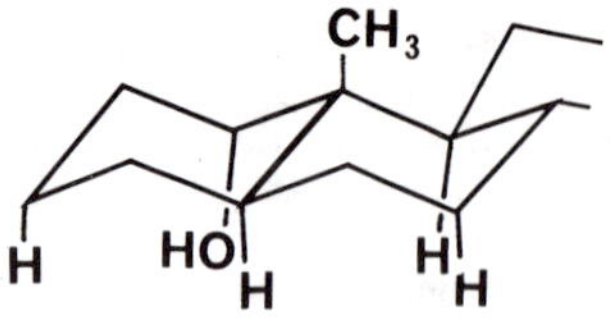

2β-ol
OH:CH$_3$, OH:H
20

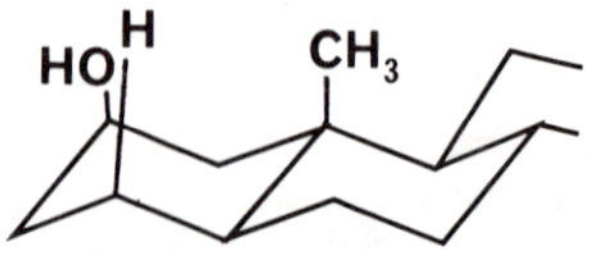

4β-ol
2 OH:H, OH:CH$_3$
35

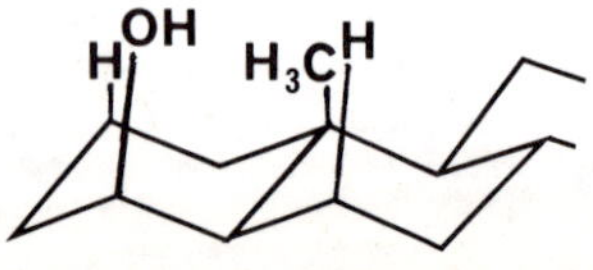

It should be added that spatial crowding can also produce unusual reactions. For example, crowded carbenium ions undergo C-H insertion reactions on nearby alkyl groups (223). An example, is the following:

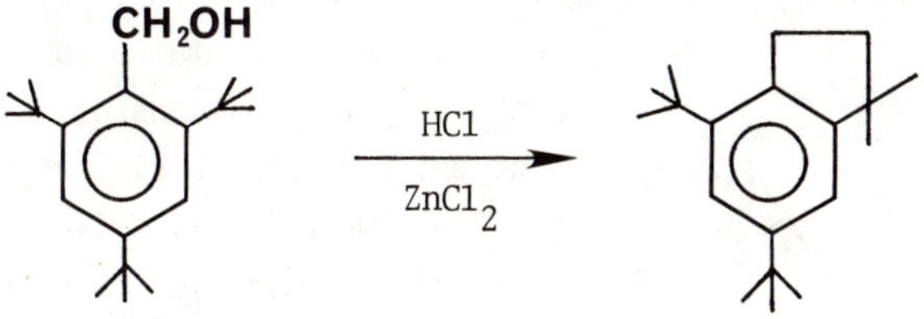

QUESTIONS

1. Explain the fact that 1,3-di-t-butylbutadiene reacts with maleic anhydride in a Diels-Alder reaction but 2,3-di-t-butyl-butadiene does not react.
2. Account for the observation that N,N-dimethylaniline couples with diazonium ions at the ortho and para positions whereas 2,6-N,N-tetramethylaniline is inert toward diazonium ions.
3. The δ_{OH} of isomeric hydroxy-acetylnaphthalenes are (G. O. Dudek, Spectrochim. Acta 19: 691 (1963)):

1-hydroxy-2-acetylnaphthalene	δ_{OH} = 13.98 ppm
1-acetyl-2-hydroxynaphthalene	13.42
2-acetyl-3-hydroxynaphthalene	11.52

    Comment on this trend in δ_{OH} values.

ANSWERS

1. The 1,3-isomer can assume a planar s-cis conformation but the 2,3-isomer is nonplanar (see Chapter 11).

2. The transition state for aromatic substitution is resonance stabilized and resembles S, where Z is the electrophile. Such resonance can occur in the transition states for reaction at the ortho or para positions of N,N-dimethylaniline. However, the ortho positions are blocked in the 2,6-dimethyl derivative and resonance is sterically blocked for resonance at the para position. This demonstrates the importance of resonance in aromatic electrophilic substitution.

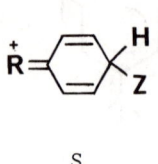

S

3. δ_{OH} is markedly affected by H bonding. Since the C_1-C_2 bond has greater double-bond character than the C_2-C_3 bond, the intramolecular H bond is much stronger for the 1,2-substituted isomers. That of the 1-hydroxy-2-acetyl isomer is the stronger because in the 1-acetyl-2-hydroxy isomer, the CH_3CO group cannot be coplanar with the ring owing to steric hindrance from the peri hydrogen on C_8.

REFERENCES
Part IV

1. H.B. Burgi and L.S. Bartell, *J. Am. Chem. Soc.* 94: 5236 (1972).
2. D.H.R. Barton and O. Hassel, *Topics in Stereochem.* 6: 1 (1972).
3. J.D. Dunitz, *Pure Appl. Chem.* 25: 495 (1971); J. Dale, *Pure Appl. Chem.* 25: 469 (1971).
4. L.N. Ferguson, *Highlights of Alicyclic Chemistry*, Palisades, N.J.: Franklin, 1973.
5. For leading references see: J.P. Lowe, *Science* 179: 527 (1973); L. Radom, W.A. Lathan, W.J. Hehre, and J.A. Pople, *J. Am. Chem. Soc.* 95: 693 (1973).
6. See W.J. Adams, H.J. Geise, and L.S. Bartell, *J. Am. Chem. Soc.* 92: 5013 (1970) for pertinent references.
7. W.S. Johnson, et al., *J. Am. Chem. Soc.* 83: 606 (1961).
8. M.S. Newman, *J. Org. Chem.* 30: 5554 (1965).
9. N.S. Zefirov and N.M. Shekhiman, *Russ. Chem. Revs.* 40: 315 (1971).
10. L.A. Cohen and S. Takahashi, *J. Am. Chem. Soc.* 95: 443 (1973); A.J. Bowles, W.O. George, and D.B. Cunliffe-Jones, *J. Chem. Soc.* (B) 1070 (1970).
10a. K. Hagen and K. Hedberg, *J. Am. Chem. Soc.* 95: 8263 (1973).
11. J.T. Waldron and W.H. Snyder, *J. Am. Chem. Soc.* 95: 5491 (1973).
12. A. Liberles, A. Greenberg, and J.E. Eilers, *J. Chem. Educ.* 50: 676 (1973).
13. N.D. Epiotis and W. Cherry, *Chem. Commun.* 278 (1973); *J. Am. Chem. Soc.* 95: 3087 (1973).
14. H.G. Viehe, J. Dale, and E. Franchimont, *Ber.* 97: 244 (1963).
15. R.B. Turner, A.D. Jarrett, P. Goebel, and B.J. Mallon, *J. Am. Chem. Soc.* 95: 790 (1973).
16. G.T. Armstrong and S. Marantz, *J. Chem. Phys.* 38: 169 (1963).
17. O. Bastiansen, *Acta Chem. Scand.* 4: 926 (1950).
18. S. Wolfe, *Accts. Chem. Res.* 5: 102 (1972).
19. K.L. Servis and D.J. Bowler, *J. Am. Chem. Soc.* 95: 3392 (1973).
20. V. Boekelheide and T. Miyasaka, *J. Am. Chem. Soc.* 89: 1709 (1967).
21. S.F. Nelsen and J.P. Gillespie, *J. Am. Chem. Soc.* 94: 6238 (1972).
22. J.F. Chiang, *J. Am. Chem. Soc.* 93: 5044 (1971).
23. G.M. Lampman, *Amer. Chem. Soc. Petroleum Res. Fund Annual Rept.* 33 (1972).
24. G. Fodor and K. Nádor, *Nature* 169: 462 (1952).
25. R.D. Stolow, *J. Am. Chem. Soc.* 86: 2165, 2170 (1964).

26. J.B. Hendrickson, R.K. Boeckman, Jr., J.D. Glickson, and E. Grunwald, *J. Am. Chem. Soc.* 95: 494 (1973); E.S. Glazer, R. Knorr, C. Ganter, and J.D. Roberts, *J. Am. Chem. Soc.* 94: 6026 (1972); M. Christl and J.D. Roberts, *J. Org. Chem.* 37: 3443 (1972).
27. See J.B. Hendrickson, *J. Am. Chem. Soc.* 90: 7036-7061 (1967).
28. F.A.L. Anet and J.J. Wagner, *J. Am. Chem. Soc.* 93: 5266 (1971).
29. See E.A. Noe and J.D. Roberts, *J. Am. Chem. Soc.* 94: 2020 (1972) and references cited there.
30. O. Ermer, H. Eser, and J.D. Dunitz, *Helv. Chim. Acta* 54: 2469 (1971).
31. See *Determination of Organic Structures by Physical Methods*, Vols. 3-5, eds. F.C. Nachod and J.J. Zuckerman, New York: Academic, 1971-1973.
32. C.W. Jefford, R.C. McCreadie, P. Muller, and J. Pfyffer, *J. Chem. Educ.* 50: 181 (1973); D.S. Tarbell and J.R. Hazen, *J. Am. Chem. Soc.* 91: 7657 (1969).
33. Numerous examples of such assignments are made by D.S. Noyce and B.E. Johnston, *J. Org. Chem.* 34: 1252 (1969).
34. H.S. Aaron, G.E. Wicks, Jr., and C.P. Radar, *J. Org. Chem.* 29: 2248 (1964).
35. See N.L. Allinger and S.P. Jindal, *J. Org. Chem.* 37: 1042 (1972).
36. O.L. Chapman, *J. Am. Chem. Soc.* 85: 2014 (1963); P. Laszlo and P.v.R. Schleyer, *J. Am. Chem. Soc.* 85: 2017 (1963); G.V. Smith and H. Kriloff, *J. Am. Chem. Soc.* 85: 2016 (1963).
37. R.A. Bell and J.K. Saunders, *Topics in Stereochem.* 7: 1 (1973); P.D. Kennewell, *J. Chem. Educ.* 47: 278 (1970); C.L. Van Antwerp, *J. Chem. Educ.* 50: 638 (1973).
38. D.H. Geske, *Progr. Phys. Org. Chem.* 4: 125 (1967).
39. G.A. Russell and G.R. Stevenson, *J. Am. Chem. Soc.* 93: 2432 (1971); J.K. Kochi, P.J. Krusic, and D.R. Eaton, *J. Am. Chem. Soc.* 91: 1877 (1969).
40. G.A. Russell and H. Malkus, *J. Am. Chem. Soc.* 89: 160 (1967).
41. L.M. Stock and P.E. Young, *J. Am. Chem. Soc.* 94: 7686 (1972).
42. C.E. Hudson and N.L. Bauld, *J. Am. Chem. Soc.* 94: 1158 (1972).
43. V.I. Minkin, O.A. Osipov, and Y.A. Zhdanov, *Dipole Moments in Organic Chemistry*, New York: Plenum, 1970.
44. D.D. Faulk and A. Fry, *J. Org. Chem.* 35: 364 (1970).
45. D.M. Grant and B.V. Cheney, *J. Am. Chem. Soc.* 89: 5315 (1967).
46. H. Brouwer and J.B. Stothers, *Canad. J. Chem.* 50: 601 (1972).
47. F.A.L. Anet, C.H. Bradley, and G.W. Buchanan, *J. Am. Chem. Soc.* 93: 258 (1971).
48. A. Liberles and B. Matlosz, *J. Org. Chem.* 36: 2710 (1971).
49. J.F. Tocanne, *Tetrahedron* 28: 363 (1972).
50. See B.A. Pawson, H.-C. Cheung, S. Gurbaxani, and G. Saucy, *J. Am. Chem. Soc.* 92: 336 (1970) and references cited there.
51. N. Harada and K. Nakanishi, *Chem. Commun.* 310 (1970) and earlier publications.

52. J.H. Brewster, *Topics in Stereochem.* 2: 33 (1967); R.D. Stolow and K. Sachdev, *Tetrahedron* 21: 1889 (1965); R.G. Bergman, *J. Am. Chem. Soc.* 91: 7405 (1969).

53. See K. Mislow, *Introduction to Stereochemistry*, New York: Benjamin, 1965.

54. G. Lowe, *Chem. Commun.* 411 (1965); W.R. Moore, H.W. Anderson, S.D. Clark, and T.M. Ozretich, *J. Am. Chem. Soc.* 93: 4932 (1971).

55. See J.C.P. Schwarz, *Physical Methods in Organic Chemistry*, San Francisco: Holden-Day, 1964, Chap. 6; N. Harada and K. Nakanishi, *Accts. Chem. Res.* 5: 257 (1972).

56. J. Fried, M.J. Green, and G.V. Nair, *J. Am. Chem. Soc.* 92: 4137 (1970); B. Rimiker, *et al.*, *J. Am. Chem. Soc.* 76: 313 (1954).

57. H. Akimoto and S. Yamada, *Tetrahedron* 27: 5999 (1971).

58. K. Mislow, *et al.*, *J. Am. Chem. Soc.* 84: 1455 (1962).

59. K. Mislow, *et al.*, *J. Am. Chem. Soc.* 85: 1342 (1963).

60. P.A. Browne, M.M. Harris, R.Z. Mazengo, and S. Singh, *J. Chem. Soc.* (C) 3990 (1971).

61. A.W. Burgstahler and N.C. Naik, *Helv. Chim. Acta* 54: 2920 (1971).

62. H.L. Goering, *et al.*, *J. Org. Chem.* 37: 3019 (1972).

63. V. Prelog, *et al.*, *Helv. Chim. Acta* 36: 325, 320, 308 (1953).

64. A. Horeau and H.B. Kagan, *Tetrahedron* 20: 2431 (1964).

65. M. Miyamoto, *et al.*, *Tetrahedron* 23: 411 (1967).

66. A.J. Bowles, W.O. George, and D.B. Cunliffe-Jones, *Chem. Commun.* 103 (1970).

67. J.G. Aston, *Determinations of Organic Structures by Physical Methods*, ed. E.A. Braude and F. Nachod, New York: Academic, 1955, Chap. 13.

68. C.H. Bushweller, G.E. Yesowitch, and F.H. Bissett, *J. Org. Chem.* 37: 1449 (1972).

69. R.J. Abraham and K. Parry, *J. Chem. Soc.* (B) 539 (1970).

70. J.D. Kemp and K.S. Pitzer, *J. Chem. Phys.* 4: 749 (1936); *J. Am. Chem. Soc.* 59: 276 (1937).

71. E.B. Wilson, *Chem. Soc. Revs.* (London) 1: 293 (1972); R.H. Larkin and R.C. Lord, *J. Am. Chem. Soc.* 95: 5129 (1973); J.P. Lowe, *Progr. in Phys. Org. Chem.* 6, eds. A. Streitwieser, Jr., and R.W. Taft, New York: Interscience, 1968, p. 1; J.R. Durig, J. Bragin, S.M. Craven, C.M. Player, Jr., and Y.S. Li, *Developments in Applied Spectroscopy* 9, eds. A.L. Perkins and E.L. Grove, New York: Plenum, 1971, pp. 23-71.

72. G. Binsch, *Topics in Stereochem.* 3: 97 (1968).

73. H. Kessler, *Angew. Chem. Int. Ed.* 9: 219 (1970).

74. A. Greenberg, *J. Chem. Educ.* 49: 575 (1972).

75. For more exact methods, see references cited by F.H. Suydam and C.H. Yoder, *J. Chem. Educ.* 48: 849 (1971). See also F.B. Mallory, S.L. Manatt, and C.S. Wood, *J. Am. Chem. Soc.* 87: 5433 (1965); S. Alexander, *J. Chem. Phys.* 40: 2741 (1964).

76. B.L. Hawkins, W. Bremser, S. Borcić, and J.D. Roberts, *J. Am. Chem. Soc.* 93: 4472 (1971).
77. S.I. Miller, *J. Chem. Educ.* 41: 421 (1964); J. Lielmezs and J.P. Morgan, *Nature* 202: 1106 (1964).
78. D.Y. Curtin, P.E. Bender, and D.S. Hetzel, *J. Org. Chem.* 36: 565 (1971).
79. I.R. Epstein and W.N. Lipscomb, *J. Am. Chem. Soc.* 92: 6094 (1970).
80. C.H. Bushweller and W.G. Anderson, *Tetra. Letters* (18) 1811 (1972).
81. R.E. Gall, D. Landman, G.P. Newsoroff, and S. Sternhell, *Australian J. Chem.* 25: 109 (1972).
82. R. Price, G. Schilling, L. Ernst, and A. Mannschreck, *Tetra. Letters* (17) 1689 (1972).
83. N.M. Sergeyev, K.F. Abdulla, and V.R. Skvarchenko, *Chem. Commun.* 368 (1972).
83a. J.E. Anderson and D.I. Rawson, *Chem. Commun.* 830 (1973).
84. F.A.L. Anet and M. Ahmad, *J. Am. Chem. Soc.* 86: 119 (1964).
85. T. Drakenberg, *Tetra. Letters* (18) 1743 (1972).
86. H.S. Gutowsky, J. Jonas, and T.H. Siddall, III, *J. Am. Chem. Soc.* 89: 4300 (1967).
87. J.J. Bergman and W.D. Chandler, *Canad. J. Chem.* 50: 353 (1972).
88. L.N. Ferguson, *Highlights of Alicyclic Chemistry*, Palisades, N.J.: Franklin, 1973, Chap. 1.
89. V.I. Minkin, O.A. Osipov, and Y.A. Zhdanov, *Dipole Moments in Organic Chemistry*, New York: Plenum, 1970.
90. R.A. Beaudet, *J. Am. Chem. Soc.* 88: 1390 (1966).
91. V. Baliah and M. Uma, *Tetrahedron* 19: 455 (1963).
92. W.F. Anzilotti and B.C. Curran, *J. Am. Chem. Soc.* 65: 607 (1943).
93. G.F. Longster and E.E. Walker, *Trans. Farad. Soc.* 49: 228 (1954).
94. P.H. Cureton, C.G. LeFevre, and R.J.W. LeFevre, *J. Chem. Soc.* 4447 (1961).
95. W.E. Heyd and C.A. Cupas, *J. Am. Chem. Soc.* 93: 6086 (1971).
96. F.A.L. Anet and J. Sandstrom, *Chem. Commun.* 1558 (1971).
97. J.E. Anderson, *J. Am. Chem. Soc.* 91: 6374 (1969).
98. J.C. Jochims and F.A.L. Anet, *J. Am. Chem. Soc.* 92: 5524 (1970).
99. M.A. Robb, W.J. Haines, and I.G. Csizmadia, *J. Am. Chem. Soc.* 95: 42 (1973).
100. R.O. Hutchins, L.D. Kopp, and E.L. Eliel, *J. Am. Chem. Soc.* 90: 7174 (1968).
101. C.H. Bushweller and J.W. O'Neal, *J. Org. Chem.* 35: 276 (1970).
102. L.N. Ferguson, *J. Chem. Educ.* 47: 46 (1970).
103. On cyclobutadiene, see O.L. Chapman, C.L. McIntosh, and J. Pacansky, *J. Am. Chem. Soc.* 95: 614 (1973); A. Krantz, C.Y. Lin, and M.D. Newton, *J. Am. Chem. Soc.* 95: 2744 (1973).

104. P.E. Eaton and K. Nyl, *J. Am. Chem. Soc.* 93: 2786 (1971); D. Ginsberg, *Accts. Chem. Res.* 2: 121 (1969); D. Ermer, R. Gerdil, and J.D. Dunitz, *Helv. Chim. Acta* 54: 2476 (1971).

105. W.G. Dauben and D.L. Whalen, *J. Am. Chem. Soc.* 88: 4739 (1966).

106. P.E. Eaton and T.W. Cole, Jr., *J. Am. Chem. Soc.* 86: 3157 (1964).

107. J.H. Dopper and H. Wynberg, *Tetra. Letters* (9) 763 (1972); W.E. Barth and R.G. Lawton, *J. Am. Chem. Soc.* 93: 1730 (1971).

108. R.J. Stedman, L.D. Davis, and L.S. Miller, *J. Org. Chem.* 33: 1280 (1968).

109. M. Tichý, L. Kruezo, and J. Hapala, *Tetra. Letters* (8) 699 (1972); H.W. Whitlock, Jr., and M.W. Siefken, *J. Am. Chem. Soc.* 90: 4929 (1968).

110. P.A. Waitkus, E.B. Sanders, L.I. Peterson, and G.W. Griffin, *J. Am. Chem. Soc.* 89: 6318 (1967).

111. U. Biethan, U.V. Gizycki, and H. Musso, *Tetra. Letters* (20) 1477 (1965).

112. J.L. Ripoll, J.C. Limasset, and J.M. Conia, *Tetrahedron* 27: 2431 (1971); A.P. Krapcho and F.J. Waller, *J. Org. Chem.* 37: 1079 (1972).

113. T.J. Katz and N. Acton, *J. Am. Chem. Soc.* 95: 2738 (1973).

114. N.L. Allinger, M.T. Tribble, M.A. Miller, and D.H. Wertz, *J. Am. Chem. Soc.* 93: 1637 (1971); N.C. Baird, *Tetrahedron* 26: 2185 (1970); W.K. Bratton, J. Szilard, and C.A. Cupas, *J. Org. Chem.* 32: 2019 (1967).

115. J. Coops, et al., *Rec. Trav. Chim.* 79: 1226 (1960).

116. P.v.R. Schleyer, J.E. Williams, and K.R. Blanchard, *J. Am. Chem. Soc.* 92: 2377 (1970); S. Chang, et al., *J. Am. Chem. Soc.* 92: 3109 (1970); M. Mansson, N. Rapport, and E.F. Westrum, Jr., *J. Am. Chem. Soc.* 92: 7296 (1970).

117. B.D. Kybett, et al., *J. Am. Chem. Soc.* 88: 626 (1966).

118. S.S. Wong and E.F. Westrum, Jr., *J. Am. Chem. Soc.* 93: 5317 (1971).

119. G.J. Gleicher and P.v.R. Schleyer, *J. Am. Chem. Soc.* 89: 582 (1968).

120. In contrast, see R.H.-W. Wong and G.J. Gleicher, *J. Org. Chem.* 38: 1957 (1973).

121. W.C. Danen, T.J. Tipton, and D.G. Saunders, *J. Am. Chem. Soc.* 93: 5186 (1971).

122. J.D. Roberts and V.C. Chambers, *J. Am. Chem. Soc.* 73: 5030 (1951).

123. P.v.R. Schleyer and R.D. Nicholas, *J. Am. Chem. Soc.* 83: 182 (1961).

124. W. McFarlane, *Quart. Revs.* 23: 187 (1969).

125. T. Axenrod, et al., *J. Am. Chem. Soc.* 93: 6536 (1971).

126. S.L. Manatt, M.A. Cooper, C.W. Mallory, and F.B. Mallory, *J. Am. Chem. Soc.* 95: 975 (1973); F.G. Weigert and J.D. Roberts, *J. Am. Chem. Soc.* 94: 6021 (1972).

127. K. Frei and H.J. Bernstein, *J. Chem. Phys.* 38: 1216 (1963).

128. M.D. Newton and J.M. Schulman, *J. Am. Chem. Soc.* 94: 767 (1972).

129. G.L. Closs and R.B. Larrabee, *Tetra. Letters* (4) 287 (1965).

130. A. Streitwieser, Jr., R.A. Caldwell, and W.R. Young, *J. Am. Chem. Soc.* 91: 529 (1969).

131. D.G. Farnum, J. Chickos, and P.E. Thurston, *J. Am. Chem. Soc.* 88: 3075 (1966).

132. J.F. Chaing, C.F. Wilcox, Jr., and S.H. Bauer, *J. Am. Chem. Soc.* 90: 3149 (1968).

133. J.W. Wilt, H.F. Dabek, Jr., J.P. Berliner, and C.A. Schneider, *J. Org. Chem.* 35: 2406 (1970).

134. K.B. Wiberg and V.Z. Williams, Jr., *J. Org. Chem.* 35: 369 (1970); J.W. Wilt, H.F. Dabek, Jr., J.P. Berliner, and C.A. Schneider, *J. Org. Chem.* 35: 2402 (1970); F.W. Baker, R.C. Parish, and L.M. Stock, *J. Am. Chem. Soc.* 89: 5677 (1967).

135. S.E. Bales and R.D. Reike, *J. Org. Chem.* 37: 3866 (1972).

136. J.H. Markgraf and W.I. Scott, *Chem. Commun.* 296 (1967).

137. P.G. Gassman, *Accts. Chem. Res.* 4: 128 (1971). For an MO explanation, see F.S. Collins, J.K. George, and C. Trindle, *J. Am. Chem. Soc.* 94: 3732 (1972).

138. H.K. Hall, *et al.*, *J. Am. Chem. Soc.* 93: 121, 137 (1971).

139. M. Pomerantz and E.W. Abrahamson, *J. Am. Chem. Soc.* 88: 3970 (1966).

140. W.R. Moore and C.R. Costin, *J. Am. Chem. Soc.* 93: 4910 (1971).

141. G. Montaudo and C.G. Overberger, *J. Org. Chem.* 38: 804 (1973).

142. K.B. Wiberg, *et al.*, *J. Am. Chem. Soc.* 94: 7396, 7402 (1972).

143. P.G. Gassman and F.J. Williams, *J. Am. Chem. Soc.* 93: 2704 (1971).

144. See description in L.N. Ferguson, *Highlights of Alicyclic Chemistry*, Palisades, N.J.: Franklin, 1973, p. 210.

145. F.R. Jensen and B.E. Smart, *J. Am. Chem. Soc.* 91: 5686 (1969).

146. A.P. Krapcho and R.G. Johanson, *J. Org. Chem.* 36: 146 (1971), T.G. Traylor, *et al.*, *J. Am. Chem. Soc.* 93: 5715 (1971); G.A. Olah and G. Liang, *J. Am. Chem. Soc.* 95: 3792 (1973).

147. J.E. Nordlander, S.P. Jindal, and D.J. Kilko, *Chem. Commun.* 1136 (1969).

148. H.C. Brown, R.S. Fletcher, and R.B. Johannessen, *J. Am. Chem. Soc.* 73: 212 (1951).

149. H.C. Brown and M. Barkowski, *J. Am. Chem. Soc.* 74: 1894 (1952).

150. H.C. Brown and K. Ichikawa, *Tetrahedron* 1: 221 (1957); R. Heck and V. Prelog, *Helv. Chim. Acta* 38: 1541 (1955).

151. C.D. Gutsche, *Org. Reactions* 8: 364 (1954).

152. P.v.R. Schleyer, *et al.*, *J. Am. Chem. Soc.* 93: 3189 (1971) and earlier papers.

153. W.G. Dauben and J.L. Chitwood, *J. Am. Chem. Soc.* 92: 1624 (1970).

154. J.P. Lorand, S.D. Chodroff, and R.W. Wallace, *J. Am. Chem. Soc.* 90: 5266 (1968); R.C. Fort, Jr., and R.E. Franklin, *J. Am. Chem. Soc.* 90: 5267 (1968).

155. R.W. Fessenden and R.H. Schuler, *J. Chem. Phys.* 39: 2147 (1963).

156. P.J. Krusic, T.A. Rettig, and P.v.R. Schleyer, *J. Am. Chem. Soc.* 94: 995 (1972).

157. J.M. Landesberg and J. Sieczkowski, *J. Am. Chem. Soc.* 93: 973 (1971).

158. C.S. Foote, *J. Am. Chem. Soc.* 86: 1853 (1964); P.v.R. Schleyer, *J. Am. Chem. Soc.* 86: 1854 (1964).

159. J.E. Nordlander, F.Y.-H. Wu, S.P. Jindal, and J.B. Hamilton, *J. Am. Chem. Soc.* 91: 3962 (1969).

160. E.W. Garbisch, Jr., S.M. Schildcrout, D.B. Patterson, and C.M. Sprecher, *J. Am. Chem. Soc.* 87: 505, 2932 (1965).

161. M.A. Mühs and F.T. Weiss, *J. Am. Chem. Soc.* 84: 4697 (1962).

162. A.J. Fry and W.B. Farnham, *J. Org. Chem.* 34: 2314 (1969).

163. J.A. Marshall and H. Fauble, *J. Am. Chem. Soc.* 92: 948 (1970); J.R. Wiseman and W.A. Pletcher, *J. Am. Chem. Soc.* 92: 956 (1970). For a review, see G. Köbrich, *Angew. Chem. Int. Ed.* 12: 464 (1973).

164. R. Keese and E.-P. Krebs, *Angew. Chem. Int. Ed.* 10: 262 (1971), who argue that their trapping reactions provide evidence for the transient formation of 1-norbornene as a reactions intermediate. See also J.A. Chong and J.R. Wiseman, *J. Org. Chem.* 37: 8627 (1972).

165. H. Pracejus, *Chem. Ber.* 92: 988 (1959).

166. H.K. Hall, Jr., and R.C. Johnson, *J. Org. Chem.* 37: 697 (1972).

167. R. Srinivasan, *J. Am. Chem. Soc.* 90: 2752 (1968).

168. N.J. Turro, *et al.*, *Accts. Chem. Res.* 5: 92 (1972); P.A. Leermakers and G.F. Vesley, *J. Chem. Educ.* 41: 535 (1964); J.S. Swenton, *J. Chem. Educ.* 46: 7 (1969).

169. Review: H.J. Hageman and U.E. Wiersum, *Chemistry in Britain* 9: 206 (1973).

170. R. Bloch, R.A. Marty, and P. de Mayo, *J. Am. Chem. Soc.* 93: 3071 (1971).

171. A. Brown, M.J.S. Dewar, and W. Schoeller, *J. Am. Chem. Soc.* 92: 5517 (1970).

172. G. Schroder and J.F.M. Oth, *Angew. Chem. Int. Ed.* 6: 414 (1967); J.A. Berson, *Accts. Chem. Res.* 1: 152 (1968).

173. L.G. Greifenstein, J.B. Lambert, M.J. Broadhurst, and L.A. Paquette, *J. Org. Chem.* 38: 1210 (1973); R. Hoffman and W.-D. Stohrer, *J. Am. Chem. Soc.* 93: 6941 (1971).

174. J.S. McKennes, L. Brener, J.S. Ward, and R. Pettit, *J. Am. Chem. Soc.* 93: 4957 (1971).

175. M.J. Jorgenson and T.J. Clark, *J. Am. Chem. Soc.* 90: 2188 (1968).

176. H.J. Reich, E. Ciganek, and J.D. Roberts, *J. Am. Chem. Soc.* 92: 5166 (1970).

177. I.E. DenDesten, L. Kaplan, and K.E. Wilzbach, *J. Am. Chem. Soc.* 90: 5868 (1968). For analogous photo-isomerizations of strained naphthalenes, see W.L. Mandella and R.W. Franck, *J. Am. Chem. Soc.* 95: 971 (1973).

178. D.M. Lamal and L.H. Dunlap, Jr., *J. Am. Chem. Soc.* 94: 6562 (1972).

179. V. Balasubramaniyan, *Chem. Revs.* 66: 567 (1966).
180. J.E. Anderson, R.W. Franck, and W.L. Mandella, *J. Am. Chem. Soc.* 94: 4608 (1972).
181. N.L. Allinger, J.S. Maul, and M. Hickey, *J. Org. Chem.* 36: 2747 (1971).
182. S. Itô, Y. Fukazawa, and Y. Iitako, *Tetra. Letters* (9) 741, 745 (1972).
183. H. Wynberg, A. DeGroot, and D.W. Davies, *Tetra. Letters* 1083 (1963).
184. C.R. Brundle and M.B. Robin, *J. Am. Chem. Soc.* 92: 5550 (1970).
185. K. Hagen and K. Hedberg, *J. Am. Chem. Soc.* 95: 1003 (1973).
186. C.F. Aten, L. Hedberg, and K. Hedberg, *J. Am. Chem. Soc.* 90: 2463 (1968).
187. R.W. Franck and E.G. Leser, *J. Org. Chem.* 35: 3932 (1970).
188. H. Weller-Feilchenfeld, E.D. Bergmann, and A. Hirschfield, *Tetra. Letters* 4129 (1965).
189. R. Adams, *Rec. Chem. Progr.* (Summer Issue) 91 (1949).
190. F.H. Westheimer and J.E. Mayer, *J. Chem. Phys.* 14: 733 (1946); T.L. Hill, *J. Chem. Phys.* 14: 465 (1946).
191. K. Mislow, *et al.*, *J. Am. Chem. Soc.* 84: 1455 (1962); M.M. Harris, *Proc. Chem. Soc.* (London) 367 (1959).
192. M. Rieger and F.H. Westheimer, *J. Am. Chem. Soc.* 72: 19 (1950).
193. G.L. Sloan and W.R. Vaughan, *J. Org. Chem.* 22: 750 (1957).
194. E. Müller, *et al.*, *Ber.* 74: 807 (1941) and 72: 2063 (1939).
195. H.A. Lloyd, K.S. Warren, and H.M. Fales, *J. Am. Chem. Soc.* 88: 5544 (1966).
196. H.C. Brown and B. Kanner, *J. Am. Chem. Soc.* 88: 986 (1966).
197. A. Fratiello, G.A. Vidulich, and Y. Chow, *J. Org. Chem.* 38: 2309 (1973).
198. H.C. Brown and M.D. Taylor, *J. Am. Chem. Soc.* 69: 1332 (1947).
199. E.J. Forman and D.N. Hume, *J. Phys. Chem.* 63: 1949 (1959); R.G. Pearson and F.V. Williams, *J. Am. Chem. Soc.* 76: 258 (1954); H.K. Hall, *J. Am. Chem. Soc.* 79: 5444 (1957); L. Sacconi and G. Lomarbo, *J. Am. Chem. Soc.* 82: 6266 (1960).
200. J.I. Brauman and L.K. Blair, *J. Am. Chem. Soc.* 90: 6561 (1968); M.S.B. Munson, *J. Am. Chem. Soc.* 87: 2332 (1965); W.G. Henderson, *et al.*, *J. Am. Chem. Soc.* 94: 4728 (1972).
201. J.I. Brauman, *et al.*, *J. Am. Chem. Soc.* 93: 6360 (1971).
202. J.P. Briggs, R. Yamdagni, and P. Kebarle, *J. Am. Chem. Soc.* 94: 5128 (1972).
203. M. Taagepera, *et al.*, *J. Am. Chem. Soc.* 94: 1369 (1972).
204. D.H. McDaniel, *Science* 125: 545 (1957).
205. E.M. Arnett, *et al.*, *J. Am. Chem. Soc.* 94: 4724 (1972); M.S. Nozari and R.S. Drago, *J. Am. Chem. Soc.* 94: 6877 (1972).
206. L.A. Cohen, *J. Org. Chem.* 22: 1333 (1955).
207. H.H. Lochte and P. Brown, *J. Am. Chem. Soc.* 72: 4297 (1950).

208. G.S. Hammond and D.H. Hogle, *J. Am. Chem. Soc.* 77: 338 (1955).

208a. R.T. McIver, Jr., and J.H. Silvers, *J. Am. Chem. Soc.* 95: 8462 (1973).

209. It is well known that sterically hindered acids cannot be esterified by normal procedures. For comments on various methods used for this purpose, see J. Grundy, B.G. James, and G. Pattenden, *Tetra. Letters* (9) 757 (1972).

210. C.D. Cook, *et al.*, *J. Org. Chem.* 25: 1429 (1960); E. Muller, *et al.*, *Chem. Ber.* 92: 474 (1959).

211. N.C. Yang and A.J. Castro, *J. Am. Chem. Soc.* 82: 6208 (1960).

212. M. Ballester, *et al.*, *J. Am. Chem. Soc.* 93: 2215 (1971).

213. N. Kornblum and R. Seltzer, *J. Am. Chem. Soc.* 83: 3668 (1961).

214. J.M. Pochan, J.E. Baldwin, and W.H. Flygare, *J. Am. Chem. Soc.* 91: 1896 (1969).

215. O.L. Chapman and C.L. McIntosh, *Chem. Commun.* 770 (1971).

216. See also M. Cherest, H. Felkin, and C. Frajerman, *Tetra. Letters* (5) 379 (1971).

217. D.E. Pearson and F. Greer, *J. Am. Chem. Soc.* 77: 1294 (1955).

218. G.S. Hammond and M.F. Hawthorne, *Steric Effects in Organic Chemistry*, ed. M.S. Newman, New York: Wiley, 1956, Chap. 3.

219. H.C. Brown and W.H. Bonner, *J. Am. Chem. Soc.* 76: 605 (1954); see also G.A. Olah and S.J. Kuhn, *J. Am. Chem. Soc.* 84: 3684 (1962) for steric effects of the solvent.

220. C. Hoogzand and W. Hübel, *Angew. Chem. Int. Ed.* 73: 680 (1961) and *Chem. Ber.* 93: 103 (1960); E.M. Arnett and M.E. Strem, *Chem. and Ind.* (London) 2008 (1961); H.A. Bruson, F.W. Grant, and E. Bobko, *J. Am. Chem. Soc.* 80: 3633 (1958); L.R.C. Barclay, N.D. Hall, and J.W. MacLean, *Tetra. Letters* (7) 243 (1961).

221. W.W. Kaeding, *J. Org. Chem.* 26: 4851 (1961).

222. J. Schreiber and E. Eschenmoser, *Helv. Chim. Acta* 38: 1529 (1955).

223. J.J. Looker, D.P. Maier, and T.H. Regan, *J. Org. Chem.* 37:3401 (1972).

INDEX

Ab initio calculations, 3, 14
Absolute configurations, 461
Absorbance, 95
Absorption of light energy, 6
Absorption spectroscopy, 91 (see also Spectroscopy)
Abstraction of iodine atoms, 12
Acetic acid: ir spectral data, 389; carbonyl frequency, 82
Acetophenones and substituent constants, 292; dihedral angles, 419; ir spectral data, 375; uv constituent constants, 416; table, 412
Acetylenes, basicity of, 393
Acetylmesitylene, 211
Acid dissociation and H bonding, 28; of phenols, 252
Acidic hydrocarbons, 256
Acidities of excited states, 427
Acidity and H bond strengths, 25; of hindered acids, 525
Acid strength and H. bonding, 26; effect of solvation, 524; of charge-transfer complexes, 239; of enols, 30
Activation energies for thermal rearrangements, 505
Acylium ions, heats of formation, table, 78
Adamantyl radical, 499
Additivity of covalent radii, 10
Alcohols: basicity, 82; ^{13}C nmr chemical shifts, 83, 85; dissociation constants from polar substituent constants, 51; H bonding, 29, 35; ionization potentials, 81, 242; ir absorption frequencies, 105; mass spectral characteristics, 155; ^{17}O nmr chemical shifts, 83; relative acidities, 78
Aldehyde hydrates, 48
Alicycles, classification, 485; trivial names, 484
Aliphatic acids: dissociation constants, 65; nmr chemical shifts, 66
Alkanes, relative acidities, 78
Alkylanilines, dipole moments, 331
Alkylation and dissociation energies, 12
Alkylbenzenes: charge-transfer complexes, 245; dipole moments, 80; integrated ir intensities, 80; nmr chemical shifts, 80; ionization potentials, 80; spectral data, table, 408
Alkylbenzoic acids, fragmentation in mass spectrometers, 158
Alkyl chlorides, dipole moments, 81
Alkyl groups: electrical effects of, 64, 77; summary, 89; relative electronegativity, 53
Alkyl halides, relative basicity, 394
Alkyl iodides, 12
Alkylphosphines, basicity, 524
Allene, 18
Allyl: π electron distribution, 5; resonance, 273; system, 7
Allylic fission, 156
Alternation of charges, 70
Amide resonance, 502
Amides, rotation barriers, 478
Amines: dipole moments, 60; ^{14}N chemical shifts, 83; relative basicities, 78
Amplitude of stationary waves, 2
Aniline resonance, 203, 253
Anisoles, nmr chemical shifts, 332
Anisotropic effects, 133
(14)Annulene, 226
(16)Annulene, 224
(18)Annulene, 134, 215, 224
Antiaromatic rings, 198
Antiaromaticity, 226
Antibonding orbitals, 4, 6, 118
Approximate solutions, 3
Arene-TCNE complexes, 243
Arndt, 181
Aromatic character: and magnetic properties, 216; and spectral properties, 216; of heterocycles, 217
Aromaticity: and bond lengths, 232; and J_{ortho} coupling, 219; and resonance, 215; and solvent nmr shifts, 221; definitions, 215; of chelate rings, 222; properties for, table, 182; theoretical parameter, 229
Aromatic nitration, effect on cyclopropyl groups, 192
Aromatic properties vs. olefinic properties, 181
Aromatic substitution, steric effects, 531
Arrhenius activation energy, 364

545

Arsole, 217
Aryl carbenes, 218
Aryl ethers, uv spectra, 341
β-Arylketones, 350
Assignment of uv bands, 120
Associated liquids and H bonding, 22
Associated solutions, 389
Asteranes, 484
Atomic electronegativities, 44
Atomic orbitals, 3, 296
Atomic radii, 9
Atomic susceptibilities, 97
Atrolactic acids, 462
Audrieth, 246
Auxochromes, 119
Azines, 236
Azobenzenes, uv spectral data, 365
Azo compounds, uv spectral data, 365

Badger's rule, 116
Baker-Nathan, 77, 79, 87, 186
Bar mass spectrum of cyclohexane, 158
Barrelene, 188
Barton, 452
Barriers to rotation: in ethanes, table, 474; theory, 443
Base peak, 148
Base strengths: and reference acid, 523; from nmr spectral data, 522; in gas phase, 524; of aliphatic amines, 521; solvent effects, 523; gas phase, 253; in excited states, 427; of anilines, 261; table, 262; of haloketones, 70
Basicity: and H bond strengths, 25; and ir frequency shifts, table, 394; and resonance, 230; of acetophenones, 290
Basicity of amines: and H bonding, 29; and steric effects, 521
Basketanes, 484
Beasley, 7
Benzalanilines: uv spectra, 345; table, 346
Benzaldehydes, uv spectra, 374
Bengen, 38
Benzene: π electron distribution, 7; molecular orbitals, 7; resonance energy, 197
Benzils, uv spectra, 373
Benzoate sector rule, 461
Benzobishomotropylium system, 228
Benzocarborane, 217
Benzocyclopropene, 381
Benzophenones: uv spectra, 343; table, 346
Benzotropones: ir spectra data, 376; resonance, 375

Benzoylacetone, 263; uv spectrum, 268
Benzoyl resonance, 409
Benzyl radical, 266
Beynon, 150, 160
Biadamantyl, 441
Bi-t-butyl, 441
Bicyclo bromides, reactivities, 498
Bicyclo compounds: ^{19}F nmr chemical shifts, 87; ionization potentials, 358
Bicyclononanes, 385, 451
Bicyclooctadiene, ionization potentials, 357
Bicyclooctane, 486
Bifurcate H bonds, 26
Bimesityl, uv spectral data, 373
Biphenyls: conformational analysis, 461; dihedral angles, 419, 515; rotation barriers, 479; uv spectral data, 368, 374
Biradicals, 264; and steric hindrance, 517
Birdcage compounds, 484
Bisected conformation for cyclopropyl groups, 193
Bishomoaromatic rings, 228
Bond angles, 9, 14; and electron pair repulsion, 16; in dihaloethylenes, 445; references, 167
Bond angle strain and carbonyl frequencies, 423
Bond character and ring strain, 388
Bond characteristics, table, 8
Bond dipole moments, table, 57
Bond dipoles, 8
Bond energies: and hybridization, 14; table, 13
Bond energy: terms, 12; table, 13; values, 10
Bond force constants, 114; table, 115
Bond ir frequencies, table, 100
Bond ir stretching frequencies, 8
Bond length, 8, 9; and bond type, 8; and resonance, 194; and s-character, 422; effect on ir frequencies, 109; of aromatics, 182; of H-C bonds, 46; references, 167; table, 9
Bond moments, 56; C-C bond, 58; H-C bonds, 46
Bond order, 204
Bond orbitals and dissociation energies, 11
Bond polarity: and electronegativity, 48; effect on ir frequencies, 111
Bond polarizabilities, 55
Bond strength, 8, 10; and bond type, 8
Bond type, 8
Bond uv absorption, 8

Bonding: in charge-transfer complexes, 237; in organic molecules, 1; orbitals, 4
Bonds: chemical, 1; covalent, 1
Boiling points and H bonding, 26
Boron trifluoride, 184
Branch and Calvin, 51
Bredt's rule, 286
Brewster's method of absolute configuration determination, 461
Bromine pentafluoride, 16
Bullvalene, 504, 505
Butadiene, 203; π electron distribution, 6
Butadienes: conformational analysis, 480; and Diels-Alder reaction, 533; steric effects in, 513; uv spectral data, 366
Buttressing effect, 369, 516; and restricted rotation, 511, 512; in carboxylic acids, 390; in methylphenanthrenes, 444
Butylbenzenes, 164
t-Butyl cation, 87
n-Butyl 3,5-dinitrobenzoate inclusion complexes, 40
t-Butyl groups, rotation barriers for, 476

Calculation of dipole moments, 58
Carbanions, nmr chemical shifts, 333
Carbenium ions: and cnmr chemical shifts, 84, 88; definition, 297; heats of formation, table, 78; and hyperconjugation, 78
Carbocations: chemical shifts, 298; definition, 297; in mass spectra, 153; in solvolysis, 298; nmr chemical shifts, 84
Carbon bond angles and ^{13}C-H coupling, 422
Carbon dioxide, vibrational modes, 103
Carbon tetrachloride, bond lengths in, 10
Carbonyl bond angles and ir frequencies, 422; table, 423
Carbonyl compounds, ionization potentials, 188
Carbonyl hydrates, 499
Carbonyl ionization potentials, 359
Carbonyl ir band: assignment, 401; identification of, 111
Carbonyl ir frequencies: and conformational analysis, 453; and solvolysis rates, 500
Carboxylic acids: and H bonding, 34; resonance, 253; substituent effects on dissociation, 65

α-Carotene, 164
Catalysis by inclusion complexation, 41
C-C bond: cleavages, 153; dissociation energies, 267; hybridization in cyclopropyl ring, 193; hyperconjugation, 87, 185; moments, 58
^{13}C chemical shifts, 147; for α,β-enones, 324, 325; for mesitylene, 335
Channel complexes, 38
Characterization of olefins by spectra, 321
p-Character of C-C bonds and ring strain, 495
Charge alternation, 69
Charge densities and nmr chemical shifts, 325
Charge-transfer complexes, 237; and hyperconjugation, 186; and ionization potentials, 81, 240, 242; bonding, 237; effects of substituents, 242; in H bonding, 21; of TCNE, 290; properties, table, 240; relative stabilities, table, 242, 394; uv bands, 120
^{13}C-H coupling and aromatic character, 218; and carbon bond angles, 422
C-H dissociation energies, 267
Chelate rings, aromaticity, 222
Chelates, coupled oscillations, 107
Chelation and ir spectra, 391
Chemical bonds, 1; definition, 8
Chemical properties and H bonding, 33
Chemical shielding, 132
Chemical shift, 132; aromatic range, 134
Chloromethyltriptycene, 477, 478
Choleic acids, 38
Chromate oxidations, 532; of alcohols, 463
Chromatography and H bonding, 31
Chromophores, 119, 355; and uv spectra, 342
cis-trans Conformers, phenols, 392
Cis-trans-isomers: dipole moments, 60, 61; heats of hydrogenation, table, 197; uv spectral data, 365
Cis-trans-enones, ir spectral data, 376
Classical carbocations, 299
Classical mechanics methods, 1
Clathrates, 38
Cleavage of aliphatic ketones in mass spectra, 157
Cleavages in mass spectra, 153
CNDO methods, 1
CNMR: of alcohols, 83, 85; of carbocations, 84

548   INDEX

^{13}C nmr chemical shifts: and H bonding, 396; in aliphatic acids, 66; in carboxylic acids, 397; of carbenium ions, 88; of ketones, 86; of phenol in different solvents, table, 397; of α,β-unsaturated esters and acids, 86
C-O-H vibrations, 101
Color and H bonding, 31
Combination tones, 106
Complementary resonance, 208, 342
Complexes, inclusion, 37
Complex formation, enthalpy of, 109
Configurations of hybrid bond orbitals, table, 15
Conformational analysis, 441; and ring inversions, 472; by dipole moments, 458, 480; by esr spectroscopy, 457; cis-trans-enones by ir, 377; cyclobutanoids, 448; cyclodecanoids, 452; cycloheptanoids, 450; cyclononanoids, 451; cyclooctanoids, 450-451; cyclopentanoids, 448-449; dynamic, 469; empirical correlations, 453; of free radicals, 499; static, 441
Conformational energies, table, 471
Conformational equilibria, 469
Conjugated chromophores, spectral characteristics, 339
Conjugated enones, pe spectra, 129
Conjugation, 296; and pe spectra, 128; of cyclopropyl groups, 495; transmission of, 301
Consistent force field method, 167
Constituent constants, 198
Coordination numbers, 1
Cope rearrangement, 503
Coronene, 484
Correlations, for covalent bonds, 8
Cotton effect, 461
Coupled nuclei, 146
Coupled oscillations, 106
Coupling, 136
Coupling constants, 136; and charge alternation, 70
Coupling of ir bands, 106
Covalent bonding, in H bonds, 21
Covalent bonds, 2; correlations, 8; energies, table, 13; theory, 1
Covalent radii; additivity of, 10; table, 10
Criteria for aromaticity, 218
Cross-conjugated chromophores, 345
Crown ethers, 42
Crystal field theory, 308
Crystal violet, uv spectral data, 344
Cubanes, 484

Cumulated chromophores, 338
Cyanogen chloride, 201
Cyclanes: ir frequencies, 489; reactivity toward radicals, 489; strain energies, table, 485
Cyclanones: ir frequencies, 489; coupled oscillations, 107; relative basicities, 394; spectral data, 387
Cyclenes, ir spectral data, 387
Cyclic dienes: pe spectral data, 357; uv spectral data, 328
Cycloalkenones, spectral data, 368
Cycloalkyl bromides, dipole moments, 489
Cyclobutadiene, 210; paramagnetic ring current, 232
Cyclobutanoids, preferred conformations, 448
Cyclobutanones, spectral data, 350
Cyclobutenones, uv spectral data, 337
Cyclobutyl derivatives, 299
1,6-Cyclodecadiene, 128
Cyclodecane, transannular strain, 442
Cyclodecanoids, conformational analysis, 452
Cyclodecene, 188; addition of bromine, 442
Cyclodecenones, 351
Cyclodextrins, 38; bibliography, 169
1,3-Cyclodienes, uv spectral data, 367
Cycloheptanoids, preferred conformations, 450
Cycloheptatriene, 450
1,4-Cyclohexadiene, dihedral angle, 419
Cyclohexadione, 114
Cyclohexane, conformational analysis, 443
Cyclohexanediols, conformational analysis, 450
Cyclohexanes, conformational analysis, table, 471
Cyclohexanoids, preferred conformations, 449
Cyclohexanones: conformational analysis, 447, 450, 459; ir frequencies, 104; mass spectral data, 158
Cyclononanoids, conformational analysis, 451
1,4,7-Cyclononatriene, 355
Cyclooctatetraene, 203, 451; magnetic susceptibility, 97
Cyclooctene, 450; optically active, 488; dipole moment, 488
Cyclooct-3-enone, 351
Cyclopentadiene, uv spectral data, 328
Cyclopentane, strain, 443

Cyclopentanoids, preferred conformations, 448
Cyclophanes, spectral data, 382
Cyclopropane, hybridization, 430
Cyclopropene, 388; ir absorption bands, 107
Cyclopropyl carbanions, 227
Cyclopropylcarbinyl cation, 297
Cyclopropylcarbinyl derivatives: nmr spectral data, 193; solvolysis, 299
Cyclopropyl groups: effect on esr spectra, 458; effect on spectra, 190; ir frequencies, 105; stabilization of positive charges, 300; substituent constant, 189
Cyclopropylethylenes, hyperconjugation, 186
Cyclopropyl participation, 510
Cyclotron resonance spectroscopy, 35

Decalins, 487
Deductive thought, 43
Degenerate orbitals, 5, 8
Degenerate rearrangements, 504, 505; and H bonding, 34
Delocalization: and ionization potentials, 360; by alkyl groups, 64; by cyclopropyl groups, 189, 300; by phenyl groups, 301; effects, in H bonds, 20; stabilization, 7
Deshielding zones, 135
Deuteration and dipole moments, 54
Deuterium, relative electronegativity, 26, 53
Dewar resonance, 198
Dialkylethylenes, conformational analysis, 445
Dialkylpyrenes, 447
Diamagnetic ring currents, 133
Diamagnetism, 96; table, 219
Diaroylmethanes, 33
Diarylacetylenes, conformational analysis, 460
Diarylmethanes, 350; spectral data, 354
Diaryl-4-pyrones, 289; ir spectra, 112
N,N'-Diarylureas, 306
Diazonium ions, coupling reactions, 533
Di-t-Butylbutadiene, 367
Di-t-butyl ether, 12
1,4-Dienes, 350
Dienes: pe spectral data, 128; spectral data, 354; uv constituent constants, table, 410
o-Diethylbenzene, 182
Diffraction techniques, 452-453

Difluorocyclooctane, 451
Difluorodiimide, 445
Dihalobiphenyls, conformational analysis, 446
Dihalocyclohexanes, conformational analysis, 443
Dihaloethanes, conformational analysis, 444, 446
1,1-Dihaloethylenes, conformational analysis, 445
Dihedral angles: and nmr coupling, 431, 456; and spectra, 418; and Taft substituent constants, 421; and uv spectral data, 419; in biphenyls, 515; in nitrobenzenes, 370
Dimethylacetylene, 147
Dimethylenecyclohexane, 114
Dinitrobenzenes, uv spectral data, 369
Diols, H bonding in, 34
Di-o-t-butylphenols, 385
m-Dioxane, 483
p-Dioxane, inversion, 482
Diphenylbutadienes, uv spectral data, 366
3,3-Diphenylcyclopropyl-1,2-dicarboxylic acid, 31
Diphenylpolyenes, 340
Dipole-dipole forces, 36; in H bonds, 21, 24
Dipole-dipole interaction 76
Dipole-dipole repulsion, in dihalocyclohexanes, 444
Dipole moments: and absorption intensities, 124; of aldehydes, 188; and bond type, 8; and conformational analysis, 458, 464, 480, 487; and electronegativity, 49; and H bonding, 31; and ir spectra, 113; and resonance, 206, 230; and symmetry, 56, 61; calculation of, 58; electric, 55; from microwave, 54; from polar substituent constants, 51; in excited states, 427; measurement of, 55; of alkylbenzenes, 80; of alkyl chlorides, 81; of amines, 60; of bonds, table, 57; of cis-trans isomers, 60, 61; of cyclopropyl compounds, 195; of charge-transfer complexes, 239; of disubstituted benzenes, table, 59, 62, 212; of fluorinated benzenes, 62, 63; of halogen compounds, table, 207; of H-C bonds, 46; of ketones, 89, 189; of methylacetylenes, 88; of phosphinobenzenes, 331; of some anilines, 331; steric hindrance, 212; trimethylamine oxide, 334
Di-t-butylethylene, 188

Dispersion forces, in H bonds, 21, 24
Dispirodecane, 114
Dissociation of hexaarylethanes, 264
Dissociation constants: and electronegativity, 53; and field effects, 72; and temperature effects, 67
Dissociation energies, 10, 267; and steric effects, 12; and structural effects, table, 11; effect on ir frequencies, 109; from polar substituent constants, 51; of aliphatic acids, 65; of H-C bonds, 46; of substituted acetic acids, 65
Dissymmetric chromophores, 461
Disubstituted benzenes, dipole moments of, table, 59, 62
Disubstituted ethanes, conformational analysis, table, 471
Ditolyl, 388
Divinylcyclopropane, 503
1,8-Divinylnaphthalene, 447
Dodecapentaenic acid, 127
Doering, 7
d-Orbitals, 3, 18; shapes, 248
Double bond character, 204; from ir spectra, 326
Dynamic conformational analysis, 469

Effects of homoconjugation, 350
Electric dipole moments, 55
Electrical effects: of alkyl groups, 77; summary, 89; of substituents, 64; transmission coefficient, 72
Electromagnetic radiation, 91
Electromagnetic spectrum, table, 92
Electron delocalization, 9; and uv spectra, 119, 322, 328
Electron densities: and ^{13}C and ^{17}O nmr chemical shifts, 83, 335; and H bond strength, 35
Electron diffraction, 9
Electron-donor ability and H bond strengths, 25
Electronegativity, 43; and basicity, 43; and bond angles, 15, 17; and bond lengths, 18, 19, 46; and bond polarity, 48; and dipole moments of ketones, 89; and dissociation constants, 53, 76; and dissociation energies, 11; and H bonding, 21, 25, 35; and hybridization, 46, 56, 509; and H abstraction, 489; and internuclear distances, 9; and ionization of acids, 69; and ir intensities, 51; and oxidation state, 48; and thermochemistry, 44; and nmr chemical shifts, 66, 270;
atomic, 44; by Kharasch, 49; definition, 43; effect on ir frequencies, 109; from dipole moments, 49; from polar substituent constants, 49; of groups, table, 54
Electronegativity effects on spectra, 405
Electronic configurations, 5
Electronic excitation, 6, 116
Electronic transitions, 93
Electronic transmission, 231
Electron-pair repulsion: and bond angles, 14, 483; theory, 16
Electron spin-sets, 1
Electron transfer uv bands, 120
Electron-withdrawing groups and H bond strengths, 25
Electrophilic substituent constants, table, 278
Electrostatic attraction in H bonds, 20
Electrostatic effect on acid dissociation, 30
Electrostatic repulsion, 1
Electrostatics, 1
Empirical correlations, 43; for conformational analysis, 453; for bond types, 8
Energies of H bonds, table, 20
Energy: to bend a bond, 116; to stretch a bond, 116
Energy barriers to rotations: in ethanes, table, 474; in various molecules, table, 475
Energy conversion chart, 95
Energy levels, diagrams, 5, 93
Enol-keto equilibria, 262; tautomers, 12
Enols: acid strength, 30; ir spectra, 391; nmr chemical shifts, 395
Enone resonance, spectral data, 367
Enones, 350; homoconjugation, 352; ir spectral data, 374; spectral data, 352; table, 322; s-trans vs. s-cis, 376; uv constituent constants, 415; table, 411
Enone systems, and spectra, table, 351
Enthalpy of ionization of carboxylic acids, 68
Entropy effects, on dissociation constants, 67, 90; table, 69
Enzymatic deamination of keto acids, 33
Equilibria and resonance, 252
ESR spectroscopy and conformational analysis, 457
Esters, conformational analysis, 444, 469, 480
Ethanol, nmr spectral data, 142

INDEX 551

Ethers: ^{17}O nmr chemical shifts, 84; relative basicities, 79; spectral data, 325
Ethylene, π-electron distributions, 5; torsional force constant, 116
E_T values, 123, 126
Exaltation, 97, 219
Excited states, 5; acidities, 427; and uv spectra, 341
Extinction coefficient, 95

Fahlberg, 246
^{19}F chemical shifts, pentafluorobenzyl fluorides, 334
Fermi resonance, 106, 268
Ferrocene, 246; charge-transfer complexes, 243; pmr spectral data, 251
Ferromagnetism, 99
Field effects, 64, 72; on solvolysis, 76
First overtone for OH bonds, 105
First primary uv bands, 119
Flavone, mass spectral data, 164
Fluorene, 227, 231
Fluorine hyperconjugation, 285
Fluorine resonance, 280
Fluxionalism, 503
^{19}F nmr chemical shifts: and substituent constants, 279; in bicyclo compounds, 87; of fluoropyridines, 124
f-orbitals, 3
Force: to stretch a bond, 115; to bend a bond, 116
Force constants, 108; of aromatics, 182; and Raman spectra, 114; table, 115; and bond length, 116; of H-C bonds, 46
Force-field methods, 1
Forces of attraction: covalent bonding, 8; hydrogen bonding, 19; interatomic, 1; van der Waals, 36
Formamide, 206
Fragmentation: generalizations, 152; modes, 153
Free rotation, 472
Frequency of light, definition and units, 95
Frontal strain, 521
^{19}F scs values, 258
F strain, 521
Fumaric acid, 31
Fundamental frequencies, 103
Fundamental modes of vibration, 102

Gamma effect on nmr chemical shifts, 460
Gas hydrates, 38
Gas phase: acidities, 79; basicities, 79, 524
Gauche effect, 446
Glycols: H bond distances, 424; and H bonding, 27
Ground states, 5
Group electronegativities, 49; and ir intensities, 51; and ir frequencies, 52; and nmr spectra, 52; by Kharasch, 49; from dipole moments, 49; from polar substituent constants, 49; solution vs. gas phase, 53; table, 54
Group moments, 56
Guanidine, 258
Guest molecules, 38

Half-wave potentials, 243
Halobenzenes: uv spectral data, 147; table, 406, 407
Halogen compounds, dipole moments, table, 207
α-Halohydrins, 392
Haloketones, basicities, 70
o-Halophenols, 28, 385, 392
Hammett: constants and ir frequencies, 409; equation, 276; substituent constants of cyclopropyl, 189; treatment, criticisms, 283
Hammett-type constants, application, 283
Hammett-type treatments, 276; and ir spectra, 287; and uv spectra, 288
Hansch partition coefficients, table, 278
Harmonic oscillator, 107
Harmonic overtones, 104
H-C bond moment, 56
Heats of atomization, 12; table, 13; of benzene, 196
Heats of combustion, 12, 92; and H bonding, 31; and resonance energies, 196
Heats of hydrogenation, 92; and hyperconjugation, 186; and resonance energy, 196; and strain, 384; cyclononene, 356; of cis-trans isomers, table, 197
Heat of vaporization and H bonding, 31
Heavy water, H bonding, 26
Heteronins, 217
Hexaarylethanes, 264, 369
Hexafluorobutadiene, 128
Hexafluoroisopropanol, H bonding, 26
Hexafluorothioacetone, 48
Hexamethylethane, 11

Highest occupied orbital, 6
Hindered acids, acidity, 525
Hindered phenols: pK_a values, 525; radicals from, 526, 528
Hindered rotations, 269; in a variety of compounds, 516
Homoaromatic rings, 228
Homoconjugated chromophores, 350
Homoconjugation, 350, 362; and reaction mechanisms, 305; enones, table, 351; in carbocations, table, 303; in para-cyclophanes, 381
Homoenamines, spectral data, 353
Homotropylidine, 503
Hooke's law, 107
Horeau's procedure of absolute configuration determination, 462
Host molecules, 38
Hückel MO method, 1
Hückel's rule, 215
Hybrid, 181; bonds, 18; bond orbitals, 9; configurations, table, 15
Hybridization, 9; and bond energies, 14; and dissociation energies, 11; and electronegativity, 46, 56, 63, 509; and internuclear distances, 9; in cyclopropane, 430; of cyclopropyl C-C bonds, 193; of N in anilines, 409
Hydrocarbon acids, 256
Hydrocarbons, ^{13}C nmr chemical shifts, 85
Hydrogen bond distances, 22; and ir frequencies, 424
Hydrogen bond forces, 19
Hydrogen bonding: and acid dissociation, 28; and acid strength, 26; and a degenerate rearrangement, 34; and basicity of amines, 29; and boiling points, table, 27; and chemical properties, 33; and chromatography, 31; and cis-trans conformers, 392; and color, 31; and conformational analysis, 392; and dipole moments, 31; and enol content, 271; and heat of combustion, 31; and heat of vaporization, 31; and hyperconjugation, 77; and ir absorption, 35, 104, 389, 402, 520; and mass spectra, 31; and nmr chemical shifts, 395, 401, 533; and nmr shielding, 404; and physical properties, table, 32; and polarographic reduction potentials, 31; and photoreactivity, 31; and refractive index, 31; and relative basicities, 392, 394; and spectra, 389; and steric effects, 520; and thermodynamic data, 403; and transition temperatures, 26; and uv spectral changes, 395, 403; and viscosity, 31; and water solubilities, table, 27; and wet-melting points, 31; effect on ir frequencies, 109; effects on properties, 389; in alcohols, 29; in carboxylic acids, 34, 390; in enols, 30; in excited states, 404; in heavy water, 26; in o-phenols, 26; in proteins, 27; steric effects, 392; in sulfur compounds, 390; in tropolone, 33; of acetic acid in different solvents, 399

Hydrogen bonds, 19; and molecular properties, 26; and titration curves, 31; energies, table, 20; historical development, reference, 168; in peroxides, 26; model, 21; near ir absorption, 105; pK_as vs. ir frequency shifts, 393; symmetrical, 22; theory of, reference, 168

Hydrogen bond strength: and basicities, 22; and electron density, 35; and electronegativity, 25, 35; and ir frequencies, 22, 393; and nmr chemical shifts, 25; and types of forces, 25; excited states, 429; for some alcohols, 396; with π electrons, 23

Hydrogen maleate ion, 22
1,3-Hydrogen shifts, 506
1,5-Hydrogen transfers, 506
Hydronium ion, 18
Hydrophobic substituent constants, 292
Hydroxybiphenyls, uv spectra, 335
4-Hydroxycyclohexanone, 104
6-Hydroxy-2-formylfulvene, 22
Hydroxytropylium ion, H bonding, 26
Hyperconjugation, 185, 245; and bond lengths, 194; and heats of hydrogenation, 187; and H bonding, 77; and ir carbonyl frequencies, 81; and nmr chemical shifts, 86; and stability of carbocations, 78; and ring strain, 496; and solvolysis, 496; C-F, 210, 285; in carbocations, 85; in methylacetylenes, 88; of alkyl groups, 77; of C-C bonds, 87; of crotonaldehyde, 188; of cyclopentadiene, 221, 328
Hypostrophene, 505

Imides, ir absorption, 107
Inclusion complexes, 37; catalytic effect, 41; dimensional characteristics, 38; general properties, 40; uses, 41

Increment values, 486
Indene, 231
Indenes, pe spectra, 349
Induced dipoles, 37, 55
Induced magnetism, 99
Induced polarization, 55
Induction, 65; and alkyl groups, 77; substituent constant, 75
Inductive effect, 50; and nmr chemical shifts, 82, 86; and polar substituent constants, 82; in alkyl groups, 64; on solvation, 87; transmission coefficient, 73
Inductive thought, 43
Infrared radiation, 91
Infrared spectra, 99; and bond lengths, 9 (*See also* IR spectra)
Inner shell electrons, 21
Insulated chromophores, 338, 374
Integrated intensities, 133; and polarizability, 80
Interaction moments, 514
Interaction resonance, 203; moments, table, 209
Interatomic distances, references, 167
Intermolecular forces, 36
Intermolecular H bonds, 19; uv spectra, 400
Internal energies, 91
Internal strain, 497
Internuclear distances, 9
Intramolecular forces: and bond angles, 15; and molecular properties, 441
Intramolecular geometry, and spectra, 418
Intramolecular H bonding, 19, 26, 391, 396, 400; in diols, 34; in glycols, 27
Intramolecular strain from π clouds, 447
Inversion barriers, paracyclophanes, 481
Ionization constants of alkylacetic acids, 68
Ionization potentials: and H bond strength, 25; and hyperconjugation, 186; and ir spectra, 327; and homoconjugation, 350; and uv spectra, 365; of acetophenones, 289; of alcohols, 81, 242; of alkylbenzenes, 80; of arenes, 243; of azo compounds, 365; of benzenes, 290; of bicyclo hydrocarbons, 358; of carbonyl compounds, 188; of CT complexes, table, 240, 242; of nitro compounds, 269; of olefins, table, 187; of vinyl silanes, 349; pes, 126, 309

IR absorption and H bonding, 104
IR absorption frequencies, and group electronegativities, 52
IR absorption intensities, and electronegativities, 51
IR bands, coupling, 106
IR carbonyl band, 111
IR carbonyl frequency, 81
IR constituent constants, ketones, table, 413
IR frequencies: and amide resonance, 502; and carbonyl bond angles, 422; table, 423; and complex stability, 109; and conformational analysis, 453, 464; and electronegativity, 51, 405; and Hammett constants, 409; and H bond distances, 424; and H bonding, 23; and H bond strengths, 393; and reduced mass, 108; and ring size, 431; and steric hindrance, 383; cyclanes, 489; effect of bond polarity, 111; effect of force constant, 109; effect of reduced mass, 110; in benzamides, 409; isotope effect, 110; of ketones and solvolysis rates, 500; of octadecane derivatives, table, 108; of substituted acetic acids, 82; solvent effects, 111; ranges, table, 100
IR frequency shifts: and basicity, table, 394; and pK_as, 393
IR group frequencies, in different solvents, 112
IR intensities, for conformational analysis, 377
IR spectra: and conformational analysis, 469; and double bond character, 326; and Hammett-type treatments, 287; and H bonding, 22, 389; buttressing effect, 390; chelation effect, 391; ionization potentials, 327; of cyclopropyl derivatives, 191; of ethers, 325; pK_as, 327; of 4-pyrones, 112; of strained benzene rings, 379; of tropolones, 112; table, 113
IR spectral data: acetophenones, 375; benzotropones, 376; carboxylic acids, 389; cis-trans enones, 376; cyclohexanol, 389; cyclenes, 387; enones, 374; halophenols, 393; ketosteroids, 378; of metal π complexes, 250
IR spectroscopy, and H bonding, 35
IR stretching frequencies and bond type, 8
I strain, 497
Isolated chromophores, 338
Isolated π orbitals, 6
Isotopic distributions, table, 150

554   INDEX

Isotopic effect on ir frequencies, 110
Isotopes, effect on mass spectra, 149
Isovalent resonance, 184, 201

J_{ortho} coupling, 219
J values, 136

Kekulé rings, and resonance stability, 234
Ketimines, 350; spectral data, 353
β-Ketocyclopropanes, 350
Keto-enol tautomers, 12, 262
Ketones: basicity, 496; cmnr chemical shifts, 86; conformational analysis, 453-455; dipole moments, 89; ir constituent constants, table, 413; ir frequencies and bond angle, table, 423; ir spectral data, 408; ^{17}O nmr chemical shifts, 86; pK_a values, 259; pmr chemical shifts, 86; resonance, 71; spectral data, table, 327
Ketone complexes, ir frequencies, 110
Ketone hydrates, 48
Ketosteroids, 378
Kinetics and resonance, 252
Kirkwood-Westheimer, 77
Kolbe electrolysis, 274

Lewis' covalent bond theory, 1
Light quanta, 93
Linear free-energy equations, 277
Linearly conjugated chromophores, 339
Lithium alkoxides, 24
Lithium bonds, 24
Local excitation uv bands, 120
Lone pair electrons: and bond angles, 16; and H bonding, 23; effective size, 483; repulsion, 447
Lowest unoccupied orbital, 6
Lutidine borontrifluoride, 522

Magnetic: equivalency, 136, 141; fields, 96; moments, 131; nuclei, 130; properties, of aromatic systems, 216; susceptibility, 96
Magnetogyric ratio, 130
Malachite Green, uv data, 344
Maleic acid, 31
Mass spectra, 148; and H bonding, 31
Mass spectroscopy, structural problems, 160
McLafferty rearrangement, 160
Melting points of phenols, 28
Mesitylene, cmr spectral data, 335
Mesityl oxide, 377; uv spectra in different solvents, 126
Mesomeric-field effects, 280

Mesomerism, 181
Metal π complexes, 237, 246; ir spectral data, table, 250
Metallocenes, 246; bonding, 248
Metaparacyclophane, 147
Meta polyphenyls, 338
Metastable ions, 163
Metastable peaks, 149
Meta substituent constants, 280
Methyl acetylenes, dipole moments, 88
Methyl alcohol, dipole moment, 58
Methyl chloride, vibrational modes, 103
Methylene vibrations, 101
Methylnorbornene, 147
1-Methylpentalene, 503
Michler's hydrol, uv spectral data, 344
Microwave spectra: and bond lengths, 9; and dipole moments, 54; and H bond distances, 22
Modern structural theory, 2, 179
Modes of fragmentation, 153
Modes of vibration, 101
Molar extinction coefficient, 95
Molecular absorption of energy, 99
Molecular energy diagram, 216
Molecular forces, 1
Molecular geometry, 1; and resonance, 200
Molecular ion, 148; peak, 148
Molecular mechanics, 1
Molecular orbitals: of benzene, 7; of allyl systems, 5
Molecular orbital energy diagrams, 117, 127
Molecular orbital methods, 1; theory, 2; theory on metallocenes, 248
Molecular polarization, 43
Molecular properties and H bonding, 26
Mulliken self-consistent field MO method, 1
Multiple correlation indexes, 124
Multiplicity, 137
Mutual polarization of dipoles, 63

Naphthol, 254
Natural abundances of isotopes, table, 150
^{14}N chemical shifts, 82
Neopentyl chloride, 179
Neopentyl p-nitrobenzoates, 497
N-ethylaniline, mass spectral data, 152
Neutron diffraction, 9; and H bond distances, 22
Nickel tetracarbonyl, 250
Nitroaniline resonance, 202, 371
Nitroanilines: ionization potentials, 129; resonance, 202, 203

Nitrobenzenes: dihedral angles, 420; nmr chemical shifts, 332; uv bands, 122
3-Nitrocatechol, 28
Nitrogen inversion, 482
Nitrogen rule for mass spectra, 152
Nitro group, shielding zones, 135, 370
p-Nitrophenylacetylene, dipole moment, 63
2-Nitroresorcinol, 28
NMR chemical shifts: and conformational analysis, 453, 464, 470; and electronegativity, 66, 270; and H bonding, 395, 401; and H bond strengths, 25; and inductive effects, 82; and π electron densities, 333; carbanions, 333; for selected groups, table, 139, 140; in 1,3-diketones, 267; in metal π complexes, 251; in nitrobenzenes, 370; methylene groups, table, 412, 414; of aliphatic acids, 66; of alkylbenzenes, 80; of esters, 399; of ketones, 86; of substituted benzenes, 332; substituent effects on, 71; table, 417
NMR coupling: and conformational analysis, 453, 463; and dihedral angles, 418, 456; and Hammett constants, 430; and ring size, 431; and ring strain, 379, 490; bicycloheptanes, 430; for disubstituted benzenes, 209
NMR parameters for carbocations, table, 298
NMR shielding: abutting protons, 383; and H bonding, 404; by phenyl groups, 519; by ring currents, 432; by various groups, 459; of ^{13}C chemical shifts, 460; paracyclophanes, 382
NMR spectra: and Hammett-type treatments, 288; of cyclopropyl compounds, 191; of strained benzene rings, 379; phospholenes, 330; solvent effects, 333; spin-spin coupling, 136
NMR spectral data: and base strengths, 522; and Q values, 295; enones, 367; for cyclopropylcarbinyl cation, 193; for α,β-enones, 323
NMR spectroscopy, 129; and group electronegativity, 52
^{15}N nmr chemical shifts of substituted anilines, 205
Nodal plane, 2

Nonbonding energy levels, 6
Nonclassical carbocations, 350
Nonclassical resonance, 296
Nonmagnetic nuclei, 130
Norbornadiene, ionization potentials, 357
Norbornanes, conformational analysis, 457
Norbornene: heats of hydrogenation, 384, 490; ir spectral data, 490
2-Norbornyl cation, 297, 298, 360
7-Norbornyl cation, 298
Norbornyl derivatives, absolute configurations, 462
Norbornyl-type substrates, solvolysis of, table, 302
Nuclear Overhauser effects, 457
Nucleophilic assisted ionization, 304
Nucleophilic substituent constants, table, 278

3-Octanol, mass spectral data, 154
Octant rule, 461
Olefinic properties vs. aromatic properties, 182
Olefins: characterization by spectra, 321; ionization potentials, table, 187; strained, 363
^{17}O nmr chemical shifts, 86; of alcohols, 83; of ethers, 84
Optical density, 95
Optical isomerism and steric hindrance, 515
Orbital hybridization: and bond lengths, 9; and inductive effects, 64
Orbital lobes, 3
Orbital overlap, 3, 350
Orbitals, 2
Orbital s character of strained bonds, 490
Orbital type and bond angles, 15
Organomercurials, 49
Ortho coupling constants, 210
Ortho phenols, H bonding, 26
Ortho substituent constants, 282
Oscillator strength, 122
Overtones, 104; table, 105
Oxidation state and electronegativity, 48
Ozonolysis, 182

Para coupling constants, 210
(2,2)Paracyclophane, 424; properties, 381; spectral data, 380; strain, 447
Paracyclophanes, 386; inversion barriers, 481

Paramagnetic ring currents, 222
Paramagnetism, 99
Para polyphenyls, 340
Parrison-Pople method, 1
Pascal's constants, 97; table, 98
Pauling: scale of electronegativities, 44; table, 47; van der Waals radii, 37
Peak area ratios, 137
Pentafluorobenzyl fluoride, 334
Pentalenyl dianion, 219
Perchloroarylmethanes, 260
Perchlorobishomocubanone, 423
Perchloro radicals, 529
Perchlorotriptycene, 260
Perfluoroamines, 46
Perfluorocarbon acids, 257
Perfluoronorbornyl ion, 286
Phases of a wave, 2
o-Phenanthroline, 29
Phenols: acidities in excited states, table, 427; cis-trans conformers, 392; melting points, 28; resonance, 252
Phenonium ions, 304
Phenylacetylene, dipole moment, 62
β-Phenylethanol, dihedral angle, 425
Phenyl isonitrile, 185
Phenyl radicals, 12
Phosphinobenzenes, 330
Phosphole, 217
Phospholenes, 330
Phosphones, 51
Phosphorus substituent constants, table, 278
Photochemical nitration, 430
Photochemical reactivity and homoconjugation, 350
Photoelectron spectra: and conjugation, 128; spiro compounds, 349
Photoelectron spectral data, hydrocarbons, 356
Photoelectron spectroscopy, 126, 309; and homoconjugation, 355; and spiroconjugation, 349; of 2-norbornyl cation, 360
Photolytic isomerization, 506
Photoreactivity: and H bonding, 31
Phthalic anhydride, mass spectral data, 154
Physical properties, and H bonding, table, 32
Picramides, dihedral angles, 420
Picric acid, 254
Picryl iodide, 204
π electron densities, and nmr chemical shifts, 333
π electron distributions, 5

π electrons and H bonding, 23
π induction, 64, 72; and electronegativity, 70; in aromatic acids, 70; in aryl rings, 71
π orbitals, 4
pK data and resonance energies, 199
pK_a: and ir frequency shifts, 393; and ir spectra, 327; and resonance, 255; of phenols, table, 255
pK_a values: excited states, 427; of aromatic amines, 268; of ketones, 259; of pyridines, 189
pK_R+ values, 226
Planarity of nitroanilines, 203
PMR chemical shifts: of protonated alcohols, 88; of protonated ethers, 88
Polar character: and acidity, 490; and ^{13}C-H coupling, 491; and resonance, 205
Polarizabilities: and base strengths of amines, 79; and CT complex stability, 240; and dissociation energies, 11; and integrated intensities, 80; and internuclear distances, 9; and ionization potentials of alcohols, 243; and ir carbonyl frequencies, 81; and uv absorption, 79, 122, 343; and Raman spectra, 113; and relative electronegativity, 53; and spectra, 405; and stability of acylium ions, 78; of alkylbenzenes, 408; of alkyl groups, 53; table, 55
Polarization, 55: and uv spectra, 119; of alkyl groups, 77; of bonding electrons, 64; of C-H bonds, 78; of π electrons, 70
Polar substituent constants, 82, 279; and basicity of alcohols, 82; and electronegativity, 49; ir absorption, 106; table, 50
Polarographic reduction potentials and H bonding, 31
Polycyanines, 340, 341
Polyethers, 42
Polyfluorobenzenes, dipole moments of, 62
p-orbitals, 3
Potential energy, 91; definition, 92; diagram, 94
Precession, 130
Prelog's method of absolute configuration determination, 462
Prismanes, 484
Propellanes, 484
n-Propyl iodide, nmr spectral data, 142
Proteins and H bonding, 27; spectra of, 322

Proton affinities, 195; of amines, 254
Protonated cyclopropane, 195, 301
Protonated ethers, pmr chemical shifts, 88
$PtCl_3$-ethylene complex, 249
Pyracyclene, 224
Pyrene, 222
Pyridazine, mass spectral data, 508
Pyridinium-N-phenolbetaine, 123
4-Pyrone, 106
Pyrrole carboxylic acid, nmr spectral data, 143

Q parameter, 293, 373
Quadricyclene, 486
Quantum mechanics, 3
Quarter-wave potentials, 243
Quinoidal resonance, 429
Quinols, 38
Q values, tables, 294

Radio waves, 92
Raman frequencies: and force constants, 114; and reduced mass, 114
Raman spectra, 113; and bond lengths, 9
Reaction constants, 277; and reaction mechanisms, 283
Reaction mechanisms: and reaction constants, 283; and resonance, 273
Redox potentials, and substituent constants, 291
Reduced mass, 108; and Raman spectra, 114; effect on ir frequencies, 110
Refractive index and H bonding, 31
Rehybridization energies, 10
Relative acid strengths: and electronegativities, 69; of carbon acids, 65; of carboxylic acids, 65; strongest monobasic acid, 66
Relative basicities: from spectra, 392; of aliphatic amines, table, 523; of pyridines, table, 522; of some solvents, table, 397
Relative electronegativity of cyclopropyl ring, 195
Relative group electronegativities, 49
Relative intensities of mass lines, 151
Relative polarizabilities of alkyl troups, 77
Relative reaction rates, 1
Resolution: by clathration, 41; by differential complexation, 42
Resonance: and acidity, 254; and aromaticity, 215; and base strengths of amines, 230, 261; and bond lengths, 194; and dipole moments, 206, 230; and dissociation of hexaarylethanes, 264; and dynamics, 252; and equilibria, 252; and ir frequencies, 201; and keto-enol tautomerism, 262; and kinetics, 252; and molecular geometry, 200; and ^{15}N chemical shifts in anilines, 205; and planarity, 202, 212; and polar character, 205; and reaction mechanisms, 273; and rotation barriers, 202; and spectra, 336; and steric hindrance, 365; concept, 181; in excited states, 429; of benzotropones, 375; of benzoylacetone, 263; of charge-transfer complexes, 237; of α,β-enones, 322; of ketones, 71; of triarylmethyl radicals, 265; rules, 184
Resonance energies: definition, 196; table, 183, 198; of allyl system, 274
Resonance hybrid, 182
Resonance interaction, 207
Resonance moments, table, 208
Resonance structures, 183
Resonance theory, 2; and spectra, 322; applications, 200; origin, 180
Resonance vs. MO theory, 308
Repulsive forces in H bonds, 21, 24
Restoring force constant, 108
Restricted rotations, 472
Retro Diels-Alder reactions, in mass spectroscopy, 156, 165, 166
Ring currents, 133, 222
Ring inversions, and conformation analysis, 472
Ring strain: and addition reactions, 500, 509; and bonding, 489; and Bredt's rule, 501; and C-H acidity, 491; and electronegativities, 494; and hyperconjugation, 496; and isomerization, 502; and molecular properties, 484; and nmr coupling, 379, 490; and nmr shielding, 432; and OH acidity, 492; and olefinic C-C bonds, 507; and reactivity, 497, 507; and $\overline{s}$ character, 493; and solvolysis rates, 508; and spectra, 388; and thermochemical data, 507; effect on dissociation constants, 493; energies, 485
Rochow, 45
Rocking vibrations, 101
Rotanes, 484
Rotation barriers, 269, 373, 472, 516; amides, 478; and resonance, 202; aryl ethers, 479; benzaldehydes, 478; biphenyls, 479; by nmr spectroscopy, 472; in benzamides, 409; theory, 443

558   INDEX

Rotational energy, 91
Rotational transitions, 93
Rules for writing resonance forms, 184

Sandwich compounds, 247
s-character: and acid strengths, 492; and bond angles, 492; and ^{13}C-H coupling, 422
Schrödinger equation, 2
Scissor bend vibrations, 101
SCS values, 71, 73
Secondary uv bands, 120
Semibulvalene, 505
Semitheoretical correlations, 43
Shielding effects, of valence electrons, 132
Shielding zones, 133; of nitro group, 370
Σ delocalization, 301
Σ induction, 64, 72; and ionization of acids, 69
Σ orbitals, 4
Sigmatropic rearrangements, 503
Silica gel, effect on uv spectra, 125
Skeletal breathing, 101
Solvation: and relative electronegativity, 53; energies, 79
Solvent effects: on base strengths, 523; on ir frequencies, 111; on nmr chemical shifts, 398, 399; on spectra, 333; on uv spectra, 118, 123, 401
Solvent parameters, 123
Solvent polarity, and uv spectra, 124
Solvent shifts, and aromaticity, 221
Solvolysis, effect of cyclopropyl groups, 192; rates, table, 302; via carbocations, 298
s orbitals, 3
Spectra: and constituent constants, 410; and electron delocalization, 322; and electronegativity, 405; and H bonding, 389; and intramolecular geometry, 418; and polarizability, 405; and steric effects, 363; of conjugated chromophores, 339; of α,β-enones, table, 322; of proteins, 322
Spatial crowding and spectral effects, 383
Spectral data: cyclanones, 387; cyclic ketones, 351; cyclobutanones, 350; cyclophanes, 382; diarylmethanes, 354; dienes, 354; enone systems, 351; homoenamines, 353; ketimines, 353
Spectral effects: from spatial crowding, 383; from benzene ring strain, 379

Spectral properties of aromatic systems, 216
Spectra-structure correlations, 321
Spectroscopy, 94; and conformational analysis, 453; uses, 96
Spin-coupled nuclei, 139
Spin orientations, 136
Spin-spin coupling constants, table, 138; and chemical shifts, 146
Spin-spin interactions, 136
Splitting patterns, 140
Spirane carboxylic acids, 76
Spiroaromaticity, 349
Spiro compounds, pe spectra, 349
Spiroconjugated chromophores, 347
Spiroconjugation, 329, 349, 362; and pe spectroscopy, 349
Splitting patterns, 140
Squalene, 41
Stable free radicals, 526, 528, 529
Static conformational analysis, 442
Stationary waves, 2
Sterically crowded carbenium ions, 533
Steric conditions and bond length, 9
Steric effects: and base strengths, 521; and chromate oxidations, 532; and dissociation energies, 12; and spectra, 363; and uv enhancement, 379; miscellaneous, 511; on C-C bond lengths, 441; of peri H atoms, 511; on uv spectra, 513, 514, 365
Steric enhancement, 370
Steric hindrance: and biradicals, 517; and ir frequencies, 383; and resonance, 365; effect on properties, 441; in di-t-butyl-benzenes, 444; to addition reactions, 530; to aromatic substitution, 531; to bond contraction, 77; to esterification, 179; to H bonding, 520; to hydrolysis reactions, 527; to resonance, 71, 212, 261; to solvation, 77
Steric stabilization, 527, 530
Steric strain, in nitrobenzenes, 371
Stilbenes: dihedral angles, 419; uv spectral data, 365
Strain: and bond character, 388; and heats of hydrogenation, 384; and uv spectra, 388; cyclophanes, 381; in benzene rings, 379
Strain energies: additivity, 488; cyclanes, table, 485; of some alicycles, table, 487
Strained olefins, 363
Structural parameter constants, 97
Structure elucidation by mass spectra, 160
Substituent chemical shift values, 73

Substituent constants, 277; and bioactivity, 292; and ionization potentials of acetophenones, 289; and ionization potentials of benzenes, 291; and redox potentials, 291; for aliphatic compounds, 282; for heterocyclic rings, 282; for organophosphorus compounds, 282; for phenols, 282; for pyridine rings, 283; meta, 280; ortho, 282; table, 278

Substituent effects: on nmr chemical shifts, 71; table, 417; on uv bands, table, 121, 329

Substituent electrical effects, 64

Substituent induced chemical shifts, 71

Substituted adamantanes, solvolysis, 87

Substituted benzaldehydes, rotation barriers, 478

S values for solvents, 124

Symmetrical H bonds, 22

Taft, 50, 78, 82; σ_I constants, table, 278; σ_R constants, table, 278; steric constants, 292

Taft substituent constants, and dihedral angles, 421

Tautomers, 12

T values, 132

TCNE-charge transfer complexes, 290

Temperature effects: on dissociation constants, 67; on uv spectra, 126

Tetracyclones, uv spectra, 343

Tetrafluoropropanol, 125

Tetramethylsilane, 132

Thermal rearrangements, cyclic olefins, 364

Thermochemistry and electronegativity, 44

Thermodynamic data and H bonding, 403

Thermodynamic quantities, for ring inversions, table, 473

Thiourea complexes, 38

Torsional strain, 443

Transannular strain, 442; in cyclodecanes, 452

trans,cis-1,3,5-heptatriene, ir spectrum, 102

Transconjugation, 350

trans-fused cyclopropanes, 495

Transition temperatures, and H bonding, 26

Translational energy, 91

Transmission coefficient, of inductive effects, 73

Trifluoromethylaniline, 287

Trimethylamine oxide, dipole moment, 334

Triphenyl carbocations, uv spectra, 343

Triphenylmethyl, 264

Triptycene, 260

Trishomoaromatic ring, 356

Tri-t-butylamine, 441

Trivial names of alicycles, 484

Tropone, 33, 220

Tropolone, 33, 220, ir spectra, 112

Twistanes, 484

Twisted double bonds, 367

Ultraviolet. See UV.

α,β-Unsaturated acids and esters, cnmr chemical shifts, 86

Urea, 200; complexes, 38

UV absorption, and bond type, 8

UV bands, substituent effects, table, 121

UV constituent constants: for acetophenones, table, 412; for dienes, 410; for α,β-enones, 411

UV enhancement, and steric effects, 379

UV partials, 373

UV spectra, 116; and assignment of C=O ir band, 401; and buttressing effect, 369; and conformational analysis, 453; and dihedral angles, 419; and excited states, 427; and Hammett-type treatments, 288; and H bond strengths, 396; and intramolecular H bonds, 400; and MO energy diagrams, 117; and polarizability, 122, 343, 405; and resonance, 328; and solvent effects, 334; and steric hindrance, 513, 514; benzalanilines, 345; cross-conjugated chromophores, 345; cumulated chromophores, 339; effect of cyclopropyl groups, table, 194; meta polyphenyls, 338; of benzene derivatives, 119; of conjugated chromophores, 339; of cyclopropyl compounds, 190; of insulated chromophores, 338; of ketones in different solvents, 125; of linearly conjugated molecules, table, 340; of nitrobenzenes, 122; of strained benzene rings, 379; oscillator strength, 122; solvent effects, 118, 123, 401; spiroconjugated chromophores, 347; substituent effects, table, 329; temperature effects, 126

UV spectral data: and H bonding, 403; alkylbenzenes, table, 408; alkylbutadienes, 366; azo compounds, 365; biphenyls, 368, 374; butadienes, 366; cis-trans isomers, 365; 1,3-cyclodienes, 367; dinitrobenzenes,

369; diphenylbutadienes, 366; stilbenes, 365; strained olefins, 363; substituted naphthalenes, 361
UV wavelength shifts, and H bonding, 395

Valence-bond: structures, 181; theory, 308
Valence electron-repulsion theories, 1
Valence isomers, of benzene, 507
Valence shell electrons, 1
Valence tautomerism, 503
van der Waals: attraction, 36; forces, 36; radii, table, 37
van't Hoff equation, 469
Vertical ionization, 127
Vibrational: modes, 101; coordinates, 102; quantum number, 103; transitions, 93
Vinyl bromide, 183
Vinyl cations, 274
Vinyl chloride: ionization potentials, 127; nmr spectral data, 145
Vinyl halides, dipole moments, 207
Vinyl ketones, dihedral angles, 419
Vinyl silanes, ionization potentials, 349

Viscosity and H bonding, 31
Visible spectra, 116
Vitamin A palmitate, 41

Wagging vibrations, 101
Water solubility, and H bonding, 26
Wave equation, 2
Wave function, 2
Wavelength of light, definition and units, 95
Wave Mechanical methods, 1
Werner complexes, 38, 40
Wet-melting points and H bonding, 31
Woodward-Hoffmann rules, 308

X-bands, 343
X-ray diffraction 9; on charge transfer complexes, 239
X-rays, 91
$\underline{X}$ values for solvents, 123

Y-bands, 343
$\underline{Y}$ values for solvents, 123

Zero-bridge bonds, 495
Ziegler, 246
$\underline{Z}$ values for solvents, 123, 126
Zwischenstufen, 181

AUTHOR INDEX

Aaron, H.S.   536
Abdulla, K.F.   538
Abraham, R.J.   537
Abrahamson, E.W.   540
Abruscato, G.J.   435
Abushanab, E.   438
Acton, N.   539
Adams, R.   542
Adams, W.J.   535
Adcock, W.   172, 316
Adolph, H.G.   175, 435
Ahmad, M.   311, 538
Akiyama, S.   433
Alden, R.A.   439
Alder, K.   174
Alder, R.W.   169
Alexander, S.   537
Allan, E.A.   437
Allen, G.   315
Allen, L.C.   168, 172
Allinger, N.L.   167, 175, 177, 315, 434, 439, 536, 539, 542
Allred, A.L.   170, 510
Ammon, H.L.   312
Anastassiou, A.G.   312, 313
Anderson, A.G., Jr.   313
Anderson, G.L.   171, 172, 174, 315
Anderson, G.R.   314
Anderson, H.W.   537
Anderson, J.E.   434, 538, 542
Anderson, J.K.   311, 435
Anderson, W.G.   538
Andreades, S.   317
Andrews, L.J.   314
Andrews, T.G., Jr.   312
Andrist, A.H.   435
Anet, F.A.L.   177, 311, 313, 433, 436, 536, 538
Angelelle, J.M.   317
Anzilotte, W.F.   538
Ariens, E.J.   318
Armstrong, G.T.   535
Arnett, E.M.   171, 172, 315, 318, 433, 437, 542, 543
Arnold, R.T.   435, 436
Arrington, J.P.   434
Ash, M.L.   318
Ashe, A.J., III   312
Astle, M.J.   169
Aston, J.G.   537

Aten, C.F.   542
Axenrod, T.   311, 539
Aue, D.H.   508

Baba, H.   167, 310
Badger, G.M.   312
Badger, R.M.   175, 437
Bailes, H.M.   437
Bain, A.D.   176
Baird, N.C.   173, 311, 435, 539
Baitinger, W.F.   317
Baker, A.D.   176
Baker, A.W.   437
Baker, C.   176
Baker, F.W.   171, 172, 316, 540
Baker, J.W.   173, 310, 318
Balasubramaniyan, V.   433, 435, 437, 542
Baldt, J.H.   487
Baldwin, J.E.   435, 543
Bales, S.E.   540
Baliah, V.   214, 538
Ballester, M.   315, 435, 543
Balsubramanian, A.   35
Bank, S.   433, 439
Barclay, L.R.C.   543
Barcza, S.   174
Barfield, M.   438
Barker, C.C.   433
Barkowski, M.   540
Bailey, D.S.   175
Bartell, L.S.   18, 167, 535
Bartlett, P.D.   168, 319, 439
Bartman, B.   317
Barton, D.H.R.   535
Barton, T.J.   171, 313
Barth, W.E.   539
Bassler, G.C.   174, 177
Bastiansen, O.   169, 535
Bates, K.B.   433
Batich, C.   176, 438
Bauer, R.H.   175
Bauer, S.H.   312, 540
Bauer, W., Jr.   439
Bauld, N.L.   536
Baumann, H.   313
Bavin, P.M.G.   168
Beak, P.   313
Beasley, G.H.   167

Beauchamp, J.L.   169
Beaudet, R.A.   538
Beavers, W.A.   434
Becker, E.D.   177
Becker, E.I.   433
Becker, G.   311
Beeman, C.P.   176
Bekárek, V.   317
Bell, J.V.   170
Bell, R.A.   536
Bell, R.P.   171
Bellamy, L.J.   175, 437
Bellettini, A.G.   172
Bender, M.L.   170
Bender, P.E.   538
Benfey, O.T.   169
Bengen, F.   246
Bennett, C.   314
Benson, R.C.   313
Benson, S.W.   167, 316
Bent, H.A.   167
Bentley, F.F.   175
Bentley, K.W.   433
Bentley, T.W.   318
Berg, E.   314
Bergman, J.J.   538
Bergman, R.G.   316, 537
Bergmann, E.D.   542
Berlenback, B.E.   433
Berliner, E.   172
Berliner, J.P.   540
Bernett, W.A.   310
Bernstein, H.J.   539
Berson, A.   176
Berson, J.A.   541
Bertelli, D.J.   234, 312
Berwin, H.J.   172, 318
Beveridge, D.L.   167, 175
Beynon, J.H.   177, 178
Bhatnager, V.M.   169
Bienvenue, A.   436
Bier, A.   314
Bierl, B.A.   175
Biethan, U.   539
Bigeleisen, J.   433
Billups, W.E.   436
Binder, R.G.   435
Bingham, R.C.   167
Binsch, G.   537
Birks, J.B.   314
Bischof, P.   176, 320, 434, 435
Bissett, F.H.   537
Bjerrum, N.   99

561

Blagdon, D.E. 319
Blair, L.K. 168, 172, 173, 542
Blakeney, A.J. 436
Blanchard, E.P. 174
Blanchard, K.R. 539
Blatz, P. 433
Bloch, R. 541
Blomquist, A.T. 435
Bloomfield, J.J. 434
Bloor, J.E. 439
Bloser, D. 168
Blout, E.R. 435
Bobko, E. 543
Bock, B. 313
Boeckman, R.K. 536
Boekelheide, V. 313, 436, 535
Bohlmann, R. 176
Bohm, H. 434
Bohme, D.K. 173
Boikess, R.S. 177, 436
Bolles, T.F. 175
Bolton, P.D. 171
Bondi, A. 169
Bonner, T.G. 318
Bonner, W.H. 543
Booth, J. 169
Borcic, S. 538
Boschi, R. 434
Bouis, P.A. 172
Boulton, A.J. 63
Bourn, A.J.R. 436
Bouwhuis, E. 440
Bovey, F.A. 173, 177
Bowden, K. 172
Bowler, D.J. 535
Bowles, A.J. 535, 537
Bowman, P.S. 169
Boyd, R.J. 435
Boykin, D.W., Jr. 317, 318
Boyland, E. 169
Brachel, H.v. 174
Bradley, L.H. 536
Bradley, R.H. 168
Brady, J.D. 437
Brady, S.F. 311
Bragin, J. 438, 537
Branch, G.E.K. 433, 170, 171
Brand, J.D.C. 176
Bratoz, S. 437
Bratton, W.K. 539
Braude, E.A. 312, 439, 537
Brauman, J.I. 172, 173, 177, 436, 542

Brauman, J.L. 168
Braun, R.L. 313
Breen, J.J. 433
Bremser, W. 538
Brener, L. 541
Breslow, R. 170, 311, 313
Brewster, J.H. 537
Brier, P.N. 168
Briggs, J.P. 315, 542
Broadhurst, M.J. 541
Brockway, L.O. 18
Brookhart, N. 310, 404
Brouwer, H. 173, 433, 438
Brown, A. 541
Brown, C.L. 160
Brown, D.G. 175
Brown, H.C. 170, 310, 316, 437, 497, 540, 542, 543
Brown, J. 311
Brown, M.G. 167
Brown, P. 542
Brown, R.S. 319
Brown, T.L. 168, 170, 171, 173, 317, 438
Brown, W.G. 435
Browne, P.A. 537
Brownlee, R.T.C. 173, 317
Brownstein, S. 176
Broxton, T.J. 173
Bruck, P. 436
Brundle, C.R. 176, 542
Bruson, H.A. 543
Brutcher, F.V., Jr. 439
Buchanan, G.L. 435, 439
Buchanan, G.W. 313, 536
Buchs, A. 318
Buckles, R.E. 168
Budzikiewicz, H. 177, 178
Bunzli, J.C. 176, 435
Burawoy, A. 171, 172, 342, 438, 439
Burgi, H.B. 535
Burgstahler, A.W. 537
Burlingame, A.L. 177
Bursay, J.T. 169
Bursey, M.M. 165, 169, 173, 318, 509
Burton, D.J. 315, 317
Burton, F.G. 317
Buss, V. 172, 310, 319
Buter, J. 310
Bushweller, C.H. 537, 538

Cabana, A. 170
Calder, G.V. 171
Caldwell, R.A. 540
Calvin, M. 169, 170, 171
Campbell, A.D. 317
Campbell, M.M. 178
Campbell, P. 170
Campbell, W.J. 436
Canady, W.J. 168
Cargioli, J.D. 171, 315
Carter, J.V. 318
Carter, P. 436
Castellucci, N.T. 174
Castro, A.J. 543
Catalano, E. 168
Cava, M.P. 177
Cavestri, R. 310
Chadwick, D. 176
Chaing, J.F. 540
Chakrabarti, B. 438
Chambers, V.C. 539
Chan, R.K. 314
Chandler, W.D. 538
Chang, D.S.G. 434
Chang, S. 539
Chantooni, M.K., Jr. 317, 436
Chapman, O.L. 433, 536, 538, 543
Charette, J.J. 315
Charton, M. 317, 318
Cheney, B.V. 176, 536
Chenier, P.J. 319
Cherest, M. 543
Cherry, W. 535
Cheung, H.-C. 536
Chiang, J.F. 312, 433, 535
Chickos, J. 169, 315, 540
Childs, R.F. 313
Chitwood, J.L. 540
Chodroff, S.D. 540
Chong, J.A. 541
Chong, S.-L. 172, 195
Chow, H.S. 314
Chow, L.W. 315
Chow, S.W. 436
Chow, W.Y. 436
Chow, Y. 176, 434, 542
Christ, H.A. 173, 437
Christian, C.G. 169
Christiansen, G.A. 169
Christl, M. 536
Christman, D.L. 170
Christoffersen, R.E. 314
Ciganek, E. 541
Clar, E. 180
Clark, J. 315
Clark, R.A. 433

AUTHOR INDEX

Clark, S.D.   537
Clark, T.J.   541
Clarke, P.A.   437
Clemens, L.M.   314
Clementi, S.   319
Cleveland, J.D.   310
Clevenger, J.V.   319
Closs, G.L.   174, 539
Closson, W.D.   170, 176, 433, 434, 439
Coan, S.B.   433
Coates, G.E.   315
Coates, R.M.   319
Coblentz, W.W.   100
Coetzee, J.F.   175
Cohen, L.A.   170, 317, 318, 434, 535, 542
Cole, A.R.H.   35
Cole, T.W., Jr.   539
Collen, J.   310
Collins, F.S.   540
Collman, J.P.   313
Colpa, J.P.   440
Colter, A.K.   314
Colthup, N.B.   174
Comisarow, M.B.   334
Conant, J.B.   310
Conia, J.M.   539
Cook, C.D.   543
Cook, D.   318
Cooner, T.M.   173
Cooper, M.A.   436, 539
Coops, J.   539
Corbin, T.F.   310
Corver, H.A.   313
Costin, C.R.   540
Cottee, F.H.   436
Cotter, R.J.   314
Coulson, C.A.   168, 311, 312
Cowan, D.O.   434
Cowling, S.A.   362
Cox, J.D.   167, 168
Craig, R.R.   174
Cram, D.J.   170, 175, 319, 436, 439
Cramer, F.D.   169, 170
Crandall, J.K.   434
Craven, S.M.   537
Crawford, B.L.   174
Cresswell, W.T.   311
Creswell, C.J.   174
Crews, P.O.   312
Cross, F.J.   315
Csizmadia, I.G.   538
Csizmadia, V.M.   320
Cullison, D.A.   434
Cunliffe-Jones, D.B., 535, 537
Cupas, C.A.   538, 539
Cureton, P.H.   538
Curran, B.C.   538
Curtin, D.Y.   538
Custard, H.C.   174, 439

Dabek, H.F., Jr.   540
Dabrowska, U.   168
Dabrowski, J.   168
Dack, M.R.J.   175
Dahlgren, G., Jr.   169
Dahn, H.   173
Daintith, J.   176
Dale, J.   535
Danen, W.C.   168, 539
D'Antonio, G.P.   313
Darby, N.   313
Dauben, H., Jr.   312
Dauben, W.G.   539, 540
Daudel, R.   167
Davidson, R.B.   435
Davies, D.W.   542
Davies, M.   172, 318
Davis, L.D.   539
Day, N.   177
Dayal, S.K.   317
Deady, L.W.   173
Dearden, J.C.   438
Deavenport, D.L.   319
deBie, D.A.   440
De Franco, R.J.   434
De Groot, A.   542
de Hoffmann, E.   315
De Jongh, D.   177
de Jongh, R.O.   440
de la Mare, P.B.D.   315, 435
Delay, F.   335
DelPesco, T.W.   440
de Mayo, P.   435, 541
DenDesten, I.E.   541
Deno, N.C.   316
De Puy, C.H.   319
Dermer, O.C.   319
De Selms, R.C.   335
De Tar, D.F.   177
De Vries, L.   436, 440
Dewar, M.J.S.   167, 172, 176, 310, 311, 316, 320, 439, 541
Diaz, A.   319
Dickinson, C.   439
Diehl, P.   173, 318, 437
Digenus, G.A.   269, 409

Dimroth, K.   176
Dischler, B.   439
Dixon, W.T.   312
Djerassi, C.   177, 178
Doak, G.O.   317
Dodson, R.M.   436
Doering, W.v.E.   167
Dollish, F.R.   175
Donohue, J.   311
Dopper, J.H.   539
Dorman, D.E.   177
Dorsay, W.S.   169
Doub, L.   175
Dowd, P.   434, 438
Drago, R.S.   173, 175, 318, 437, 542
Drakenberg, T.   538
Dreiding, A.S.   434
Duchesne, J.   314
Dudek, G.O.   533
Dunitz, J.D.   535, 536, 539
Dunlap, L.H.   541
Durig, J.R.   537
Dwek, R.A.   315
Dyer, J.R.   177
Dyke, T.   438

Eastes, J.W.   438
Eaton, D.R.   536
Eaton, P.E.   539
Eglinton, G.   436
Ehrenson, S.   317
Eilers, J.E.   535
El-Bayoumi, M.A.   434
Elbermani, M.F.   175
Eliel, E.L.   538
Ellerhorst, R.H.   439
Ellis, P.D.   177
El-Sayed, M.F.A.   311
Elwood, J.K.   320
Emerson, M.E.   168
Engel, P.   440
Epiotis, N.D.   535
Epstein, I.R.   538
Epstein, M.J.   510
Epstein, W.W.   169
Erickson, K.C.   438
Erickson, R.E.   168
Ermer, D.   539
Ermer, O.   438, 536
Ernst, L.   538
Eschenmoser, A.   436, 543
Eser, H.   536
Evans, H.D.   439
Evans, T.R.   176
Eyring, E.M.   169

AUTHOR INDEX

Fainberg, A.H. 176
Fales, H. 168, 436, 542
Farnham, W.B. 541
Farnum, D.G. 169, 540
Farrar, T.C. 177
Fassel, V.A. 170
Fateley, W.G. 175
Fauble, H. 541
Faulk, D.D. 436, 536
Fay, R.C. 313
Featon, R.A. 433
Felkin, H. 543
Fellows, R. 316
Ferguson, L.N. 168, 169, 175, 176, 310, 312, 314, 315, 316, 317, 318, 319, 433, 434, 436, 437, 438, 440, 535, 538, 540
Ferrand, E.F. 433
Ferraris, G. 168
Ferretti, J.A. 177
Fessenden, R.W. 540
Fickes, G.N. 435
Filipescu, N. 434
Filler, R. 315, 316
Fink, R. 434
Firestone, R.A. 167
Fischer, E. 246
Fisera, L. 317
Fish, R.W. 314
Fisher, W.F. 172
Fitzgerald, R. 172
Fitzpatrick, J.D. 312
Fleming, I. 433
Fleming, K.A. 171
Fletcher, R.S. 540
Fliszar, S. 173
Flood, S.H. 437
Florence, T.M. 318
Flurry, R.L., Jr. 440
Flygare, W.H. 312, 313, 543
Fodor, G. 535
Folsom, T.K. 438
Foos, J.S. 362
Foote, C.S. 439, 440, 541
Forbes, W.F. 434, 435, 438
Ford, R.A. 315
Foreman, M.I. 314
Forman, E.J. 542
Fort, R.C., Jr. 540
Förster, T. 439
Foster, R.G. 147, 314
Frajerman, C. 543
France, H. 435

Franchimont, E. 535
Franck, R.W. 177, 436, 541, 542
Frank, J. 169
Franklin, J.L. 172, 195, 311
Franklin, R.E. 540
Fratiello, A. 542
Freeman, D.E. 175
Frei, K. 539
Fried, J. 537
Friedel, R.A. 173, 178
Frost, D.C. 176
Fry, A. 436, 536, 541
Frijita, T. 316
Fukazawa, Y. 542
Fukunaga, T. 434
Fuson, N. 438
Fyfe, C.A. 314

Gac, N.A. 316
Gajewski, J.H. 311
Gall, R.E. 538
Ganter, C. 536
Ganz, C.R. 314
Garbisch, E.W., Jr. 541
Garner, A.Y. 316
Garratt, P.J. 312
Gassman, P.G. 174, 176, 540
Gatos, G.C. 439
Gaunt, J. 318
Geise, H.J. 535
Geiseler, G. 175
Gentilli, B. 246
Geroge, J.K. 540
George, W.O. 535, 537
Gerdil, R. 539
Gerloch, M. 320
Geske, D.H. 536
Giam, C.S. 176
Gielen, M. 176
Giles, C.H. 169
Gilles, J.-M. 313
Gillespie, J.P. 535
Gillespie, R.J. 167
Gilligan, W.H. 437, 438
Ginsberg, D. 539
Gizycki, U.V. 539
Glazer, E.S. 536
Gleicher, G.J. 173, 539
Gleiter, R. 434, 435
Glew, D.N. 436
Glickson, J.D. 536
Goan, J.C. 314
Godfrey, M. 317
Godycke, L.E. 437
Goebel, C.V. 168

Goebel, P. 535
Goering, H.L. 319, 434, 537
Gohlke, R.S. 178
Golden, D.M. 316
Golden, R. 172, 316
Goldenson, J. 170
Goldfish, E. 311
Goldstein, A. 435
Goldstein, J.H. 312
Goldstein, M. 437
Goldstein, M.J. 319
Gompper, R. 312
Goodman, M. 319
Goodyear, C. 246
Gordon, M. 320
Gorman, A.A. 362
Gorodetsky, M. 315
Goss, F.R. 312
Götz, H. 175
Gould, R.F. 318
Gove, J.L. 438
Grant, D.M. 176, 536
Grant, F.W. 543
Grant, M.W. 172
Green, M.J. 537
Green, M.L.H. 315
Greenberg, A. 535, 537
Greene, G.H., Jr. 172
Greer, F. 543
Greifenstein, L.G. 541
Greig, C.C. 440
Griffin, G.W. 539
Griffith, O.H. 170
Grinter, R. 440
Grisdale, P.J. 172
Grob, C.A. 75, 172, 316
Grohman, K.K. 313
Gross, J.C. 434
Gross, M.L. 174, 317
Grove, E.L. 537
Groves, J.K. 436, 439
Groves, L.G. 173
Grubbs, E.J. 172
Grubbs, R. 313
Grundy, J. 543
Grunwald, E. 168, 171, 536
Grutzner, J.B. 313, 319
Gund, P. 315
Günther, H. 176, 232
Gurbaxani, S. 536
Gutowsky, H.S. 538
Gutsche, C.D. 540

Habermehl, G. 174
Haddon, R.C. 409
Hadzi, D. 168, 437

AUTHOR INDEX 565

Hafner, K. 232
Hagan, M.M. 169
Hagden, E.L. 319
Hageman, H.J. 541
Hagen, E.L. 195
Hagen, K. 535, 542
Hahn, R.C. 177, 310, 440
Haines, W.J. 538
Hajek, G. 433
Halevi, E.V. 171
Hall, F.M. 171
Hall, H.K. 487, 540, 541, 542
Hall, N.D. 543
Hallas, G. 433
Halpern, Y. 310, 433
Halton, B. 320
Hamada, Y. 318, 409, 431
Hamilton, J.B. 541
Hamlet, Z. 176
Hammett, L.P. 276, 316
Hammond, G.S. 543
Hamon, D.P.G. 434
Hanack, M. 316
Hancock, C.K. 316, 440
Handloser, L. 434
Hanna, M.W. 314
Hanson, R.L. 436
Hanstein, W. 172, 318
Hapala, J. 539
Harada, N. 536, 537
Harget, A.J. 311
Hariharan, P.C. 320
Harrell, S.A. 168
Harris, C.L. 172
Harris, J.M. 316
Harris, M.M. 537, 542
Harrison, I.T. 169
Hart, H. 176, 319
Hartzler, H.D. 174
Haruta, M. 318
Haselbach, E. 176
Hashmall, J.A. 435
Haslam, J.L. 169
Hassel, O. 452, 535
Hastie, A. 311
Hata, Y. 174
Haug, P. 176
Haugen, W. 311
Hauser, K.W. 176
Havinga, E. 440
Hawkins, B.L. 538
Hawthorne, M.F. 310, 543
Hay, J.M. 167
Hazen, J.R. 536
Heap, N. 434
Heck, R. 540
Hedberg, K. 535, 542

Hedberg, L. 542
Hehre, W.J. 172, 313, 535
Heidberg, J. 311, 435
Heilbronner, E. 175, 176, 314, 320, 434, 435, 438, 440
Heisler, J. 170
Helgeson, R.C. 170, 439
Helm, D.v.D. 434
Henderson, W.G. 542
Hendrickson, J.B. 536
Hepfinger, N.F. 437
Herbrandson, H.F. 176
Hermann, R.B. 172
Hermans, J., Jr. 169
Herndon, W.C. 167, 320
Herrmann, R.I. 177, 435
Herschback, D.R. 175
Herscovitch, R. 315
Herzog, H.L. 435
Hess, B.A. 311
Hess, L.D. 176
Hess, R.E. 318
Hetzel, D.S. 538
Heyd, W.E. 538
Heymann, M.L. 435
Hickey, M. 437, 542
Higasi, K. 167, 310
Hill, E.A. 317
Hill, R.K. 434
Hill, T.L. 542
Hinrichs, H.H. 232
Hiraoka, K. 171
Hirota, M. 318, 409, 431
Hirotsu, T. 313
Hirsch, J.A. 315
Hirschfield, A. 542
Hixon, S.S. 310
Hobey, W.D. 311, 313
Hoffmann, J.M., Jr. 313
Hoffmann, R. 319, 320, 434, 435, 541
Hoffsommer, J.C. 175, 435
Hofmann, A. 246
Hogle, D.H. 542
Holden, J.R. 439
Holland, J.M. 434
Hollenberg, J.L. 174
Hollins, R.A. 436
Holsboer, F. 312
Holtz, D. 317
Holtzclaw, H.F. 315
Honour, R.J. 175
Hood, F.P. 173
Hoogzand, C. 543

Hopkinson, A.C. 440
Horeau, A. 537
Hornung, V. 435
Horowitz, R. 246
Horsley, J.A. 320
Hota, N.K. 312
House, H.O. 436, 519
Howard, P.H. 177, 310, 440
Howell, B.A. 319
Hruska, F. 318
Hsu, C.-M. 319
Huang, H.H. 62
Hübel, W. 543
Hückel, E. 215, 312
Hückel, W. 169
Hudec, J. 320
Hudson, C.E. 536
Huggins, M.L. 168
Huheey, J.E. 172
Huisgen, R. 439
Hume, D.N. 542
Hunsberger, I.M. 437
Huse, G. 311
Hutchins, R.O. 538
Hutchinson, R.E.J. 173
Hutton, H.M. 318
Hyden, S. 439

Ichikawa, K. 540
Ichikawa, M. 315
Idoux, J.P. 440
Iffland, D.C. 435
Ihrig, A.M. 319
Iitako, Y. 542
Ikemoto, I. 314
Imamura, A. 434
Imanishi, S. 438
Ingold, C.K. 312
Ingold, K.U. 436
Inoue, T. 176
Inuzuka, K. 438
Ishibe, N. 437
Ishida, H. 314
Ito, M. 176, 438
Ito, S. 542
Iwama, A. 388
Iwamura, H. 439
Iyoda, J. 171

Jackman, L.M. 173, 409, 438
Jackson, G. 440
Jacobus, N.C. 318
Jaeger, W. 519
Jaffe, H.H. 167, 170, 175, 317, 439
James, B.G. 543
James, R.V. 437

AUTHOR INDEX

Jameson, D.A. 409
Jamieson, G. 435, 439
Jamkowski, W.C. 177
Janusonis, G.A. 311, 435
Januszewski, J. 173
Jarrett, A.D. 535
Jautelat, M. 177
Jeffery, G.A. 169
Jeffery, G.H. 311
Jefford, C.W. 536
Jeffries, P.R. 35
Jensen, F.R. 540
Jensen, M.B. 171
Jesaitis, R.G. 314
Jeuell, C.L. 173, 310, 318
Jewett, J.G. 319
Jindal, S.P. 536, 540, 541
Jochims, J.C. 538
Johannsen, R.B. 540
Johanson, R.G. 540
Johnson, C.D. 316, 317, 440
Johnson, C.R. 177
Johnson, R.C. 541
Johnson, S.M. 313
Johnson, W.S. 438, 535
Johnston, B.E. 172, 536
Johnstone, R.A.W. 318, 362
Jonas, J. 538
Jonathan, N. 175
Jones, A.J. 176, 312
Jones, D.W. 168, 434
Jones, E.S. 439
Jones, H.L. 317, 439
Jones, M., Jr. 388
Jones, N. 436, 439
Jones, R.A. 175
Jones, R.N. 436
Jones, W.M. 312, 317, 318
Jorgensen, W.L. 167
Jorgenson, M.J. 541
Joris, L. 168, 318, 439
Josien, M.-L. 438
Jouanne, J.v. 311
Joyner, B.L. 172
Jurinski, N.B. 173, 314

Kabachnik, M.I. 317
Kabakoff, D.S. 310
Kaeding, W.W. 437, 543
Kagan, H.B. 537
Kagarise, R.E. 170
Kagiya, T. 176
Kaiser, A. 172

Kalzitzky, A.R. 173
Kamienski, B. 317
Kamlet, M.J. 175, 315, 435, 437, 438, 439
Kane, V.V. 388
Kanner, B. 542
Kaplan, L. 541
Kaplan, M.L. 168
Karabatsos, G.J. 175
Karle, I. 167
Karle, J. 167, 313
Karmack, W.D. 312
Karplus, M. 438
Kasai, M. 318, 409, 431
Kato, T. 436
Katritzky, A.R. 175, 316, 317
Katz, S. 362
Katz, T.J. 539
Katzenellenbogen, E.R. 433
Kavanagh, T.E. 409
Kayser, E.G. 438
Kealy, T.J. 246, 315
Kebarle, P. 171, 315, 542
Keefer, R.M. 314
Keepsell, D.G. 436
Keese, R. 541
Kelby, D.P. 173
Keldner, L. 311
Kellogg, R.M. 310
Kelly, D.P. 173, 310
Kelly, I. 169, 438
Kelsey, D.R. 316
Kelvin, W. 96
Kemp, J.D. 472, 537
Kemp-Jones, A.V. 176, 312
Kerber, R.C. 319
Kennewell, P.D. 536
Keprianov, A.I. 433
Kerlinger, H.O. 169
Kerr, J.A. 168
Kessler, H. 537
Ketcham, R. 310
Ketelaar, J.A.A. 171
Khalil, O.S. 176, 269
Kharasch, M.S. 49, 170
Kheifets, G.M. 315
Kilko, D.J. 540
Killian, F.L. 170
King, G.S.D. 313
King, W.T. 174
Kingston, D.G.I. 169
Kirby, G.W. 433
Kirkpatrick, J.L. 319
Kirkwood, J.G. 171

Kiser, R.W. 177
Kistiakowsky, G.B. 310
Kivelson, D. 436
Klaboe, P. 314
Klabunds, K.J. 315, 317
Klamperer, W. 438
Klein, H.S. 171
Kleinfclter, D.C. 404
Klemperer, W. 175
Klopfenstein, C.E. 313
Kloster-Jensen, E. 438
Klyne, W. 315, 435
Kmet, T.J. 312
Knorr, R. 536
Ko, E.C.F. 195
Kübrich, G. 541
Kochi, J.K. 536
Koepsell, D. 519
Koga, K. 170
Kogon, I.C. 174
Kolattukudy, P.E. 178
Kollman, P.A. 168
Koltoff, I.M. 317, 436
Kol'tsov, A.I. 315
König, C. 232
Königshofen, H. 176
Kopp, L.D. 538
Kornblum, N. 543
Kosower, E.M. 176, 434
Kovac, J. 317
Kozima, K. 175
Kraihanzel, C.S. 168
Krantz, A. 314, 538
Krapcho, A.P. 539, 540
Kraut, J. 439
Krebs, E.-P. 541
Kriloff, H. 536
Kromhout, R.A. 168
Kronenberg, M.E. 440
Kross, R.D. 170
Krueger, P.J. 409
Kruezo, L. 539
Krusic, P.J. 536, 541
Kruszewski, J. 312
Krutosikova, A. 317
Krygowski, T.M. 312, 317
Kuhn, R. 176, 315
Kuhn, S.J. 437, 543
Kulkarni, S.V. 316
Kundiger, D.G. 439
Kuroda, H. 314
Kurylo, M.J. 173, 314
Kwart, H. 439
Kwiram, A.L. 170
Kybett, B.D. 539

Laane, J. 438
LaBar, R.A. 312

AUTHOR INDEX

Laidler, K.J. 168
Laity, J.L. 312
Lallemand, J.-Y. 177
Lamal, D.M. 541
Lambert, J.B. 175, 541
Lampman, G.M. 535
Lancelot, C.J. 434
Landesberg, J.M. 541
Landman, D. 538
Lane, C.A. 172
Langsdorf, W.P. 173
Lankamp, H. 316
Lansbury, P.T. 318
Lao, S.C. 314
Larkin, R.H. 537
Larrabee, R.B. 539
Larsen, E. 178
Larsen, J.W. 171, 172, 315, 433
Laszlo, P. 174, 433, 439, 536
Lathan, W.A. 535
Laurie, V.W. 171, 175, 185
Lauterbur, R.C. 318
Law, P.A. 319
Lawlah, R.G. 173
Lawton, R.G. 539
Leal, G. 307
Lebel, N.A. 177
Lecomte, J. 100
Lederberg, J. 177
Lee, C.C. 195
Lee, D.J. 172
Leermakers, P.A. 176, 434, 541
Lee-Ruff, E. 173
Leey, G.C. 171
LeFevre, C.G. 538
LeFevre, R.J.W. 61, 167, 538
Lehn, J.M. 434
Leicester, J. 311
Leonard, N.J. 435
Lepore, G. 319
Leser, E.G. 436, 542
Letsinger, R.L. 440
Leung, C.S. 171, 172
Levasseur, L.A. 314
Levin, I.W. 175
Levin, R.H. 176, 177, 388
Levitt, B.W. 170, 173, 318
Levitt, L.S. 170, 173, 318
Levy, G.C. 177, 310, 315, 404

Levy, M.N. 315
Lewis, G.N. 65, 433
Lewis, I.C. 316
Lewis, T.P. 173
Ley, H. 175
Lhomme, J. 319
Li, Y.S. 537
Liang, G. 319, 433, 540
Liberles, A. 535, 536
Liddell, V. 174
Liebman, J.F. 168
Lielmezs, J. 538
Lifson, S. 167
Lightner, D.A. 165, 434
Lillien, I. 434
Limasset, J.C. 539
Lin, C.Y. 538
Lingefelter, E.C. 313
Linke, R. 437
Linnett, J.W. 167, 175
Lintvedt, R.L. 315
Liotta, C.L. 172, 312
Lipinski, C.A. 317
Lippert, J.L. 314
Lippincott, E.R. 437
Lipscomb, W.N. 174, 176, 538
Lister, D.G. 311
Little, E.J., Jr. 170
Liu, A. 173
Lloyd, D. 311
Lloyd, H.A. 436, 542
Lochmüller, C.H. 169
Lochte, H.H. 542
Loew, L.M. 319
Lomarbo, G. 542
Lombardi, J.R. 175, 440
Long, F.A. 169
Longone, D.T. 314, 436
Longster, G.F. 538
Looker, J.J. 543
Lorand, J.P. 540
Lord, R.C. 438, 537
Lorenco, R.J. 319
Lorenzo, G.A. 310, 440
Lossing, F.P. 310
Lovering, E.G. 168
Lowe, G. 537
Lowe, J.P. 535, 537
Lowe, J.U., Jr. 315, 437
Lowrey, A.H. 313
Luder, W.F. 167
Luft, R. 316
Lum, K.K. 317
Lupton, E.C. 172, 317
Lustig, E. 177
Lutskii, A.E. 169

Luttke, W. 174, 311
Luz, Z. 315
Lykos, P.G. 440
Lyle, J.L. 176
Lynch, B.M. 409

MacDonald, A.A. 313
MacDonald, B.C. 409
Machmer, P. 314
Maciel, G.E. 171, 173, 177, 437, 438
Mack, J.L. 316
Mackensen, G. 170
Mackenzie, R.K. 311
Mackor, E.L. 440
MacLean, C. 316, 440
MacLean, J.W. 543
MacLeay, R.E. 318
MacNicol, D.D. 311
Maguire, M.M. 168
Maier, D.P. 543
Maksimova, I.N. 171
Malkus, H. 536
Mallon, B.J. 535
Mallory, C.W. 539
Mallory, F.B. 537, 539
Manatt, S.L. 436, 537, 539
Mancini, V. 319
Mandella, W.L. 541, 542
Manion, M. 317
Mannschreck, A. 538
Mansson, M. 539
Marantz, S. 535
Marchand, A.P. 172
Marcus, S.H. 318
Margolin, Z. 434
Marion, G. 319
Markgraf, J.H. 540
Marshall, D.R. 311
Marshall, J.A. 311, 438, 541
Martell, A.E. 174
Martin, H.D. 320, 434
Martin, J. 436
Martin, J.C. 310
Martinelli, L. 310
Marty, R.A. 541
Marx, G.S. 317
Masamuni, S. 174, 176, 232, 312, 313
Maskornick, M.J. 171
Mason, S.F. 311
Mastryukova, T.A. 317
Mateescu, G.D. 319, 433, 435
Matlosz, B. 536
Matsui, Y. 174

AUTHOR INDEX

Matsuoka, H. 318, 409, 431
Matteson, D.S. 312
Maul, J.J. 437, 542
Mayeda, E.A. 314
Mayer, C.F. 434
Mayer, J.E. 516, 542
Mayer-Pitsch, E. 342
Mazengo, R.Z. 537
Mazur, S. 313, 315
McBee, E.T. 170, 434
McCain, J.H. 440
McCarthy, P.J. 174
McCarville, A.R. 313
McClellan, A.L. 169, 171
McConnell, W.V. 169
McCreadie, R.C. 536
McDaniel, D.H. 168, 542
McDaniel, J.C. 319
McDowell, B.L. 235, 315
McFarlane, W. 539
McGlynn, S.P. 176, 269
McIntosh, C.L. 538, 543
McIver, J.W., Jr. 171
McIver, M.C. 147
McIver, R.T., Jr. 35, 172, 543
McKennes, J.S. 541
McLafferty, F.W. 151, 165, 177, 178, 318
McNeil, R.L. 311
McQuillin, F.J. 310
Mecke, R. 174
Meeks, J.L. 176, 269
Megerle, G.H. 314
Melzer, M.S. 317
Menkin, V.I. 169
Merrifield, R.E. 314
Metcalf, B.W. 313
Metz, W.D. 314
Meyers, E.A. 316
Meyers, T.J. 312
Michel, R.H. 317
Migdal, S. 319
Miles, M.H. 169
Milewich, L. 168
Miller, A.L. 439
Miller, H.K. 170
Miller, J. 316
Miller, L.L. 314, 320
Miller, L.S. 539
Miller, M.A. 167, 434, 539
Miller, S.I. 318, 320, 437, 538
Mills, J.M. 174
Mills, J.W. 169
Milne, G.W.A. 177

Milun, M. 311
Minesinger, R.R. 435, 437
Minkin, V.I. 171, 311, 536, 538
Mislow, K. 434, 439, 461, 537, 542
Misume, S. 388, 435
Mitchell, R.H. 313
Mitsky, J. 318
Miyamoto, M. 537
Miyasaka, T. 535
Mizogami, S. 436
Mo, Y.K. 310
Modena, G. 316
Moncur, M.V. 319
Montana, A.F. 313
Montaudo, G. 540
Moodie, R.B. 173
Moon, S. 314, 434
Moore, W.R. 537, 540
Morgan, J.P. 538
Mori, K. 315
Moriarty, T.C. 171
Moricka, M. 313
Moriconi, E.J. 434
Morita, T. 320
Moritani, I. 310
Moser, C. 167
Moser, R.J. 269, 409
Mosser, S. 171
Mueller, W.A. 176
Muenter, J.S. 171
Muhs, M.A. 541
Mukherjee, T.K. 314
Müller, E. 542, 543
Muller, P. 536
Mulliken, R.S. 314
Mullins, C.B. 437
Munson, M.S.B. 173, 542
Murahashi, S.-I. 313
Murata, I. 313
Murayama, D.R. 313
Murphy, J. 168
Murphy, R.B. 172, 438
Murrell, J.N. 175
Murty, T.S.S.R. 317
Musher, J.I. 176
Musso, H. 539
Mutha, S.C. 310
Myers, D.K. 433

Nachod, F.C. 312, 433, 536, 537
Nador, K. 535
Naik, N.C. 537
Nair, G.V. 537
Nakagawa, M. 433

Nakagawa, T.W. 314
Nakamoto, K. 174
Nakamura, A. 315
Nakamura, N. 232, 312
Nakanishi, K. 536, 537
Namanworth, E. 174, 310
Namikawa, K. 177, 436
Nasielski, J. 176
Nathan, W.S. 310
Natowsky, S. 319
Natterstad, J.J. 437
Nauta, W.Th. 316, 433
Neckers, D.C. 433
Neely, S.C. 434
Nelsen, S.F. 535
Nelson, D.A. 176
Nelson, G.L. 177
Neuenschwander, M. 312
Neufield, F.R. 176
Newman, A.C.D. 170
Newman, M.S. 170, 316, 535, 543
Newman, R.M. 438
Newsoroff, G.P. 538
Newton, M.D. 174, 538, 539
Nicholas, R.D. 439, 539
Nichols, R.W. 317
Nickol, S.L. 317
Nickon, A. 169, 437
Nicoletti, R. 165
Niederhauser, A. 312
Nishida, S. 310
Nishida, T. 169
Nixon, W.B. 173
Nnadi, J.C. 434
Noack, K. 436
Nobel, A. 246
Noe, E.A. 536
Nordblom, G.D. 314
Nordlander, J.E. 540, 541
Norman, R.O.C. 315
Norris, C.L. 313
Noyce, D.S. 172, 317, 536
Nozari, M.S. 173, 542
Nye, M.J. 438
Nyl, K. 539
Nyquist, R.A. 168, 174, 318, 437

O'Brien, F.L. 318
O'Brien, D.H. 174
O'Connell, E.J., Jr. 169
O'Connor, W.F. 434
Ogawa, I.A. 319
Ogoshi, H. 437
O'Hare, M.J. 438
Oi, N. 175

AUTHOR INDEX

Okamoto, Y. 316
Okamura, W.H. 235
Olah, G.A. 173, 174, 177, 310, 313, 318, 319, 334, 433, 434, 435, 437, 540, 543
Oma, H. 232
Omura, I. 310
Ona, H. 312
O'Neal, J.W. 538
Orchin, M. 167, 175, 311, 320, 439
Orr, S.F.D. 169
Osawa, E. 436
Osipov, O.A. 171, 311, 536, 538
Ossip, P.S. 314
Ostercamp, D.L. 438
Ostlund, N.S. 171
O'Sullivan, D.G. 317
Oth, J.F.M. 313, 541
Otsu, T. 316
Otsubo, T. 388, 436
Overberger, C.G. 540
Overman, L.E. 170
Owens, P.H. 172, 173
Ozoe, H. 437
Ozretich, T.M. 537

Pacansky, J. 538
Pace, R.J. 437
Pape, P.G. 176
Paquette, L.A. 510, 541
Parish, R.C. 171, 172, 315, 316, 540
Parker, W. 436
Parry, K. 537
Parry, K.A.W. 438
Pasto, D.J. 177
Pattenden, G. 543
Patterson, D.B. 541
Paul, I.C. 313, 439
Pauling, L. 169, 170
Pauson, P.L. 246, 315
Pavoz, H.J. 317
Pawson, B.A. 536
Payling D.W. 318
Peake, E.G. 170
Pearson, D.E. 543
Pearson, R.G. 173, 542
Pearson, W.B. 168
Pedersen, C.J. 170
Pedersen, E. 178
Pelah, Z. 178
Perjessy, A. 317
Perkins, A.L. 537
Perkins, W. 246
Perrin, D.D. 315

Person, W.B. 314
Peters, A.W. 434
Peterson, L.I. 539
Peterson, M.R. 174
Peterson, P.E. 172
Petro, A.J. 171
Pettit, R. 307, 312, 541
Pews, R.G. 439
Pfyffer, J. 536
Phelan, N.F. 311, 320
Phillips, J. 318
Phillips, W.D. 314
Pickett, H.M. 168
Pickett, L.W. 435
Pier, E. 173
Pietra, F. 313
Piette, L.H. 311
Pilato, L.A. 436
Pilcher, G. 167
Pimentel, G.C. 169
Pinchas, S. 175
Pine, S.H. 171
Pinkus, A.G. 174, 439
Pinzelli, R.F. 316, 317
Pippert, D. 433
Pitha, J. 174
Pitts, J.N., Jr. 176
Pitzer, K.S. 168, 537
Player, C.M., Jr. 537
Pletcher, W.A. 431, 541
Ploss, G. 232
Pochan, J.M. 312, 543
Podall, H.E. 314
Politzer, P. 316
Pomerantz, M. 540
Pople, J.A. 167, 171, 172, 310, 319, 320, 535
Porter, C. 440
Porter, R.D. 173, 310, 319
Poulter, C.D. 177, 436
Powell, H.M. 170, 311
Pracejus, H. 541
Prelog, V. 537, 540
Price, C.C. 317
Price, M.J. 172
Price, R. 538
Price, W.C. 173
Prinzbach, H. 320, 434
Pritchard, H.O. 170
Pritchard, G.O. 175
Purcell, K.F. 437
Purdue, N. 174

Quin, L.D. 433
Quirk, R.P. 171, 315, 318, 433

Racela, W. 171, 315
Radar, C.P. 536
Radom, L. 172, 310, 311, 319, 320, 535
Rae, I.D. 177, 435
Ragelis, E.P. 177
Ralph, P.D. 315
Raman, C.V. 113
Ramsay, O.B. 440
Ramsey, B.G. 433
Rao, C.N.R. 35, 318, 437
Rapala, R.T. 435
Rapiejko, R.J. 170
Rapoport, H. 235, 315
Rapport, N. 539
Rath, N.S. 436
Rau, H. 435
Rawson, D.I. 538
Reagen, H. 435
Recker, K. 176
Reddy, G.S. 312
Redmore, D. 433
Ree, B.R. 310
Reed, L.L. 313
Reed, R.I. 177
Reel, H. 313
Reeve, W. 435
Reeves, L.W. 437
Reeves, R.A. 175
Regan, T.H. 543
Reich, H.J. 541
Reichardt, C. 175, 176
Reichel, D.M. 435
Reike, R.D. 540
Rekker, R.F. 433
Rembraum, A. 167
Remington, W.R. 435
Renk, E. 172
Retcofsky, H.L. 173
Rettig, T.A. 541
Rewicki, D. 315
Reynolds, W.F. 318
Rhodes, Y.E. 170, 310, 319
Richards, W.G. 320
Richter, P. 175
Ridge, D.P. 169
Rieger, M. 542
Riemenschneider, J.L. 319, 435
Rimiker, B. 537
Ripoll, J.L. 539
Ritchie, C.D. 175
Riveros, J.M. 35, 172, 173
Rivers, J.M. 168
Roark, J.L. 319
Robb, M.A. 538
Roberts, C.W. 439
Roberts, J.D. 167, 176, 177, 536, 538, 539

Robiette, A.G. 168
Robin, M.B. 176, 542
Robins, J. 172, 438
Robinson, C.H. 168
Robinson, L. 317
Robinson, R. 312
Rochow, E.G. 170
Rodebush, W.H. 435
Rogasch, P.E. 175
Rogers, F.E. 170
Rogers, L.B. 440
Rogers, M.T. 170, 317
Rondestvedt, E. 435
Rose, C.B. 435
Rosenblum, M. 314
Rosenfeld, J. 195, 319
Rossetti, G.P. 318
Rossmy, G. 437
Roth, R.W. 313
Roth, W.R. 435
Ruby, A. 174
Ruchardt, C. 168, 316
Rudolph, J.P. 171
Runquist, O. 174, 178
Russell, G.A. 536
Russo, T.J. 433
Ryason, P.R. 436

Sacconi, L. 542
Sachdev, K. 434, 537
Sadekov, I.D. 169
Sadler, P.S. 317
Saito, H. 437
Sakai, M. 176
Sakata, Y. 436
Salem, L. 167
Sakurai, H. 318
Sam D.J. 170
Sanchez, R.A. 439
Sanderd, E.B. 539
Sanderson, R.T. 45, 168 170
Sandorfy, C. 170
Sandstrom, J. 538
Sano, H. 171
Santelli, M. 438
Saucy, G. 536
Saunders, D.G. 168, 539
Saunders, J.K. 536
Saunders, M. 195, 319
Saunders, R.A. 178
Savitsky, G.B. 177, 436, 437
Schaad, L.J. 311
Schaar, B.E. 315
Schaefer, H. 439
Schaefer, J.P. 313
Schaefer, W. 312

Schaeffer, T. 318
Schaeffer, W.D. 169
Schenck, G.E. 313
Schenck, G.E. 433
Scheraga, H.A. 169
Scherer, K.V., Jr. 312
Schildcrout, S.M. 541
Schilling, G. 538
Schindlbauer, H. 433
Schleyer, P.v.R. 167, 168, 172, 173, 310, 316, 317, 319, 320, 433, 434, 437, 439, 486, 509, 536, 539, 540, 541
Schmeising, H.N. 310, 439
Schmelzer, A. 176
Schmickler, H. 176
Schmidt, H. 310
Schmidt, W. 312
Schmiedel, K. 175
Schneider, C.A. 540
Schneider, H.R. 173
Schneider, W.G. 311
Schoeller, W. 541
Schofield, R.T.C. 317
Schreck, J.O. 317
Schreiber, J. 543
Schroder, G. 176, 313, 541
Schroeder, R. 437
Schrumpf, G. 311
Schubert, W.M. 172, 438
Schuler, R.H. 540
Schilman, J.M. 539
Schure, R. 316
Schwarz, J.C.P. 537
Schweig, A. 310
Schweig, S. 434
Scott, A.I. 314, 438
Scott, J.A. 35, 172
Scott, K.N. 433
Scott, W.I. 540
Searles, S., Jr. 437, 439
Sease, J.W. 317
Sebastian, J.F. 172
Sekigawa, K. 317
Seltzer, R. 543
Semmelhack, M.F. 362, 434
Sergeyev, N.M. 538
Serpone, N. 313
Servis, K.L. 535
Seybold, G. 312
Shambhu, M.B. 269, 409

Sharkey, A.G. 178
Sharpe, T. 310
Shechter, H. 310
Sheehan, M. 436
Shekhiman, N.M. 535
Sheline, R.K. 311
Shelton, E.M. 438
Shen, K.W. 319, 320, 510
Sheppard, W.A. 171
Sheratte, M.B. 438
Sherman, L. 433
Sherrod, S.A. 316
Sherry, A.D. 437
Shilton, R. 435
Shiner, V.J., Jr. 172
Shorter, J. 316
Shoulders, B.A. 438
Shrader, S.R. 177
Shriver, D.F. 315
Shulgin, A.T., 169, 437
Shultz, J.L. 178
Sicher, J. 435
Siddall, T.H. 538
Sieczkowski, J. 541
Siefken, M.W. 539
Siegel, H. 435
Siepmann, T. 176
Silber, E. 320
Silberman, R.G. 177
Silverstein, R.M. 174, 177
Silvers, J.H. 543
Simkin, D.J. 169
Simmons, H.E. 170, 174, 434
Simpson, W.T. 172
Sinanoglu, O. 167
Singh, S. 537
Sinnott, M.V. 316
Sisler, H.H. 169
Skell, P.S. 316
Skinner, D.A. 169
Skinner, H.A. 170
Skvarchenko, V.R. 538
Slade, R.C. 320
Slejko, F.L. 437
Slifkin, M.A. 314
Slightom, E.L. 172
Sloan, G.L. 542
Smakula, A. 176
Small, L.E. 171
Smith, C.D. 487
Smith, D.F., Jr. 312
Smith, G.G. 317
Smith, G.V. 536
Smith, J.W. 312
Smith, L.M. 173
Smith, P.G. 362
Smith, S.G. 176
Smith, W.B. 312, 316, 319, 320

AUTHOR INDEX

Smitherman, H.C., Jr. 175, 312, 318
Smyth, C.P. 171, 311
Snyder, J.P. 312, 435
Snyder, W.H. 535
Sobotka, Z. 311
Sondheimer, F. 235, 313, 439
Songstad, J. 173
Sorenson, T. 172
Soto, H. 314
Soucy, K.T. 318
Soulen, R.L. 439
Souma, Y. 171
Souter, R.W. 169
Spencer, J.N. 168
Spinna, E. 172
Spoerri, P.E. 317
Sprecher, C.M. 541
Srinivasan, R. 541
Stals, J. 167
Stang, P.J. 174
Stasiewicz, M. 317
Stavely, L.A.K. 169
Stedman, D.E. 171
Stedman, R.J. 539
Steele, D. 171
Steele, D.R. 171
Steele, W.R.S. 169
Sternhell, S. 438, 538
Stevenson, G.R. 536
Stevens, C.L. 177
Stewart, R.F. 172
Stewart, R. 173
Stidham, H.D. 438
Stiglianai, W.M. 195
Stiles, M. 439
Stock, J. 434
Stock, L.M. 171, 172, 174 310, 315, 316, 536, 540
Stohere, W.D. 541
Stolow, R.D. 174, 535, 537
Story, P. 319
Stote, D.S. 433
Stothers, J.B. 173, 177, 433, 438, 536
Strait, L.A. 310
Streitwieser, A., Jr. 167 171, 172, 314, 317, 318, 537, 540
Strem, M.E. 543
Stromme, K.O. 437
Suda, M. 232, 312
Sulzberg, T. 314
Sumida, T. 176
Sunners, B. 311
Susz, B.P. 318

Sutton, L.E. 61, 312
Suydam, F.H. 537
Susuki, H. 175, 439
Swain, C.G. 172, 173, 317
Sweigert, D.A. 176
Sweitel, G. 436
Swenson, J.R. 435
Swenton, J.S. 175, 541
Swistun, Z. 168
Szilard, J. 539

Taagepera, M. 542
Tabit, C.Y. 173
Taft, R.W., Jr. 169, 170, 316, 317, 318, 433 439, 537
Takahashi, H. 314
Takahashi, S. 170, 434, 535
Takeda, K. 172
Takino, T. 319
Tamres, M. 437
Tanaka, I. 174
Tanaka, Y. 320
Tang, W.P. 438
Tanida, H. 172, 319
Tanida, H.T. 174
Tannenbaum, H. 170
Tarbell, D.S. 536
Taylor, D.R. 436
Taylor, G.R. 172
Taylor, M.D. 168, 542
Taylor, P.F. 169
Taylor, T.G. 172
Taylor, T.W.J. 61
Teraji, T. 310
Thomas, H.T. 434
Thompson, A.R. 171
Thompson, H.B. 171
Thompson, H.W. 168, 409
Thompson, J.A. 319
Thurston, P.E. 169, 540
Tichy, M. 436, 539
Tidwell, T.T. 435
Timberlake, J.W. 316
Tlmka, J.M. 170
Tipson, T.J. 168
Tipton, T.J. 539
Toby, F.S. 175
Toby, S. 175
Tocanne, J.F. 536
Tokumitsu, T. 437
Tolbert, B.M. 433
Tomson, M.B. 174
Tonellato, L. 316
Topson, R.D. 173, 316, 317

Toyoda, T. 388
Traetteberg, M. 311
Traficante, D.D. 438
Traylor, T.G. 167, 318, 319, 439, 540
Traynham, J.G. 317, 319
Trentwith, A.B. 316
Tribble, M.T. 167, 317, 539
Trinajstic, N. 311
Trindle, C. 540
Trotter, P.J. 314
Trucker, D.E. 433
Trueblood, K.N. 311
Tsang, W. 316
Tsubomura, H. 314
Tsutsui, M. 315
Tuck, R.H. 318
Tufariello, J.J. 319
Turner, D.W. 176
Turner, R.B. 310, 535
Turro, N.J. 541
Tyler, J.K. 311

Uetrecht, J.P. 313
Uma, M. 214, 538
Usselman, M.C. 435

Van Antwerp, C.L. 536
van Bekkum H. 439
Van-Catledge, F.A. 310
Vandenbelt, J.M. 175
Vander Jagt, D.L. 170
Vargas, L. 170, 310
Vassie, S. 195
Vaughan, W.R. 542
Veracini, C.A. 313
Verkade, J.G. 313
Verkade, P.E. 439
Verloop, A. 318
Vervanic, C.J. 172
Vesley, G.F. 541
Vidulich, G.A. 542
Viehe, H.G. 535
Virgilio, J.A. 317
Vogel, A.I. 311
Vogel, E. 176, 313
Vogel, G.C. 437
Vogel, P. 195, 319
Voorhees, K.J. 510
Vopel, K.H. 232

Wade, K. 315
Wagner, E.L. 171
Wagner, J. 434
Wagner, J.J. 536
Wahl, G.H., Jr. 174
Waitkus, P.A. 539
Waldron, J.T. 535

AUTHOR INDEX

Wallace, R.W. 540
Walker, E.E. 538
Waller, F.J. 539
Walsh, T.D. 314
Walter, G.F. 435
Wang, S.S. 314
Wang, S.Y. 434
Ward, J.S. 541
Warren, C.H. 434
Warren, C.L. 436
Warren, K.S. 436, 542
Warren, R.G. 176, 434
Warren, R.N. 312
Watts, L. 312
Wayland, B.B. 318
Weaver, H.E. 173
Weber, W.P. 168
Webers, V.J. 436
Wechter, W.J. 175
Wehry, E.L. 440
Weidner, U. 434
Weigert, F.G. 539
Weil, J.A. 311, 435
Weiler, L. 176
Weis, L.D. 176
Weiss, F.T. 541
Weiss, K. 434
Weleb, J.G.K. 409
Weller-Feilchenfeld, H. 542
 542
Wells, P.R. 170, 171, 172, 316, 317
Wepster, B.M. 315, 317, 435, 439
Wertz, D.H. 167, 539
West, R. 168, 437
Westenberg, A.A. 19
Westerman, P.W. 177, 319
Westheimer, F.H. 171, 516, 542
Westrum, E.F., Jr. 539
Wettermark, G. 434
Whalen, D.L. 539
Wheatley, P.J. 167, 174
Wheeler, G.L. 312
Wheelwright, W.L. 436
Wheland, G.W. 310, 311
White, A.M. 173, 318
White, E.H. 169
Whitehead, M.A. 170
Whitham, G.H. 434
Whitlock, H.W., Jr. 539
Wiberg, K.B. 167, 540
Wicks, G.E., Jr. 536
Wiersum, U.E. 541
Wilcox, C.F. 172, 174, 313, 319, 433, 540
Wiley, G.R. 318, 437

Wilkins, C.L. 174
Williams, A.E. 177, 178
Williams, D.H. 177, 178, 433
Williams, F.J. 540
Williams, F.V. 169, 542
Williams, J.E. 172, 539
Williams, L.L. 174
Williams, R.L. 437
Williams, V.Z., Jr. 540
Williamson, B. 435
Williamson, D.G. 176
Williamson, K.L. 171, 318, 438
Williamson, M.A. 177
Willis, H.A. 437
Wilmshurst, J.K. 170
Wilson, E.B., Jr. 19, 537
Wilson, J.D. 312
Wilt, J.W. 319, 540
Wilzbach, K.E. 541
Winstein, S. 176, 177, 310, 313, 319, 404, 434, 436
Winter, R.E. 315
Winterman, D.R. 169
Wiseman, J.R. 431, 438, 501, 541
Wiskott, E. 509
Witanowski, M. 173
Wojnarowski, W. 434
Wolf, A.D. 388
Wolf, R. 172
Wolfe, S. 535
Wollrabe, V. 178
Wong, R.H.W. 539
Wong, S.S. 539
Wood, C.S. 537
Woodward, A.J. 175
Woodward, R.B. 320
Woodworth, C.W. 173
Worley, S.D. 176
Worman, J.J. 176
Wrixton, A.D. 314
Wu, F.Y.H. 541
Wulf, O.R. 174
Wyatt, P.A.H. 440
Wynberg, H. 539, 542

Yager, B.J. 316
Yamada, H. 175
Yamamoto, H. 313
Yamamoto, K. 313
Yamamoto, S. 172
Yamamoto, T. 316
Yamdagni, R. 171, 315, 542

Yang, N.C. 543
Yanovskaya, L.A. 318
Yates, P. 174
Yerkess, J. 168
Yesowitch, G.E. 537
Yoder, C.H. 318, 537
Yoshida, Z. 318, 436, 437
Young, L.B. 173
Young, P.E. 310, 536
Young, W.R. 540

Zadorozhnyi, B.A. 440
Zalar, F.V. 174
Zefirov, N.S. 535
Zeiss, G.D. 434
Zhdanov, Y.A. 171, 311, 536, 538
Ziffer, H. 175
Zimmerman, H.E. 440
Zook, D. 433
Zuckerman, J.J. 433, 536
Zumwalt, L.R. 437